ORGANIC SYNTHESES

ADVISORY BOARD

Kay Brummond	Clayton Heathcock	John A. Ragan
Robert M. Coates	Louis S. Hegedus	Viresh H. Rawal
Elias J. Corey	Andrew B. Holmes	William R. Roush
Dennis P. Curran	David L. Hughes	Dieter Seebach
Huw M. L. Davies	Carl R. Johnson	Martin F. Semmelhack
Scott E. Denmark	Sarah E. Kelly	Masakatsu Shibasaki
Jonathan Ellman	Mark Lautens	Bruce E. Smart
Albert Eschenmoser	Dawei Ma	Amos B. Smith III
Margaret M. Faul	Steven F. Martin	Brian M. Stoltz
Ian Fleming	David Mathre	Keisuke Suzuki
Tohru Fukuyama	Marvin J. Miller	Kenneth B. Wiberg
Alois Fürstner	Koichi Narasaka	Peter Wipf
Leon Ghosez	Wayland E. Noland	Steven Wolff
Edward J. J. Grabowski	Ryoji Noyori	John Wood
David J. Hart	Larry E. Overman	Hasashi Yamamoto
	Andreas Pfaltz	

FORMER MEMBERS OF THE BOARD NOW DECEASED

Roger Adams	Nathan L. Drake	Melvin S. Newman
Homer Adkins	William D. Emmons	C. R. Noller
C. F. H. Allen	L. F. Fieser	Leo A. Paquette
Richard T. Arnold	Jeremiah P. Freeman	W. E. Parham
Werner E. Bachmann	R. C. Fuson	Charles C. Price
Henry E. Baumgarten	Henry Gilman	Norman Rabjohn
Richard E. Benson	Cliff S. Hamilton	John D. Roberts
A. H. Blatt	W. W. Hartman	Gabriel Saucy
Robert K. Boeckman Jr.	E. C. Horning	R. S. Schreiber
Virgil Boekelheide	Herbert O. House	John C. Sheehan
Ronald Breslow	Robert E. Ireland	William A. Sheppard
Arnold R. Brossi	John R. Johnson	Ichiro Shinkai
George H. Büchi	William S. Johnson	Ralph L. Shriner
T. L. Cairns	Oliver Kamm	Lee Irvin Smith
Wallace H. Carothers	Andrew S. Kende	H. R. Snyder
James Cason	Kenji Koga	Robert V. Stevens
Orville L. Chapman	Nelson J. Leonard	Max Tishler
H. T. Clarke	C. S. Marvel	Edwin Vedejs
David E. Coffen	Satoru Masamune	James D. White
J. B. Conant	B. C. McKusick	Frank C. Whitmore
Arthur C. Cope	Albert I. Meyers	Ekkehard Winterfeltd
William G. Dauben	Wataru Nagata	Peter Yates

ORGANIC SYNTHESES

AN ANNUAL PUBLICATION OF SATISFACTORY
METHODS FOR THE PREPARATION OF
ORGANIC CHEMICALS

VOLUME 98

2021

CHRIS SENANAYAKE
VOLUME EDITOR

BOARD OF EDITORS

RICK L. DANHEISER, *Editor-in-Chief*

KEVIN R. CAMPOS	CRISTINA NEVADO
CATHLEEN CRUDDEN	SARAH E. REISMAN
KUILING DING	RICHMOND SARPONG
NEIL K. GARG	CHRIS H. SENANAYAKE
MASAYUKI INOUE	DIRK TRAUNER
NUNO MAULIDE	TEHSHIK YOON

CHARLES K. ZERCHER, *Associate Editor*
DEPARTMENT OF CHEMISTRY
UNIVERSITY OF NEW HAMPSHIRE
DURHAM, NEW HAMPSHIRE 03824

The procedures in this article are intended for use only by persons with prior training in experimental organic chemistry. These procedures must be conducted at one's own risk. *Organic Syntheses, Inc.*, its Editors, and its Board of Directors do not warrant or guarantee the safety of individuals using these procedures and hereby disclaim any liability for any injuries or damages claimed to have resulted from or related in any way to the procedures herein.

Copyright © 2021 by Organic Syntheses, Inc. All rights reserved.

Published by John Wiley & Sons, Inc., Hoboken, New Jersey.

Published simultaneously in Canada.

No part of this publication may be reproduced, stored in a retrieval system or transmitted in any form or by any means, electronic, mechanical, photocopying, recording, scanning, or otherwise, except as permitted under Section 107 or 108 of the 1976 United States Copyright Act, without either the prior written permission of the Publisher, or authorization through payment of the appropriate per-copy fee to the Copyright Clearance Center, Inc., 222 Rosewood Drive, Danvers, MA 01923, (978) 750-8400, fax (978) 750-4470, or on the web at www.copyright.com. Requests to the Publisher for permission should be addressed to the Permissions Department, John Wiley & Sons, Inc., 111 River Street, Hoboken, NJ 07030, (201) 748-6011, fax (201) 748-6008, or online at http://www.wiley.com/go/permission.

Limit of Liability/Disclaimer of Warranty: While the publisher and author have used their best efforts in preparing this book, they make no representations or warranties with respect to the accuracy or completeness of the contents of this book and specifically disclaim any implied warranties of merchantability or fitness for a particular purpose. No warranty may be created or extended by sales representatives or written sales materials. The advice and strategies contained herein may not be suitable for your situation. You should consult with a professional where appropriate. Neither the publisher nor author shall be liable for any loss of profit or any other commercial damages, including but not limited to special, incidental, consequential, or other damages.

For general information on our other products and services or for technical support, please contact our Customer Care Department within the United States at (800) 762-2974, outside the United States at (317) 572-3993 or fax (317) 572-4002.

Wiley also publishes its books in a variety of electronic formats. Some content that appears in print may not be available in electronic formats. For more information about Wiley products, visit our web site at www.wiley.com.

"John Wiley & Sons, Inc. is pleased to publish this volume of Organic Syntheses on behalf of Organic Syntheses, Inc. Although Organic Syntheses, Inc. has assured us that each preparation contained in this volume has been checked in an independent laboratory and that any hazards that were uncovered are clearly set forth in the write-up of each preparation, John Wiley & Sons, Inc. does not warrant the preparations against any safety hazards and assumes no liability with respect to the use of the preparations."

Library of Congress Catalog Card Number:

ISBN: 978-1-119-90394-9

Printed in the United States of America

ORGANIC SYNTHESES

VOLUME	VOLUME EDITOR	PAGES
I*	† Roger Adams	84
II*	† James Bryant Conant	100
III*	† Hans Thacher Clark	105
IV*	† Oliver Kamm	89
V*	† Carl Shipp Marvel	110
VI*	† Henry Gilman	120
VII*	† Frank C. Whitmore	105
VIII*	† Roger Adams	139
IX*	† James Bryant Conant	108
Collective Vol. I	A revised edition of Annual Volumes I-IX †Henry Gilman, Editor-in-Chief 2nd Edition revised by † A. H. Blatt	580
X*	† Hans Thacher Clarke	119
XI*	† Carl Shipp Marvel	106
XII*	† Frank C. Whitmore	96
XIII*	† Wallace H. Carothers	119
XIV	† William W. Hartman	100
XV*	† Carl R. Noller	104
XVI*	† John R. Johnson	104
XVII*	† L. F. Fieser	112
XVIII*	† Reynold C. Fuson	103
XIX*	† John R. Johnson	105
Collective Vol. II	A revised edition of Annual volumes X-XIX † A. H. Blatt, Editor-in-Chief	654
20*	† Charles F. H. Allen	113
21*	† Nathan L. Drake	120
22*	† Lee Irvin Smith	114
23*	† Lee Irvin Smith	124
24*	† Nathan L. Drake	119
25*	† Werner E. Bachmann	120
26*	† Homer Adkins	124
27*	† R. L. Shriner	121
28*	† H. R. Snyder	121
29*	† Cliff S. Hamilton	119
Collective Vol. III	A revised edition of Annual Volumes 20-29 † E. C. Horning, Editor-in-Chief	890
30*	† Arthur C. Cope	115
31*	† R. S. Schreiber	122
32*	† Richard Arnold	119
33*	† Charles Price	115
34*	† William S. Johnson	121
35*	† T. L. Cairns	122

*Out of print.
†Deceased.

VOLUME	VOLUME EDITOR	PAGES
36*	† N. J. Leonard	120
37*	† James Cason	109
38*	† James C. Sheehan	120
39*	† Max Tishler	114
Collective Vol. IV	A revised edition of Annual Volumes 30-39 † Norman Rabjohn, *Editor-in-Chief*	1036
40*	† Melvin S. Newman	114
41*	† John D. Roberts	118
42*	† Virgil Boekelheide	118
43*	† B. C. McKusick	124
44*	† William E. Parham	131
45*	† William G. Dauben	118
46*	E. J. Corey	146
47*	† William D. Emmons	140
48*	† Peter Yates	164
49*	Kenneth B. Wiberg	124
Collective Vol. V	A revised edition of Annual Volumes 40-49 † Henry E. Baumgarten, *Editor-in-Chief*	1234
Cumulative Indices to Collective Volumes, I, II, III, IV, V	† Ralph L. and † Rachel H. Shriner, *Editors*	
50*	† Ronald Breslow	136
51*	† Richard E. Benson	209
52*	† Herbert O. House	192
53*	† Arnold Brossi	193
54*	† Robert E. Ireland	155
55*	† Satoru Masamune	150
56*	† George H. Büchi	144
57*	Carl R. Johnson	135
58*	† William A. Sheppard	216
59*	Robert M. Coates	267
Collective Vol. VI	A revised edition of Annual Volumes 50-59 Wayland E. Noland, *Editor-in-Chief*	1208
60*	† Orville L. Chapman	140
61*	† Robert V. Stevens	165
62*	Martin F. Semmelhack	269
63*	† Gabriel Saucy	291
64*	† Andrew S. Kende	308
Collective Vol. VII	A revised edition of Annual Volumes 60-64 † Jeremiah P. Freeman, *Editor-in-Chief*	602
65*	† Edwin Vedejs	278
66*	Clayton H. Heathcock	265
67*	Bruce E. Smart	289
68*	† James D. White	318
69*	† Leo A. Paquette	328

Out of print.
†*Deceased.*

VOLUME	VOLUME EDITOR	PAGES
Reaction Guide to Collective Volumes I-VII and Annual Volumes 65-68		854
	DENNIS C. LIOTTA AND MARK VOLMER, *Editors*	
Collective Vol. VIII	A revised edition of Annual Volumes 65-69	696
	† JEREMIAH P. FREEMAN, *Editor-in-Chief*	

Cumulative Indices to Collective Volumes, I, II, III, IV, V, VI, VII, VIII

70*	† ALBERT I. MEYERS	305
71*	LARRY E. OVERMAN	285
72*	† DAVID L. COFFEN	333
73*	† ROBERT K. BOECKMAN, JR.	352
74*	† ICHIRO SHINKAI	341
Collective Vol. IX	A revised edition of Annual Volumes 70-74	840
	† JEREMIAH P. FREEMAN, *Editor-in-Chief*	
75	AMOS B. SMITH, III	257
76	STEPHEN MARTIN	340
77	DAVID S. HART	312
78	WILLIAM R. ROUSH	326
79	LOUIS S. HEGEDUS	328
Collective Vol. X	A revised edition of Annual Volumes 75-79	810
	† JEREMIAH P. FREEMAN, *Editor-in-Chief*	
80	STEVEN WOLFF	259
81	RICK L. DANHEISER	296
82	EDWARD J. J. GRABOWSKI	195
83	DENNIS P. CURRAN	221
84	MARVIN J. MILLER	410
Collective Vol. XI	A revised edition of Annual Volumes 80-84	1138
	CHARLES K. ZERCHER, *Editor-in-Chief*	
85	SCOTT E. DENMARK	321
86	JOHN A. RAGAN	403
87	PETER WIPF	403
88	JONATHAN A. ELLMAN	472
89	MARK LAUTENS	561
Collective Vol. XII	A revised edition of Annual Volumes 85-89	1570
	CHARLES K. ZERCHER, *Editor-in-Chief*	
90	DAVID L. HUGHES	376
91	KAY M. BRUMMOND	365
92	VIRESH H. RAWAL	386
93	ERICK M. CARREIRA	421
94	MARGARET FAUL	398
95	BRIAN M. STOLTZ	511
96	JOHN L. WOOD	598
97	MOHAMMAD MOVASSAGHI	367
98	CHRIS SENANAYAKE	552

*Out of print.
†Deceased.

NOTICE

Beginning with Volume 84, the Editors of *Organic Syntheses* initiated a new publication protocol, which is intended to shorten the time between submission of a procedure and its appearance as a publication. Immediately upon completion of the successful checking process, procedures are assigned volume and page numbers and are then posted on the Organic Syntheses website (www.orgsyn.org). The accumulated procedures from a single volume are assembled once a year and submitted for publication. The annual volume is published by John Wiley and Sons, Inc., and includes an index. The hard cover edition is available for purchase through the publisher. Incorporation of graphical abstracts into the Table of Contents began with Volume 77. Annual volumes 70–74, 75–, 80–84 and 85–89 have been incorporated into five-year versions of the collective volumes of *Organic Syntheses*. Collective Volumes IX, X, XI and XII are available for purchase in the traditional hard cover format from the publishers.

Beginning with Volume 88, a new type of article, referred to as Discussion Addenda, appeared. In these articles submitters are provided the opportunity to include updated discussion sections in which new understanding, further development, and additional application of the original method are described.

Organic Syntheses, Inc., joined the age of electronic publication in 2001 with the release of its free web site (www.orgsyn.org). The site is accessible through internet browsers using Macintosh and Windows operating systems, and the database can be searched by key words and sub-structure. John Wiley & Sons, Inc., and Accelrys, Inc., partnered with Organic Syntheses, Inc., to develop a database (www.mrw.interscience.wiley.com/osdb) that is available for license with internet solutions from John Wiley & Sons, Inc. and intranet solutions from Accelrys, Inc.

Both the commercial database and the free website contain all annual and collective volumes and indices of *Organic Syntheses*. Chemists can draw structural queries and combine structural or reaction transformation queries with full-text and bibliographic search terms, such as chemical name, reagents, molecular formula, apparatus, or even hazard warnings or phrases. The contents of individual or collective volumes can be browsed by lists of titles, submitters' names, and volume and page references, with or without structures.

The commercial database at www.mrw.interscience.wiley.com/osdb also enables the user to choose his/her preferred chemical drawing package, or to utilize several freely available plug-ins for entering queries. The user is also able to cut and paste existing structures and reactions directly into the structure search query or their preferred chemistry editor, streamlining workflow. Additionally, this database contains links to the full text of primary literature references via CrossRef, ChemPort, Medline, and ISI Web of Science. Links to local holdings for institutions using open url technology can also be enabled. The database user can limit his/her search to, or order the search results by, such factors as reaction type, percentage yield, temperature, and publication date, and can create a customized table of reactions for comparison. Connections to other Wiley references are currently made via text search, with cross-product structure and reaction searching to be added in the near future. Incorporations of new preparations will occur as new material becomes available.

INFORMATION FOR AUTHORS OF PROCEDURES

Organic Syntheses welcomes and encourages submissions of experimental procedures that lead to compounds of wide interest or that illustrate important new developments in methodology. Proposals for *Organic Syntheses* procedures will be considered by the Editorial Board upon receipt of an outline proposal as described below. A full procedure will then be invited for those proposals determined to be of sufficient interest. These full procedures will be evaluated by the Editorial Board, and if approved, assigned to a member of the Board for checking. In order for a procedure to be accepted for publication, each reaction must be successfully repeated in the laboratory of a member of the Editorial Board at least twice, with similar yields (generally $\pm 5\%$) and selectivity to that reported by the submitters.

Organic Syntheses Proposals

A cover sheet should be included providing full contact information for the principal author and including a scheme outlining the proposed reactions (an *Organic Syntheses* Proposal Cover Sheet can be downloaded at orgsyn.org). Attach an outline proposal describing the utility of the methodology and/or the usefulness of the product. Identify and reference the best current alternatives. For each step, indicate the proposed scale, yield, method of isolation and purification, and how the purity of the product is determined. Describe any unusual apparatus or techniques required, and any special hazards associated with the procedure. Identify the source of starting materials. Enclose copies of relevant publications (attach pdf files if an electronic submission is used).

Submit proposals by mail or as e-mail attachments to:

Professor Charles K. Zercher
Associate Editor, *Organic Syntheses*
Department of Chemistry
University of New Hampshire
23 Academic Way, Parsons Hall
Durham, NH 03824

Electronic submissions through the website (www.orgsyn.org) is strongly encouraged.

Submission of Procedures

Authors invited by the Editorial Board to submit full procedures should prepare their manuscripts in accord with the Instructions for Authors, which are described below or may be downloaded at orgsyn.org. Submitters are also encouraged to consult this volume of *Organic Syntheses* for models with regard to style, format, and the level of experimental detail expected in *Organic Syntheses* procedures. Manuscripts should be submitted to the Associate Editor. Electronic submissions are encouraged; procedures will be accepted as e-mail attachments in the form of Microsoft Word files with all schemes and graphics also sent separately as ChemDraw files.

Procedures that do not conform to the Instructions for Authors with regard to experimental style and detail will be returned to authors for correction. Authors will be notified when their manuscript is approved for checking by the Editorial Board, and it is the goal of the Board to complete the checking of procedures within a period of no more than six months.

Additions, corrections, and improvements to the preparations previously published are welcomed; these should be directed to the Associate Editor. However, checking of such improvements will only be undertaken when new methodology is involved.

NOMENCLATURE

Both common and systematic names of compounds are used throughout this volume, depending on which the Volume Editor felt was more appropriate. Systematic Chemical Abstracts nomenclature, used in the Collective Indexes for the title compound and a selection of other compounds mentioned in the procedure, is provided in an appendix at the end of each preparation. Chemical Abstracts Registry numbers, which are useful in computer searching and identification, are also provided in these appendices.

ACKNOWLEDGMENT

Organic Syntheses wishes to acknowledge the contributions of Merck and AstaTech Biopharmaceutical to the success of this enterprise through their support, in the form of time and expenses, of members of the Board of Editors.

INSTRUCTIONS FOR AUTHORS

All organic chemists have experienced frustration at one time or another when attempting to repeat reactions based on experimental procedures found in journal articles. To ensure reproducibility, *Organic Syntheses* requires experimental procedures written with considerably more detail as compared to the typical procedures found in other journals and in the "Supporting Information" sections of papers. In addition, each *Organic Syntheses* procedure is carefully "checked" for reproducibility in the laboratory of a member of the Board of Editors.

Even with these more detailed procedures, the experience of *Organic Syntheses* editors is that difficulties often arise in obtaining the results and yields reported by the submitters of procedures. To expedite the checking process and ensure success, we have prepared the following "Instructions for Authors" as well as a **Checklist for Authors** and **Characterization Checklist** to assist you in confirming that your procedure conforms to these requirements. Please include a completed Checklist together with your procedure at the time of submission. Procedures submitted to *Organic Syntheses* will be carefully reviewed upon receipt and procedures lacking any of the required information will be returned to the submitters for revision.

Scale and Optimization

The appropriate scale for procedures will vary widely depending on the nature of the chemistry and the compounds synthesized in the procedure. However, some general guidelines are possible. For procedures in which the principal goal is to illustrate a synthetic method or strategy, it is expected, in general, that the procedure should result in at least 2 g and no more than 50 g of the final product. In cases where the point of the procedure is to provide an efficient method for the preparation of a useful reagent or synthetic building block, the appropriate scale also should not exceed 50 g of final product. Exceptions to these guidelines may be granted in special circumstances. For example, procedures describing the preparation of reagents employed as catalysts will often be acceptable on a scale of less than 2 g.

In considering the scale for an *Organic Syntheses* procedure, authors should also take into account the cost of reagents and starting materials. In general, the Editors will not accept procedures for checking in which the cost of any one of the reactants exceeds **$500** for a single full-scale run. Authors are requested to identify the most expensive reagent or starting material on the procedure submission checklist and to estimate its cost per run of the procedure.

It is expected that all aspects of the procedure will have been optimized by the authors prior to submission, and it is required that each reaction will have been carried out at least twice on exactly the scale described in the procedure, and with the results reported in the manuscript.

It is appropriate to report the weight, yield, and purity of the product of each step in the procedure as a range. In any case where a reagent is employed in significant excess, a Note should be included explaining why an excess of that reagent is necessary. If possible, the Note should indicate the effect of using amounts of reagent less than that specified in the procedure.

The Checking Process

A unique feature of papers published in *Organic Syntheses* is that each procedure and all characterization data is carefully checked for reproducibility in the laboratory of a member of the Board of Editors. In the event that an editor finds it necessary to make any modifications in an experimental procedure, then the published article incorporates the modified procedure, with an explanation and mention of the original protocol often included in a Note. The yields reported in the published article are always those obtained by the checkers. In general, the characterization data in the published article also is that of the checkers, unless there are significant differences with the data obtained by the authors, in which case the author's data will also be reported in a Note.

Reaction Apparatus

Describe the size and type of flask (number of necks) and indicate how *every* neck is equipped.

> "A 500-mL, three-necked, round-bottomed flask equipped with an 3-cm Teflon-coated magnetic stirbar, a 250-mL pressure-equalizing addition funnel fitted with an argon inlet, and a rubber septum is charged with ...."

Indicate how the reaction apparatus is dried and whether the reaction is conducted under an inert atmosphere. Note that in general balloons are not acceptable as a means of maintaining an inert atmosphere unless warranted by special circumstances. The description of the reaction apparatus can be incorporated in the text of the procedure or included in a Note.

> "The apparatus is flame-dried and maintained under an atmosphere of argon during the course of the reaction."

In the case of procedures involving unusual glassware or especially complicated reaction setups, authors are encouraged to include a photograph or drawing of the apparatus in the text or in a Note (for examples, see *Org. Syn.*, Vol. 82, 99 and Coll. Vol. X, pp 2, 3, 136, 201, 208, and 669).

Use of Gloveboxes

When a glovebox is employed in a procedure, justification must be provided in a Note and the consequences of carrying out the operation without using a glovebox should be discussed.

Reagents and Starting Materials

All chemicals employed in the procedure must be commercially available or described in an earlier *Organic Syntheses* or *Inorganic Syntheses* procedure. For other compounds, a procedure should be included either as one or more steps in the text or, in the case of relatively straightforward preparations of reagents, as a Note. In the latter case, all requirements with regard to characterization, style, and detail also apply. Authors are encouraged to consult with the Associate Editor if they have any question as to whether to include such steps as part of the text or as a Note.

Authors are encouraged to consider the use of "substitute solvents" in place of more hazardous alternatives. For example, the use of *t*-butyl methyl ether (MTBE) should be considered as a substitute for diethyl ether, particularly in large scale work. Authors are referred to the articles "Sanofi's Solvent Selection Guide: A Step Toward More Sustainable Processes" (Prat, D.; Pardigon, O.; Flemming, H.-W.; Letestu, S.; Ducandas, V.; Isnard, P.; Guntrum, E.; Senac, T.; Ruisseau, S.; Cruciani, P. Hosek, P. *Org. Process Res. Dev.* **2013**, *17*, 1517-1525) and "Solvent Replacement for Green Processing" (Sherman, J.; Chin, B.; Huibers, P. D. T.; Garcia-Valis, R.; Hatton, T. A. *Environ. Health Perspect.* **1998**, *106* (Supplement I, 253-271) as well as the references cited therein for discussions of this subject. In addition, a link to a "solvent selection guide" can be accessed via the American Chemical Society Green Chemistry website at http://www.acs.org/content/acs/en/greenchemistry/research-innovation/tools-for-green-chemistry.html.

In one or more Notes, indicate the purity or grade of each reagent, solvent, etc. It is highly desirable to also indicate the source (company the chemical was purchased from), particularly in the case of chemicals where it is suspected that the composition (trace impurities, etc.) may vary from one supplier to another. In cases where reagents are purified, dried, "activated" (e.g., Zn dust), etc., a detailed description of the procedure used should be included in a Note. In other cases, indicate that the chemical was "used as received".

"Diisopropylamine (99.5%) was obtained from Aldrich Chemical Co., Inc. and distilled under argon from calcium hydride before use. THF (99+%) was obtained from Mallinckrodt, Inc. and distilled from sodium benzophenone ketyl. Diethyl ether (99.9%) was purchased from Aldrich Chemical Co., Inc. and purified by pressure filtration under argon through activated alumina. Methyl iodide (99%) was obtained from Aldrich Chemical Co., Inc. and used as received."

The amount of each reactant must be provided in parentheses in the order mL, g, mmol, and equivalents with careful consideration to the correct number of **significant figures**. Avoid indicating amounts of reactants with more significant figures than makes sense. For example, "437 mL of THF" implies that the amount of solvent must be measured with a level of precision that is unlikely to affect the outcome of the reaction. Likewise, "5.00 equiv" implies that an amount of excess reagent must be controlled to a precision of 0.01 equiv.

The concentration of solutions should be expressed in terms of molarity or normality, and not percent (e.g., 1 N HCl, 6 M NaOH, not "10% HCl").

Reaction Procedure

Describe every aspect of the procedure clearly and explicitly. Indicate the order of addition and time for addition of all reagents and how each is added (via syringe, addition funnel, etc.).

Indicate the temperature of the reaction mixture (preferably internal temperature). Describe the type of cooling (e.g., "dry ice-acetone bath") and heating (e.g., oil bath, heating mantle) methods employed. Be careful to describe clearly all cooling and warming cycles, including initial and final temperatures and the time interval involved.

Describe the appearance of the reaction mixture (color, homogeneous or not, etc.) and describe all significant changes in appearance during the course of the reaction (color changes, gas evolution, appearance of solids, exotherms, etc.).

Indicate how the reaction can be monitored to determine the extent of conversion of reactants to products. In the case of reactions monitored by TLC, provide details in a Note, including eluent, R_f values, and method of visualization. For reactions followed by GC, HPLC, or NMR analysis, provide details on analysis conditions and relevant diagnostic peaks.

"The progress of the reaction was followed by TLC analysis on silica gel with 20% EtOAc-hexane as eluent and visualization with *p*-anisaldehyde. The ketone starting material has $R_f = 0.40$ (green) and the alcohol product has $R_f = 0.25$ (blue)."

Reaction Workup

Details should be provided for reactions in which a "quenching" process is involved. Describe the composition and volume of quenching agent, and time and temperature for addition. In cases where reaction mixtures are added to a quenching solution, be sure to also describe the setup employed.

> "The resulting mixture was stirred at room temperature for 15 h, and then carefully poured over 10 min into a rapidly stirred, ice-cold aqueous solution of 1 N HCl in a 500-mL Erlenmeyer flask equipped with a magnetic stirbar."

For extractions, the number of washes and the volume of each should be indicated as well as the size of the separatory funnel.

For concentration of solutions after workup, indicate the method and pressure and temperature used.

> "The reaction mixture is diluted with 200 mL of water and transferred to a 500-mL separatory funnel, and the aqueous phase is separated and extracted with three 100-mL portions of ether. The combined organic layers are washed with 75 mL of water and 75 mL of saturated NaCl solution, dried over 25 g of $MgSO_4$, filtered through a 250-mL medium porosity sintered glass funnel, and concentrated by rotary evaporation (25 °C, 20 mmHg) to afford 3.25 g of a yellow oil."

> "The solution is transferred to a 250-mL, round-bottomed flask equipped with a magnetic stirbar and a 15-cm Vigreux column fitted with a short path distillation head, and then concentrated by careful distillation at 50 mmHg (bath temperature gradually increased from 25 to 75 °C)."

In cases where solid products are filtered, describe the type of filter funnel used and the amount and composition of solvents used for washes.

> " ... and the resulting pale yellow solid is collected by filtration on a Büchner funnel and washed with 100 mL of cold (0 °C) hexane."

When solid or liquid compounds are dried under vacuum, indicate the pressure employed (rather than stating "reduced pressure" or "dried *in vacuo*").

> " .... and concentrated at room temperature by rotary evaporation (20 mmHg) and then at 0.01 mmHg to provide .... "

> "The resulting colorless crystals are transferred to a 50-mL, round-bottomed flask and dried overnight in a 100 °C oil bath at 0.01 mmHg."

Purification: Distillation

Describe distillation apparatus including the size and type of distillation column. Indicate temperature (and pressure) at which all significant fractions are collected.

" .... and transferred to a 100-mL, round-bottomed flask equipped with a magnetic stirbar. The product is distilled under vacuum through a 12-cm, vacuum-jacketed column of glass helices (Note 16) topped with a Perkin triangle. A forerun (ca. 2 mL) is collected and discarded, and the desired product is then obtained, distilling at 50-55 °C (0.04-0.07 mmHg) .... "

Purification: Column Chromatography

Provide information on TLC analysis in a Note, including eluent, R_f values, and method of visualization.

Provide dimensions of column and amount of silica gel used; in a Note indicate source and type of silica gel.

Provide details on eluents used, and number and size of fractions.

"The product is charged on a column (5 x 10 cm) of 200 g of silica gel (Note 15) and eluted with 250 mL of hexane. At that point, fraction collection (25-mL fractions) is begun, and elution is continued with 300 mL of 2% EtOAc-hexane (49:1 hexanes:EtOAc) and then 500 mL of 5% EtOAc-hexane (19:1 hexanes:EtOAc). The desired product is obtained in fractions 24-30, which are concentrated by rotary evaporation (25 °C, 15 mmHg) .... "

Use of Automated Column Chromatography

Automated column chromatography should not be used for purification of products unless the use of such systems is essential to the success of the procedure. When automated column chromatography equipment is employed in a procedure, justification must be provided in a Note and the consequences of carrying out the purification using conventional column chromatography must be discussed.

Purification: Recrystallization

Describe procedure in detail. Indicate solvents used (and ratio of mixed solvent systems), amount of recrystallization solvents, and temperature protocol. Describe how crystals are isolated and what they are washed with. A photograph of the crystalline product is often valuable to indicate the form and color of the crystals.

"The solid is dissolved in 100 mL of hot diethyl ether (30 °C) and filtered through a Buchner funnel. The filtrate is allowed to cool to room temperature, and 20 mL of hexane is added. The solution is cooled at -20 °C overnight and the resulting crystals are collected by suction filtration on a Buchner funnel, washed with 50 mL of ice-cold hexane, and then transferred to a 50-mL, round-bottomed flask and dried overnight at 0.01 mmHg to provide ... "

Characterization

Physical properties of the product such as color, appearance, crystal forms, melting point, etc. should be included in the text of the procedure. Comments on the stability of the product to storage, etc. should be provided in a Note.

In a Note, provide data establishing the identity of the product. This will generally include IR, MS, ^{1}H-NMR, and ^{13}C-NMR data, and in some cases UV data. Copies of the proton and carbon NMR spectra for the products of each step in the procedure should be submitted showing integration for all resonances. Submission of copies of the NMR spectra for other nuclei are encouraged as appropriate.

In the same Note, provide analytical data establishing that the purity of the **isolated** product is at least 97%. **Note that this data should be obtained for the material on which the yield of the reaction is based**, not for a sample that has been subjected to additional purification by chromatography, distillation, or crystallization. Elemental analysis for carbon and hydrogen (and nitrogen if present) agreeing with calculated values within 0.4% is preferred. However, **quantitative** NMR, GC, or HPLC analyses involving measurements versus an internal standard will also be accepted. See *Instructions for Authors* at orgsyn.org for procedures for quantitative analysis of purity by NMR and chromatographic methods. Provide details on equipment and conditions for GC and HPLC analyses.

In procedures involving non-racemic, enantiomerically enriched products, optical rotations should generally be provided, but **enantiomeric purity must be determined by another method** such as chiral HPLC or GC analysis.

In cases where the product of one step is used without purification in the next step, a Note should be included describing how a sample of the product can be purified and providing characterization data for the pure material. Copies of the proton NMR spectra of both the product both *before* and *after* purification should be submitted.

Safety Note and Hazard Warnings

Effective in August 2017, the first Note in every article is devoted to addressing the safety aspects of the procedures described in the article. The Article Template provides the required wording and format for Note 1, which reminds readers of the importance of carrying out risk assessments and hazard analyses prior to performing all experiments:

> Prior to performing each reaction, a thorough hazard analysis and risk assessment should be carried out with regard to each chemical substance and experimental operation on the scale planned and in the context of the laboratory where the procedures will be carried out. Guidelines for carrying out risk assessments and for analyzing the hazards associated

with chemicals can be found in references such as Chapter 4 of "Prudent Practices in the Laboratory" (The National Academies Press, Washington, D.C., 2011; the full text can be accessed free of charge at http://www.nap.edu/catalog.php?record_id=12654). See also "Identifying and Evaluating Hazards in Research Laboratories" (American Chemical Society, 2015) which is available via the associated website "Hazard Assessment in Research Laboratories" at https://www.acs.org/content/acs/en/about/governance/committees/chemicalsafety/hazard-assessment.html. In the case of this procedure, the risk assessment should include (but not necessarily be limited to) an evaluation of the potential hazards associated with *(enter list of chemicals here)*, as well as the proper procedures for *(list any unusual experimental operations here)*. *(Provide additional cautions with regard to exceptional hazards here)*.

For the required list of chemicals, authors should include all reactants, solvents, and other chemicals involved in the reactions described in the article.

With regard to the list of experimental operations, this list should be limited to those operations that potentially pose significant hazards. Examples may include

- Vacuum distillations
- Reactions run at elevated pressure or in sealed reaction vessels
- Photochemical reactions

In the case of experiments that involve exceptional hazards such as the use of pyrophoric or explosive substances, and substances with a high degree of acute or chronic toxicity, authors should provide additional guidelines for how to carry out the experiment so as to minimize risk. These instructions formerly would have appeared as red "Caution Notes" in *Organic Syntheses* articles. Note that it is not essential to describe general safety procedures such as working in a hood, avoiding skin contact, using eye protection, etc., since these are discussed in the Prudent Practices reference mentioned in the "Working with Hazardous Chemicals" statement within each article. Efforts should be made to avoid the use of toxic and hazardous solvents and reagents when less hazardous alternatives are available.

Discussion Section

The style and content of the discussion section will depend on the nature of the procedure.

For procedures that provide an improved method for the preparation of an important reagent or synthetic building block, the discussion should focus on the advantages of the new approach and should describe and reference all of the earlier methods used to prepare the title compound.

In the case of procedures that illustrate an important synthetic method or strategy, the discussion section should provide a mini-review on the new methodology. The scope and limitations of the method should be discussed, and it is generally desirable to include a table of examples. Please be sure each table is numbered and has a title. Competing methods for accomplishing the same overall transformation should be described and referenced. A brief discussion of mechanism may be included if this is useful for understanding the scope and limitations of the method.

Titles of Articles

In cases where the main thrust of the article is the illustration of a synthetic method of general utility, the title of the article should incorporate reference to that method. Inclusion of the name of the final product is acceptable but not required. In the case of articles where the objective is the preparation of a specific compound of importance (such as a chiral ligand), then the name of that compound should be part of the title.

Examples

Title without name of product:

"Stereoselective Synthesis of 3-Arylacrylates by Copper-Catalyzed Syn Hydroarylation" (*Org. Synth.* **2010**, *87*, 53).

Title including name of final product (note name of product is not required):

"Catalytic Enantioselective Borane Reduction of Benzyl Oximes: Preparation of (S)-1-Pyridin-3-yl-ethylamine Bis Hydrochloride" (*Org. Synth.* **2010**, *87*, 36).

Title where preparation of specific compound is the subject:

"Preparation of (S)-3,3'-Bis-Morpholinomethyl-5,5',6,6',7,7',8,8'-octahydro-1,1'-bi-2-naphthol" (*Org. Synth.* **2010**, *87*, 59).

Heading Scheme

The title of the article should be followed by a "Heading Scheme" comprising separate equations for each step in the article. Authors should consult the article template for instructions concerning ChemDraw settings and format. In general, reaction equations should not include details such as reaction time and the number of equivalents of reagents, with the exception of reactants employed in catalytic amounts which can be labeled as "cat." or by specifying mol%.

Style and Format for Text

Articles should follow the style guidelines used for organic chemistry articles published in the ACS journals such as *J. Am. Chem. Soc.*, *J. Org.*

Chem., *Org. Lett.*, etc. as described in the ACS Style Guide (3rd Ed.). The text of the procedure should be created using the Word template available on the *Organic Syntheses* website. Specific instructions with regard to the manuscript format (font, spacing, margins) is available on the website in the "Instructions for Article Template" and embedded within the Article Template itself.

Style and Format for Tables and Schemes

Chemical structures and schemes should be drawn using the standard ACS drawing parameters (in ChemDraw, the parameters are found in the "ACS Document 1996" option) with a maximum full size width of 15 cm (5.9 inches). The graphics files should then be pasted into the Word document at the correct location and the size reduced to 75% using "Format Picture" (Mac) or "Size and Position" (Windows). Graphics files must also be submitted separately. All Tables that include structures should be entirely prepared in the graphics (ChemDraw) program and inserted into the word processing file at the appropriate location. Tables that include multiple, separate graphics files prepared in the word processing program will require modification.

Tables and schemes should be numbered and should have titles. The title for a Table should be included within the ChemDraw graphic and placed immediately above the table. The title for a scheme should be included within the ChemDraw graphic and placed immediately below the scheme. Use 12 point Palatino Bold font in the ChemDraw file for all titles. For footnotes in Tables use Helvetica (or Arial) 9 point font and place these immediately below the Table.

Photographs

Photographs illustrating key elements of procedures are required in every article published in Organic Syntheses. Authors are expected to furnish photos with their original submissions and photos may also be provided by the Checkers of procedures. Photographs should be inserted into articles at the place in the text and Notes where they are first referred to and should be numbered and labeled as Figures with descriptive titles. Particularly useful subjects for photographs include:

- Photos of reaction flasks depicting how each neck is equipped
- Photos of reaction mixtures illustrating color changes, heterogeneity, etc.
- Photos of TLC plates showing degree of resolution and the color of spots
- Photos of crystalline reaction products illustrating color and crystal type

Acknowledgments and Author's Contact Information

Contact information (institution where the work was carried out and mailing address for the principal author) should be included as footnote 1. This footnote should also include the email address for the principal author, as well as ORCID for the principal author. Acknowledgment of financial support should be included in footnote 1.

Biographies and Photographs of Authors

Photographs and 100-word biographies of all authors should be submitted as separate files at the time of the submission of the procedure. The format of the biographies should be similar to those in the Volume 84 procedures found at the orgsyn.org website. Photographs can be accepted in a number of electronic formats, including tiff and jpeg formats.

DISPOSAL OF CHEMICAL WASTE

General Reference: *Prudent Practices in the Laboratory* National Academy Press, Washington, D.C. 2011.

Effluents from synthetic organic chemistry fall into the following categories:

1. **Gases**
   1a. Gaseous materials either used or generated in an organic reaction.
   1b. Solvent vapors generated in reactions swept with an inert gas and during solvent stripping operations.
   1c. Vapors from volatile reagents, intermediates and products.
2. **Liquids**
   2a. Waste solvents and solvent solutions of organic solids (see item 3b).
   2b. Aqueous layers from reaction work-up containing volatile organic solvents.
   2c. Aqueous waste containing non-volatile organic materials.
   2d. Aqueous waste containing inorganic materials.
3. **Solids**
   3a. Metal salts and other inorganic materials.
   3b. Organic residues (tars) and other unwanted organic materials.
   3c. Used silica gel, charcoal, filter aids, spent catalysts and the like.

The operation of industrial scale synthetic organic chemistry in an environmentally acceptable manner* requires that all these effluent categories be dealt with properly. In small scale operations in a research or academic setting, provision should be made for dealing with the more environmentally offensive categories.

1a. Gaseous materials that are toxic or noxious, e.g., halogens, hydrogen halides, hydrogen sulfide, ammonia, hydrogen cyanide, phosphine, nitrogen oxides, metal carbonyls, and the like.
1c. Vapors from noxious volatile organic compounds, e.g., mercaptans, sulfides, volatile amines, acrolein, acrylates, and the like.

---
*An environmentally acceptable manner may be defined as being both in compliance with all relevant state and federal environmental regulations *and* in accord with the common sense and good judgment of an environmentally aware professional.

2a. All waste solvents and solvent solutions of organic waste.
2c. Aqueous waste containing dissolved organic material known to be toxic.
2d. Aqueous waste containing dissolved inorganic material known to be toxic, particularly compounds of metals such as arsenic, beryllium, chromium, lead, manganese, mercury, nickel, and selenium.
3. All types of solid chemical waste.

Statutory procedures for waste and effluent management take precedence over any other methods. However, for operations in which compliance with statutory regulations is exempt or inapplicable because of scale or other circumstances, the following suggestions may be helpful.

Gases

Noxious gases and vapors from volatile compounds are best dealt with at the point of generation by "scrubbing" the effluent gas. The gas being swept from a reaction set-up is led through tubing to a large trap to prevent suck-back and into a sintered glass gas dispersion tube immersed in the scrubbing fluid. A bleach container can be conveniently used as a vessel for the scrubbing fluid. The nature of the effluent determines which of four common fluids should be used: dilute sulfuric acid, dilute alkali or sodium carbonate solution, laundry bleach when an oxidizing scrubber is needed, and sodium thiosulfate solution or diluted alkaline sodium borohydride when a reducing scrubber is needed. Ice should be added if an exotherm is anticipated.

Larger scale operations may require the use of a pH meter or starch/iodide test paper to ensure that the scrubbing capacity is not being exceeded.

When the operation is complete, the contents of the scrubber can be poured down the laboratory sink with a large excess (10-100 volumes) of water. If the solution is a large volume of dilute acid or base, it should be neutralized before being poured down the sink.

Liquids

Every laboratory should be equipped with a waste solvent container in which *all* waste organic solvents and solutions are collected. The contents of these containers should be periodically transferred to properly labeled waste solvent drums and arrangements made for contracted disposal in a regulated and licensed incineration facility.**

**If arrangements for incineration of waste solvent and disposal of solid chemical waste by licensed contract disposal services are not in place, a list of providers of such services should be available from a state or local office of environmental protection.

Aqueous waste containing dissolved toxic organic material should be decomposed *in situ*, when feasible, by adding acid, base, oxidant, or reductant. Otherwise, the material should be concentrated to a minimum volume and added to the contents of a waste solvent drum.

Aqueous waste containing dissolved toxic inorganic material should be evaporated to dryness and the residue handled as a solid chemical waste.

Solids

Soluble organic solid waste can usually be transferred into a waste solvent drum, provided near-term incineration of the contents is assured.

Inorganic solid wastes, particularly those containing toxic metals and toxic metal compounds, used Raney nickel, manganese dioxide, etc. should be placed in glass bottles or lined fiber drums, sealed, properly labeled, and arrangements made for disposal in a secure landfill.** Used mercury is particularly pernicious and small amounts should first be amalgamated with zinc or combined with excess sulfur to solidify the material.

Other types of solid laboratory waste including used silica gel and charcoal should also be packed, labeled, and sent for disposal in a secure landfill.

Special Note

Since local ordinances may vary widely from one locale to another, one should always check with appropriate authorities. Also, professional disposal services differ in their requirements for segregating and packaging waste.

**Robert K. Boeckman, Jr.
1944 – 2021**

Robert K. Boeckman, Jr., Marshall D. Gates Emeritus Professor of Chemistry at the University of Rochester, died on September 5, 2021 from complications of non-alcoholic steatohepatitis (NASH). Bob is survived by Mary Delton, his wife of 45 years.

Bob had been active in *Organic Syntheses* for more than three decades. He joined the Board of Editors in 1988, edited Volume 73 in 1996, and became a member of the Board of Directors in 2001. Bob served as President of Organic Syntheses, Inc. from 2012 until his death in 2021. The organization thrived under Bob's leadership with net worth and expenditures for programs in support of organic chemistry increasing by fifty percent.

Bob was born on August 26, 1944 in Pasadena, CA while his father was stationed there during WWII. He grew up in Dayton, OH where his parents continued operation of the popular Boeckman family butcher shop business. After his B.S. degree from Carnegie Institute of Technology he went to Brandeis University and completed his Ph.D. in 1970 mentored by James Hendrickson and Ernest Grunwald. After postdoctoral finishing school with Gilbert Stork at Columbia University, in 1972 he began his academic career as an Assistant Professor at Wayne State University. He quickly advanced through the ranks becoming full Professor there in 1979. In 1980 Andy Kende persuaded him to move to the University of Rochester where he spent the next four decades. He became the Marshall D. Gates Professor in 2002 and served as chair of the department from 2003 to 2013 before retiring from the University in July 2019. Bob's research efforts focused on the development of new synthetic methodology, including asymmetric synthesis, applicable to the synthesis of complex multifunctional target structures including biologically significant natural products. He and his coworkers developed innovative total syntheses of a number of complex natural products, with particularly notable and pioneering contributions in the area of sesterterpene and alkaloid synthesis. Included among his many conquests were gascardic acid, (+/-)-pleuromutilin, (+/-)-ceroplastol, (+/-)-tirandamycin A, (-)-nakadomarin A, (+)-ikarugamycin, (+)-tetronolide, (+)-laurenyne, and (+) and (-)-saudin. After retirement he continued with a medicinal chemistry project involving bisphosphonates and bisphosphonate drug conjugates. His work resulted in more than 150 publications. To the University of Rochester, Bob bequeathed endowment for the Robert K. Boeckman Jr and Mary H. Delton Family Distinguished Professorship in Organic Chemistry.

In a scenario engineered by Hans and Ieva Reich, Bob met his wife Mary Delton at a chemistry conference in Kingston, Jamaica in January 1976. They were married the following June. Bob and Mary shared loves of science, horses, cats, and nature. Upon Bob's retirement Bob, Mary, favorite horses and cats moved from their horse farm in Honeoye Falls, NY outside of Rochester to a home of their design complete with a horse barn and riding trails in the Adirondacks. Trees on the property provided for a semi-commercial production of maple syrup. Bob was a long-term supporter and recent President of Pet Pride of New York, Inc., a no-kill, cats only, sanctuary and adoption center; the organization truly prospered under his leadership.

Bob served as an Associate Editor of the *Journal of Organic Chemistry* from 1997 until 2016. He served as Chair of the ACS Division of Organic Chemistry in 2001. He was an inaugural American Chemical Society Fellow and was also a Fellow of the American Association for the Advancement of Science and the Japanese Society for the Promotion of Science. Among Bob's other honors are an A.P. Sloan Fellowship, a Research Career Development Award from the National Institutes of Health, and an ACS Cope Scholar Award. He was twice awarded the Alexander Von Humboldt Stiftung Research Prize for Senior Scientists. At Rochester, he received the William H. Riker University Award for graduate teaching in 2009. Bob served for many years as a consultant to Novartis and a variety of other pharmaceutical companies

Bob achieved remarkable balance in his life. He was internationally recognized for his creative contributions to the art and practice of organic syntheses. He was exemplary in his service to his department, to his university, to his profession and to society. He was a praiseworthy teacher at all levels from large undergraduate classes to one-on-one with members of his research group. He was a steadfast supporter of his faculty colleagues and students and postdocs from his group. Bob had an extensive collection of sweater vests – constantly wearing them was a source of comfort for Bob and amusement for his wife, colleagues and students.

He made time to enjoy his personal life – their horses, their cats, nature, good food, fine wine and great investments. He was a caring and responsive friend. He is profoundly missed.

Carl R. Johnson

James D. White
1935 – 2020

James D. White, Emeritus Professor of Chemistry at Oregon State University, died on February 10, 2020, in Albany, Oregon, at the age of 84. He is survived by his daughters, Julie White and Amy Blake, and his grandchildren, Joanna White, Logan Blake, and Andrew Blake.

Jim White was born in Bristol, England, in 1935. He spent much of an eventful childhood living in India after his family moved there to escape Europe during the tumultuous years of World War II. Following the end of the war, the White family returned to England and Jim completed his school days in Tiverton, Devon. After post-war national service as a Pilot Officer in the Royal Air Force, he pursued higher education at Queen's College, University of Cambridge, graduating with a B.A. degree in Natural Sciences in 1959. From that time onwards, Jim would make his home in North America, venturing with his first wife Muriel initially to Vancouver, Canada, where he studied for an M.S. degree with Raymond Bonnett at the University of British Columbia. In 1961, Jim transitioned to the Massachusetts Institute of Technology to pursue a Ph.D. degree with George Büchi. The themes explored by the work that he conducted for his Ph.D. dissertation, a total synthesis of thujopsene and the isolation and structural elucidation of two metabolites from *Penicillium rubrum*, would cement Jim's life-long fascination with natural products chemistry.

Remaining in Cambridge, MA, Jim took up his first independent academic appointment as an Instructor of Chemistry at Harvard University in 1965. Within six years, he had advanced to the level of Associate Professor at Harvard before being eagerly recruited to join the chemistry faculty at Oregon State University in Corvallis, Oregon. Jim and his young family happily settled into life in Corvallis and he spent the rest of his long and productive career at OSU, retiring as Distinguished Professor Emeritus in 2003. He remained an active force on campus and within the wider chemistry community long after his retirement, publishing research papers well into his eighties, the most recent of which appeared in 2019.

Jim's great passion, and the focus of nearly all of his research endeavors over the years, was the total synthesis of complex natural product molecules. A J. D. White synthesis is unmistakable for its elegance and characterized by the execution of unusual stratagems often based on transformation types that had yet to enter the mainstream at the time of its conception. In this regard, Jim was a leader in the field of target-directed synthesis and a common reaction to seeing a new synthesis from his lab (always of an important and well-chosen molecule) would be, "I wish I had thought of that!" For example, Jim's synthesis of codeine incorporates an intramolecular Rh-catalyzed CH insertion from a diazoketone to forge the D- ring from a complete phenanthrene template. This effort occurred long before CH functionalization became a routinely explored tactic for complex molecule synthesis. Jim was likewise an early adopter of ring-closing olefin metathesis (RCM) in the arena of target-directed synthesis; here, the elaboration of alkaloids such as australine and pinnaic acid via transannular reactions of unsaturated azacyclic precursors generated by RCM is instructive. Photochemical transformations also featured prominently in Jim's work over the years. A beautiful demonstration is found in his synthesis of byssochlamic acid, one of the so-called nonadride natural products, via an ingenious [2+2] cycloaddition/cycloreversion approach to construct a nine-membered ring from a transient strained polycyclic intermediate. Similarly remarkable highlights are strewn throughout the large number of other notable syntheses that the White laboratory was able to achieve with Jim at the helm. Interested readers are encouraged to avail themselves of their preparations of euonyminol, avermectin, ibogamine, and the rutamycins, in particular.

Jim White served on the editorial boards of several important international journals and periodicals, including *Organic Syntheses* (1983-1991), for which he was Editor-in-Chief (in 1989), and *Organic Reactions* (1991- 1997). He was an Associate Editor for the *Journal of the American Chemical Society* from 1989 until 1994, and he served as U.S. Editor for *Chemical Communications* from 1996 until 2004. He was the recipient of numerous national and international awards including a Guggenheim Fellowship (1990), the Centenary Medal of the Royal Society of Chemistry (1999), the American Chemical Society Arthur C. Cope Scholar Award (2003), and the 2006 Outstanding Scientist Award of the Oregon Academy of Sciences. In 1995, he was awarded an honorary Sc.D. degree from his alma mater, the University of Cambridge.

Jim was an active sportsman who enjoyed tennis, skiing, and sailing, pursuits that he shared with his second wife Valerie Bishop, wo sadly predeceased him. He loved to travel, for both work and pleasure, and he was a common sight at scientific meetings around the world and on excursions to exotic places organized by the MIT alumnae program with his partner, Wendy McKee. Well-known and widely admired in the chemistry community, he was a friend and a respected colleague to many and he helped to shape the careers of a large number of his graduate student and post-doctoral coworkers, myself included. He is greatly missed

Paul R. Blakemore
Corvallis, Oregon

PREFACE

This 98th volume of *Organic Syntheses* provides a practical and diverse synthetic transformation handbook, leading to a cross-pollination of ideas, helping to overcome existing hurdles, and makes sustainable chemistry an integral part of the chemical society. This volume was generated by stellar worldwide academic and industrial scientists and has been independently checked by researchers in the laboratories of one of the Organic Syntheses Editorial Board Members.

In the classic work of logics and tactics of chemical synthesis, Corey and Cheng articulated the value of considering molecular complexity in synthesis planning and pointed to molecular size, element and functional group content, cyclic connectivity, stereocenter content, chemical reactivity, and structural instability as key parameters in the creation of synthetic strategies that previously did not exist. While chemical reactivity and structural instability affect the likelihood of success during the physical manipulation of compounds, the remaining features are explicitly related to the graphical representation of the target structure.

It is important to note that the above-mentioned activities have certainly improved our standard of living and the well-being of our planet as a whole. Organic synthesis influences many areas that include the production of common materials, such as plastics, polymers, basic chemicals, textiles, agricultural products and make sustainable chemistry an integral part of the chemical society. It also has an important role in energy and related fundamental chemistry such as electrochemistry, electricity, photoredox etc. and plays a huge role in the development of critical medicines.

Organic Syntheses consist of a renowned academic and industry Board of Directors and outstanding Editorial Board Members who have incredibly supported the success for the 98th volume during the ongoing COVID-19 pandemic. The stellar Associate Editor Chuck Zercher and our Editor-in-Chief, Rick Danheiser have continuously kept the Board of Editors informed and maintain a high level of transparency and provide enormous guidance to accomplish the task of the Editorial Board. I would like to thank them for their patience with me as both a member of the Board of Editors and as the editor of Volume 98. The experience of serving on the Board of Editors for the past eight years was amazing and intellectually inspiring. I am grateful to all the past and current members of the Board for providing such a memorable experience with the Organic Syntheses Organization. I would also like to thank Dr. Gopal Sirasani and Dr. Praveen Gajula for supporting and proofreading the content of the procedures before submission. I am incredibly fortunate to have an advisor, mentor and a great friend, Professor Carl Johnson. I would like to thank him for providing incredible social events and gatherings which all participants feel inclusive, while enjoying a friendship of lifetime and cherishable memories.

CHRIS SENANAYAKE

TABLE OF CONTENTS

Synthesis of Chiral Organoiodine Catalyst for Enantioselective Oxidative Dearomatization Reactions: *N,N'*-(2*S*,2'*S*)-(2-Iodo-1,3-phenylene)bis(oxy)bis(propane-2,1-diyl)bis(2,4,6-trimethylbenzamide)   1

Muhammet Uyanik, Shinichi Ishizaki and Kazuaki Ishihara

Chiral Organoiodine-catalyzed Enantioselective Oxidative Dearomatization of Phenols   28

Muhammet Uyanik, Shinichi Ishizaki and Kazuaki Ishihara

Reductive Deuteration of Ketones with Magnesium and D₂O for the Synthesis of α-Deutero-O-methyl-benzhydrol 51

Nengbo Zhu, Wen-Ming Wan and Hongli Bao

Palladium-Catalyzed Acetylation of Arylbromides 68

Milauni M. Mehta, Andrew V. Kelleghan, and Neil K. Garg

Cu-catalyzed Allylic Perfluoroalkylation of Alkenes Using Perfluoro Acid Anhydrides: Preparation of N-(5,5,5-Trifluoro-2-penten-1-yl)phthalimide 84

Yuma Aoki, Shintaro Kawamura, and Mikiko Sodeoka

Synthesis of a Phosphorous Sulfur Incorporating Reagent 97
for the Enantioselective Synthesis of Thiophosphates

Prantik Maity, Amitha S. Anandamurthy, Vijaykumar Shekarappa, Rajappa
Vaidyanathan, Bin Zheng, Jason Zhu, Michael A Schmidt, Richard J. Fox, Kyle W.
Knouse, Julien C. Vantourout, Phil S. Baran, and Martin D. Eastgate

Discussion Addendum for:
Preparation of (S)-tert-ButylPyOx and Palladium-Catalyzed 117
Asymmetric Conjugate Addition of Arylboronic Acids

Stephen R. Sardini and Brian M. Stoltz

Palladium-Catalyzed Hydrodefluorination of Fluoroarenes 131

Joseph J. Gair, Ronald L. Grey, Simon Giroux, and Michael A. Brodney

Synthesis of the Isocyanide Building Block Asmic, anisylsulfanylmethylisocyanide

Embarek Alwedi, Bilal Altundas, Allen Chao, Zachary L. Ziminsky, Maanasa Natrayan, and Fraser F. Fleming

Mild *mono*-Acylation of 4,5-Diiodoimidazole: Preparation of 1-(5-Iodo-1*H*-imidazole-4-yl)pent-4-en-1-one

Michael Morgen, Jasmin Lohbeck, and Aubry K. Miller

Enantioselective Michael-Proton Transfer-Lactamization for Pyroglutamic Acid Derivatives: Synthesis of Dimethyl-(*S,E*)-5-oxo-3-styryl-1-tosylpyrrolidine-2,2-dicarboxylate

Christian M. Chaheine, Conner J. Song, Paul T. Gladen, and Daniel Romo

**Preparation of 1-Hydrosilatrane, and Its Use in the Highly   227
Practical Synthesis of Secondary and Tertiary Amines from
Aldehydes and Ketones via Direct Reductive Amination**

Fawwaz Azam and Marc J. Adler

**Synthesis of Tetraaryl-, Pentaaryl-, and Hexaaryl-1,4-dihydropyrrolo-   242
[3,2-*b*]pyrroles**

Maciej Krzeszewski, Mariusz Tasior, Marek Grzybowski and Daniel T. Gryko

Preparation of Hindered Aniline CyanH and Application in the Allyl-Ni-Catalyzed α,β-Dehydrogenation of Carbonyls

Alexandra K. Bodnar, Aneta Turlik, David Huang, Will Butcher, Joanna K. Lew, and Timothy R. Newhouse

Synthesis of tert-Alkyl Phosphines: Preparation of Di-(1-adamantyl)phosphonium Trifluoromethanesulfonate and Tri-(1-adamantyl)phosphine

Thomas Barber and Liam T. Ball

Preparation of 1-Benzyl-7-methylene-1,5,6,7-tetrahydro-4H-benzo[d]imidazol-4-one 315

Michael Morgen, Jasmin Lohbeck and Aubry K. Miller

Catalytic Diazoalkane-Carbonyl Homologation: Synthesis of 2,2-Diphenylcycloheptanone and Other Quaternary or Tertiary Arylalkanones and Spirocycles by Ring Expansion 343

Jason S. Kingsbury, Victor L. Rendina, Jacob S. Burman, and Brittany A. Smolarski

C2 Amination of Pyridine with Primary Amines Mediated by Sodium Hydride in the Presence of Lithium Iodide 363

Jia Hao Pang, Derek Yiren Ong, and Shunsuke Chiba

Synthesis and Acylation of 1,3-Thiazinane-2-thione

Stuart C. D. Kennington, Oriol Galeote, Miguel Mellado-Hidalgo,
Pedro Romea and Fèlix Urpí

**Preparation of (Bis)Cationic Nitrogen-Ligated I(III) Reagents:
Synthesis of [(pyridine)$_2$IPh](OTf)$_2$ and [(4-CF$_3$-pyridine)$_2$IPh](OTf)$_2$**

Bilal Hoblos and Sarah E. Wengryniuk

Preparation of 6-(Triethylsilyl)cyclohex-1-en-1-yl Trifluoromethanesulfonate as a Precursor to 1,2-Cyclohexadiene 407

Ryo Nakura, Kazuki Inoue, Mayu Itoh, Atsunori Mori, and Kentaro Okano

Discussion Addendum for:
Intra- and Intermolecular Kulinkovich Cyclopropanation Reactions of Carboxylic Esters with Olefins: Bicyclo[3.1.0]-hexan-1-ol and *trans*-2-benzyl-1-methylcyclopropan-1-ol 430

Jin Kun Cha

Synthesis of Chiral Aziridine Ligands for Asymmetric Alkylation with Alkylzincs: Diphenyl((S)-1-((S)-1-phenylethyl)aziridin-2-yl)methanol

Siyuan Sun and Pavel Nagorny

Large-Scale Preparation of Oppolzer's Glycylsultam

Upendra Rathnayake, H. Ümit Kaniskan, Jieyu Hu, Christopher G. Parker, and Philip Garner

Stereoselective [2+2] Cycloadditions to Glucal Derived β-Lactams: Synthesis of Tri-O-Bn-D-glucal and Derivatives 491

Maria Varghese, Hannah E. Caputo, Ruiqing Xiao, Anant Balijepalli, Aladin Hamoud, and Mark W. Grinstaff

Preparation of 6-(Triethylsilyl)cyclohex-1-en-1-yl Trifluoromethanesulfonate as a Precursor to 1,2-Cyclohexadiene 509

Kazuki Inoue, Kengo Inoue, Atsunori Mori, and Kentaro Okano

Late-stage C–H Functionalization with 2,3,7,8–Tetrafluorothianthrene: Preparation of a Tetrafluorothianthrenium-salt

Samira Speicher, Matthew B. Plutschack, and Tobias Ritter

Synthesis of Chiral Organoiodine Catalyst for Enantioselective Oxidative Dearomatization Reactions: *N,N'*-(2*S*,2'*S*)-(2-Iodo-1,3-phenylene) bis(oxy)bis(propane-2,1-diyl)bis(2,4,6-trimethylbenzamide)

Muhammet Uyanik, Shinichi Ishizaki and Kazuaki Ishihara*[1]

Graduate School of Engineering, Nagoya University, Furo-cho, Chikusa, Nagoya 464-8603, Japan

Checked by Cayetana Zarate and Kevin Campos

Procedure (Note 1)

A. *tert-Butyl (R)-(2-hydroxypropyl)carbamate (1)*. A 250 mL, two-necked (main 24/40, side 24/40 joints), round-bottomed flask is equipped with a 25 x 8 mm, Teflon-coated magnetic stir bar, a 10-mL graduated, pressure-equalizing addition funnel fitted with a rubber septum at the top, and an inlet adapter with 3-way stopcock (Note 2) (Figure 1A). The set-up is evacuated under high vacuum (5.0 mmHg, few seconds) and filled with nitrogen (three cycles). Under a positive pressure of nitrogen, the flask is charged with (R)-1-amino-2-propanol (3.90 mL, 49.3 mmol, 1.0 equiv) (Note 3), triethylamine (7.24 mL, 52 mmol, 1.04 equiv) (Note 4), and CH_2Cl_2 (38 mL) (Note 5) via syringe in less than 1 minute each through the inlet adapter. The flask is immersed in an ice-water bath (0 °C) and the mixture is stirred at 500 rpm for 10 min under a nitrogen atmosphere while the addition funnel is closed and charged with di-*tert*-butyl dicarbonate (Boc_2O, 11.8 mL, 51.5 mmol, 1.03 equiv) (Note 6). Boc_2O is added dropwise to the stirring mixture at 0 °C over 20 min via the addition funnel (Figure 1B). The inside wall of the addition funnel is washed with CH_2Cl_2 (2 mL) via a syringe. The colorless reaction mixture is allowed to warm to 26 °C, and the resulting mixture is stirred at 400 rpm for 5 h at 26 °C (Note 7). The resulting colorless solution (Figure 1C) is concentrated by rotary evaporator (30 °C, 15 mmHg) resulting in a colorless oil. The crude material (Figure 1D) is further purified by flash column chromatography (Note 8) (Figures 1E–G) to give **1** (8.16 g, 46.6 mmol, 93%) as a colorless oil (Notes 9, 10, and 11) (Figure 1H).

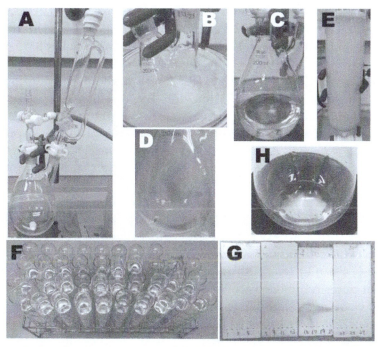

Figure 1. Synthesis of compound **1**; (A) Reaction setup; (B,C) Reaction progress; (D) Crude product; (E) Column chromatography (the checkers used a Combi-flash system); (F) Fractions; (G) TLC of fractions; (H) Pure product **1** (photos provided by submitters)

B. *Di-tert-butyl ((2S,2'S)-((2-iodo-1,3-phenylene)bis(oxy))bis(propane-2,1-diyl))dicarbamate (2)*. A 250 mL, two-necked (main 24/40, side 24/40 joints), round-bottomed flask is charged with **1** (6.86 g, 40.0 mmol, 2.5 equiv), 2-iodoresorcinol (3.78 g, 16.0 mmol, 1.0 equiv) (Note 12), and triphenylphosphine (10.5 g, 40.0 mmol, 2.5 equiv) (Note 13), and the flask is equipped with a 25x8 mm, Teflon-coated magnetic stir bar, a 60-mL graduated, pressure-equalizing addition funnel fitted with a rubber septum at the top, and a rubber septum with an inlet connected to a nitrogen Schlenk line (Note 2) (Figure 2A). The whole system is evacuated under high vacuum (5.0 mmHg, few seconds) and filled with nitrogen (three cycles). Under a positive pressure of nitrogen, THF (48.0 mL) (Note 14) is added via a syringe in less than 1 minute through the rubber septum. The flask is immersed in an ice-water bath (0 °C) and the mixture is stirred at 500 rpm for 10 min under a nitrogen atmosphere while the addition funnel is closed and charged with diisopropyl azodicarboxylate (DIAD, 1.9 M in toluene, 21.1 mL, 40.0 mmol, 2.5 equiv) (Note 15). DIAD is added dropwise to the stirring mixture over 40 min via the addition funnel at 0 °C (Figure 2B). The inside wall of the addition funnel is washed with THF (2 mL) via syringe. The reaction mixture is allowed to warm, and the mixture is stirred at 500 rpm for 13 h at 26 °C (Note 16). The resulting clear pale-yellow solution (Figure 2C) is concentrated by rotary evaporator (30 °C, 15 mmHg) for 15 min until about ca. 30 mL remains. To precipitate the byproducts (phosphine oxide and hydrazine derivatives), hexanes (30 mL) (Note 17) is added and the resulting mixture is stirred at 500 rpm for 5 min to make a slurry, which is then concentrated by rotary evaporator (30 °C, 15 mmHg) for additional 5 min. To further precipitate the byproducts, Et_2O (10 mL) (Note 18) and hexanes (30 mL) (Note 17) are added and the resulting mixture is stirred at 500 rpm for 5 min to make a slurry. The resulting mixture is filtered through a plug of tightly packed celite (3 g) pre-wetted with Et_2O (30 mL) (Note 18) on a sintered glass funnel (4 cm diameter, medium porosity) under vacuum suction (Figure 2D). The flask is rinsed with Et_2O (50 mL) (Note 16) and this rinse is used to wash the precipitate. The wet-cake is further washed with Et_2O (ca. 200 mL) (Note 18). The filtrate, collected in a 500-mL round-bottomed flask, is concentrated by rotary evaporator (30 °C, 40 mmHg). The crude residue (Figure 2E) is purified by column chromatography (Note 19) (Figure 2F–H) to give **2** as a colorless amorphous solid (8.12 g, 14.8 mmol, 93%) (Notes 20, 21, and 22) (Figure 2I), which is used in the next step.

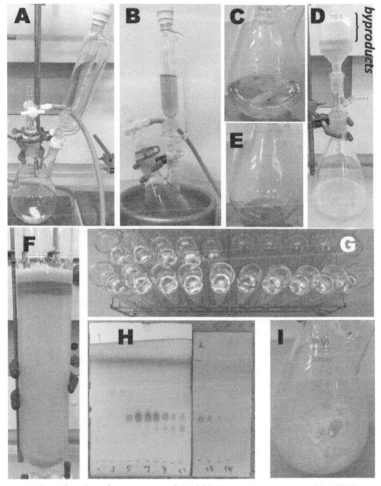

Figure 2. Synthesis of compound 2; (A) Reaction setup; (B, C) Reaction progress; (D) Filtration; (E) Crude product; (F) Column chromatography; (G) Fractions; (H) TLC of fractions; (I) Pure product 2 (photos provided by submitters)

C. *N,N'-((2S,2'S) ((2-Iodo-1,3-phenylene)bis(oxy))bis(propane-2,1-diyl))bis-(2,4,6-trimethylbenzamide)* (**4**). A 500 mL, two-necked (24/40), round-bottomed flask is charged with **2** (as obtained from step B) and the flask is equipped with a 25 x 8 mm, Teflon-coated magnetic stir bar, an inlet adapter with 3-way stopcock fitted with a nitrogen inlet (Note 2) (Figure 3A). The set-up is evacuated under high vacuum (5.0 mmHg, few seconds) and filled with nitrogen (three cycles). Dichloromethane (75 mL) (Note 5) (Figure 3B) and trifluoroacetic acid (11.0 mL, 143 mmol, 10.1 equiv) (Note 23) (Figure 3C) are added via syringe over ca. 1 min and 5 min, respectively, through the rubber septum at 23 °C. The resulting mixture is stirred at 500 rpm for 6 h at 26 °C (Note 24). The resulting pale brown suspension (Figure 3D) is cooled to 0 °C in an ice-bath and quenched with 2 M NaOH (80 mL, pH 13) (Note 25) (Figure 3E). The mixture is transferred to a 200 mL separatory funnel. The aqueous layer is separated and extracted with CH_2Cl_2 (8 x 80 mL) (Notes 26 and 27) (Figure 3F). The combined organic layers are transferred to a 1 L separatory funnel and washed with saturated brine (150 mL) (Note 28), dried over anhydrous Na_2SO_4 (60 g) (Note 29), and filtered through a sintered glass funnel (6.5 cm diameter, medium porosity) under vacuum suction. The filtrate is concentrated by rotary evaporation (30 °C, 15 mmHg) and dried under vacuum (23 °C, 5 mmHg, 13 h) to give **3** as an orange oil (4.99 g, including some impurities) (Figure 3G), which is used in the next step without further purification (Note 30).

Figure 3. Synthesis of compound 3; (A) Reaction setup; (B–D) Reaction progress with color changing; (E) Quenching; (F) Work-up; (G) Crude product 3 (photos provided by submitters)

A 250 mL, two-necked (main 24/40, side 15/25 joints), round-bottomed flask equipped with a 3.5 cm, Teflon-coated magnetic stir bar is charged with **3** (as obtained from previous reaction) and 4-dimethyl aminopyridine (3.81 g, 31.2 mmol, 2.2 equiv) (Note 31) and the flask is equipped with a 20-mL graduated, pressure-equalizing addition funnel fitted with a rubber septum at the top, and an inlet adapter with 3-way stopcock connected to a nitrogen Schlenk line (Note 2) (Figure 4A). The set-up is evacuated under high vacuum (5.0 mmHg, few seconds) and filled with nitrogen (three cycles). CH_2Cl_2 (100 mL) (Note 5) and triethyl amine (6.58 mL, 46.8 mmol, 3.0 equiv) (Note 4) are added via syringe in less than 1 minute each through the inlet adapter. The flask is immersed in an ice-water bath at 0 °C and stirred at 500 rpm for 10 min under a nitrogen atmosphere while the addition funnel is closed and charged with 2,4,6-trimethylbenzoyl chloride (MesCOCl, 54.6 mmol, 9.07 mL, 3.5 equiv) (Note 32). MesCOCl is added dropwise for ca. 7 min to the stirring mixture via the addition funnel at 0 °C (Figure 4B). The inside wall of the addition funnel is washed with CH_2Cl_2 (2 mL) via syringe. The reaction mixture is allowed to warm to 26 °C, and the resulting mixture is stirred at 500 rpm for 13 h at 26 °C (Note 33). The resulting clear yellow solution (Figure 4C) is cooled to 0 °C and quenched with 1 M HCl (70 mL, pH = 1) (Note 34) (Figure 4D). The mixture is transferred to a 300 mL separatory funnel. The aqueous layer is separated and extracted with CH_2Cl_2 (2 x 80mL) (Note 26) (Figure 4E). The combined organic layers are transferred to a 500 mL separatory funnel, washed with saturated aqueous $NaHCO_3$ (100 mL) (Note 35) (Figure 4F) and saturated brine (100 mL) (Note 28), and dried over anhydrous Na_2SO_4 (40 g) (Note 29). The combined organic layers are filtered through a sand (Note 36), celite (Note 37), silica gel (Note 38) and 3-aminopropyl-functionalized silica gel (NH silica) (Note 39) pad (Note 40) on a sintered glass funnel (6.5 cm diameter, medium porosity) under vacuum suction, and the pad is washed with CH_2Cl_2 (100 mL) (Note 26) followed by EtOAc (200 mL) (Note 41) (Figure 4G). The filtrate (Figure 4H) is concentrated by rotary evaporation (30 °C, 15 mmHg) and dried under vacuum (23 °C, 5.0 mmHg, 1 h). The crude yellow residue (Figure 4I) is recrystallized (Note 42) from toluene (Note 43) at –20 °C to provide **4** (3.90–4.59g, 6.06–7.14 mmol) as a pale white solid. The filtrate from the crystallization was evaporated and the resulting solid recrystallized (Note 44) from toluene (Note 43) to provide a second batch of **4** (1.88–2.62 g, 2.93–4.08 mmol) as a pale white solid. The two crops of recrystallized product were combined to give **4** (6.47 g, 10.07 mmol, 66%) as a pale white solid (Figure 4J) (Notes 45, 46, and 47).

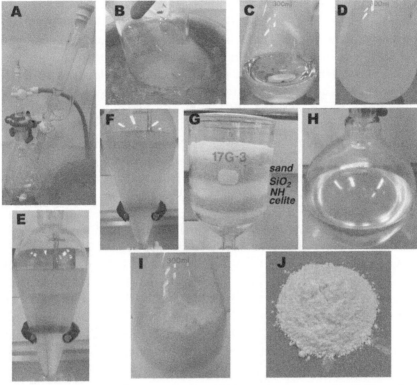

Figure 4. Synthesis of compound 4; (A) Reaction setup; (B, C) Reaction progress; (D) Quenching; (E–H) Work-up; (I) Crude; (J) Pure product 4 (photos provided by submitters)

Notes

1. Prior to performing each reaction, a thorough hazard analysis and risk assessment should be carried out with regard to each chemical substance and experimental operation on the scale planned and in the context of the laboratory where the procedures will be carried out. Guidelines for carrying out risk assessments and for analyzing the hazards associated with chemicals can be found in references such as Chapter 4 of "Prudent Practices in the Laboratory" (The National Academies Press, Washington, D.C., 2011; the full text can be accessed free of charge at https://www.nap.edu/catalog/12654/prudent-practices-in-the-laboratory-handling-and-management-of-chemical. See

also "Identifying and Evaluating Hazards in Research Laboratories" (American Chemical Society, 2015) which is available via the associated website "Hazard Assessment in Research Laboratories" at https://www.acs.org/content/acs/en/about/governance/committees/chemicalsafety/hazard-assessment.html. In the case of this procedure, the risk assessment should include (but not necessarily be limited to) an evaluation of the potential hazards associated with (*R*)-1-amino-2-propanol, triethyl amine, dichloromethane, nitrogen, di-*tert*-butyl dicarbonate, silica gel, hexanes, ethyl acetate, chloroform, 2-iodoresorcinol, triphenylphosphine, tetrahydrofuran, diisopropyl azodicarboxylate, toluene, hexanes, ethyl acetate, diethyl ether, trifluoroacetic acid, sodium hydroxide, sodium chloride, sodium sulfate, 4-dimethyl aminopyridine, 2,4,6-trimethylbenzoyl chloride, hydrogen chloride and sodium hydrogen carbonate as well as the proper procedures for experimental operations.
2. The reaction was performed under a positive pressure of nitrogen gas by using a Schlenk line.
3. (*R*)-1-Amino-2-propanol (>97.0%) was purchased from Angene and used as received.
4. Triethylamine (98.0%) was purchased from Apollo Scientific and used as received.
5. Dichloromethane (anhydrous, 99.5%) was purchased from Acros Organics and used as received.
6. Di-*tert*-butyl dicarbonate (>95.0%) was purchased from Apollo Scientific and used as received.
7. The reaction progress was monitored by TLC analysis (TLC Silica gel 60 F254, pre-coated plates (0.25 mm) purchased from Merck) (visualized with ninhydrin stain solution in EtOH/AcOH) with EtOAc/hexanes (1:1) as eluent. Product R_f = 0.31, starting material R_f = 0.00.
8. The product **1** was purified by flash column chromatography in a Teledyne ISCO CombiFlash instrument using a RediSep® RF 120 g gold silica column, a gradient of EtOAc/hexanes from 15% to 50% and a flow rate of 85 mL/min. Hexanes and EtOAc were purchased from Acros Organics and used as received. The reaction crude oil was added to a RediSep® RF 12 g silica cartridge and the residue in the flask was transferred to the cartridge with hexanes (1 mL). The cartridge was then connected to the column. The product was eluted with 1.0 L of hexanes/EtOAc (15% EtOAc) followed by 1.0 L of hexanes/EtOAc (50% EtOAc). The column chromatography was monitored by TLC analysis

(TLC Silica gel 60 F254, pre-coated plates (0.25 mm) purchased from Merck) (visualized with ninhydrin stain solution in EtOH/AcOH) with EtOAc/hexanes (1:1) as eluent. Product R_f = 0.31. Fractions (25 x 150 mm, 50 mL) 27–39 were collected and concentrated by rotary evaporation (30 °C, 15 mmHg) and dried under high vacuum (23 °C, 5.0 mmHg, 40 h). Product **1** was used in step B.

9. *tert*-Butyl (*R*)-(2-hydroxypropyl)carbamate (**1**): ^{1}H NMR (CDCl$_3$, 600 MHz): δ: 1.17 (d, *J* = 6.3 Hz, 3H), 1.44 (s, 9H), 2.30 (br s, 1H), 2.99 (dd, *J* = 14.0, 7.5 Hz, 1H), 3.15 – 3.34 (m, 1H), 3.89 (ddq, *J* = 9.5, 6.4, 3.2 Hz, 1H), 4.95 (br s, 1H); ^{13}C NMR (CDCl$_3$, 151 MHz) δ: 20.8, 28.5, 48.2, 67.8, 79.8, 156.9; IR (film): 3362, 2974, 1685, 1159, 761, 691 cm^{-1}; [α]$^{20}_D$ = –359 (*c* 1.11, CHCl$_3$); Anal. Calcd. For C$_8$H$_{17}$NO$_3$: C, 54.84; H, 9.78; N, 7.99. Found: C, 53.47; H, 9.71; N, 7.73.

10. The purity of compound **1** (98%) was determined by quantitative ^{1}H NMR analysis using compound **1** (29.33 mg, 0.167 mmol) and 1,2,4,5-tetramethylbenzene (20.66 mg, 0.154 mmol) as an internal standard.

11. A second reaction on the same scale provided 8.32 g (95%) of the product with identical purity.

12. 2-Iodoresorcinol (>97.0%) was purchased from Combi-Blocks. This compound was purified prior use to provide high yield. To this compound (5.13–5.27 g) was added CHCl$_3$ (6 mL) at –10 °C, and the solids were collected by suction filtration on a 60 mL sintered glass funnel (24/40 frit, fine porosity), washed with cold (–10 °C) CHCl$_3$ (6 mL) and dried under vacuum (23 °C, 5.0 mmHg, 5 h) to give a pure compound.

13. Triphenylphosphine (>97.0%) was purchased from Combi-Blocks and used as received.

14. THF (anhydrous, 99.5%) was purchased from Acros Organics and used as received.

15. Diisopropyl azodicarboxylate (DIAD) was purchased from eMolecules, Inc. and a 1.9 M solution in toluene was prepared dissolving DIAD (11.1 mL, 56.3 mmol) in 30 mL toluene. Toluene (anhydrous) was purchased from Acros Organics and used as received.

16. The reaction progress was monitored by TLC analysis (TLC Silica gel 60 F254, pre-coated plates (0.25 mm) purchased from Merck) (visualized with ninhydrin) with EtOAc/hexanes (1:2) as eluent. Product R_f = 0.49, starting material R_f = 0.29.

17. Hexanes (≥95.0%) was purchased from Acros Organics and used as received.

18. Diethyl ether (≥99.5%) was purchased from Sigma-Aldrich and used as received.
19. Product **2** was purified by flash column chromatography in a Teledyne ISCO CombiFlash instrument using a RediSep® RF 120 g gold silica column, a gradient of EtOAc/hexanes from 10% to 20%, and a flow rate of 85 mL/min. The reaction crude was charged to a RediSep® Rf 12 g silica cartridge that later was connected to the column. The product was eluted with 1.5 L of hexanes/EtOAc (10% EtOAc) followed by 1.5 L of hexanes/EtOAc (20% EtOAc). The column chromatography was monitored by TLC analysis (TLC Silica gel 60 F254, pre-coated plates (0.25 mm) purchased from Merck) (visualized with 254 nm UV lamp and phosphomolybdic acid stain solution) with EtOAc/hexanes (1:2) as eluent. Product R_f = 0.49. The fractions (25x150 mm, 50 mL) 32-51 were collected and concentrated by rotary evaporation (30 °C, 15 mmHg), and dried under high vacuum (23 °C, 5.0 mmHg, 40 h) to give compound **2**.
20. Di-*tert*-butyl ((2S,2'S)-((2-iodo-1,3-phenylene)bis(oxy))bis(propane-2,1-diyl))dicarbamate (**2**): ^{1}H NMR (CDCl$_3$, 600 MHz) δ: 1.33 (d, *J* = 6.2 Hz, 6H), 1.43 (s, 18H), 3.31 (dt, *J* = 13.5, 6.2 Hz, 2H), 3.59 – 3.46 (m, 2H), 4.58 – 4.43 (m, 2H), 5.09 (br s, 2H), 6.50 (d, *J* = 8.3 Hz, 2H), 7.20 (t, *J* = 8.2 Hz, 1H); ^{13}C NMR (CDCl$_3$, 151 MHz) δ: 17.3, 28.5, 45.8, 75.5, 79.6, 82.4, 107.3, 129.9, 156.3, 158.1; LC/MS analysis (Agilent 1200 instrument with an Inductively Coupled Plasma (IPC)): *m/z* (relative intensity): 573 (97.35%); [α]^{20}$_D$ = +87.0 (*c* 0.02, CHCl$_3$); Anal. Calcd. For C$_{22}$H$_{35}$IN$_2$O$_6$: C, 48.01; H, 6.41; N, 5.09. Found: C, 47.59; H, 6.06; N, 4.72.
21. The purity of compound **2** (96%) was determined by quantitative ^{1}H NMR analysis using compound **2** (6.5 mg, 11.81 µmol) and 1,2,4,5-tetramethylbenzene (9.6 mg, 71.56 µmol) as an internal standard. SFC CHIRALCEL® OD-3 (3.0 µm, 150 x 4.6 mm) column, MP A: CO$_2$, MP B: 25 mM isobutylamine (IBA) in ethanol, column temperature: 40 °C, wavelength: PDA, pressure: 2600 psi, flow rate: 2.5 mL/min, gradient as below, t_R = 2.10 min (*R*), t_S = 2.51 min (*S*), small impurity detected (<4% by UV analysis).
22. A second reaction on the same scale provided 8.44 g (15.3 mmol, 95%) of the product.
23. Trifluoroacetic acid (>99.0%) was purchased from Oakwood Chemical and used as received. Equivalencies were based upon the amount of starting material, corrected for purity (Note 21).

24. The reaction progress was monitored by TLC analysis (TLC Silica gel 60 F254, pre-coated plates (0.25 mm) purchased from Merck) (visualized with 254 nm UV lamp and molybdatophosphoric acid) with EtOAc/hexanes (1:2) as eluent. Product R_f = 0.02, starting material R_f = 0.49.
25. NaOH (≥97.0%) was purchased from Fisher Chemical and used as received. 2M NaOH solution was prepared using deionized water. Control of pH was crucial for extracting compound **3** from aqueous phase during work-up.
26. Dichloromethane (99.0%) for work-up was purchased from Acros Organics and used as received.
27. The aqueous layer was extracted eight times to remove completely the diamine product from the aqueous layer.
28. Sodium chloride (>99.0%) was purchased from Fisher Chemical and used as received. Saturated brined solution was prepared using deionized water.
29. Sodium sulfate, anhydrous (>99%) was purchased from Fisher Chemical and used as received.
30. Compound **3** containing small impurities was obtained as a pale orange oil: ^{1}H NMR (CDCl$_3$, 600 MHz) δ: 1.18 (d, J = 6.2 Hz, 6H), 2.73 – 2.85 (m, 4H), 3.58 (s, 3H (integration value of NH$_2$ is low), 4.25 (h, J = 6.0 Hz, 2H), 6.36 (d, J = 8.3 Hz, 2H), 7.06 (t, J = 8.2 Hz, 1H); ^{13}C NMR (CDCl$_3$, 151 MHz) δ: 17.3, 47.7, 77.8, 82.3, 106.8, 129.6, 158.4 (a resonance arising from an impurity appears at approximately 28 ppm); [α]$^{30.3}_D$ = 84.0 (c 0.50, CHCl$_3$); Anal. Calcd. For C$_{12}$H$_{19}$IN$_2$O$_2$: C, 41.16; H, 5.47; N, 8.00. Found: C, 41.02; H, 5.84; N, 7.25.
31. 4-Dimethylaminopyridine (>99.0%) was purchased from Oakwood Chemical and used as received.
32. 2,4,6-Trimethyl benzoyl chloride (>80%) was purchased from Angene and used as received.
33. The reaction progress was monitored by TLC analysis (TLC Silica gel 60 F254, pre-coated plates (0.25 mm) purchased from Merck) (visualized with 254 nm UV lamp) with MeOH/CHCl$_3$ (1:30) as eluent. Product R_f = 0.72, starting material R_f = 0.39.
34. Concentrated HCl (35.0~37.0%) was purchased from Sigma-Aldrich and used as received. 1 M HCl solution was prepared using deionized water.
35. NaHCO$_3$ (99.5~100.3%) was purchased from Fisher Chemical and used as received. Saturated aqueous NaHCO$_3$ solution was prepared using deionized water.

36. Sand was purchased from Fisher Chemical and used as received.
37. Celite was purchased from Fisher Chemical and used as received.
38. Silica gel was purchased from Fisher Chemical and used as received.
39. 3-Aminopropyl-functionalized silica gel (NH silica) was purchased from Sigma-Aldrich and used as received.
40. From bottom to top, 10 g of celite, 10 g of 3-aminopropyl-functionalized silica gel (NH silica), 10 g of silica gel, and 60 g of sand were added to a sintered glass funnel and wetted with CH_2Cl_2 (50 mL) (Figure 4G).
41. EtOAc (99%) was purchased from Acros Organics and used as received.
42. To a 250 mL round-bottomed flask containing the crude residue and a 2.5 cm, Teflon-coated magnetic stir bar was added toluene (80 mL), and the resulting mixture was heated to 110 °C with 500 rpm stirring in a silicone oil bath (Figure 5A). After 20 min, toluene (2 mL x 5 times) was added to the stirring cloudy solution at 110 °C every 5 minutes until the solids were dissolved completely (total toluene = 110 mL) (Figure 5B). The resulting clear yellow solution was removed from the oil bath,

Figure 5. Recrystallization of compound 4; 1st Round: (A) Stirring suspension with 100 mL toluene at 110 °C; (B) Clear solution (total amount of toluene added: 110 mL); (C) Before cooling; (D) After cooling at –20 °C; (E) Collected solids from 1st round; (F) Filtrate; 2nd Round: (G) Stirring suspension with 30 mL toluene at 110 °C; (H) After cooling at –20 °C; (I) Collected solids from 2nd round (photos provided by submitters)

stirring bar was removed, and the mixture was cooled to 23 °C for 30 min (Figure 5C). Crystal seeds of compound **4** were then added to the solution and the solution was cooled to –20 °C in a freezer for 16 h (Figure 5D). The resulting white solid was collected by suction filtration on a Büchner funnel (7 cm diameter) and washed with cold (–20 °C) toluene (40 mL) (Figure 5E). The resulting pale white solid was then transferred to a 200 mL round-bottomed flask and dried under vacuum for 16 h (110 °C, 5.0 mmHg) to give **4** (3.90 g, 6.06 mmol) as a pale white solid.

43. Toluene (99%) was purchased from Fisher Chemical and used as received.

44. The filtrate (Figure 5F) was concentrated by rotary evaporation (40 °C, 15 mmHg) and dried under vacuum (23 °C, 5.0 mmHg, 0.5 h). To a 100 mL round-bottomed flask containing the residue and a 2.5 cm, Teflon-coated magnetic stir bar was added toluene (30 mL), and the resulting mixture was heated to 110 °C with 500 rpm stirring (Figure 5G). After 20 min, additional toluene (2 mL x 5 times) was added to the stirring cloudy solution at 110 °C every 5 minutes until the solids were dissolved completely (total toluene = 40 mL). The stirring bar was removed, and the clear yellow solution was cooled to 23 °C for 30 min. Crystal seeds of compound **4** were added and the solution was cooled to –20 °C in a freezer for 16 h (Figure 5H). The resulting white solid was collected by suction filtration on a Büchner funnel (7 cm diameter) and washed with cold (–20 °C) toluene (40 mL) (Figure 5I). The solids were then transferred to a 100 mL round-bottomed flask and dried under vacuum for 16 h (110 °C, 5.0 mmHg) to provide a second batch of **4** (1.88 g, 2.93 mmol) as a pale white solid.

45. *N,N'*-((2S,2'S)-((2-Iodo-1,3-phenylene)bis(oxy))bis(propane-2,1-diyl))bis-(2,4,6-trimethylbenzamide) (**4**) has the following properties: mp 196–201 °C (decomposed); ^{1}H NMR (CDCl$_3$, 600 MHz) δ: 1.40 (d, *J* = 6.2 Hz, 6H), 2.21 (s, 12H), 2.25 (s, 6H), 3.62 – 3.51 (m, 2H), 3.92 (ddd, *J* = 13.9, 6.7, 3.2 Hz, 2H), 4.74 – 4.63 (m, 2H), 6.21 (t, *J* = 5.5 Hz, 2H), 6.53 (d, *J* = 8.3 Hz, 2H), 6.80 (s, 4H), 7.22 (t, *J* = 8.3 Hz, 1H); ^{13}C NMR (CDCl$_3$, 151 MHz) δ: 17.6, 19.3, 21.2, 44.6, 75.2, 82.4, 107.3, 128.4, 130.2, 134.2, 134.8, 138.6, 157.8, 171.0; IR (film): 3228, 2971, 2916, 2867, 1458, 1242, 1152, 879, 748 cm^{-1}; LC/MS analysis (Agilent 1200 instrument with an Inductively Coupled Plasma (IPC)): *m/z* (relative intensity): 643 (97.10%); [α]$^{20}_D$ = +216.6 (*c* 0.038, CHCl$_3$) for ~99% ee; Anal. Calcd. For C$_{32}$H$_{39}$IN$_2$O$_4$: C, 59.81; H, 6.12; N, 4.36. Found: C, 59.09; H, 6.02; N, 4.41.

46. The purity of compound **4** (99%) was determined by quantitative ^{1}H NMR analysis using compound **4** (15.4 mg, 0.024 mmol) and 1,2,4,5-tetramethylbenzene (13.0 mg, 0.097 mmol) as an internal standard. SFC CHIRALCEL® OD-3 (3.0 µm, 150 x 4.6 mm) column, MP A: CO_2, MP B: 25 mM isobutylamine (IBA) in ethanol, column temperature: 40 °C, wavelength: PDA, pressure: 2600 psi, flow rate: 2.5 mL/min, gradient as belows, t_R = 6.74 min (*R*), t_{RS} = 7.12 min (*S*), meso compound was detected in racemic sample.
47. A second run of Step C on the same scale provided 6.52 g (10.15 mmol, 68%) of the product.

Working with Hazardous Chemicals

The procedures in *Organic Syntheses* are intended for use only by persons with proper training in experimental organic chemistry. All hazardous materials should be handled using the standard procedures for work with chemicals described in references such as "Prudent Practices in the Laboratory" (The National Academies Press, Washington, D.C., 2011; the full text can be accessed free of charge at http://www.nap.edu/catalog.php?record_id=12654). All chemical waste should be disposed of in accordance with local regulations. For general guidelines for the management of chemical waste, see Chapter 8 of Prudent Practices.

In some articles in *Organic Syntheses*, chemical-specific hazards are highlighted in red "Caution Notes" within a procedure. It is important to recognize that the absence of a caution note does not imply that no significant hazards are associated with the chemicals involved in that procedure. Prior to performing a reaction, a thorough risk assessment should be carried out that includes a review of the potential hazards associated with each chemical and experimental operation on the scale that is planned for the procedure. Guidelines for carrying out a risk assessment and for analyzing the hazards associated with chemicals can be found in Chapter 4 of Prudent Practices.

The procedures described in *Organic Syntheses* are provided as published and are conducted at one's own risk. *Organic Syntheses, Inc.*, its Editors, and its Board of Directors do not warrant or guarantee the safety of individuals using these procedures and hereby disclaim any liability for any injuries or damages claimed to have resulted from or related in any way to the procedures herein.

Discussion

Despite its long history,[2] the development of enantioselective hypervalent organoiodine catalysis is one of the most challenging areas in asymmetric synthesis.[3] Representative examples of chiral hypervalent iodine reagents that were used for various asymmetric oxidations before our first report are shown in Scheme 1 (the best enantioselectivities observed with these reagents are also shown).[4] However, the enantioselectivities of these reactions were moderate (<80%), except for Kita's reagent reported in 2008

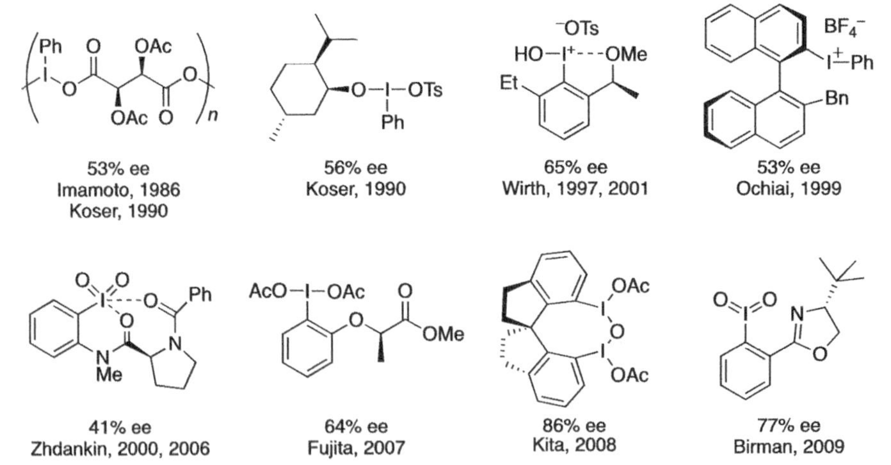

Scheme 1. Representative chiral organoiodine (III or V) reagents reported before our first work[4]

with 86% ee,[5a] which was the highest asymmetric induction using chiral hypervalent iodines reported at that time. Additionally, there have been only a few examples of *in situ*-generated chiral hypervalent organoiodine (III or V) catalysts with *m*-CPBA as a terminal oxidant (Scheme 2).[5a,6a,6b] In 2008, Kita's group designed a chiral hypervalent organoiodine catalyst with a conformationally rigid 1,1'-spirobiindane backbone, and applied this reagent to the enantioselective oxidative spirolactonization of 1-naphthol derivatives to the corresponding spirolactones with moderate enantioselectivity (up to 69% ee).[5a]

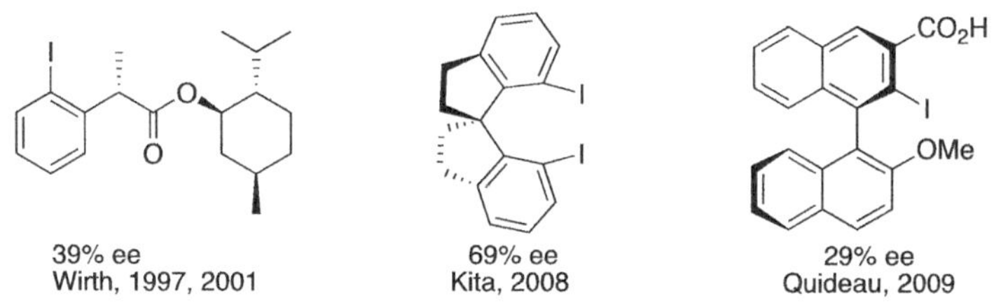

Scheme 2. Chiral organoiodine catalysts used with *m*-CPBA as an oxidant reported before our first work[5a,6a,6b]

In sharp contrast to Kita's conformationally rigid design, we reported the rational design of conformationally flexible hypervalent organoiodines as chiral catalysts based on secondary nonbonding interactions (i.e. intramolecular hydrogen-bonding interactions).[7,8] In 2010, we reported the design of conformationally flexible C_2-symmetric chiral iodoarenes **1** consisting of three units, including an iodoaryl moiety (**A**), chiral linkers (**B**), and subfunctional groups (**C**) (Scheme 3).[7] These units can be easily combined to give a wide variety of chiral iodoarenes **1**. Notably, the hypervalent iodines (III) **2** generated in situ from iodoarenes **1** were expected to exhibit intramolecular hydrogen-bonding interactions between the acidic hydrogen of **C** (NHAr) and the ligand (L, such as an acetoxy group, alkoxy

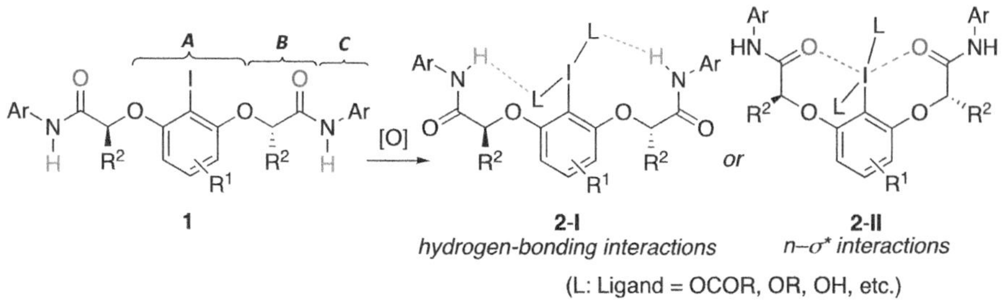

Scheme 3. Design of our 1st generation conformationally flexible chiral organoiodine, **1**

group, hydroxy group, etc.) of iodine (III) (**2-I**). Alternatively, intramolecular n–σ* interactions between the electron-deficient iodine (III) center (σ*$_{C-I}$ orbital) of **A** and the Lewis-basic group of **C** (lone pair *n*), such as carbonyl groups, might also be generated (**2-II**). We envisioned that a suitable chiral environment might be constructed around the iodine (III) center of **2** via such non-covalent bonding intramolecular interactions. Our lactate-based[4h] 1st generation catalyst **1a** (R^1, R^2, Ar = H, Me, Mes) could be successfully applied to the enantioselective oxidative dearomatization of 1-naphthol derivatives (Kita spirolactonization) to the corresponding spirolactones (up to 92% ee).

However, lactate-based catalysts **1** were found to be insufficient for the oxidation of phenols, which were less reactive than 1-naphthols, in terms of both reactivity and enantioselectivity. To overcome these limitations, we designed new chiral organoiodines **3**, 2nd generation catalysts, derived from 2-aminoalcohol instead of lactate as a chiral source (Scheme 4).[8a] As both **1** and **2** consist of 2-iodoresorcinol (**A**) and secondary amide (**C**) units, the corresponding acidic hydrogen atoms are at the same distance from the iodine center. On the other hand, because sp^2-hybridized carbonyl groups are moved to the outer sides, the 2nd generation catalysts **3** would be much more conformationally flexible. Moreover, **3** would be much more stereochemically stable than **1** as the stereocenters are far from the carbonyl groups.

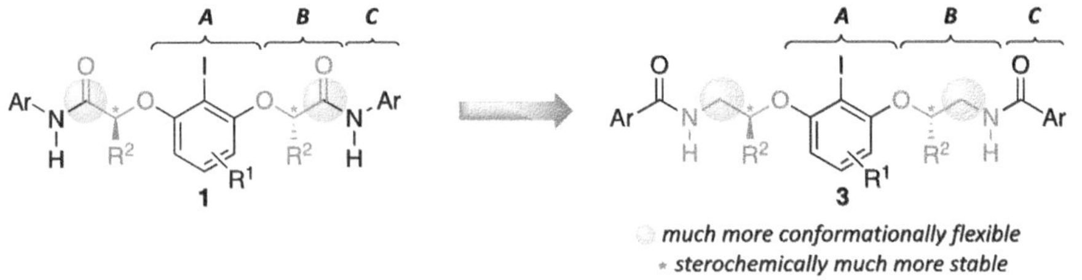

Scheme 4. Design of 2nd generation organoiodines 3 using 2-aminoalcohol as a chiral source instead of lactate

The enantioselective oxidation of not only 1-naphthols[8b] but also 2-naphthols[8b] and phenols[8a] as well as hydroquinone derivatives[8c] could be achieved by using organoiodine catalysts **3** (R^1, R^2 = H, Me) to give the corresponding cyclohexadienone spirolactones with excellent enantioselectivities up to 99% ee (Scheme 5). As an application of our enantioselective organoiodine catalysis, the asymmetric synthesis of (–)-maldoxin was achieved through the oxidative dearomatization of pestheic acid with excellent enantioselectivity.[8d] X-ray diffraction and NOE (Nuclear Overhauser Effect) NMR analyses of in situ-generated organoiodines (III) **4a** showed that a suitable chiral environment around the iodine (III) center was constructed via intramolecular hydrogen-bonding interactions (Scheme 6).[8a]

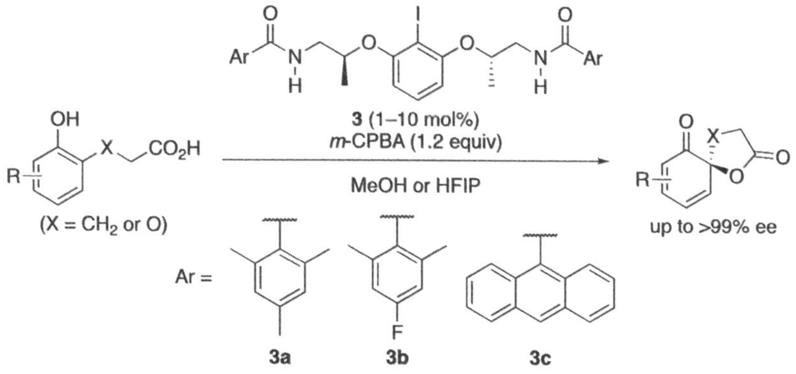

Scheme 5. Enantioselective oxidative dearomatization of a variety of arenols using 2nd generation catalysts 3[8]

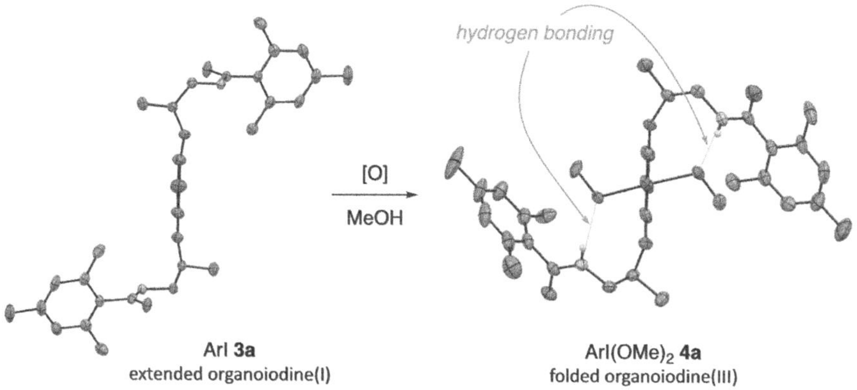

Scheme 6. X-ray structures of iodine (I) 3a and iodine (III) 4a[8]

Our conformationally flexible 1st and 2nd generation organoiodines 1–3 could be successfully applied to a variety of enantioselective oxidative transformations as a reagent or catalyst.[9–11] Wirth's group reported a series of enantioselective oxidations using our secondary bisamide reagent 2a[7] including the oxyamination of homoallylic urea derivatives to isoureas,[9a] intramolecular thioamination of alkenes to pyrrolines or indolines,[9b] oxidative rearrangements of chalcones[9c] or 1,1-disubstituted alkenes[9d] to α-aryl ketones, oxidative rearrangement of arylketones in the presence of orthoesters to α-arylesters,[9e] and intramolecular enantioselective α-functionalization of carbonyl compounds through tethers between the nucleophile and silyl enol ethers[9f] (Scheme 7).

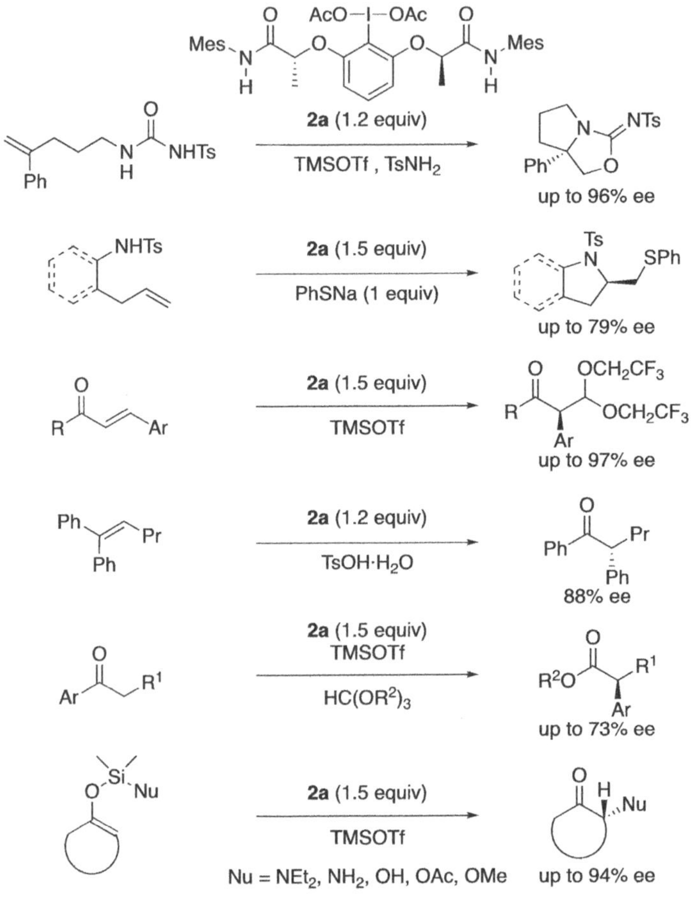

Scheme 7. Enantioselective oxidative transformations using lactate-based chiral organoiodine (III) reagent 2a reported by Wirth's group[9]

Enantioselective oxidations using catalytic amounts of iodoarenes **1** with an oxidant were also achieved (Scheme 8).[10] Gong's group reported the **1a**-catalyzed dearomatizative cyclization of 1-hydroxy-N-aryl-2-naphthamide derivatives to generate all-carbon spiro-stereocenter using *m*-CPBA as a stoichiometric oxidant.[10a] Masson's group reported the

1a-catalzyed enantioselective sulfonyl- or phosphoryl-oxylactonization of 4-pentenoic acids with *m*-CPBA to give the corresponding sulfonyloxy- or phosphoryloxy-γ-butyrolactones.[10b] On the other hand, Muñiz and colleagues reported the intermolecular vicinal diacetoxylation of terminal styrenes using structurally congested secondary bisamide **1b** as a catalyst in the presence of peracetic acid as an oxidant.[10c] The same research group also reported the **1b**-catalzyed intermolecular dearomatization of phenols at the *para*-position using *m*-CPBA as an oxidant to give the corresponding *para*-quinols with moderate stereoselectivity.[10d]

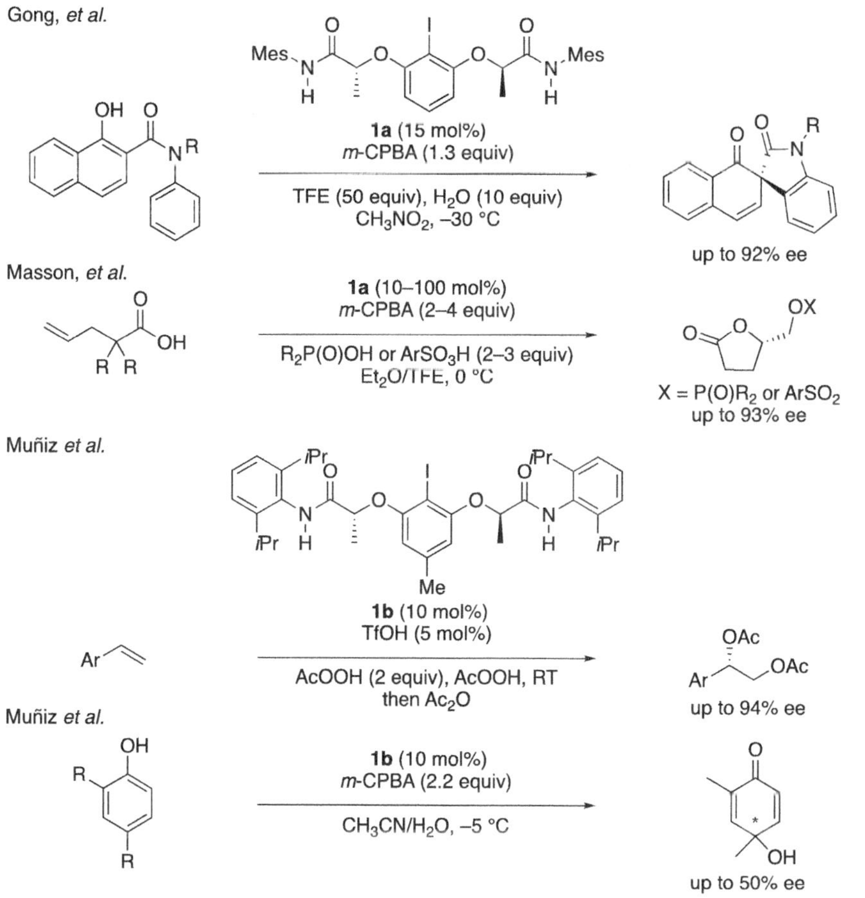

Scheme 8. Enantioselective oxidative transformations using lactate-based chiral organoiodine(I) catalysts **1**[10]

On the other hand, 2nd generation catalysts 3 and their analogues 5 have also been applied to enantioselective oxidations (Scheme 9).[11] Gilmour and colleagues reported the enantioselective vicinal difluorination or fluorocyclization of terminal alkenes using *in situ*-generated aryliodonium difluoride species from a catalytic amount of 3a with selectfluor as an oxidant in the presence of an excess amount of HF, albeit with only moderate enantioselectivity.[11a] Ciufolini and colleagues reported the enantioselective spiroetherification of 1-naphthol derivatives with *m*-CPBA as an oxidant to give the corresponding spiroethers with good to high enantioselectivities.[11b]

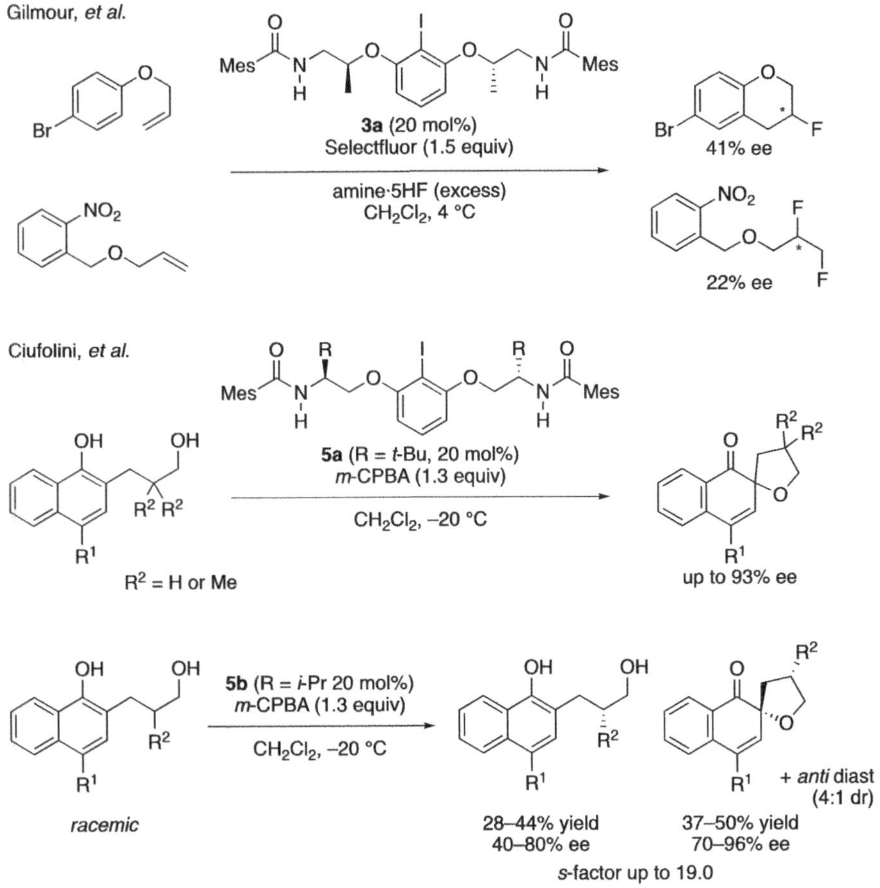

Scheme 9. Enantioselective oxidative transformations using amino alcohol-derived chiral organoiodine(I) catalysts 3 or 5[11]

Interestingly, catalyst **5a**, a structural analogue of our catalyst **3a**, was found to be superior to **3a** with respect to both stereoselectivity and reactivity for this particular reaction, especially under low catalyst-loading conditions.[11b] The same research group also reported the oxidative kinetic resolution of 1-naptholic alcohols using catalyst **5b** with an *S*-factor of up to 19.0.[11c] These examples highlight the substantial scope of conformationally flexible designer organoiodine (III) catalysis.

References

1. Graduate School of Engineering, Nagoya University, Furo-cho, Chikusa, Nagoya 464-8603, Japan. E-mail: ishihara@cc.nagoya-u.ac.jp. ORCID: 0000-0003-4191-3845. We thank JSPS.KAKENHI (15H05755 to K.I., 15H05484 to M.U., 18H01973 to M.U.), and the Program for Leading Graduate Schools: IGER Program in Green Natural Sciences (MEXT).
2. (a) Zhdankin, V. V. *Hypervalent Iodine Chemistry: Preparation, Structure and Synthetic Applications of Polyvalent Iodine Compounds*, Wiley, New York, 2014. (b) Kaiho, T., Ed.; *Iodine Chemistry and Application*, Wiley, Hoboken, New Jersey, 2015. (c) Wirth, T., Ed.; *Hypervalent Iodine Chemistry*; In Top. Curr. Chem. 373, Springer, Berlin, 2016. (d) Yoshimura, A.; Zhdankin, V. V. *Chem. Rev.* **2016**, *116*, 3328–3435.
3. (a) Lupton, D. W.; Ngatimin, M. *Aust. J. Chem.* **2010**, *63*, 653–658. (b) Liang, H.; Ciufolini, M. A. *Angew. Chem. Int. Ed.* **2011**, *50*, 11849–11851. (c) Uyanik, M.; Ishihara, K. *J. Synth. Org. Chem., Jpn.* **2012**, *70*, 1116–1122. (d) Parra, A.; Reboredo, S. *Chem. Eur. J.* **2013**, *19*, 17244–17260. (e) Harned, A. M. *Tetrahedron. Lett.* **2014**, *55*, 4681–4689. (f) Berthiol, F. *Synthesis* **2015**, 587–603. (g) Basdevant, B.; Guilbault, A.-A.; Beaulieu, S.; Lauriers, A. J.-D.; Legault, C. Y. *Pure Appl. Chem.* **2017**, *89*, 781–789.
4. (a) Imamoto, T.; Koto, H. *Chem. Lett.* **1986**, *15*, 967–968. (b) Ray, D. G.; Koser, G. F. *J. Org. Chem.* **1992**, *57*, 1607–1610. (c) Ray, D. G.; Koser, G. F. *J. Am. Chem. Soc.* **1990**, *112*, 5672–5673. (d) Wirth, T.; Hirt, U. H. *Tetrahedron: Asymmetry* **1997**, *8*, 23. (e) Hirt, U. H.; Schuster, M. F. H.; French, A. N.; Wiest, O. G.; Wirth, T. *Eur. J. Org. Chem.* **2001**, 1569–1579. (f) Ochiai, M.; Takaoka, Y.; Masaki, Y. *J. Am. Chem. Soc.* **1999**, *121*, 9233–9234. (g) Ladziata, U.; Carlson, J.; Zhdankin, V. V. *Tetrahedron Lett.* **2006**, *47*, 6301–6304. (h) Fujita, M.; Okuno, S.; Lee, H. J.; Sugimura, T.;

Okuyama, T. *Tetrahedron Lett.* **2007**, *48*, 8691–8694. (i) Boppisetti, J. K.; Birman, V. B. *Org. Lett.* **2009**, *11*, 1221–1223.
5. (a) Dohi, T.; Maruyama, A.; Takenaga, N.; Senami, K.; Minamitsuji, Y.; Fujioka, H.; Caemmerer, S. B.; Kita, Y. *Angew. Chem. Int. Ed.* **2008**, *47*, 3787–3790. (b) Dohi, T.; Takenaga, N.; Nakae, T.; Toyoda, Y.; Yamasaki, M.; Shiro, M.; Fujioka, H.; Maruyama, A.; Kita, Y. *J. Am. Chem. Soc.* **2013**, *135*, 4558–4566.
6. (a) Altermann, S. M.; Richardson, R. D.; Page, T. K.; Schmidt, R. K.; Holland, E.; Mohammed, U.; Paradine, S. M.; French, A. N.; Richter, C.; Bahar, A. M.; Witulski, B.; Wirth, T. *Eur. J. Org. Chem.* **2008**, 5315–5328. (b) Quideau, S.; Lyvinec, G.; Marguerit, M.; Bathany, K.; Ozanne-Beaudenon, A.; Buffeteau, T.; Cavagnat, D.; Chenede, A. *Angew. Chem. Int. Ed.* **2009**, *48*, 4605–4609. Recent examples for hypervalent iodine catalysis: (c) Fujita, M.; Yoshida, Y.; Miyata, K.; Wakisaka, A.; Sugimura, T. *Angew. Chem. Int. Ed.* **2010**, *49*, 7068–7071. (d) Röben, C.; Souto, J. A.; Gonzáles, Y.; Lishchynskyi, A.; Muñiz, K. *Angew. Chem. Int. Ed.* **2011**, *50*, 9478–9482. (e) Fujita, M.; Mori, K.; Shimogaki, M. Sugimura, T. *Org. Lett.* **2012**, *14*, 1294–1297. (f) Kong, W.; Feige, P. Haro, T. de; Nevado, C. *Angew. Chem. Int. Ed.* **2013**, *52*, 2469–2473. (g) Shimogaki, M.; Fujita, M.; Sugimura, T. *Eur. J. Org. Chem.* **2013**, 7128–7138. (h) Bosset, C.; Coffinier, R.; Peixoto, P. A.; El Assal, M.; Miqueu, K.; Sotiropoulos, J.-M.; Pouységu, L.; Quideau, S. *Angew. Chem. Int. Ed.* **2014**, *53*, 9860–9864. (i) Basdevant, B.; Legault, C. L. *Org. Lett.* **2015**, *17*, 4918–4921. (j) Murray, S. J.; Ibrahim, H. *Chem. Commun.* **2015**, *51*, 2376–2379. (k) Bekkaye M.; Masson, G. *Synthesis* **2016**, *48*, 302–312. (l) Ahmad, A.; Silva, L. F. *J. Org. Chem.* **2016**, *81*, 2174–2181. (m) Feng, Y.; Huang, R.; Hu, L.; Xiong, Y.; Coeffard, V. *Synthesis* **2016**, 2637–2644. (n) Banik, S. M.; Medley, J. W.; Jacobsen, E. N. *Science* **2016**, *353*, 51–54. (o) Woerly, E. M.; Banik, S. M.; Jacobsen, S. M. *J. Am. Chem. Soc.* **2016**, *138*, 13858–13861. (p) Shimogaki, M.; Fujita, M.; Sugimura, T. *Angew. Chem. Int. Ed.* **2016**, *55*, 15797–15801. (q) Muñiz, K. Barreiro, L.; Romero, R. M.; Martínez, C. *J. Am. Chem. Soc.* **2017**, *139*, 4354–4357. (r) Hempel, C.; Maichle-Mössmer, C.; Peric, M. A.; Nachtsheim, B. J. *Adv. Synth. Catal.* **2017**, *359*, 2931–2941. (s) Ogasawara, M.; Sasa, H.; Hu, H.; Amano, Y.; Nakajima. H.; Takenaga, N.; Nakajima, K.; Kita, Y.; Takahashi, T.; Dohi, T. *Org. Lett.* **2017**, *19*, 4102–4105. (t) Hashimoto, H.; Shimazaki, Y.; Omatsu, Y.; Maruoka, K. *Angew. Chem. Int. Ed.* **2018**, *57*, 7200–7204. (u) Ding, Q.; He, H.; Cai, Q. *Org. Lett.* **2018**, *20*, 4554–4557. (v) Antien, K.; Pouységu, L.; Deffieux, D.; Massip, S.; Peixoto, P.; Quideau, S. *Chem. Eur. J.* **2019**, *25*, 2852–2858.

7. (a) Uyanik, M.; Yasui, T.; Ishihara, K. *Angew. Chem. Int. Ed.* **2010**, *49*, 2175–2177. (b) Uyanik, M.; Yasui, T.; Ishihara, K. *Tetrahedron* **2010**, *66*, 5841–5851.
8. (a) Uyanik, M.; Yasui, T.; Ishihara, K. *Angew. Chem. Int. Ed.* **2013**, *52*, 9215–9218. (b) Uyanik, M.; Yasui, T.; Ishihara, K. *J.Org. Chem.* **2017**, *82*, 11946–11953. (c) Uyanik, M.; Sasakura, N.; Mizuno, M.; Ishihara, K. *ACS Catal.* **2017**, *7*, 872–876. (d) Suzuki, T.; Watanabe, S.; Uyanik, M.; Ishihara, K.; Kobayashi, S.; Tanino, K. *Org. Lett.* **2018**, *20*, 3919–3922.
9. (a) Farid, U.; Wirth, T. *Angew. Chem. Int. Ed.* **2012**, *51*, 3462–3465. (b) Mizar, P.; Niebuhr, R.; Hutchings, M.; Farooq, U.; Wirth, T. *Chem. Eur. J.* **2016**, *22*, 1614–1617. (c) Farid, U.; Malmedy, F.; Claveau, R.; Albers, L.; Wirth, T. *Angew. Chem. Int. Ed.* **2013**, *52*, 7018–7022. (d) Brown, M.; Kumar, R.; Rehbein, J.; Wirth, T. *Chem. Eur. J.* **2016**, *22*, 4030–4035. (e) Malmedy, F.; Wirth, T. *Chem. Eur. J.* **2016**, *22*, 16072–16077. (f) Mizar, P.; Wirth, T. *Angew. Chem. Int. Ed.* **2014**, *53*, 5993–5997.
10. (a) Zhang, D.-Y.; Xu, L.; Wu, H.; Gong, L.-Z. *Chem. Eur. J.* **2015**, *21*, 10314–10317. (b) Gelis, C.; Dumoulin, A.; Bekkaye, M.; Neuville, L.; Masson, G. *Org. Lett.* **2017**, *19*, 278–281. (c) Haubenreisser, S.; Wöste, T. H.; Martínez, C.; Ishihara, K.; Muñiz, K. *Angew. Chem. Int. Ed.* **2016**, *55*, 413–417. (d) Muñiz, K.; Fra, L. *Synthesis* **2017**, 2901–2096.
11. (a) Monár, I. G.; Gilmour, R. *J. Am. Chem. Soc.* **2016**, *138*, 5004–5007. (b) Jain, N.; Xu, S.; Ciufolini, M. A. *Chem. Eur. J.* **2017**, *23*, 4542–4546. (c) Jain, N.; Ciufolini, M. A. *Synthesis* **2018**, *51*, 3322.

Appendix
Chemical Abstracts Nomenclature (Registry Number)

(R)-1-Amino-2-propanol: 2-Propanol, 1-amino-, (2R)-; (2799-16-8)
Triethyl amine, Et$_3$N: Ethanamine, N,N-diethyl-; (121-44-8)
Di-*tert*-butyl decarbonate, Boc$_2$O: Dicarbonic acid, bis(1,1-dimethylethyl) ester; (24424-99-5)
2-Iodoresorcinol: 1,3-Benzenediol,2-iodo-; (41046-67-7)
Triphenylphosphine. Ph$_3$P; Phosphine, triphenyl-; (603-35-0)
Diisopropyl azodicarboxylate, DIAD: 1,2-Diazenedicarboxylic acid, 1,2-bis(1-methylethyl) ester; (2446-83-5)
Trifluoroacetic acid: Acetic acid, trifluoro-; (76-05-1)
4-Dimethyl aminopyridine, DMAP; (1122-58-3)
2,4,6-Trimethylbenzoyl chloride, MesCOCl; (938-18-1)

Prof. Muhammet Uyanik was born in Samsun, Turkey, in 1981 and received his Ph.D. from Nagoya University, Japan, in 2007 under the direction of Professor Kazuaki Ishihara. He was appointed as an Assistant Professor at Nagoya University in 2007, and he became Associate Professor in 2020. His research focused on the development of halogen-based catalysis for the oxidative transformations.

Shinichi Ishizaki was born in Mie, Japan, in 1996. He received his Bachelor's degree from Nagoya University in 2018. He is continuing his graduate studies at Nagoya University under the supervision of Professors Kazuaki Ishihara and Muhammet Uyanik. His research focused on the chiral organoiodine catalysis.

Prof. Kazuaki Ishihara received his Ph.D. from Nagoya University in 1991 under the direction of Professor Hisashi Yamamoto. He had the opportunity to work under the direction of Professor Clayton H. Heathcock at the University of California, Berkeley, for three months in 1988. After completing his postdoctoral studies with Professor E. J. Corey at Harvard University, he joined Professor H. Yamamoto's group at Nagoya University as an assistant professor in 1992, and he became associate professor in 1997. In 2002, he was appointed to his current position as a full professor. His research interest is the rational design of high-performance catalysts based on acid–base combination chemistry.

Cayetana Zarate received her B.Sc. in 2012 from the University of Valladolid (Spain). She pursued doctoral studies with Professor Ruben Martin in the area of nickel-catalyzed activation of 'inert' C–O, C–H and C–F bonds. After receiving her Ph.D. in 2017, Cayetana moved to the United States for a postdoctoral appointment with Professor Paul Chirik at Princeton University. In collaboration with Merck & Co., Inc., she designed catalysts for the isotopic radiolabeling of drug candidates. In 2019, Cayetana joined the Merck Discovery Process Chemistry Team in Boston, where her research involves the delivery of innovative chemistry that enables acceleration of drug discovery and development.

Chiral Organoiodine-catalyzed Enantioselective Oxidative Dearomatization of Phenols

Muhammet Uyanik, Shinichi Ishizaki, and Kazuaki Ishihara*[1]

Graduate School of Engineering, Nagoya University, Furo-cho, Chikusa, Nagoya 464-8603, Japan

Checked by Tao Wang and Kevin Campos

Procedure (Note 1)

A. *3-(4-Chloro-1-hydroxynaphthalen-2-yl)propanoic acid* (**2**). An oven-dried, 200 mL, two-necked (main 24/40, side 15/25 joints), round-bottomed flask is charged with 4-chloro-1-naphthol (7.14 g, 40.0 mmol, 1.0 equiv) (Note 2) under an air atmosphere, and the flask is equipped with a 3.5 cm, Teflon-coated magnetic stir bar, a reflux condenser fitted with a nitrogen inlet adapter, and a rubber septum (Note 3) (Figure 1A). Via the nitrogen inlet, the round-bottomed flask is evacuated under high vacuum (5.0 mmHg, few seconds) and refilled with nitrogen (three cycles). Toluene (40 mL) (Note 4), trifluoromethanesulfonic acid (0.35 mL, 4.0 mmol, 10 mol%) (Note 5) and acrylic acid (13.7 mL, 200 mmol, 5.0 equiv) (Note 6) are added via a syringe in less than 1 min each through the rubber septum. The reaction mixture is further degassed by bubbling N_2 via a metal needle (16 gauge) for 20 min at room temperature with stirring. The flask is immersed in a silicone oil bath and the bath is heated to 130 °C (Figure 1B). The mixture is refluxed with stirring for 24 h under a nitrogen atmosphere (Note 7). The resulting purple solution (Figure 1C) is cooled to 0 °C in an ice-bath for 10 min (Figure 1D). The reflux condenser is removed, and a 250 mL addition funnel is attached. An aqueous solution of K_2HPO_4 (120 mL, 25 wt%) (Note 8) is placed in the addition funnel and is added over 5 minutes to the reaction mixture in the round-bottomed flask. This mixture is then transferred to a 250 mL separatory funnel assisted by 60 mL of EtOAc (Note 9), and the aqueous phase is separated (Figure 1E) (Note 10). The organic phase is washed with saturated brine (80 mL) (Note 11) and dried over anhydrous Na_2SO_4 (20 g) (Note 12). The filtrate is concentrated by rotary evaporation (40 °C, 40 mmHg) and dried under vacuum (23 °C, 5.0 mmHg, 6 h) to give a purple solid (10.5–11.1 g, 56.8–58.1 wt% purity, 64.1–69.3% yield) (Figure 1F), which is used in the next step without further purification (Note 13).

Figure 1. Synthesis of compound 1. A) Reaction setup, B, C) Reaction progress, D) Before quenching, E) Work-up, F) Crude product (photos provided by authors)

A 300 mL, one-necked, round-bottomed flask (24/40 joint) equipped with a 3.5-cm, Teflon-coated magnetic stir bar is charged with **1**, THF (40 mL) (Note 14) and MeOH (40 mL) (Note 15) under an air atmosphere. 2 M NaOH (55 mL) (Note 16) is added at 23 °C and the resulting mixture is stirred for 3 h at 23 °C (Note 17) (Figure 2A). The resulting solution (Figure 2B) is concentrated using a rotary evaporator (30 °C, 30 mmHg) until about ca. 60 mL remains (Figure 2C). The concentrated mixture is transferred to a 300 mL separatory funnel using water (60 mL). The aqueous layer is extracted with Et$_2$O (3 x 80 mL) (Note 18) (Figure 2D). The Et$_2$O layers are discarded. The resulting aqueous layer is transferred to a 300 mL Erlenmeyer flask and Et$_2$O (80 mL) (Note 18) is added and cooled to 0 °C. The resulting mixture is acidified carefully (Note 19) through the addition of 2 M HCl (65 mL) (Note 20) at 0 °C over a period of 3 min. The resulting mixture is transferred to a 500 mL separatory funnel using Et$_2$O (30 mL). The aqueous layer is separated and extracted with Et$_2$O (100 mL) (Note 18). The combined organic layers are washed with saturated brine (100 mL) (Note 11), dried over anhydrous Na$_2$SO$_4$ (30 g) (Note 12), and filtered through a sintered glass funnel (4 cm diameter, medium porosity) under vacuum suction. The filtrate is concentrated by rotary evaporation (20 °C, 45 mmHg) (Note 21) to give a deep brown oil (Figure 2E). To the resulting oil, containing 3-(4-chloro-1-hydroxynaphthalen-2-yl)propanoic acid (**2**), is added CHCl$_3$ (50 mL) (Note 22), which is then concentrated by rotary evaporation (20 °C, 45 mmHg) (Note 23) and dried under vacuum (23 °C, 5.0 mmHg) to give a brown solid (Figure 2F). To the resulting solid is added CHCl$_3$ (15 mL) (Note 22) at 0 °C (Note 24), and the solids are collected by suction filtration on a Büchner funnel (7 cm diameter), and washed with cold (0 °C) CHCl$_3$ (15 mL) (Note 22) followed by hexane (20 mL) (Note 25). The resulting solid is dried under vacuum (23 °C, 5.0 mmHg, 12 h) to give **2** (3.62–3.98 g, 97.5–99.7 wt%, 36.0–38.7 % yield) as a beige solid (Figures 2G) (Notes 26 and 27).

Figure 2. Synthesis of compound 2. A) Reaction setup, B) Before evaporation, C) After evaporation, D) Work-up, E) Crude, F) Solidification of crude product, G) Product 2
(photos provided by authors)

B. *(S)-4'-Chloro-3,4-dihydro-1'H,5H-spiro[furan-2,2'-naphthalene]-1',5-dione* (**3**). An oven-dried 500 mL, three-necked (main 24/40, two sides 15/25 joints), round-bottomed flask is charged with **2** (6.27 g, 25.0 mmol, 1.0 equiv) and N,N'-(2S,2'S)-(2-iodo-1,3-phenylene)bis(oxy)bis(propane-2,1-diyl)bis(2,4,6-trimethylbenzamide (1.61 g, 2.50 mmol, 10 mol%) (Note 28) under an air atmosphere, and the flask is equipped with a 3.5 cm oval Teflon-coated magnetic stir-bar, an internal low-temperature thermometer, an inlet adapter fitted with a nitrogen inlet, and a rubber septum (Figure 3A). A nitrogen line is evacuated under high vacuum (5.0 mmHg, few seconds) and filled with nitrogen (three cycles). The septum is removed, and 1,2-dichloroethane (245 mL) (Note 29) and ethanol (8.76 mL, 150 mmol, 6.0 equiv) (Note 30) are added with a graduated cylinder under nitrogen flow. The septum is replaced. The flask is immersed in an EtOH bath in a low-temperature chamber (Note 31) that is cooled to –20 °C (Figure 3B) (Note 32). After the internal temperature is reached to –20 °C, the septum is removed, and *meta*-chloroperoxybenzoic acid (*m*-CPBA, 6.72 g, 30.0 mmol, 1.2 equiv) (Note 33) is added slowly (within 1~2 min) via the neck under nitrogen flow. The wall of the neck is washed with 1,2-dichloroethane (5 mL) (Note 29), and the septum is replaced. The reaction mixture is stirred at –20 °C for 24 h under nitrogen atmosphere (Note 34). To the resulting brown solution (Figure 3C) are added saturated aqueous Na_2SO_3 (50 mL) (Note 35) and saturated aqueous $NaHCO_3$ (50 mL) (Note 36) (Note 37), and then the reaction flask is moved to an ice-bath (0 °C) and stirred for additional 5 min. The resulting mixture is transferred to a 500 mL separating funnel and the aqueous layer is separated (Figure 3D). The organic layer is washed with saturated aqueous $NaHCO_3$ (50 mL) (Note 36) (Figure 3E). The combined aqueous layers are extracted with $CHCl_3$ (2 x 80 mL) (Note 22). The combined organic layers are washed with saturated brine (150 mL) (Note 11), dried over anhydrous Na_2SO_4 (50 g) (Note 12), and filtered through a sintered glass funnel (6.5 cm diameter, medium porosity) under vacuum suction. The filtrate is concentrated by rotary evaporation (30 °C, 30 mmHg). The crude (Figure 3F) is further purified by column chromatography (Note 38) (Figures 3G–I) to give **3** (5.33 g, 86% yield, 96% ee) (Notes 39, 40 and 41) as a pale-yellow solid (Note 42) (Figure 3J).

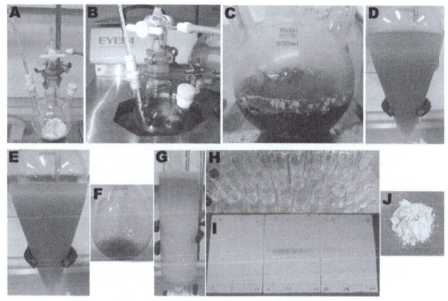

Figure 3. Synthesis of compound 3. A) Reaction setup, B, C) Reaction progress, D, E) Work-up, F) Crude product, G) Column chromatography, H) Fractions, I) TLC of fractions, J) Pure product 3
(photos provided by authors)

Notes

1. Prior to performing each reaction, a thorough hazard analysis and risk assessment should be carried out with regard to each chemical substance and experimental operation on the scale planned and in the context of the laboratory where the procedures will be carried out. Guidelines for carrying out risk assessments and for analyzing the hazards associated with chemicals can be found in references such as Chapter 4 of "Prudent Practices in the Laboratory" (The National Academies Press, Washington, D.C., 2011; the full text can be accessed free of charge at https://www.nap.edu/catalog/12654/prudent-practices-in-the-laboratory-handling-and-management-of-chemical. See also "Identifying and Evaluating Hazards in Research Laboratories" (American Chemical Society, 2015) which is available via the associated website "Hazard Assessment in Research Laboratories" at

https://www.acs.org/content/acs/en/about/governance/committees/chemicalsafety/hazard-assessment.html. In the case of this procedure, the risk assessment should include (but not necessarily be limited to) an evaluation of the potential hazards associated with 4-chloro-1-naphthol, toluene, trifluoromethanesulfonic acid, acrylic acid, nitrogen, sodium hydrogen carbonate, ethyl acetate, sodium chloride, sodium sulfate, tetrahydrofuran, methanol, sodium hydroxide, diethyl ether, hydrogen chloride, chloroform, hexane, N,N'-(2S,2'S)-2,2'-(2-iodo-1,3-phenylene)bis(oxy)bis(propane-2,1-diyl)bis(2,4,6-trimethylbenzamide, 1,2-dichloroethane, ethanol, *meta*-chloroperoxybenzoic acid, sodium sulfite and silica gel, as well as the proper procedures for experimental operations. Although no problems were encountered with the explosion of *m*-CPBA or other chemicals in the use of the oxidation procedure (step B), prudence dictates that all operations should be conducted in a hood.

2. 4-Chloro-1-naphthol (≥98%) was purchased from Sigma-Aldrich and used as received.
3. The reaction is performed under a positive pressure of nitrogen gas by using a nitrogen-filled balloon or connecting the reaction setup to nitrogen manifold via a glass adapter on top of the reflux condenser.
4. Toluene (anhydrous, 99.8%) was purchased from Sigma-Aldrich and used as received.
5. Trifluoromethanesulfonic acid (≥99%) was purchased from Sigma-Aldrich and used as received.
6. Acrylic acid (99%, anhydrous) was purchased from Sigma-Aldrich and used as received.
7. The reaction progress could be checked by TLC analysis (TLC Silica gel 60 F254, pre-coated plates (0.25 mm) purchased from Merck) (visualized with 254 nm UV lamp and molybdatophosphoric acid) with EtOAc/acetone/hexane (1:2:10) as an eluent. Product R_f = 0.42, starting material R_f = 0.32. Alternatively, the reaction can be monitored by NMR analysis of the crude reaction mixture: Take 20 μL of crude reaction mixture and dissolve into 600 μL $CDCl_3$ inside of an NMR tube. If starting material remains, a doublet peak (J = 8.0 Hz) around 6.75 ppm will appear in ^{1}H NMR spectrum.
8. K_2HPO_4 (≥98%) was purchased from Sigma-Aldrich and used as received. The 25 wt% solution is prepared by dissolving 250 g of salt into 750 mL of deionized water.
9. EtOAc (≥99.8%) was purchased from Sigma-Aldrich in an amber glass bottom and used as received.

10. This phase separation should be done without extended waiting (<15 min). A flashlight may be needed to assist in observing the separation.
11. Sodium chloride (NaCl (≥99.0%)) was purchased from Sigma-Aldrich and used as received. Saturated brined solution was prepared using deionized water.
12. Sodium sulfate (≥99.0%) was purchased from Sigma-Aldrich and used as received.
13. If desired, compound **1** can be purified on small scale (crude: 107 mg) by flash column chromatography on SiO_2. A column (diameter: 24 mm, height: 500 mm) was charged with 25 g of silica (E. Merck Art. 9385) and hexane. Sand with 1 cm minimum height (30–50 mesh particle size; purchased from FUJIFILM Wako Pure Chemical Corporation) was added to the top of the column (sand was used to assist packing). The crude residue was transferred to the column. The product was eluted with 0.42 L of hexane/EtOAc = 20:1. The product was checked by TLC analysis (TLC Silica gel 60 F254, pre-coated plates (0.25 mm) purchased from Merck) (visualized with 254 nm UV lamp and molybdatophosphoric acid) with EtOAc/acetone /hexane (1:2:10) as an eluent. Product R_f = 0.42. The fractions (18 x 130 mm) 14–20 were collected and concentrated by rotary evaporation (30 °C, 15 mmHg) and dried under vacuum (23 °C, 5.0 mmHg, 6 h). Alternatively, this compound can be purified with an automatic LC purification system (CombiFlash Rf+ from Teledyne ISCO). Mobile phase was ramped from 100% hexanes to 100% EtOAc over 30 min for the purification. When purified in this way compound **1** was obtained as an off-white solid, which has the following properties: mp 121.5–124.5 °C ; ^{1}H NMR (500 MHz, CDCl$_3$) δ: 2.90 (dd, J = 8.2, 6.6 Hz, 2H), 3.13 (dd, J = 8.3, 6.5 Hz, 2H), 7.39 (s, 1H), 7.58–7.65 (m, 2H), 8.20–8.27 (m, 2H). ^{13}C NMR (125 MHz, CDCl$_3$) δ: 23.9, 29.1, 117.5, 121.5, 124.6, 124.9, 125.4, 127.3, 127.5, 127.7, 130.6, 146.0, 167.9; IR (film): 2917, 2850, 1754, 1368, 1252, 1172, 1099, 872 cm^{-1} Anal. Calcd. For $C_{13}H_9ClO_2$: C, 67.11; H, 3.90. Found: C, 67.43; H, 3.83.
14. THF (≥99.9%) was purchased from Sigma Aldrich in a Sure/Seal bottle and used as received.
15. MeOH (99.8%) was purchased from Sigma-Aldrich in a Sure/Seal bottle and used as received.

16. NaOH (≥97.0%) was purchased from Sigma-Aldrich and used as received. 2 M NaOH solution was prepared by dissolving 20 g of solid NaOH into 250 mL of deionized water.
17. The reaction progress was checked by TLC analysis (TLC Silica gel 60 F254, pre-coated plates (0.25 mm) purchased from Merck) (visualized with 254 nm UV lamp and molybdatophosphoric acid) with EtOAc/acetone/hexane (1:2:10) as an eluent. Product R_f = 0.03, starting material R_f = 0.41.
18. Et$_2$O (≥99.7%) was purchased from Sigma-Aldrich and used as received.
19. To prevent the acid-catalyzed intramolecular re-lactonization of **2** to **1**.
20. Concentrated HCl (36.5–38.0%) was purchased from Sigma-Aldrich and used as received. 2 M HCl solution was prepared using deionized water.
21. If the concentration was performed at higher temperature (>25 °C), **2** would be converted gradually to lactone **1**.
22. CHCl$_3$ (≥99.8.0%) was purchased from Sigma-Aldrich and used as received.
23. CHCl$_3$ was used to promote solidification of the crude product. In some cases the residue remains as an oil. Sonication can assist in crashing out the product.
24. To obtain consistent product quality, big chunks of solid at this stage should be broken up with a spatula and the mixture stirred at 0 °C for at least 20 min with magnetic stir bar before filtration.
25. *n*-Hexane (≥95.0%) was purchased from Sigma-Aldrich and used as received.
26. 3-(4-Chloro-1-hydroxynaphthalen-2-yl)propanoic acid (**2**) has the following properties: mp 107.5 °C (decomposed); ^{1}H NMR (DMSO-d_6, 500 MHz) δ: 2.57 (t, *J* = 7.7 Hz, 2H), 2.99 (t, *J* = 7.6 Hz, 2H), 7.49 (s, 1H), 7.57 (ddd, *J* = 8.1, 6.8, 1.3 Hz, 1H), 7.61 (ddd, *J* = 8.1, 6.8, 1.3 Hz, 1H), 8.01–8.06 (m, 1H), 8.25–8.30 (m, 1H), 9.47 (s, 1H), 12.17 (s, 1H); ^{13}C NMR (DMSO-d_6, 125MHz) δ: 25.0, 33.9, 120.5, 122.5, 122.6, 123.4, 125.7, 126.5, 126.8, 128.4, 129.4, 149.1, 174.1; IR (film): 3362, 3284, 3055, 2944, 2652, 2573, 1689, 1447, 1379, 1235, 880 cm^{-1}; Anal. Calcd. For C$_{13}$H$_{11}$ClO$_3$: C, 62.29; H, 4.42. Found: C, 62.42; H, 4.26.
27. Purity (97.5–99.7 wt%) for the three runs was established by qNMR using 1,3,5-trimethoxybenzene as an internal standard.
28. Chiral aryliodine catalyst was prepared in an *Org. Synth.* procedure[2] that directly precedes this manuscript. *N,N'*-[(2*S*,2'*S*)-[(2-Iodo-1,3-phenylene)bis-(oxy)]bis(propane-2,1-diyl)]bis(mesitylamide) (Cas No:

1399008-27-5) can be also purchased from Tokyo Chemical Industry Co., Ltd. (TCI).
29. 1,2-Dichloroethane (anhydrous, 99.8%) was purchased from Sigma-Aldrich and used as received.
30. Ethanol (>99.5%) was purchased from Sigma-Aldrich and used as received.
31. EYELA Low Temp. Pairstirrer PSL-1400 was used.
32. This reaction can also be carried out in a jacketed 500 mL 3-necked, round bottomed flask (all 24/40 joints) connected to a chiller unit. Jacket coolant temperature was set at –24.5 °C to be able to reach –20.0 °C internal reaction temperature.
33. *m*-CPBA (77 %) was purchased from Sigma-Aldrich and used as received.
34. The reaction progress was checked by TLC analysis (TLC Silica gel 60 F254, pre-coated plates (0.25 mm) purchased from Merck) (visualized with 254 nm UV lamp and molybdatophosphoric acid) with AcOH/EtOAc/hexane (1:50:50) as an eluent. Product R_f = 0.55, starting material R_f = 0.24.
35. Na_2SO_3 (≥98.0%) was purchased from Sigma-Aldrich and used as received. Saturated aqueous Na_2SO_3 solution was prepared using deionized water.
36. $NaHCO_3$ (≥99.7%) was purchased from Sigma-Aldrich and used as received. Saturated $NaHCO_3$ solution was prepared using deionized water.
37. Saturated aqueous Na_2SO_3 and $NaHCO_3$ solutions were pre-cooled in an ice bath (0 °C).
38. The product **3** was purified using column chromatography with a gradient of EtOAc/hexane from 20% to 25%. A column (diameter: 55 mm, height: 700 mm) was charged with 130 g of silica (E. Merck Art. 9385) and hexane. Sand with 2 cm minimum height (30–50 mesh particle size; purchased from FUJIFILM Wako Pure Chemical Corporation) was added to the top of the column (sand was used to assist packing). The crude residue was transferred to the column. The product was eluted with 0.5 L of hexane/EtOAc = 4:1 followed by 2.4 L of hexane/EtOAc = 3:1. The product was checked by TLC analysis (TLC Silica gel 60 F254, pre-coated plates (0.25 mm) purchased from Merck) (visualized with 254 nm UV lamp and molybdatophosphoric acid) with EtOAc/hexane (1:1) as an eluent. Product R_f = 0.53. The fractions (30 x 200 mm) 10–28 were collected and concentrated by rotary evaporation (30 °C, 15 mmHg) and

dried under vacuum (23 °C, 5.0 mmHg, 14 h). Alternatively, this compound can be purified with an automatic LC purification system (CombiFlash Rf+ from Teledyne ISCO). Mobile phase was ramped from 100% hexanes to 100% EtOAc over 30 mins for the purification. The fractions containing product was combined and concentrated by rotary evaporation (30 °C, 100-40 mmHg) and dried under vacuum (23 °C, 5.0 mmHg, 2 h) with N_2 flow on top.

39. (S)-4'-Chlorospiro[tetrahydrofuran-2,2'-(1'H-naphthaline)]-1',5-dione (**3**) has the following properties: mp 101.5–103.6 °C; ^{1}H NMR (CDCl$_3$, 400 MHz) δ: 2.24 (ddd, *J* = 13.5, 11.1, 9.7 Hz, 1H), 2.45 (ddd, *J* = 13.5, 9.6, 2.3 Hz, 1H), 2.62 (ddd, *J* = 17.7, 9.7, 2.3 Hz, 1H), 2.89 (ddd, *J* = 17.7, 11.1, 9.6 Hz, 1H), 6.39 (s, 1H), 7.51 (td, *J* = 7.5, 1.6 Hz, 1H), 7.72–7.78 (m, 2H), 8.05 (ddd, *J* = 7.7, 1.3, 0.6 Hz, 1H); ^{13}C NMR (CDCl$_3$, 100 MHz) δ: 26.7, 31.6, 83.6, 126.2, 127.5, 128.2, 129.3, 130.2, 131.9, 134.6, 135.9, 175.9, 194.8; IR (film): 3055, 2950, 1780, 1687, 1592, 1291, 1188, 1160, 1025, 941 cm^{-1}; Anal. Calcd. For C$_{13}$H$_9$ClO$_3$: C, 62.79; H, 3.65. Found: C, 62.96; H, 3.41.

40. SFC-Conditions: Column: CHIRALPAK IA-3 (3.0 μm, 150 x 4.6 mm) MP A: CO$_2$; MP B: 25 mM isobutylamine (IBA) in methanol. Column temperature: 40 °C. Wavelength: PDA. Pressure: 2600 psi. Flow rate: 2.5 mL/min, Hexane/EtOH = 10:1 as eluent, 1.0 mL/min, t_R = 15.6 min (R), t_R = 17.4 min (S); $[α]^{25.0}_D$ = –127.2 (*c* 1.00, CHCl$_3$) for 96% ee;

41. Purity (98.4 wt%) was established by qNMR using 1,3,5-trimethoxybenzene as an internal standard.

42. A half-scale run was performed, and the product (**3**) was obtained 2.91 g (82% yield, 97% purity) with 94% ee.

Working with Hazardous Chemicals

The procedures in *Organic Syntheses* are intended for use only by persons with proper training in experimental organic chemistry. All hazardous materials should be handled using the standard procedures for work with chemicals described in references such as "Prudent Practices in the Laboratory" (The National Academies Press, Washington, D.C., 2011; the full text can be accessed free of charge at http://www.nap.edu/catalog.php?record_id=12654). All chemical waste should be disposed of in accordance with local regulations. For general

guidelines for the management of chemical waste, see Chapter 8 of Prudent Practices.

In some articles in *Organic Syntheses*, chemical-specific hazards are highlighted in red "Caution Notes" within a procedure. It is important to recognize that the absence of a caution note does not imply that no significant hazards are associated with the chemicals involved in that procedure. Prior to performing a reaction, a thorough risk assessment should be carried out that includes a review of the potential hazards associated with each chemical and experimental operation on the scale that is planned for the procedure. Guidelines for carrying out a risk assessment and for analyzing the hazards associated with chemicals can be found in Chapter 4 of Prudent Practices.

The procedures described in *Organic Syntheses* are provided as published and are conducted at one's own risk. *Organic Syntheses, Inc.*, its Editors, and its Board of Directors do not warrant or guarantee the safety of individuals using these procedures and hereby disclaim any liability for any injuries or damages claimed to have resulted from or related in any way to the procedures herein.

Discussion

The enantioselective oxidative dearomatizative coupling of arenols is an useful strategy for the synthesis of various biologically active compounds.[3] The development of enantioselective oxidative dearomatization of arenols using chiral hypervalent iodine catalysis is one of the most challenging areas in asymmetric organocatalysis.[3g,4] Especially, several researchers have pointed out that it is quite difficult to induce asymmetry via a chiral hypervalent iodine reagent because of the exclusive formation of phenoxenium ion via the dissociation of a chiral organoiodine(III) fragment during the reaction (Scheme 1).[3g,5]

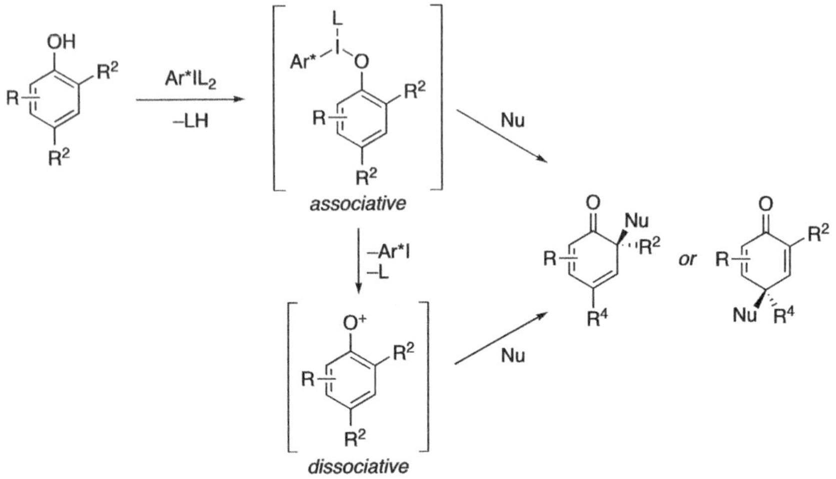

Scheme 1. General concept of the hypervalent iodine-mediated enantioselective oxidative dearomatizative coupling of arenols

In 2008, Kita and colleagues overcame this difficulty for the first time.[6] They designed a chiral organoiodine with a conformationally rigid 1,1'-spirobiindane backbone, and applied it to the enantioselective oxidative spirolactonization of 1-naphthol derivatives to the corresponding spirolactones with moderate enantioselectivity (up to 69% ee[6a]) (Scheme 2).[6] In contrast to Kita's conformationally rigid catalysts, our group introduced the rational design of conformationally flexible hypervalent organoiodines(III) as chiral catalysts based on secondary nonbonding interactions (i.e., intramolecular hydrogen-bonding interactions between the acidic amido protons and the iodine(III) ligands) (Scheme 2).[7,7]

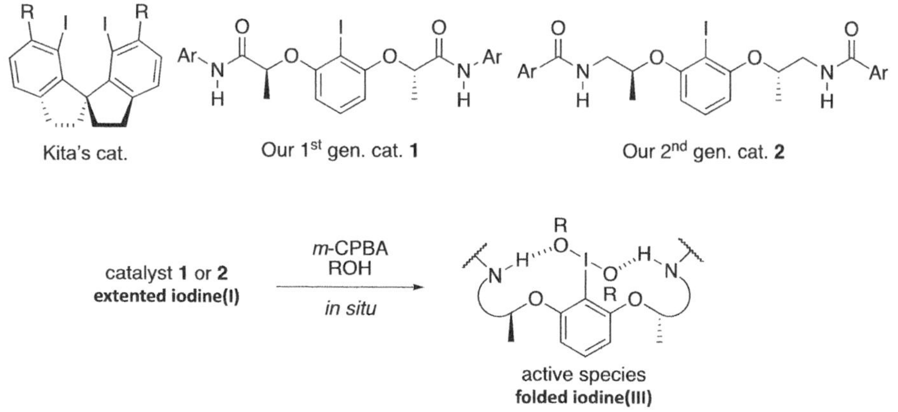

Scheme 2. Kita's and our chiral organoiodine catalysts

In 2010, we designed C_2-symmetric organoiodine **1a** (1st-generation catalyst, Ar = Mes) derived from lactate as a chiral source for the catalytic enantioselective oxidative spirolactonization of 1-naphthol derivatives **3** to the corresponding spirolactones **4** with high enantioselectivity of up to 92% ee (Scheme 3).[7]

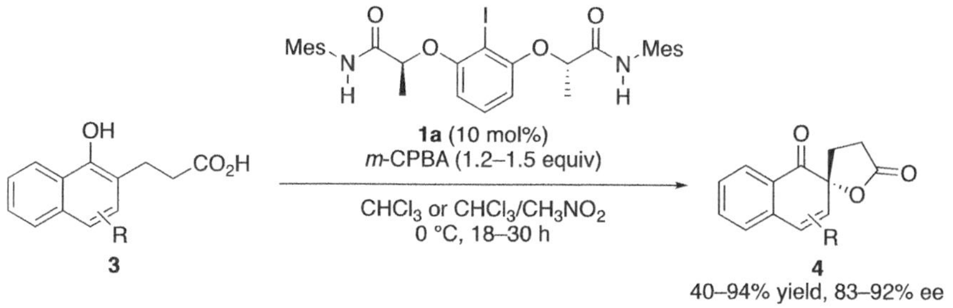

Scheme 3. Enantioselective oxidative spirolactonization of 1-naphthols using 1st-generation catalyst 1a

However, lactate-based **1** was found to be insufficient for the oxidation of phenols **5**, which were less reactive than 1-naphthols with respect to not only reactivity but also enantioselectivity.[8a] To overcome these limitations, we designed a new chiral iodoarene **2a** (2nd-generation catalyst, Ar = Mes) derived from 2-aminoalcohol instead of lactate as a chiral source (Scheme 4).[8a] A catalyst loading of 1 to 10 mol% was enough to give the desired cyclohexadienone spirolactones **6** and the subsequent Diels–Alder adducts **7** with excellent enantioselectivities (up to 99% ee).

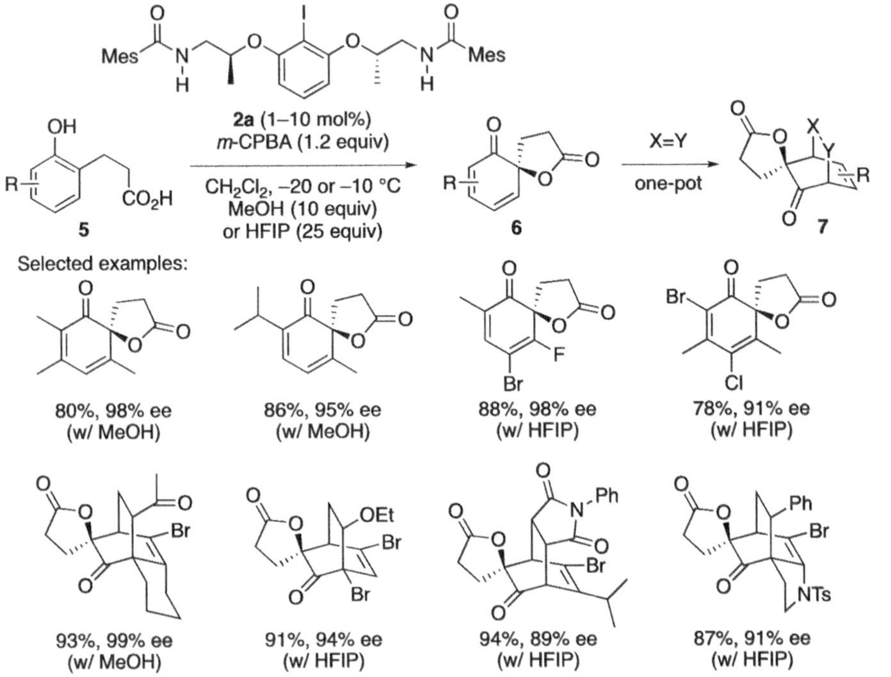

Scheme 4. Enantioselective oxidative spirolactonization of phenols **5** using 2nd-generation catalysts **2a**

Moreover, we succeeded in rationally controlling the desired associated pathway using alcohol additives such as methanol or ethanol for electron-rich phenols (Schemes 4 and 5).[8a] On the other hand, the use of 1,1,1,3,3,3-hexafluoroisopropanol (HFIP) was crucial to induce high reactivity for the electron-deficient phenols, which seems to disfavor a dissociative path (Scheme 4).[8a,9] This can be explained by the disfavoring of the dissociative path of electron-deficient phenols under these conditions. Furthermore, X-ray diffraction and NOE (Nuclear Overhauser Effect)–NMR analyses of an situ-generated organoiodine(III) showed that a suitable chiral environment around the iodine(III) center was constructed via intramolecular hydrogen-bonding interactions.[7a]

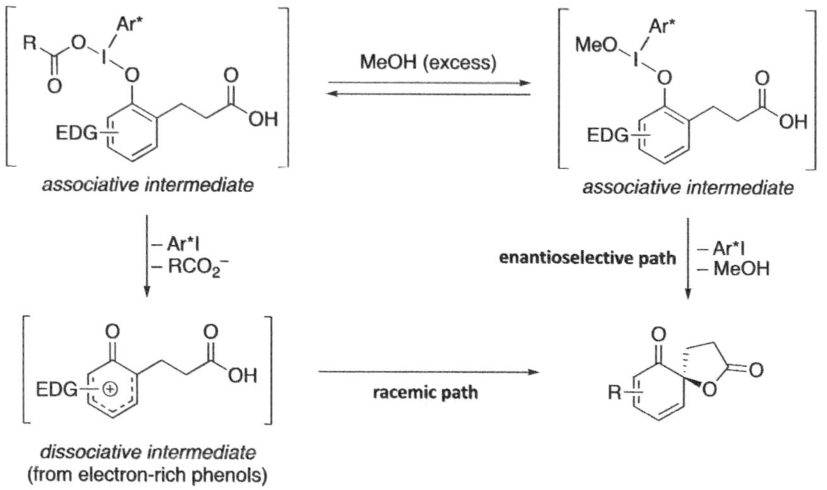

[Scheme 5 diagram showing associative intermediates with Ar*, MeO, EDG substituents, leading via enantioselective path and racemic path to products, with dissociative intermediate from electron-rich phenols]

Scheme 5. Additional methanol effect on the enantioselective dearomatization of electron-rich phenols

By using our flexible designer organoiodine catalysts **1** or **2**, we also achieved the first enantioselective oxidative dearomatization of *ortho*- and *para*-hydroquinone derivatives **9** or **10** to give the corresponding masked benzoquinones **11** or **12** with high enantioselectivities (Scheme 6).[8b] A tether strategy in which phenols were O-tethered to an acetic acid or ethanol unit at the *ortho*- or *para*-position realized rapid intramolecular cyclization enantioselectively prior to dissociation[8a] of the chiral iodine moiety. Interestingly, the use of slightly modified catalyst **2b** was found to be superior to the use of **2a**. Especially, this remote electronic effect was found to enhance enantioselectivity for the oxidation of phenols in a toluene–buffer (*pH* 7.0) biphasic solvent. Additionally, to achieve high enantioselectivity for the highly challenging *para*-cyclization reaction, we designed new lactate-derived catalysts **1b** bearing a deeper chiral cavity.

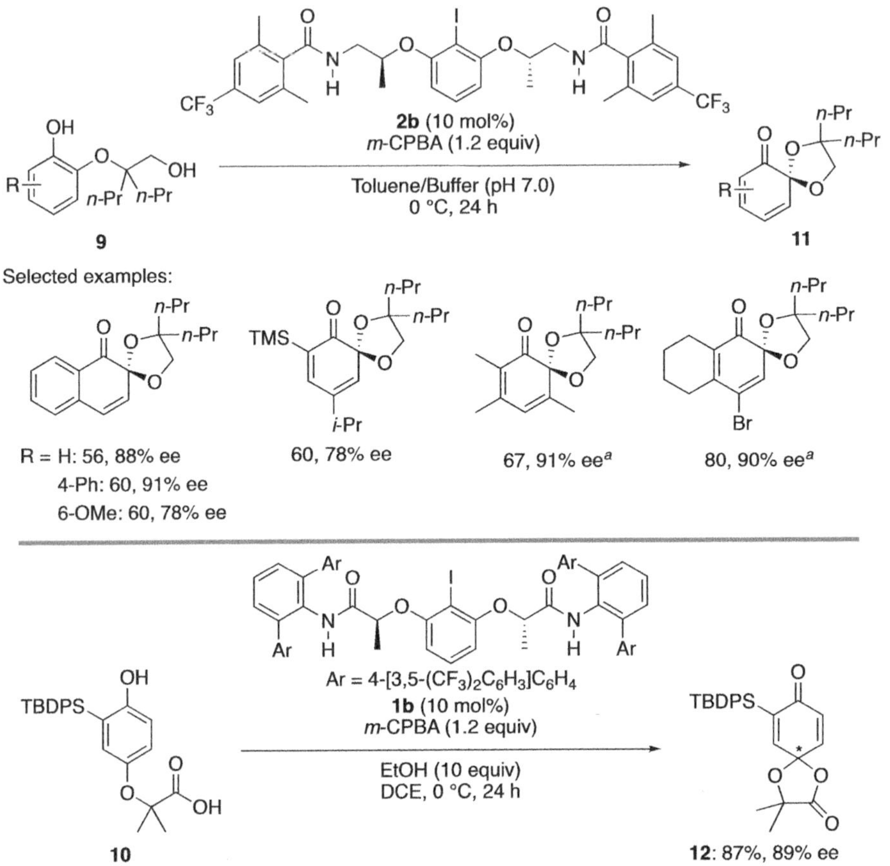

Scheme 6. Enantioselective oxidative spirolactonization of *ortho*- (9) and *para*-hydroquinones (10) using designer organoiodine catalysts 2b and 1b, respectively. [a] 1a was used instead of 1b in $CHCl_3/H_2O$

Recently, we also achieved an enantioselective oxidative dearomatization of 2-naphthol derivatives 9 by using our conformationally flexible organoiodine catalyst **2a** (Scheme 7).[8c] Moreover, excellent enantioselectivities were achieved for 1-naphthol derivatives 3, which had previously[7] been obtained with lower enantioselectivities with lactate-based organoiodine 1.[8c] Interestingly, the use of HFIP and methanol as additives was crucial to induce high enantioselectivity for 2-naphthol and 1-napthols, respectively.

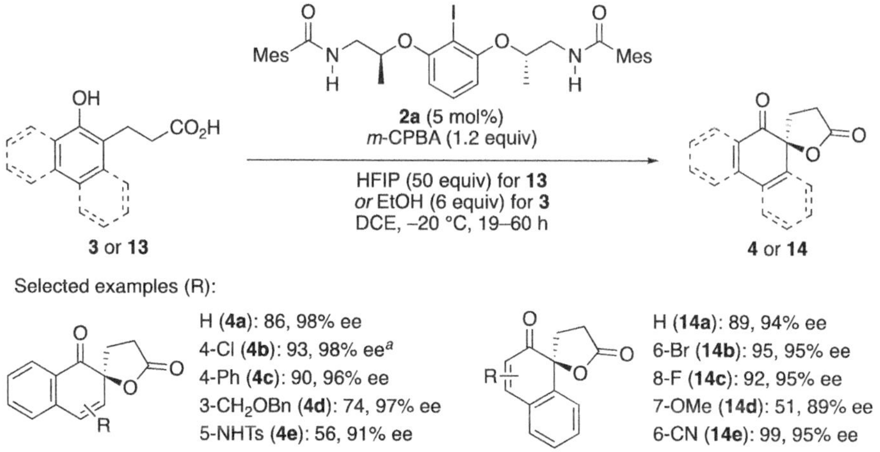

Scheme 7. Enantioselective oxidative spirolactonization of 1- (**3**) and 2-naphthols (**13**) using 2nd-generation catalyst **2a**. [a] **2a** (10 mol%)

Considering the safety issues and practicality of the large-scale experiments, here, we investigated the reaction conditions and oxidant purity for the oxidative spirolactonization of 1-naphthol **3b** (Table 1). To avoid any reproducibility problems, we used purified *m*-CPBA (>99%) as an oxidant under diluted conditions (0.02 M) for small scale-reactions.[7,8] Although no problems were encountered with the oxidant during purification, the use of commercially available *m*-CPBA (77%) would be preferred in large-scale reactions. Additionally, to make the reaction practical, the use of large amounts of solvents should be avoided in large-scale experiments. First, the oxidation of **3b** was performed on 0.5 mmol-scale to give **4b** with the same enantioselectivity albeit in slightly lower yield as in the original report,[8c] which was performed on 0.1 mmol-scale (entry 2 versus entry 1). We next investigated several parameters in 0.5 mmol-scale reactions and compared the result. To our delight, the same results were obtained with the use of commercially available *m*-CPBA (77%) instead of the purified compound (>99%) as an oxidant (entry 3). Next, the amount of solvent used was investigated and 0.1 M was found to the optimal concentration of starting material **3b** (entries 4 and 5). Interestingly, the reaction could be performed even at room temperature without any significant loss of enantioselectivity, however, the chemical yield of **4b** was decreased due to undesired intramolecular dehydrative lactonization (entries 6–8). Finally, the 50-fold scale-up reaction (25-mmol) under optimized conditions (as in entry 4) could be achieved with almost the same results (entry 9, see procedure section for details).

Table 1. Investigation of the Reaction Parameters for Scale-up

Entry	y (M)	T (°C)	t (h)	Yield (%)	ee (%)
1[a,b]	0.02	−20	26	93	98
2[a]	0.02	−20	24	86	98
3	0.02	−20	24	86	98
4	0.1	−20	24	86	96
5	0.2	−20	24	84	93
6	0.1	−10	20	84	96
7	0.1	0	16	83	95
8	0.1	25	2.5	79	94
9[c]	0.1	−20	24	87	96

[a] Reaction was performed under the original conditions using purified m-CPBA (>99%).[8c]
[b] 0.1 mmol-scale. [c] 25 mmol-scale.

References

1. Graduate School of Engineering, Nagoya University, Furo-cho, Chikusa, Nagoya 464-8603, Japan. E-mail: ishihara@cc.nagoya-u.ac.jp. ORCID: 0000-0003-4191-3845. We thank JSPS.KAKENHI (15H05755 to K.I., 15H05484 to M.U., 18H01973 to M.U.), and the Program for Leading Graduate Schools: IGER Program in Green Natural Sciences (MEXT).
2. Uyanik, M.; Ishizaki, S.; Ishihara, K. *Org. Synth.* **2021**, *98*, 1–27.
3. (a) Magdziak, D.; Meek, S. J.; Pettus, T. R. R. *Chem. Rev.* **2004**, *104*, 1383–1430. (b) Roche, S. P.; Porco, J. A., Jr. *Angew. Chem. Int. Ed.* **2011**, *50*, 4068–4093. (c) Bartoli, A.; Rodier, F.; Commeiras, L.; Parrain, J.-L.; Chouraqui, G. *Nat. Prod. Rep.* **2011**, *28*, 763–782. (d) Zhuo, C.-X.; Zhang, W.; You, S.-L. *Angew. Chem. Int. Ed.* **2012**, *51*, 12662–12686. (e) Uyanik,

M.; Ishihara, K. *J. Synth. Org. Chem., Jpn.* **2012**, *70*, 1116–1122. (f) Quideau, S.; Pouységu, L.; Peixoto, P. A.; Deffieux, D. In *Hypervalent Iodine Chemistry*; Wirth, T. Ed.; Springer: Switzerland, 2016, pp 25–74. (g) Uyanik, M.; Ishihara, K. In *Asymmetric Dearomatization Reactions*; You, S.-L., Ed.; John Wiley & Sons: Weinheim, 2016; pp 126–152.

4. (a) Quideau, S.; Lyvinec, G.; Marguerit, M.; Bathany, K.; Ozanne-Beaudenon, A.; Buffeteau, T.; Cavagnat, D.; Chenede, A. *Angew. Chem. Int. Ed.* **2009**, *48*, 4605–4609. (b) Boppisetti, J. K.; Birman, V. B. *Org. Lett.* **2009**, *11*, 1221–1223. (c) Volp, K. A.; Harned, A. M. *Chem. Commun.* **2013**, *49*, 3001–3003. (d) Bosset, C.; Coffinier, R.; Peixoto, P. A.; El Assal, M.; Miqueu, K.; Sotiropoulos, J.-M.; Pouységu, L.; Quideau, S. *Angew. Chem., Int. Ed.* **2014**, *53*, 9860–9864. (e) Murray, S. J.; Ibrahim, H. *Chem. Commun.* **2015**, *51*, 2376–2379. (f) Zhang, D. Y.; Xu, L.; Wu, H.; Gong, L.-Z. *Chem. Eur. J.* **2015**, *21*, 10314–10317. (g) Coffinier, R.; El Assal, M.; Peixoto, P. A.; Bosset, C.; Miqueu, K.; Sotiropoulos, J.-M.; Pouységu, L.; Quideau, S. *Org. Lett.* **2016**, *18*, 1120–1123. (h) Bekkaye, M.; Masson, G. *Synthesis* **2016**, *48*, 302–312. (i) Yoshida, Y.; Magara, A.; Mino, T.; Sakamoto, M. *Tetrahedron Lett.* **2016**, *57*, 5103–5107. (j) Jain, N.; Xu, S.; Ciufolini, M. A. *Chem. Eur. J.* **2017**, *23*, 4542–4546. (k) Muñiz, K.; Fra, L. *Synthesis* **2017**, *49*, 2901–2906. (l) El Assal, M.; Peixoto, P. A.; Coffinier, R.; Garnier, T.; Deffieux, D.; Miqueu, K.; Sotiropoulos, J.-M.; Pouységu, L.; Quideau, S. *J. Org. Chem.* **2017**, *82*, 11816–11828. (m) Companys, S.; Peixoto, P. A.; Bosset, C.; Chassaing, S.; Miqueu, K.; Sotiropoulos, J.-M.; Pouységu, L.; Quideau, S. *Chem. Eur. J.* **2017**, *23*, 13309–13313. (n) Hempel, C.; Maichle-Mössmer, C.; Peric, M. A.; Nachtsheim, B. J. *Adv. Synth. Catal.* **2017**, *359*, 2931–2941. (o) Ogasawara, M.; Sasa, H.; Hu, H.; Amano, Y.; Nakajima. H.; Takenaga, N.; Nakajima, K.; Kita, Y.; Takahashi, T.; Dohi, T. *Org. Lett.* **2017**, *19*, 4102–4105. (p) Dohi, T.; Sasa, H.; Miyazaki, K.; Fujitake, M.; Takenaga, N.; Kita, Y. *J. Org. Chem.* **2017**, *82*, 11954–11960. (q) Hashimoto, H.; Shimazaki, Y.; Omatsu, Y.; Maruoka, K. *Angew. Chem. Int. Ed.* **2018**, *57*, 7200–7204. (r) Ding, Q.; He, H.; Cai, Q. *Org. Lett.* **2018**, *20*, 4554–4557. (s) Jain, N.; Ciufolini, M. A. *Synthesis* **2018**, *51*, 3322–3332. (t) Jain, N.; Hein, J. E.; Ciufolini, M. A. *Synlett* **2019**, *30*, 1222–1227. (u) Antien, K.; Pouységu, L.; Deffieux, D.; Massip, S.; Peixoto, P.; Quideau, S. *Chem. Eur. J.* **2019**, *25*, 2852–2858. (v) Yoshida, Y.; Kanashima, Y.; Mino, T.; Sakamoto, M. *Tetrahedron* **2019**, *75*, 3840–3849.

5. (a) Kürti, L.; Herczegh, P.; Visy, J.; Simonyi, M.; Antus, S.; Pelter, A. *J. Chem. Soc., Perkin Trans. 1* **1999**, 379–380. (b) Pelter, A.; Ward, R. S. *Tetrahedron* **2001**, *57*, 273–282.

6. (a) Dohi, T.; Maruyama, A.; Takenaga, N.; Senami, K.; Minamitsuji, Y.; Fujioka, H.; Caemmerer, S. B.; Kita, Y. *Angew. Chem. Int. Ed.* **2008**, *47*, 3787–3790. (b) Dohi, T.; Takenaga, N.; Nakae, T.; Toyoda, Y.; Yamasaki, M.; Shiro, M.; Fujioka, H.; Maruyama, A.; Kita, Y. *J. Am. Chem. Soc.* **2013**, *135*, 4558–4566.
7. (a) Uyanik, M.; Yasui, T.; Ishihara, K. *Angew. Chem. Int. Ed.* **2010**, *49*, 2175–2177. (b) Uyanik, M.; Yasui, T.; Ishihara, K. *Tetrahedron* **2010**, *66*, 5841–5851. We also achieved the enantioselective oxidative dearomatization of 1-naphthols using chiral quaternary ammonium hypoiodite catalysis using aqueous hydrogen peroxide as an oxidant: Uyanik, M.; Sasakura, N.; Kaneko, E.; Ohori, K.; Ishihara, K. *Chem. Lett.* **2015**, *44*, 179–181.
8. (a) Uyanik, M.; Yasui, T.; Ishihara, K. *Angew. Chem. Int. Ed.* **2013**, *52*, 9215–9218. (b) Uyanik, M.; Sasakura, N.; Mizuno, M.; Ishihara, K. *ACS Catal.* **2017**, *7*, 872–876. (c) Uyanik, M.; Yasui, T.; Ishihara, K. *J. Org. Chem.* **2017**, *82*, 11946–11953. (d) Suzuki, T.; Watanabe, S.; Uyanik, M.; Ishihara, K.; Kobayashi, S.; Tanino, K. *Org. Lett.* **2018**, *20*, 3919–3922.
9. Dohi, T.; Yamaoka, N.; Kita, Y. *Tetrahedron* **2010**, *66*, 5775–5785.

Appendix
Chemical Abstracts Nomenclature (Registry Number)

4-Chloro-1-naphthol: 1-Naphthalenol, 4-chloro-; (604-44-4)
Trifluoromethanesulfonic acid, TfOH: Methanesulfonic acid, 1,1,1-trifluoro-; (1493-13-6)
Acrylic acid: 2-Propenoic acid; (79-10-7)
N,N'-(2*S*,2'*S*)-(2-Iodo-1,3-phenylene)bis(oxy)bis(propane-2,1-diyl)bis(2,4,6-trimethylbenzamide: Benzamide, *N,N'*-[(2-Iodo-1,3-phenylene)bis[oxy[(2*S*)-2-methyl-2,1-ethanediyl]]]bis[2,4,6-trimethyl-; (1399008-27-5)
meta-Chloroperoxybenzoic acid, *m*-CPBA: Benzenecarboperoxoic acid, 3-chloro-; (937-14-4)

Prof. Muhammet Uyanik was born in Samsun, Turkey, in 1981 and received his Ph.D. from Nagoya University, Japan, in 2007 under the direction of Professor Kazuaki Ishihara. He was appointed as an Assistant Professor at Nagoya University in 2007, and became Associate Professor in 2020. His research focused on the development of halogen-based catalysis for the oxidative transformations.

Shinichi Ishizaki was born in Mie, Japan, in 1996. He received his Bachelor's degree from Nagoya University in 2018. He is continuing his graduate studies at Nagoya University under the supervision of Professors Kazuaki Ishihara and Muhammet Uyanik. His research focused on chiral organoiodine catalysis.

Prof. Kazuaki Ishihara received his Ph.D. from Nagoya University in 1991 under the direction of Professor Hisashi Yamamoto. He had the opportunity to work under the direction of Professor Clayton H. Heathcock at the University of California, Berkeley, for three months in 1988. After completing his postdoctoral studies with Professor E. J. Corey at Harvard University, he joined Professor H. Yamamoto's group at Nagoya University as an Assistant Professor in 1992, and became Associate Professor in 1997. In 2002, he was appointed to his current position as a full Professor. His research interest is the rational design of high-performance catalysts based on acid–base combination chemistry.

Tao Wang joined the Process Research Department of Merck & Co., Inc. in 2018. His research focuses on developing green, sustainable and safe process using state-of-art organic chemistry. Tao obtained his B. S. degree from Peking University in Beijing, China. In 2011, he joined the research group of Thomas Hoye in University of Minnesota to study new benzyne chemistry and develop biomimetic synthesis of natural products. After completion of his Ph.D., Tao joined the laboratory of David MacMillan as a Postdoctoral Scholar at Princeton University. While at Princeton, he explored photoredox-based methodology for small molecules synthesis and pioneered the development of a proximity-based labelling technology in biological systems.

Reductive Deuteration of Ketones with Magnesium and D₂O for the Synthesis of α-Deutero-o-methyl-benzhydrol

Nengbo Zhu, Wen-Ming Wan, and Hongli Bao[1*]

State Key Laboratory of Structural Chemistry, Key Laboratory of Coal to Ethylene Glycol and Its Related Technology, Center for Excellence in Molecular Synthesis, Fujian Institute of Research on the Structure of Matter, Chinese Academy of Sciences, 155 Yangqiao Road West, Fuzhou, Fujian 350002, P. R. of China.

Checked by Yuuki Watanabe, Koichi Hagiwara, and Masayuki Inoue

Procedure (Note 1)

A. *α-Deutero-o-methyl-benzhydrol* (**2**). A 500-mL, three-necked round-bottomed flask (central neck 29/32 joint, side necks 15/25 joint) is equipped with a Teflon-coated magnetic stir bar (oval shaped, 5.0 × 2.0 cm) (Note 2). The side necks are fitted with a 15/25 rubber septa. The central neck is equipped with a connecting adapter (upper outer joint 29/42, lower inner joint 15/25) and a water-cooled reflux Dimroth condenser (20.0 cm height, 15/25 joint), which is topped with a 15/25 three-way cock connected to a Schlenk line (Note 2). All junctures of the glassware and the rubber septa of the side-neck of the flask are sealed with Teflon tape (Figure 1A). The rubber septum of the side-neck of the flask is removed and the magnesium turning (2.40 g, 98.0 mmol, 4.9 equiv) (Note 3) (Figure 1B) is added through the side neck. The rubber septum is reattached and sealed with Teflon tape. The resultant magnesium is dried by heating with a heat gun under vacuum (3.2 mmHg) for 3 min, after which the flask is backfilled with argon (Figure 1C). This process is repeated three times. The reflux condenser is kept at room temperature (23 °C) with a constant flow of water. Following the addition of THF (30 mL) (Note 4) by a syringe through the rubber septum on

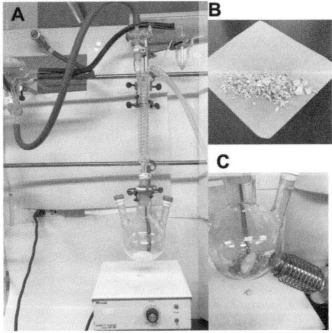

Figure 1. (A) Reaction set-up; (B) Magnesium turnings; (C) magnesium turnings dried by heating gun under vacuum

the side-neck of the flask (Figure 2A), the reaction vessel is placed in a pre-heated 70 °C silicone oil bath and stirred for 25 min. 1,2-Dibromoethane (3.46 mL, 40 mmol, 2.0 equiv) (Note 5) is then added to the flask via syringe over 20 sec (Note 6) through the rubber septum on the side-neck of the flask (Figure 2B) and the reaction mixture is stirred for 1 min (Figure 2C). At that point, a mixture (Note 7) (Figure 3) of THF (10 mL) (Note 4), 2-methylbenzophenone (3.6 mL, 20 mmol, 1.0 equiv) (Note 8) and D$_2$O (541 µL, 30 mmol, 1.5 equiv) (Note 9) is added via syringe over 1 min (Note 1) through the rubber septum on the side-neck of the flask. After the reaction

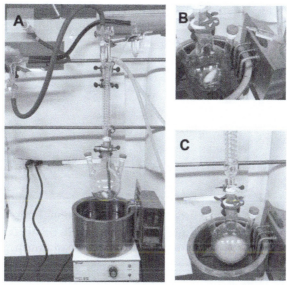

Figure 2. (A) Mixture after the addition of THF; (B) Mixture during the addition of 1,2-dibromoethane; (C) Reaction mixture immediately after the addition of 1,2-dibromoethane

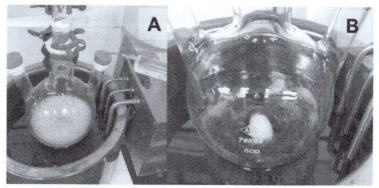

Figure 3. (A) Reaction mixture during the addition of the mixture of THF, 2-methylbenzophenone and D₂O; (B) Reaction mixture after the addition of the mixture of THF, 2-methylbenzophenone and D₂O

mixture is stirred at 70 °C for 2 h in the oil bath, the reaction vessel is removed from the oil bath and cooled to room temperature over 30 min (Note 10) (Figure 4A). The solution is then quenched with sequential addition of dichloromethane (100 mL) (Note 11) (Figure 4B) and saturated aqueous ammonium chloride (60 mL) (Note 12) under stirring. The saturated aqueous ammonium chloride is added in three 20 mL portions over 30 min (Figure 4C). Water (100 mL) is added to the flask and the solution is stirred

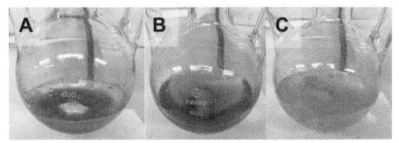

Figure 4. (A) Reaction mixture after cooling to room temperature; (B) Reaction mixture after the addition of dichloromethane; (C) Reaction mixture after the addition of saturated aqueous ammonium chloride

at approximately room temperature (23 °C) for 1 h (Figure 5A). The resultant mixture is then transferred to a 1000-mL separatory funnel. The flask is rinsed with dichloromethane (50 mL) and water (100 mL). The layers are separated, and the aqueous solution is extracted with dichloromethane (3 × 50 mL). The combined organic layers are dried for 10 min over anhydrous MgSO$_4$ (50 g) (Note 13) and filtered through a cotton plug into a 1000 mL round-bottomed flask (Figure 5B). The filter cake is rinsed with dichloromethane (3 × 30 mL) and the filtrate is concentrated on a rotary evaporator under reduced pressure (37 °C, 450 mmHg, then 69 mmHg) to give the crude mixture as a colorless oil (Figure 5C). Silica gel (4.5 g) (Note 14) and dichloromethane

  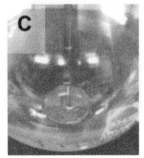

Figure 5. (A) Reaction mixture after the addition of water and stirred at 1 h; (B) Cotton filtration of MgSO$_4$ after extraction; (C) Crude colorless oil

(50 mL) are added to the resulting crude oil. The mixture is concentrated by rotary evaporation (37 °C, 450 mmHg, then 300 mmHg) to afford a white solid (Figure 6A). The resulting solid is purified by column chromatography on silica gel using petroleum ether and dichloromethane as eluents (Notes 11, 14, 15, 16, and 17). The fractions containing the product are combined and concentrated by rotary evaporation (37 °C, 400 mmHg, then 60 mmHg). The purified material is placed under vacuum (3.2 mmHg) for 5 h to afford α-deutero-*o*-methyl-benzhydrol (**2**) (2.90 g, 73% yield, 98% D) (Notes 18 and 19) as a white solid with 99.3% purity determined by qNMR analysis (Notes 20 and 21) (Table 1) (Figure 6C).

Figure 6. (A) Crude white solid; (B) Purification by column chromatography; (C) Purified product

Notes

1. Prior to performing each reaction, a thorough hazard analysis and risk assessment should be carried out with regard to each chemical substance and experimental operation on the scale planned and in the context of the laboratory where the procedures will be carried out. Guidelines for carrying out risk assessments and for analyzing the hazards associated with chemicals can be found in references such as Chapter 4 of "Prudent Practices in the Laboratory" (The National Academies Press, Washington, D.C., 2011; the full text can be accessed free of charge at https://www.nap.edu/catalog/12654/prudent-practices-in-the-laboratory-handling-and-management-of-chemical. See also "Identifying and Evaluating Hazards in Research Laboratories"

(American Chemical Society, 2015) which is available via the associated website "Hazard Assessment in Research Laboratories" at https://www.acs.org/content/acs/en/about/governance/committees/chemicalsafety/hazard-assessment.html. In the case of this procedure, the risk assessment should include (but not necessarily be limited to) an evaluation of the potential hazards associated with magnesium, 1,2-dibromoethane, tetrahydrofuran, 2-Methylbenzophenone, deuterium oxide, saturated aqueous ammonium chloride, anhydrous MgSO$_4$, silica gel, phosphomolybdic acid stain, 1,4-dimethoxybenzene and deuterotrichloromethane. It should be noted that there is a large amount of gas generation after the 1,2-dibromoethane was added to the solution. The time over which the 1,2-dibromoethane is added is 15-40 seconds; the 1,2-dibromoethaneshould not be added too quickly or too slowly. The time required for the addition of the mixture of THF, 2-methylbenzophenone and D$_2$O is about 1 minute; the mixture should not be added too quickly or too slowly. This is a fast and exothermic reaction. It is advisable to conduct these experiments in a large flask fitted with long Dimroth (checkers) or Allihn (submitters) condenser and adequately vented to a standard inert gas manifold.

2. All glassware and Teflon-coated magnetic stir bar are oven-dried prior to reaction set-up. The atmosphere of the reaction system is argon (checkers). The atmosphere of the reaction system is nitrogen (submitters).

3. The freshly peeled magnesium was peeled from magnesium rod by using planer. Magnesium rod (99.8%, metal basis) was purchased from Alfa Aesar (submitters). In this reaction, magnesium as a reducing agent loses electrons on the metal surface. It needs enough metal surface area to contact the ketone, and there is still some magnesium left after the reaction. Therefore, an excessive amount of magnesium is required. Magnesium turning (99.8%, metal basis) was purchased from Alfa Aesar and used as received (checkers).

4. Various qualities of tetrahydrofuran were investigated by the submitters (Table 1). The THF used by the checkers corresponded to entry 5 on Table 1, although the submitters indicate that THF prepared in alternate fashions provide similar results. Run 1 and 2: Tetrahydrofuran (99%) was purchased from Shanghai Titan Scientific Co., Ltd. and freshly distilled over sodium metal (submitters). Run 3: Tetrahydrofuran (anhydrous, 99.8+%, unstab., packaged under Argon in resealable ChemSeal bottles) was purchased from Alfa Aesar and

used as received (submitters). Run 4: Tetrahydrofuran (purity GC 99.9%) was purchased from Oceanpak and dried by Innovative Technology Solvent Purification System (PS-MD-5) (submitters). Run 5: Tetrahydrofuran (>99.0%) was purchased from Kanto Chemical Co., Inc. and purified by Glass Contour solvent dispensing system (Nikko Hansen & Co., Ltd.) (checkers).

Table 1. Tetrahydrofuran scope

Entry	THF (note 4)	Yield (%)	2 weight (g)	Deuterium incorporation (%)	Purity (%)
Run 1	Freshly distilled over sodium metal	70	2.78	96	98.4
Run 2	Freshly distilled over sodium metal	61	2.42	96	98.5
Run 3	Purchased from Alfa Aesar and used as received	73	2.90	92	98.2
Run 4	Dried by Innovative Technology Solvent Purification System	62	2.46	96	99.3
Run 5	Dried by Glass Contour solvent dispensing system (checkers)	73	2.90	98	99.3

5. 1,2-Dibromoethane (99%) was purchased from Adamas-beta and used as received (submitters). 1,2-Dibromoethane (99%) was purchased from Tokyo Chemical Industry Co., Ltd. and used as received (checkers).
6. It should be noted that a large amount of gas is generated after the 1,2-dibromoethane is added to the solution. The time of addition for the 1,2-dibromoethane is about 15-40 seconds, and users are cautioned to not add this material too quickly or too slowly. It is advisable to conduct these experiments in a large flask fitted with long Dimroth (checkers) or Allihn (submitters) condenser and adequately vented to a standard inert gas manifold.
7. The checkers pre-mixed THF (11 mL), 2-methylbenzophenone (4.0 mL) and D_2O (600 µL), and the resultant mixture (14.2 mL) is added to the

reaction flask. The mixture of THF, 2-methylbenzophenone and D₂O should be prepared in advance before adding the 1,2-dibromoethane to flask. The addition time of the mixture of THF, 2-methylbenzophenone and D₂O should be approximately 1 minute, and users are cautioned to not add this mixture too quickly or too slowly. This is a fast and exothermic reaction.

8. 2-Methylbenzophenone (98%) was purchased from Bide Pharmatech Ltd. and used as received (submitters). 2-Methylbenzophenone (98%) was purchased from Sigma-Aldrich and used as received (checkers).

9. Deuterium oxide (99.9% D, for NMR) was purchased from Enengy Chemical and used as received (submitters). Deuterium oxide (99.9% D) was purchased from Sigma-Aldrich and used as received (checkers).

10. TLC analysis of the mixture is shown below. The R_f value of the product and 2-methylbenzophenone (**1**) in petroleum ether/dichloromethane (3/5, v/v) was 0.15 and 0.45, respectively. The thin-layer chromatography plate (TLC silica gel 60 F_{254}, purchased from Merck KGaA) was then visualized using UV (254 nm) (Figure 7).

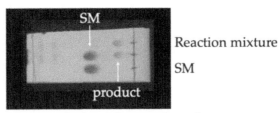

Figure 7. TLC analysis of the reaction mixture

11. Dichloromethane (99%) was purchased from Shanghai Titan Scientific Co., Ltd. and used as received (submitters). Dichloromethane (>99.5%) was purchased from Kanto Chemical Co., Inc. and used as received (checkers).

12. Ammonium chloride (>98.5%) was purchased from Nacalai Tesque, Inc. and used as received (checkers).

13. Anhydrous MgSO₄ (98%) was purchased from Sinopharm Chemical Reagent co., Ltd. and used as received (submitters). Anhydrous MgSO₄ (98%) was purchased from Tokyo Chemical Industry Co., Ltd. and used as received (checkers).

14. Silica gel (Silica gel 60N, spherical and neutral, 0.100–0.210 mm) was purchased from Kanto Chemical Co., Inc. and used as received (checkers).

15. Petroleum ether (99%) was purchased from Shanghai Titan Scientific Co., Ltd. and used as received (submitters). Petroleum ether (>90%) was purchased from Kanto Chemical Co., Inc. and used as received (checkers).
16. The column of silica (111 g) is wet packed in a 6 cm diameter × 30 cm height column using petroleum ether (400 mL) (Note 14). Then, the crude solid is loaded onto the column. At that point, fraction collection (100 mL fractions, 27 × 200 mm) is begun, and elution is continued with 800 mL of petroleum ether/dichloromethane (1/1, v/v), then 1000 mL of petroleum ether/dichloromethane (1/1.2, v/v), then 600 mL of petroleum ether/dichloromethane (1/3, v/v) and then 600 mL of petroleum ether/dichloromethane (1/4, v/v) (Figure 6B). The desired product is obtained in fractions No. 9 through No. 29 (Note 17).
17. The product was visualized by thin-layer chromatography using petroleum ether/dichloromethane (3/5, v/v). The thin-layer chromatography plate (TLC silica gel 60 F$_{254}$, purchased from Merck KGaA) was then visualized using UV (254 nm) and phosphomolybdic acid stain (Note 21) (Figure 8). Once the TLC plate was stained using phosphomolybdic acid, it is needed to heat the TLC plate by hot-plate (350 °C) before visualizing the spot of the product.

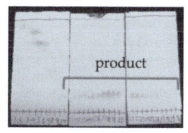
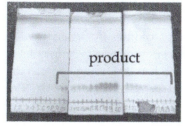

Figure 8. TLC analysis of column fractions

18. All NMR analysis was performed using deuterotrichloromethane (99.8%) as the solvent, which was purchased from Meryer Chemical Technology Co., Ltd. and used as received (submitters). Deuterotrichloromethane (99.8%) was purchased from Kanto Chemical Co., Inc. and used as received (checkers).
19. Characterization of **2**: ^{1}H NMR (400 MHz, CDCl$_3$) δ: 2.05 (brs, 1H), 2.26 (s, 3H), 7.15 (d, J = 7.3 Hz, 1H), 7.19–7.29 (m, 3H), 7.32–7.34 (m, 4H), 7.53 (dd, J = 7.8, 1.4 Hz, 1H) (a resonance corresponding to the residual non-deuterated product appears at 6.01 ppm). ^{13}C NMR (100 MHz,

CDCl$_3$) δ: 19.4, 72.9 (t, J = 22.0 Hz), 126.1, 126.2, 127.1, 127.5, 127.6, 128.5, 130.5, 135.3, 141.3, 142.7. HRMS (ESI) calcd for [C$_{14}$H$_{13}$DONa]$^+$ ([M+Na]$^+$): 222.1000, found: 222.1000. mp 91.9–92.6 °C. IR (KBr): 3330, 3059, 3025, 1486, 1452, 1181, 1035, 748, 701 cm^{-1}. The deuterium ratio was determined by ^{1}H NMR.

20. The qNMR was conducted using 1,4-dimethoxybenzene as an internal standard, with relaxation time (D$_1$) set to 40 seconds. 1,4-Dimethoxybenzene (99%) was purchased from Energy Chemical and used as received (submitters). 1,4-Dimethoxybenzene (>99%) was purchased from Kanto Chemical Co., Inc. and used as received (checkers).
21. When the reaction was carried out on a half-scale, 1.42 g (72% yield, 98% D) of product **2** was obtained with 98.4% purity as determined by qNMR.
22. Phosphomolybdic acid was purchased from Bide Pharmatech Ltd. and used as received (submitters). Phosphomolybdic acid, purchased from FUJIFILM Wako Pure Chemical Corp., was used as received (checkers).

Working with Hazardous Chemicals

The procedures in *Organic Syntheses* are intended for use only by persons with proper training in experimental organic chemistry. All hazardous materials should be handled using the standard procedures for work with chemicals described in references such as "Prudent Practices in the Laboratory" (The National Academies Press, Washington, D.C., 2011; the full text can be accessed free of charge at http://www.nap.edu/catalog.php?record_id=12654). All chemical waste should be disposed of in accordance with local regulations. For general guidelines for the management of chemical waste, see Chapter 8 of Prudent Practices.

In some articles in *Organic Syntheses*, chemical-specific hazards are highlighted in red "Caution Notes" within a procedure. It is important to recognize that the absence of a caution note does not imply that no significant hazards are associated with the chemicals involved in that procedure. Prior to performing a reaction, a thorough risk assessment should be carried out that includes a review of the potential hazards associated with each chemical and experimental operation on the scale that is planned for the procedure. Guidelines for carrying out a risk assessment and for analyzing the hazards associated with chemicals can be found in Chapter 4 of Prudent Practices.

The procedures described in *Organic Syntheses* are provided as published and are conducted at one's own risk. *Organic Syntheses, Inc.*, its Editors, and its Board of Directors do not warrant or guarantee the safety of individuals using these procedures and hereby disclaim any liability for any injuries or damages claimed to have resulted from or related in any way to the procedures herein.

Discussion

Since deuterium was discovered in 1931 by Harold C. Urey,[2] who received the 1934 Nobel Prize for his contributions, there has been significant interest in the use of deuterium and the number of deuterium-labeled compounds reported has increased dramatically.[3] Among those deuterium-labeled compounds, α-deuterated alcohols[4] enjoy a range of uses. They work as deuterium sources or key intermediates in the synthesis of deuterium-containing biologically active compounds including deuterium-containing drugs.[5] Reductive deuteration of carbonyl compounds is the most straightforward process to deliver α-deuterated alcohols.[4a-c] In principle, several different strategies can be used for their formation, including nucleophilic addition of a deuteride anion (D$^-$) to a carbonyl group,[3j] Meerwein-Ponndorf-Verley reduction,[6] single electron reduction (SER) and the umpolung strategies.[7] In terms of a practical and general synthesis of α-deuterated alcohols in large quantities, the umpolung strategies which use D$_2$O as the deuterium source have great advantages in price, supply, and safety. For general comparison, the price of D$_2$O is only 2% of that for NaBD$_4$ and 0.2% of that for LiAlD$_4$. However, the umpolung strategies have rarely been utilized in the reductive deuteration of ketones because these methods would require a large excess amount of D$_2$O or CD$_3$OD (typically more than 100 equiv).[8] We reported a practical umpolung strategy for the reductive deuteration of ketones with a Mg/BrCH$_2$CH$_2$Br/D$_2$O system, which affords α deuterated alcohols in good yields and with near quantitative incorporation of deuterium.[9] Only 1.5 equiv. of D$_2$O are required for the deuteration, while alternative methods require a large excess amount of D$_2$O or CD$_3$OD.

A variety of substituted benzophenones can be transformed into the desired products in good yields and with >89% deuterium incorporation

(Figure 9). Substrates with a moderate electron-withdrawing group, such as fluorine or chlorine, afford the corresponding products (**3a-11a**) in moderate to good yields. Strong electron-withdrawing groups, such as the cyano (**12**) and ester groups (**13**), which are sensitive to Grignard conditions, are also tolerated. Moreover, a terminal olefin (**14**) that could be reduced with

Figure 9. Substrate Scope. Reaction conditions: ketone (1.0 mmol), BrCH$_2$CH$_2$Br (2.0 equiv.), Mg (freshly peeled, 5.0 equiv.), and D$_2$O (1.5 equiv.) in THF (2 mL); isolated yield

magnesium / methanol remained unaffected under the reaction conditions.[10] The reactions occurred smoothly with substrates containing an electron-donating group, such as Me, ^{t}Bu, OMe, Ph, and NMe$_2$, and delivered the corresponding products in 48-86% yields (**15a–31a**).

Substrates with a five-, six-, or seven-membered ring underwent the reaction smoothly to produce the corresponding products (**32a–35a**, Figure 10) in 57–88% yields with up to 98% deuterium incorporation. The reactions also proceeded well with pyridines, affording the target products in 61%–82% yields (**36a–39a**). A pyridine-fused tricyclic compound participated in the reaction and afforded the desired product (**40a**) in 88% yield. The two symmetric diketones that were tested offered an opportunity for the selective reduction of one of the carbonyl groups, and afforded the monodeuterated products (**41a** and **42a**) in moderate yields with the second carbonyl group untouched.

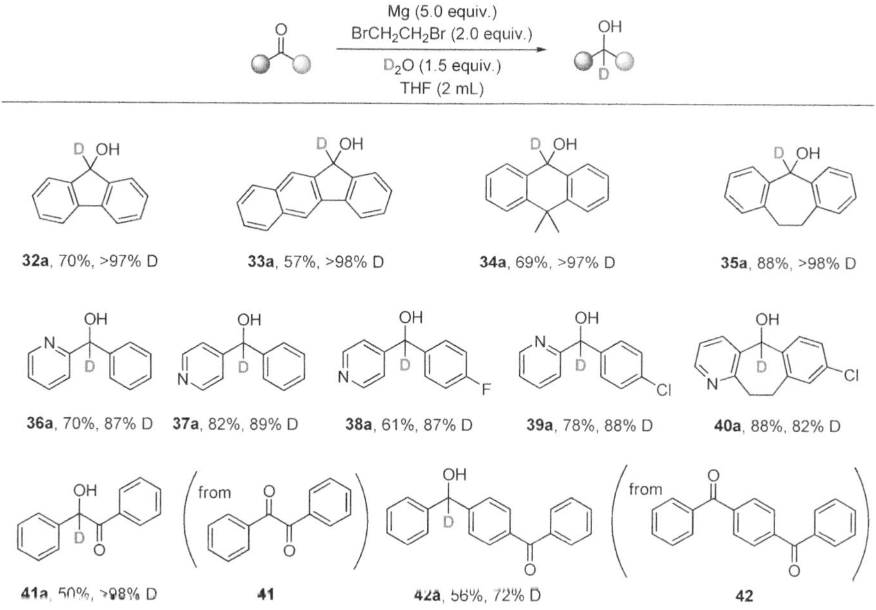

Figure 10. Other substrate Scope. Reaction conditions: ketone (1.0 mmol), BrCH$_2$CH$_2$Br (2.0 equiv.), Mg (freshly peeled, 5.0 equiv.), and D$_2$O (1.5 equiv.) in THF (2 mL); isolated yield

Fenofibrate,[11] a drug for the treatment of hyperlipidemia, mixed dyslipidemia and hypertriglyceridemia, can be selectively reduced to the corresponding deuterated alcohol (**43a**, Figure 11).

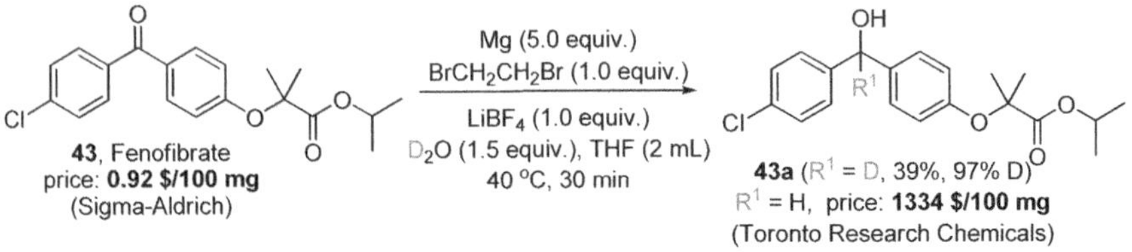

Figure 11. Reductive deuteration of Fenofibrate

In conclusion, a practical method for the reductive deuteration of carbonyl compounds to α-deuterated alcohols with excellent deuterium incorporation is reported. Only 1.5 equivalents of D_2O were required for this highly efficient transformation. This method is economical, and features good substrate scope and excellent functional group tolerance.

References

1. State Key Laboratory of Structural Chemistry, Key Laboratory of Coal to Ethylene Glycol and Its Related Technology, Center for Excellence in Molecular Synthesis, Fujian Institute of Research on the Structure of Matter, Chinese Academy of Sciences, 155 Yangqiao Road West, Fuzhou, Fujian 350002, P. R. of China. E-mail: hlbao@fjirsm.ac.cn. (ORCID: 0000-0003-1030-5089). We thank the National Key R&D Program of China (Grant No. 2017YFA0700103), the NSFC (Grant Nos. 21672213, 21871258, 21922112), the Strategic Priority Research Program of the Chinese Academy of Sciences (Grant No. XDB20000000), the Haixi Institute of CAS (Grant No. CXZX-2017-P01) for financial support.
2. Urey, H. C.; Brickwedde, F. G.; Murphy, G. M. *Phys. Rev.* **1932**, *40* (1), 1–15.
3. (a) Puleo, T. R.; Strong, A. J.; Bandar, J. S. *J. Am. Chem. Soc.* **2019**, *141* (4), 1467–1472; (b) Geng, H.; Chen, X.; Gui, J.; Zhang, Y.; Shen, Z.; Qian, P.; Chen, J.; Zhang, S.; Wang, W. *Nat. Catal.* **2019**, *2* (12), 1071–1077; (c) Liu, W.; Zhao, L.-L.; Melaimi, M.; Cao, L.; Xu, X.; Bouffard, J.; Bertrand, G.; Yan, X. *Chem* **2019**, *5* (9), 2484–2494; (d) Kerr, W. J.; Mudd, R. J.; Reid, M.; Atzrodt, J.; Derdau, V. *ACS Catal.* **2018**, *8* (11), 10895–10900; (e)

Soulard, V.; Villa, G.; Vollmar, D. P.; Renaud, P. *J. Am. Chem. Soc.* **2018**, *140* (1), 155–158; (f) Liu, C.; Chen, Z.; Su, C.; Zhao, X.; Gao, Q.; Ning, G.-H.; Zhu, H.; Tang, W.; Leng, K.; Fu, W.; Tian, B.; Peng, X.; Li, J.; Xu, Q.-H.; Zhou, W.; Loh, K. P. *Nat. Commun.* **2018**, *9* (1), 80; (g) Valero, M.; Weck, R.; Güssregen, S.; Atzrodt, J.; Derdau, V. *Angew. Chem. Int. Ed.* **2018**, *57* (27), 8159–8163; (h) Zhang, M.; Yuan, X.-A.; Zhu, C.; Xie, J. *Angew. Chem. Int. Ed.* **2019**, *58* (1), 312–316; (i) Loh, Y. Y.; Nagao, K.; Hoover, A. J.; Hesk, D.; Rivera, N. R.; Colletti, S. L.; Davies, I. W.; MacMillan, D. W. C. *Science* **2017**, *358* (6367), 1182–1187; (j) Yang, J. *Deuterium: Discovery and Applications in Organic Chemistry*: Elsevier: 2016; (k) Epstein, R. I.; Lattimer, J. M.; Schramm, D. N. *Nature* **1976**, *263* (5574), 198–202; (l) Olah, G. A.; Prakash, G. K. S.; Arvanaghi, M.; Bruce, M. R. *Angew. Chem. Int. Ed. Engl.* **1981**, *20* (1), 92–93.

4. (a) Khaskin, E.; Milstein, D. *ACS Catal.* **2013**, *3* (3), 448–452; (b) Midland, M. M.; Greer, S.; Tramontano, A., Zderic, S. A. *J. Am. Chem. Soc.* **1979**, *101* (9), 2352–2355; (c) Parry, R. J.; Trainor, D. A., *J. Am. Chem. Soc.* **1978**, *100* (16), 5243–5244; (d) Tashiro, M.; Mataka, S.; Nakamura, H.; Nakayama, K. *J. Chem. Soc. Perkin Trans. 1* **1988**, (2), 179–181.

5. (a) Borukhova, S.; Noël, T.; Hessel, V. *ChemSusChem* **2016**, *9* (1), 67–74; (b) Schmidt, F.; Stemmler, R. T.; Rudolph, J.; Bolm, C. *Chem. Soc. Rev.* **2006**, *35* (5), 454–470; (c) Wang, Q.; Sheng, X.; Horner, J. H.; Newcomb, M. *J. Am. Chem. Soc.* **2009**, *131* (30), 10629–10636; (d) Atzrodt, J.; Derdau, V.; Fey, T.; Zimmermann, J. *Angew. Chem. Int. Ed.* **2007**, *46* (41), 7744–7765; (e) Schlatter, A.; Kundu, M. K.; Woggon, W. D. *Angew. Chem. Int. Ed.* **2004**, *43* (48), 6731–6734; (f) Miyazaki, D.; Nomura, K.; Ichihara, H.; Ohtsuka, Y.; Ikeno, T.; Yamada, T. *New J. Chem.* **2003**, *27* (8), 1164–1166.

6. Williams, E. D.; Krieger, K. A.; Day, A. R. *J. Am. Chem. Soc.* **1953**, *75* (10), 2404–2407.

7. (a) Rosales, A.; Muñoz-Bascón, J.; Roldan-Molina, E.; Castañeda, M. A.; Padial, N. M.; Gansäuer, A.; Rodríguez-García, I.; Oltra, J. E, *J. Org. Chem.* **2014**, *79* (16), 7672–7676; (b) Barrero, A. F.; Rosales, A.; Cuerva, J. M.; Gansäuer, A.; Oltra, J. E. *Tetrahedron Lett.* **2003**, *44* (5), 1079–1082; (c) Szostak, M.; Spain, M.; Procter, D. J. *Org. Lett.* **2014**, *16* (19), 5052–5055.

8. Bordoloi, M. *Tetrahedron Lett.* **1993**, *34* (10), 1681–1684.

9. Zhu, N.; Su, M.; Wan, W.-M.; Li, Y.; Bao, H. *Org. Lett.* **2020**, *22* (3), 991–996.

10. Lee, G. H.; Youn, I. K.; Choi, E. B.; Lee, H. K.; Yon, G. H.; Yang, H. C.; Pak, C. S. *Curr. Org. Chem.* **2004**, *8* (13), 1263–1287.

11. McKeage, K.; Keating, G. M. *Drugs* **2011**, *71* (14), 1917–1946.

Appendix
Chemical Abstracts Nomenclature (Registry Number)

Magnesium; (7439-95-4)
1,2-Dibromoethane; (106-93-4)
Tetrahydrofuran; (109-99-9)
2-Methylbenzophenone; (131-58-8)
Deuterium oxide; (7789-20-0)
Ammonium chloride; (12125-02-9)
Petroleum ether; (8032-32-4)
1,4-Dimethoxybenzene; (150-78-7)
Deuterotrichloromethane; (865-49-6)
α-Deutero-*o*-methyl-benzhydrol; (339551-20-1)

Nengbo Zhu received his Ph.D. degree in organic chemistry at University of Chinese Academy of Sciences in 2018 under the supervision of Prof. Hongli Bao. He is currently an assistant research fellow at Fujian Institute of Research on the Structure of Matter, Chinese Academy of Sciences. He is interested in developing the new organic synthesis methodology and introducing organic reactions into polymer chemistry for the development of novel polymerization methodologies, including radical cascade polymerization and alkene functionalization polymerization.

Wen-Ming Wan received his Ph.D. degree in Polymer Chemistry and Physics at University of Science and Technology of China in 2010 and worked on polymerization-induced self-assembly under the supervision of Prof. Cai-Yuan Pan during his PhD period. He is currently a professor at Fujian Institute of Research on the Structure of Matter, Chinese Academy of Sciences, focusing on the introduction of organic reactions into polymer chemistry for the development of novel polymerization methodologies, including but not limited to Barbier polymerization, radical cascade polymerization, alkene functionalization polymerization and polymerization-induced emission.

Hongli Bao received her B.S. degree in chemistry from the University of Science & Technology of China in 2002. She obtained her Ph.D. from the joint program of the Shanghai Institute of Organic Chemistry (China) and the University of Science & Technology of China in 2008 with Professor Kuiling Ding and Professor Tianpa You. She joined the Tambar lab in 2009 and received the UT Southwestern Chilton Fellowship in Biochemistry in 2012. She started her independent career in 2014 at Fujian Institute of Research on the Structure of Matter, Chinese Academy of Science. She is interested in developing new metal-catalyzed reactions and asymmetric catalysis.

Yuuki Watanabe was born in Kanagawa, Japan. He graduated from the University of Tokyo in 2020 with B.S. in Pharmaceutical Sciences. He is continuing his graduate studies at the University of Tokyo under the supervision of Prof. Masayuki Inoue. His research interests are in the area of the total synthesis of complex natural products.

Koichi Hagiwara was born in Kanagawa, Japan., in 1989. He received his B.Sc. degree in 2013 from the University of Tokyo, and his Ph.D. degree in Pharmaceutical Sciences in 2019 under the supervision of Prof. Masayuki Inoue. He was appointed as an Assistant Professor in the Graduate School of Pharmaceutical Sciences at the University of Tokyo in 2017. His research interests include the total synthesis of bioactive and highly complex natural products.

Palladium-Catalyzed Acetylation of Arylbromides

Milauni M. Mehta, Andrew V. Kelleghan, and Neil K. Garg[*1]

Department of Chemistry and Biochemistry, University of California, Los Angeles, California 90095, United States

Checked by Martina Drescher, Daniel Kaiser, Nuno Maulide, Zach Ariki, Yuki Maekawa, and Cathleen Crudden

Procedure (Note 1)

6-Acetylbenzothiophene (1). A single-necked (24/40 joint) 250 mL round-bottomed flask is equipped with a Teflon-coated magnetic stir bar (4.0 × 1.5 cm, football-shaped). The apparatus is flame-dried under vacuum, then cooled to 23 °C under an atmosphere of argon (Note 2). The flask is charged sequentially with 6-bromobenzothiophene (8.00 g, 37.5 mmol, 1 equiv) (Note 3), cesium fluoride (22.8 g, 150 mmol, 4 equiv) (Note 4), and tetrakis(triphenylphosphine)palladium(0) (2.17 g, 1.88 mmol, 0.05 equiv) (Note 4) through the neck of the flask in singular portions. The neck of the flask is then fit with a rubber septum. An argon inlet needle and a purge needle are placed in the rubber septum, and the flask is purged for 5 min (Figure 1A). After 5 min, the vent needle is removed, and acetyltrimethylsilane (10.8 mL, 75 mmol, 2 equiv) (Note 5) is added in one portion over 1 min via a plastic syringe fit with an 18 G × 1.5″ needle. 1,2-Dichloroethane (38 mL, 1 M) (Note 6) is then added to the flask via a plastic syringe fit with an 18 G × 6″ needle in a single portion over 1 min

(Figure 1B). The rubber septum is quickly removed and replaced with a separately flame-dried air condenser with a 24/40 joint (Note 7). The reaction apparatus is then placed in an oil bath preheated to 75 °C. The reaction mixture is stirred vigorously (800 RPM) for 24 h under positive argon pressure (Figure 1C).

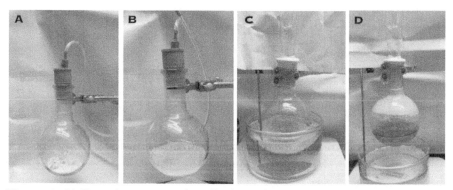

Figure 1. A) Reaction setup after flask is charged with solid reactants; B) Reaction setup after acetyltrimethylsilane; (2) and 1,2-dichloroethane addition; C) Reaction setup with air condenser and oil bath; D) Reaction mixture after stirring at 75 °C for 24 h (photos provided by submitters)

After 24 h (Note 8), the reaction flask is removed from the oil bath and allowed to cool to 23 °C (Figure 1D). Once the reaction mixture is cooled to 23 °C, the reflux condenser is removed, and the heterogeneous mixture is diluted with heptane (75 mL) (Note 9). The solution is then filtered through a plug of silica gel (50 g, pre-wetted with 100 mL ethyl acetate) (Note 10) in a fritted Büchner funnel (Note 11) into a 1000 mL round-bottomed flask using ethyl acetate as eluent (500 mL) (Notes 12 and 13). The filtrate is then concentrated under reduced pressure (31 °C, from 100 mmHg to 50 mmHg).

The resultant brown solid is purified via column chromatography using an OD 7.5 × 12 cm column of 250 g silica gel (Note 14) and eluted sequentially with 1400 mL 14:1 heptane:EtOAc and 2000 mL 9:1 heptane:EtOAc. The eluate is collected using 25 mL test tubes to provide the product in fractions 78–106 (Note 15). The fractions are combined in a collection flask and concentrated by rotary evaporation under reduced pressure (31 °C, 90 mmHg). The material is then transferred to an 8-dram vial and dried under high vacuum for 30 min (<1 mmHg) to afford 6-acetylbenzothiophene (**1**) as a yellow powder (4.15 g, 63% yield) (Notes 16, 17 and 18).

Figure 2. Isolated 6-acetylbenzothiophene (1)
(photo provided by submitters)

Notes

1. Prior to performing each reaction, a thorough hazard analysis and risk assessment should be carried out with regard to each chemical substance and experimental operation on the scale planned and in the context of the laboratory where the procedures will be carried out. Guidelines for carrying out risk assessments and for analyzing the hazards associated with chemicals can be found in references such as Chapter 4 of "Prudent Practices in the Laboratory" (The National Academies Press, Washington, D.C., 2011; the full text can be accessed free of charge at https://www.nap.edu/catalog/12654/prudent-practices-in-the-laboratory-handling-and-management-of-chemical. See also "Identifying and Evaluating Hazards in Research Laboratories" (American Chemical Society, 2015) which is available via the associated website "Hazard Assessment in Research Laboratories" at https://www.acs.org/content/acs/en/about/governance/committees/chemicalsafety/hazard-assessment.html. In the case of this procedure, the risk assessment should include (but not necessarily be limited to) an evaluation of the potential hazards associated with acetyltrimethylsilane, 6-bromobenzothiophene, tetrakis(triphenyl-

phosphine) palladium (0), 1,2-dichloroethane, hexanes, ethyl acetate, silica gel, and cesium fluoride.
2. The authors performed the reaction under an atmosphere of nitrogen.
3. 6-Bromobenzothiophene (97%) was purchased from Combi-Blocks and used after crushing the material into a fine powder with a mortar and pestle. Checkers purchased 6-bromobenzothiophene (98%) from Fluorochem and the material was used as received.
4. Cesium fluoride (99%) and Pd(PPh$_3$)$_4$ (99%) were purchased from Strem Chemicals and used as received. Checkers purchased cesium fluoride (99%) from Fluorochem and the reagent was used as received. A third run was conducted with Pd(PPh$_3$)$_4$ (98%), purchased from Fluorochem and used as received.
5. Acetyltrimethylsilane (97%) was purchased from Sigma Aldrich and used as received.
6. 1,2-Dichloroethane (99%) was purchased from Fischer Scientific and passed through an activated alumina column with argon before use. Checkers purchased 1,2-dichloroethane (99.5%) from Acros Organics and the solvent was used as received.
7. The air condenser is equipped with a rubber septum with a nitrogen inlet needle to maintain an inert atmosphere and positive pressure. The joint of the round-bottomed flask and reflux condenser is sealed with Teflon tape (Figure 3) before placing the reaction flask in an oil bath.

Figure 3. Full reaction setup with air condenser and nitrogen inlet needle (photo provided by submitters)

8. Reaction progress was monitored after 24 h using TLC analysis on silica gel with 9:1 hexanes:EtOAc as eluent. Visualization of the TLC plate was performed with UV irradiation (254 nm) and *p*-anisaldehyde. The starting material has $R_f = 0.73$ (no color, UV active), the desired ketone product has $R_f = 0.32$ (pink, UV active).

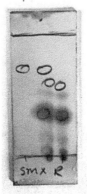

Figure 4. TLC of the crude reaction mixture (SM = starting material, X = co-spot of SM and R, R = reaction mixture) (photo provided by submitters)

9. Checkers used heptane, which was purchased from Donauchemie and used after distillation. The authors used hexanes (98.5%), which were purchased from Fisher Scientific and used as received.
10. SiliaFlash P60 (particle size 0.040–0.063 mm) was purchased from SiliCycle and used as received. The checkers used silica gel purchased from Macherey-Nagel (particle size 0.040–0.063 mm), which was used as received.
11. Filtration used a 150 mL medium porosity fritted Büchner funnel under vacuum.
12. Ethyl acetate (99.5%) was purchased from VWR and used as received. The checkers used ethyl acetate, which was purchased from Donauchemie and used after distillation.
13. The reaction flask was rinsed with ethyl acetate until all of the material was transferred onto the pad of silica on the fritted funnel. Filtrate is a light brown color as shown in Figure 5.

Figure 5. Filtration apparatus (photo provided by submitters)

14. The column is wet-packed using 250 g of silica and 600 mL of 14:1 heptane:EtOAc. The crude material is dissolved in benzene (20 mL) and loaded on the column with subsequent rinses of the round-bottomed flask using benzene (3 × 1 mL) to ensure quantitative transfer. The authors used hexanes instead of heptane.
15. Fractions containing the product were identified by TLC analysis (9:1 heptane:EtOAc as eluent). Fractions 65–77 contained the desired product and two impurities that can be visualized by UV irradiation and by *p*-anisaldehyde stain as a blue spot (R_f = 0.54, Figure 6) and a faint brown spot (R_f = 0.46, Figure 6). These fractions were not collected. Fractions 98–106 contained an impurity that can be visualized by UV irradiation and by *p*-anisaldehyde stain as a faint black spot (R_f = 0.37, Figure 6). This impurity does not impact the purity of the desired compound as judged by qNMR, so these fractions were also collected. Test tubes containing the desired product were each rinsed with EtOAc (3 × 1 mL), and the rinses were transferred into the collection flask.

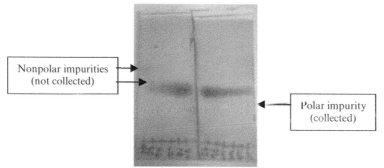

Figure 6. Representative TLC analysis showing impurities in column fractions (photo provided by submitters)

16. 6-Acetylbenzothiophene: mp 72.4–74.6 °C. ^{1}H NMR (400 MHz, CDCl$_3$) δ: 2.68 (s, 3H), 7.39 (d, J = 5.6 Hz, 1H), 7.67 (d, J = 5.3 Hz, 1H), 7.86 (d, J = 8.4 Hz, 1H), 7.96 (dd, J = 8.4, 2.3 Hz, 1H), 8.51 (br s, 1H); ^{13}C NMR (100 MHz, CDCl$_3$) δ: 26.9, 123.7, 123.7, 123.9, 124.1, 131.0, 133.4, 139.8, 143.0, 197.8; IR (film): 3106, 1677, 1593, 1391, 1356, 1269, 1249, 1237, 827, 762 cm^{-1}; HRMS–ESI (m/z) [M + H]+ calcd for C$_{10}$H$_9$OS$^+$ 177.0369; found, 177.0369.

17. The purity of **1** was determined to be >95 wt% by qNMR using 1,3,5-trimethoxybenzene (Alfa-Aesar, 99%) as the external standard. Elemental analysis performed by the checkers provided the following data: Elemental analysis: Calcd for C$_{10}$H$_8$OS: C, 68.15; H, 4.58; N, 0.0; S, 18.19; O, 9.08. Found: C, 67.75; H, 4.44; N, <0.05; S, 17.96; O, 9.21.

18. A second run performed at full scale provided 4.51 g (68%) of **1** at >97% purity. Three reactions were performed at approximately half-scale and provided product **1** in yields that ranged between 65% and 73%.

19. Another run on half-scale was performed using Pd(PPh$_3$)$_4$ (98%) purchased from Fluorochem. For this reaction, TLC analysis showed remaining starting material after 24 h, for which reason an additional 2.5% of catalyst were added (total: 7.5%) and stirring was continued for 5 h. After this time, the reaction was treated as above to yield 2.51 g (76%) of **1**.

Working with Hazardous Chemicals

The procedures in *Organic Syntheses* are intended for use only by persons with proper training in experimental organic chemistry. All hazardous materials should be handled using the standard procedures for work with chemicals described in references such as "Prudent Practices in the Laboratory" (The National Academies Press, Washington, D.C., 2011; the full text can be accessed free of charge at http://www.nap.edu/catalog.php?record_id=12654). All chemical waste should be disposed of in accordance with local regulations. For general

guidelines for the management of chemical waste, see Chapter 8 of Prudent Practices.

In some articles in *Organic Syntheses*, chemical-specific hazards are highlighted in red "Caution Notes" within a procedure. It is important to recognize that the absence of a caution note does not imply that no significant hazards are associated with the chemicals involved in that procedure. Prior to performing a reaction, a thorough risk assessment should be carried out that includes a review of the potential hazards associated with each chemical and experimental operation on the scale that is planned for the procedure. Guidelines for carrying out a risk assessment and for analyzing the hazards associated with chemicals can be found in Chapter 4 of Prudent Practices.

The procedures described in *Organic Syntheses* are provided as published and are conducted at one's own risk. *Organic Syntheses, Inc.*, its Editors, and its Board of Directors do not warrant or guarantee the safety of individuals using these procedures and hereby disclaim any liability for any injuries or damages claimed to have resulted from or related in any way to the procedures herein.

Discussion

Aryl methyl ketones and heteroaryl methyl ketones are versatile building blocks in the syntheses of fragrances,[2] resins,[3] and drug candidates.[4] Traditionally, Friedel–Crafts acylation[5] or addition of organometallic reagents into carboxylic acid derivatives[6] have been employed to access alkyl–aryl ketones. Despite the synthetic utility of these methods, several drawbacks exist, such as the necessity to use stoichiometric reagents, poor functional group tolerance, and competitive over-alkylation in the case of organometallic additions.[5a,6c–d,7]

To overcome these challenges, transition metal catalysis has made available new mechanistic paradigms that offer improved chemo- and regioselectivity in the formation of the desired aryl–acyl linkage.[8–12] Several cross-coupling approaches to form methyl aryl ketones have been reported, including carbonylative cross-couplings employing CO or CO_2 as the carbonyl source.[8,9] Although this strategy has proved effective, the necessity of manipulating toxic, gaseous reagents remains a limitation.[13] Alternative cross-coupling approaches include Heck reactions of enol ethers, followed by subsequent hydrolysis[10] and cross-couplings of α–alkyoxyvinyl metal reagents with ensuing hydrolysis.[11,12] Though these strategies eliminate the necessity to employ CO or CO_2, multiple steps are required to furnish the desired alkyl-aryl ketone. Therefore, a methodology that forms the desired aryl–acyl linkage in a single step, while avoiding the use of gaseous reagents, would be useful. The one-step procedure reported here circumvents these limitations by providing a mild, catalytic alternative that relies on commercially available reagents (i.e., acetyltrimethylsilane (2) and (hetero)aryl bromides) to construct aryl–acetyl linkages.[14]

This methodology is tolerant of a variety of substitution patterns on the aryl bromide coupling partner as summarized in Table 1. Substrates with electron-donating groups at the *ortho* position underwent the coupling smoothly, as demonstrated by products **3** and **4**. Amine and carboxylic ester functionalities were employed as shown by the formation of products **5** and **6**. Furthermore, vinyl, chloro, and trifluoromethyl moieties at the *para* position are well tolerated in this transformation. Notably, an aryl chloride was not disturbed in this methodology, as demonstrated by the formation of ketone **8**.

Table 1. Coupling of acetyltrimethylsilane and aryl-bromides

$$\text{Me-C(O)-SiMe}_3 \; (\textbf{2}) + \text{Br-Aryl} \xrightarrow[\text{1,2-dichloroethane, 75 °C}]{\text{Pd(PPh}_3)_4, \text{CsF}} \text{Me-C(O)-Aryl}$$

3 (2-OMe-C6H4), 73% yield[a]
4 (2-Me-C6H4), 56% yield[a]
5 (3-NMe2-C6H4), 89% yield[a]
6 (3-CO2Me-C6H4), 60% yield[a]
7 (4-vinyl-C6H4), 81% yield[a]
8 (4-Cl-C6H4), 73% yield[a]
9 (4-CF3-C6H4), 53% yield[a]

Conditions: Pd(PPh$_3$)$_4$ (5.0 mol%), CsF (4.0 equiv), substrate (1.0 equiv), **2** (2.0 equiv), 6 h.[15] [a] The yield of the product was determined by ^{1}H NMR analysis of crude reaction using 1,3,5-trimethoxybenzene as an internal standard (0.1 equiv).

Heteroaryl bromides are valuable test substrates because of the ubiquity of heterocycles in bioactive molecules and pharmaceutical targets. Gratifyingly, heteroaryl bromides proved to be viable cross-coupling partners as depicted in Table 2. Sulfur-containing substrates afforded desired compounds **1** and **10** in excellent yields. Quinoline and indole motifs furnished ketones **11** and **12** in good yields, respectively. This methodology is also tolerant of oxygen-containing heterocycles as indicated by the formation of ketones **13** and **14**.

Table 2. Coupling of acetyltrimethylsilane and heteroaryl-bromides

Conditions: Pd(PPh$_3$)$_4$ (5.0 mol%), CsF (4.0 equiv), substrate (1.0 equiv), **2** (2.0 equiv). [a]**10**, **11**, **12** were stirred for 6 h[15] and the yield of the product was determined by ^{1}H NMR analysis of crude reaction using 1,3,5-trimethoxybenzene as an internal standard (0.1 equiv). [b]For **12**, Pd(PPh$_3$)$_4$ (10.0 mol%) was used. [c]**1**, **13**, and **14** are isolated yields.

In summary, this methodology provides facile access to acetylated arenes from commercially available acetyltrimethylsilane (**2**) and (hetero)aryl bromides. Its operational simplicity and wide substrate scope render it an attractive alternative protocol to traditional methods for the construction of aryl–carbonyl linkages.

References

1. Department of Chemistry and Biochemistry, University of California, Los Angeles, California 90095, United States. E-mail: neilgarg@chem.ucla.edu. The authors are grateful to the University of California, Los Angeles and the Trueblood Family. These studies were also supported by shared instrumentation grants from the NSF (CHE-1048804) and the National Center for Research Resources (S10RR025631).
2. (a) Gautschi, M.; Plessis, C.; Derrer, S. Fragrance precursors. US Patent 6,939,845. 6 Sep. 2005. (b) Behan, J. M.; Clements, C. F.; Hooper, D. C.; Martin, J. R.; Melville, J. B.; Perring, K. D. Perfume compositions. US Patent 5,554,588. 10 Sep. 1996.
3. Siegel, H.; Eggersdorf, M. Ketones. Ullmann's Encyclopedia of Industrial Chemistry; Wiley-VCH Verlag GmbH: Germany, **2000**, 187–207.
4. (a) Qian, Y.; Corbett, W. L.; Berthel, S. J.; Choi, D. S.; Dvorozniak, M. T.; Geng, W.; Gillespie, P.; Guertin, K. R.; Haynes, N.-E.; Kester, R. F.; Mennona, F. A.; Moore, D.; Racha, J.; Radinov, R.; Sarabu, R.; Scott, N. R.; Grimsby, J.; Mallalieu, N. L. *ACS Med. Chem. Lett.* **2013**, *4*, 414–418. (b) Beaulieu, P. L.; Bös, M.; Cordingley, M. G.; Chabot, C.; Fazal, G.; Garneau, M.; Gillard, J. R.; Jolicoeur, E.; LaPlante, S.; McKercher, G.; Poirier, M.; Poupart, M.-A.; Tsantrizos, Y. S.; Duan, J.; Kukolj, G. *J. Med. Chem.* **2012**, *55*, 7650–7666. (c) Norman, M. H.; Andrews, K. L.; Bo, Y. Y.; Booker, S. K.; Caenepeel, S.; Cee, V. J.; D'Angelo, N. D.; Freeman, D. J.; Herberich, B. J.; Hong, F.-T.; Jackson, C. L. M.; Jiang, J.; Lanman, B. A.; Liu, L.; McCarter, J. D.; Mullady, E. L.; Nishimura, N.; Pettus, L. H.; Reed, A. B.; Miguel, T. S.; Smith, A. L.; Stec, M. M.; Tadesse, S.; Tasker, A.; Aidasani, D.; Zhu, X.; Subramanian, R.; Tamayo, N. A.; Wang, L.; Whittington, D. A.; Wu, B.; Wu, T.; Wurz, R. P.; Yang, K.; Zalameda, L.; Zhang, N.; Hughes, P. E. *J. Med. Chem.* **2012**, *55*, 7796–7816. (d) Liu, T.; Nair, S. J.; Lescarbeau, A.; Belani, J.; Peluso, S.; Conley, J.; Tillotson, B.; O'Hearn, P.; Smith, S.; Slocum, K.; West, K.; Helble, J.; Douglas, M.; Bahadoor, A.; Ali, J.; McGovern, K.; Fritz, C.; Palombella, V. J.; Wylie, A.; Castro, A. C.; Tremblay, M. R. *J. Med. Chem.* **2012**, *55*, 8859–8878. (e) Akritopoulou-Zanze, I.; Wakefield, B. D.; Gasiecki, A.; Kalvin, D.; Johnson, E. F.; Kovar, P.; Djuric, S. W. *Bioorg. Med. Chem. Lett.* **2011**, *21*, 1480–1483. (f) Kung, P.-P.; Sinnema, P.-J.; Richardson, P.; Hickey, M. J.; Gajiwala, K. S.; Wang, F.; Huang, B.; McClellan, G.; Wang, J.; Maegley, K.; Bergqvist, S.; Mehta, P. P.; Kania, R. *Bioorg. Med. Chem. Lett.* **2011**, *21*, 3557–3562.

5. For an early review on the traditional Freidel–Crafts acylation: (a) Calloway, N. O. *Chem. Rev.* **1935**, 327–392. For recent developments and applications in total syntheses: (b) Sartori, G.; Maggi, R. *Chem. Rev.* **2006** 1077–1104. (c) Schmidt, N. G.; Pavkov-Keller, T.; Richter, N.; Wiltschi, B.; Gruber, K.; Kroutil, W. *Angew. Chem. Int. Ed.* **2017**, 7615–7619. (d) Heravi, M. M.; Zadsirjan, V.; Saedi, P.; Momeni, T. *RSC Adv.* **2018**, 40061–40163.

6. For overviews of organometallic reagents in the syntheses of alkyl-aryl ketones: (a) Larock, R. C. *Comprehensive Organic Transformations*, VCH, New York, 1989. (b) Dieter, R. K. *Tetrahedron*, **1999**, 55, 4177–4236. For selected developments in alkyl-aryl ketone synthesis from organometallic reagents: (c) Gowda, M. S.; Pande, S. S.; Ramakrishna, R. A.; Prabhu, K. R. *Org. Biomol. Chem.* **2011**, 5365–5368. (d) Wang, X.; Zhang, L.; Sun, X.; Xu, Y.; Krishnamurthy, D.; Senanayake, C. H. *Org. Lett.* **2005**, 5593–5595.

7. Seyferth, D. *Organometallics*, **2009**, 1598–1605.

8. (a) Tanaka, M. *Synthesis* **1981**, 1, 47–48. (b) Hwang, K.-J.; O'Neil, J. P.; Katzenellenbogen, J. A. *J. Org. Chem.* **1992**, 57, 1262–1271. (c) Shi, Y.; Koh, J. T. *J. Am. Chem. Soc.* **2002**, 124, 6921–6928.

9. For examples of three-component carbonylative approaches utilizing organoboron species, see: (a) Ishiyama, T.; Kizaki, H.; Hayashi, T.; Suzuki, A.; Miyaura, N. *J. Org. Chem.* **1998**, 63, 4726–4731. (b) Jafarpour, F.; Rashidi-Ranjbar, P.; Kashani, A. O. *Eur. J. Org. Chem.* **2011**, 2128–2132. For an example using organosilicon reagents, see: (c) Hatanaka, Y.; Fukushima, S.; Hiyama, T. *Tetrahedron* **1992**, 48, 2113–2126. For an example involving organotin reagents, see: (d) Echavarren, A. M.; Stille, J. K. *J. Am. Chem. Soc.* **1988**, 110, 1557–1565.

10. (a) Hallberg, A.; Westfelt, L.; Holm, B. *J. Org. Chem.* **1981**, 46, 5414–5415. (b) Russell, C. E.; Hegedus, L. S. *J. Am. Chem. Soc.* **1983**, 105, 943–949.

11. Denmark, S. E.; Neuville, L. *Org. Lett.* **2000**, 2, 3221–3224.

12. (a) Kosugi, M.; Sumiya, T.; Obara, Y.; Suzuki, M.; Sano, H.; Migita, T. *Bull. Chem. Soc. Jpn.* **1987**, 60, 767–768. (b) Kwon, H. B.; McKee, B. H.; Stille, J. K. *J. Org. Chem.* **1990**, 55, 3114–3118.

13. Morimoto, T.; Kakiuchi, K. *Angew. Chem., Int. Ed.* **2004**, 43, 5580–5588.

14. Ramgren, S. D.; Garg, N. K. *Org. Lett.* **2014**, 16, 824–827.

15. Tables 1 and 2 were adapted from the original report by Ramgren and coworkers (reference 14). The reaction time in the original report was 6 h for reactions run on 0.05 mmol scale, however scaling up to 37.5 mmol required longer reaction times to ensure full consumption of starting material, thus a reaction time of 24 h is reported for this procedure.

Appendix
Chemical Abstracts Nomenclature (Registry Number)

6-Bromobenzothiophene; (17347-31-9)
Acetyltrimethylsilane; (13411-48-8)
Cesium fluoride; (13400-13-0)
Tetrakis(triphenylphosphine)palladium (0); (14221-01-3)

Milauni Mehta was born in Mumbai, India and raised in Princeton, New Jersey. In 2018, she received her B.A. in Chemistry from The Ohio State University, where she carried out research under the direction of Professor T. V. RajanBabu. In 2018, she began graduate studies at the University of California, Los Angeles, where she is currently a third-year graduate student in Professor Neil K. Garg's laboratory. Her studies primarily focus on developing transition metal-catalyzed cross-coupling reactions.

Andrew Kelleghan was born in Santa Monica, California. In 2018, he received his B. S. in Chemical and Biomolecular Engineering from University of California, Berkeley where he carried out research under the advisement of Professor Phillip Messersmith. He then pursued graduate studies at the University of California, Los Angeles, where he is currently a third-year graduate student in Professor Neil K. Garg's laboratory. His studies primarily focus on developing synthetic methods utilizing strained cyclic allenes.

Neil Garg is a Professor of Chemistry and the Kenneth N. Trueblood Endowed Chair at the University of California, Los Angeles. His laboratory develops novel synthetic strategies and methodologies to enable the total synthesis of complex bioactive molecules.

Martina Drescher is the lead technician of the Maulide group at the University of Vienna, where she has worked with several group leaders over the course of 38 years.

Daniel Kaiser received his Ph.D. at the University of Vienna in 2018, completing his studies under the supervision of Prof. Nuno Maulide. After a postdoctoral stay with Prof. Varinder K. Aggarwal at the University of Bristol, he returned to Vienna in 2020 to assume a position as senior scientist in the Maulide group. His current research focusses on the chemistry of destabilized carbocations and related high-energy intermediates.

Zach Ariki was born in Colorado, USA. He obtained his B.A. in Chemistry from Boston University in 2014 and completed his Ph.D. (2021) under the supervision of Professor Cathleen Crudden at Queen's University. His thesis focused on the development of transition metal catalyzed cross-coupling methodologies using alkyl sulfone electrophiles. Following the completion of his graduate studies, he joined GreenCentre Canada as a Development Scientist.

Yuuki Maekawa received his Ph.D. from Gifu University in 2017 under the supervision of Professor Toshiaki Murai. In 2017 he joined the Crudden group as a postdoctoral fellow, where he studied the use of sulfur-containing compounds in metal-catalyzed cross-coupling reactions. In 2019, he was awarded a JSPS Research Fellowship for Young Scientists. He is currently a Scientist at Paraza Pharma in Montreal, Canada.

Cu-catalyzed Allylic Perfluoroalkylation of Alkenes Using Perfluoro Acid Anhydrides: Preparation of N-(5,5,5-Trifluoro-2-penten-1-yl)phthalimide

Yuma Aoki, Shintaro Kawamura, and Mikiko Sodeoka[1*]

Catalysis and Integrated Research Group, RIKEN Center for Sustainable Resource Science, and Synthetic Organic Chemistry Laboratory, RIKEN Cluster for Pioneering Research, 2-1 Hirosawa, Wako, Saitama 351-0198, Japan

Checked by Martina Drescher, Daniel Kaiser, and Nuno Maulide

Procedure (Note 1)

N-(5,5,5-Trifluoro-2-penten-1-yl)phthalimide (**2**). A stirring bar (7 × 50 mm) is placed in a three-necked round-bottomed flask (500-mL size) (Note 2). The flask is fitted with two glass stoppers and a reflux condenser is connected to a vacuum-argon manifold with a KOH trap in-between (Figure 1). After addition of urea·H_2O_2 (4.85 g, 51.6 mmol, 1.2 equiv) (Note 3) to the flask, it is evacuated and backfilled with argon gas (>99.999%) three times and purged with the gas very slowly during the reaction. Then, CH_2Cl_2 (100 mL) (Note 3) is added from a 100-mL syringe fitted with PFA tubing (Note 2), after replacing the two glass stoppers with a rubber septum and an internal thermometer equipped with a Teflon® adapter. The slurry is cooled with an ice-salt bath and is slowly stirred (150 rpm) (Note 4) until the temperature falls to 2±1 °C (for ca. 1 h).

Figure 1. Reaction set-up (photo provided by submitters)

After cooling, trifluoroacetic anhydride (59.8 mL, 430 mmol, 10 equiv) (Note 3) is added to the mixture from a syringe fitted with PFA tubing (100-mL size) over ca. 5 min, temporarily leading to an increase of the internal temperature of 5 °C. The rubber septum is replaced with a glass cap, and the mixture is stirred at 2±1 °C for 1 h (Figure 2a). Then, *N*-(3-buten-1-yl)phthalimide **1** (8.65 g, 43 mmol, 1.0 equiv) (Note 3) is added in one portion *via* a funnel, followed by the addition of [Cu(CH$_3$CN)$_4$]PF$_6$ (3.21 g, 8.6 mmol, 20 mol %) (Note 3) in one portion. After the addition of the copper catalyst, the reaction mixture is warmed to 40 °C. The color of the solution changes from colorless to aqua blue, and small bubbles of carbon dioxide appear (Figure 2b) (Note 5). The mixture is stirred for 4 h and completion of the reaction is confirmed by TLC analysis (Figure 3a) (Note 6).

Figure 2. (a) Reaction of urea·H_2O_2 with trifluoroacetic anhydride (b) Color change after addition of Cu catalyst (photos provided by submitters)

The mixture is cooled to 2±1 °C and diluted with CH_2Cl_2 (200 mL) and transferred to a 1 L round-bottomed flask. While further cooling, sat. K_2CO_3 aqueous solution (300 mL) is added while monitoring the appearance of bubbles due to neutralization. Initially, this addition is performed drop wise, using a Pasteur pipette. After ~50 mL have been added in this manner, the remaining solution is slowly poured into the vigorously stirred mixture over the course of 2 min. The mixture is further stirred at ambient temperature (22–25 °C) for 1 h and transferred to a separatory funnel (1-L size). The aqueous layer is extracted with CH_2Cl_2 (3 × 200 mL), allowing the mixture to stand for 10 min each time to ensure adequate separation of the phases. The organic phases are combined in an Erlenmeyer flask (2-L size). The combined organic phase is washed with brine (50 mL) and dried over 50 g of anhydrous Na_2SO_4. The absence of peroxides in the organic phase is confirmed by using a peroxide detection strip (Figure 3b) (Notes 7 and 8). The organic phase is filtered into a flask (2-L size) through a sintered glass filter (100-mL size), and the filtrate is evaporated in a rotary evaporator (80 mmHg, 30 °C). The obtained crude product (12.4 g) is purified by means of chromatography on silica (Note 9). The product, containing a small amount of impurities, is obtained as a white solid (8.08 g) after concentration of the combined fractions by removal of the solvent with a rotary evaporator (80 mmHg, 30 °C) and drying in vacuo (<6 mmHg) for 3 h.

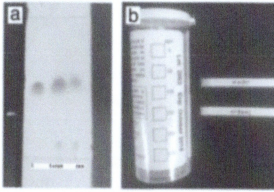

Figure 3. (a) TLC analysis of the reaction mixture (rxn) after 4 h (b) Detection of peroxide content in organic and aqueous phases after extraction (photos provided by submitters)

Heptane (15 mL) is added to the white solid in a 50 mL round-bottomed flask, and the white solid is roughly crushed with a spatula into pieces smaller than ca. 0.5 cm (Figure 4). The liquid phase is decanted with the aid of a glass Pasteur pipette (ca. 1.5 mL, 9-inch size; Iwaki IK-PAS-9P). The resulting white solid is further purified by repeating the crush-decantation process three more times, and then drying the residue in vacuo (<6 mmHg) for 3 h to give product **2** (5.66 g, 49% yield, 97% purity, E/Z = 88.5/11.5) (Notes 10, 11, 12, and 13) (Figure 5), which can be stored under ambient conditions.

Figure 4. Crushing lumps of the white solid with a spatula (photo provided by submitters)

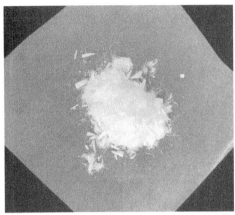

Figure 5. Photograph of pure product 2 (photo provided by submitters)

Notes

1. Prior to performing each reaction, a thorough hazard analysis and risk assessment should be carried out with regard to each chemical substance and experimental operation on the scale planned and in the context of the laboratory where the procedures will be carried out. Guidelines for carrying out risk assessments and for analyzing the hazards associated with chemicals can be found in references such as Chapter 4 of "Prudent Practices in the Laboratory" (The National Academies Press, Washington, D.C., 2011; the full text can be accessed free of charge at https://www.nap.edu/catalog/12654/prudent-practices-in-the-laboratory-handling-and-management-of-chemical. See also "Identifying and Evaluating Hazards in Research Laboratories" (American Chemical Society, 2015) which is available via the associated website "Hazard Assessment in Research Laboratories" at https://www.acs.org/content/acs/en/about/governance/committees/chemicalsafety/hazard-assessment.html. In the case of this procedure, the risk assessment should include (but not necessarily be limited to) an evaluation of the potential hazards associated with urea hydrogen peroxide, trifluoroacetic anhydride, *N*-(3-buten-1-yl)phthalimide, tetrakis(acetonitrile)copper(I) hexafluorophosphate, dichloromethane, potassium carbonate, heptane, and sodium sulfate. *(Caution!) There is a possibility of explosions when handling bis(trifluoroacetyl)peroxide (BTFAP)*

generated as a reactive intermediate in the present reaction. In particular, the phase transition from the liquid to the solid state must be avoided in accordance with the literature.[2] The peroxide must be carefully quenched with sat. K_2CO_3 aqueous as described in the procedure, and the peroxide content in the organic phase after the extraction must be checked with a test paper as described in the procedure before evaporation of the organic solvent.

2. All glassware used is dried in the oven before use. Very small amounts of grease or Teflon rings are used to seal glass joints and stopcocks. PTFE tape is used to fix and seal connections between the flask and each PVC tubing, septum, and thermometer. A syringe fitted with PFA tubing instead of a metal-needle is used to avoid the possibility of contamination of the peroxide solution with metals (see Figure 6). The vacuum/argon gas manifold system is equipped with a bubbler containing liquid paraffin at the inlet and outlet, as shown in Figure 7.

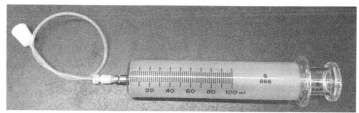

Figure 6. Syringe fitted with PFA tubing (photo provided by submitters)

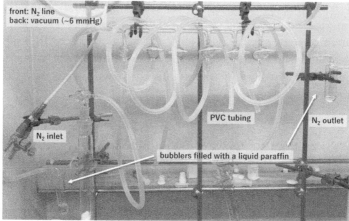

Figure 7. Vacuum/nitrogen gas manifold system (photo provided by submitters)

3. Urea·H₂O₂ (97%) was obtained from Aldrich Chemical Co., Inc. and used as received. Anhydrous CH_2Cl_2 (<10 ppm of water and <1 ppm O_2 contents) was obtained from Kanto Chemical Co., Inc. and further purified by using a Glass Contour Solvent Filtration System equipped with columns of alumina and a supported copper catalyst prior to use. The checkers purchased dichloromethane (99.6%) from Acros Organics and the solvent was used as received. Trifluoroacetic anhydride (>98.0%), N-(3-buten-1-yl)phthalimide (>98%), and $[Cu(CH_3CN)_4]PF_6$ (>97.0%) were obtained from Tokyo Chemical Industry Co., Ltd and used as received. The checkers purchased $[Cu(CH_3CN)_4]PF_6$ (>97.0%) from Sigma Aldrich and the chemical was used as received.

4. Particles of urea·H₂O₂ are scattered and adhere to the surface of the flask if stirring is vigorous. Thus, gentle stirring (ca. 150 rpm) is necessary.

5. The checkers observed the transient formation of a light blue precipitate. Carbon dioxide (CO_2) gas is released through the argon line under a slight positive flow of nitrogen gas.

6. The reaction is monitored by TLC analysis on Merck silica-gel 60 F_{254} plates with CH_2Cl_2 as the eluent. Spots on the TLC plate are visualized with a UV lamp (254 nm). The R_f values of **1** and **2** are 0.54 and 0.57, respectively.

7. XploSens PS™ was used to test for peroxides. Bis(trifluoroacetyl)peroxide (^{19}F NMR signal at −70.8 ppm in $CDCl_3$) could not be detected in the organic and water phases by ^{19}F NMR analysis. As shown in Figure 3, the test paper indicated no peroxide content in the organic phase. Thus, we can conclude that no other potentially hazardous species, such as trifluoroacetoxy peracid, is present. In addition, ^{19}F NMR analysis of the water phase shows only the signal derived from trifluoroacetic acid, which suggests that the basic workup sufficiently hydrolyzes diacyl peroxide to afford hydrogen peroxide.

8. Peroxides in the water phase are quenched with sat. $Na_2S_2O_3$ aqueous solution.

9. Silica gel chromatography was performed using a column (5 × 30 cm) containing 170 g of silica eluted with CH_2Cl_2 (3 L). Fraction sizes of approximately 95 mL were obtained using 100-mL test tubes. The product (R_f = 0.57) (Note 6) was found in fractions 6–15.

10. Characterization of **2**: mp (mixture of E- and Z-isomers, E/Z = 91/9) 83–84 °C; ^1H NMR (600 MHz, $CDCl_3$) δ: E-isomer: 2.75–2.87 (m, 2H), 4.30

(d, *J* = 5.9 Hz, 2H), 5.65–5.69 (m, 1H), 5.81 (dt, *J* = 15.3, 5.8 Hz, 1H), 7.69–7.76 (m, 2H), 7.82–7.89 (m, 2H); Z-isomer: 3.10–3.22 (m, 2H), 4.30 (overlap, 2H), 5.60–5.74 (overlap, 1H), 5.81 (overlap, 1H), 7.69–7.76 (m, 2H), 7.82–7.89 (m, 2H); ^{13}C NMR (151 MHz, CDCl$_3$) δ: E-isomer: 37.1 (q, *J* = 30 Hz), 39.2, 122.3 (q, *J* = 3.9 Hz), 123.5 (2C), 125.7 (q, *J* = 276 Hz), 131.0, 132.2 (2C), 134.2 (2C), 167.9 (2C); Z-isomer: 32.5 (q, *J* = 30 Hz), 34.5, 122.2 (q, *J* = 3.9 Hz), 123.5 (2C), 126.1 (q, *J* = 277 Hz), 129.4, 132.2 (2C), 134.2 (2C), 167.9 (2C); ^{19}F NMR (565 MHz, CDCl$_3$): δ: E-isomer: −66.29 (t, *J* = 10.6 Hz), Z-isomer: −65.97 (t, *J* = 10.4 Hz). IR (neat, cm^{-1}) 2934, 1771, 1711, 1489, 1430, 1397, 1350, 1337, 1293, 1264, 1245, 1172, 1147, 1128, 1098, 1047, 942, 721, 712; HRMS (ESI+) calc. for C$_{13}$H$_{10}$F$_3$NO$_2$ [M+Na]$^+$ 292.0556, found 292.0550.

11. The purity was determined to be 97% by quantitative ^{1}H NMR spectroscopy in CDCl$_3$ using 53.8 mg of the compound **2** together with 13.9 μL of CH$_2$Br$_2$ as an internal standard.
12. A second run on half scale gave 2.79 g of an identical product (48%, 98% purity, E/Z = 89/11).
13. The submitters reported a yield of 41% (4.80 g, 99% purity, E/Z = 91/9) when using hexane in the crush-decantation process. Additional pure product **2** (2%, 0.21 g, 99% purity, E/Z = 85/15) was obtained by evaporation of the hexane washings followed by drying in vacuo (<6 mmHg) for 3 h. The total yield of **2** was 5.01 g (43%). The checkers obtained no additional product from the residual heptane solutions.

Working with Hazardous Chemicals

The procedures in *Organic Syntheses* are intended for use only by persons with proper training in experimental organic chemistry. All hazardous materials should be handled using the standard procedures for work with chemicals described in references such as "Prudent Practices in the Laboratory" (The National Academies Press, Washington, D.C., 2011; the full text can be accessed free of charge at http://www.nap.edu/catalog.php?record_id=12654). All chemical waste should be disposed of in accordance with local regulations. For general guidelines for the management of chemical waste, see Chapter 8 of Prudent Practices.

In some articles in *Organic Syntheses*, chemical-specific hazards are highlighted in red "Caution Notes" within a procedure. It is important to recognize that the absence of a caution note does not imply that no significant hazards are associated with the chemicals involved in that procedure. Prior to performing a reaction, a thorough risk assessment should be carried out that includes a review of the potential hazards associated with each chemical and experimental operation on the scale that is planned for the procedure. Guidelines for carrying out a risk assessment and for analyzing the hazards associated with chemicals can be found in Chapter 4 of Prudent Practices.

The procedures described in *Organic Syntheses* are provided as published and are conducted at one's own risk. *Organic Syntheses, Inc.*, its Editors, and its Board of Directors do not warrant or guarantee the safety of individuals using these procedures and hereby disclaim any liability for any injuries or damages claimed to have resulted from or related in any way to the procedures herein.

Discussion

Fluoroalkyl group-containing molecules are of interest as drug candidates, agrochemicals, and chemical probe molecules in chemical biology research, because the fluorine-containing group often improves pharmaceutical efficacy and can be used for bioisosteric replacement.[3] Therefore, a practical fluoroalkylation method employing user-friendly fluoroalkyl sources is highly desirable. Among various fluoroalkyl sources, fluorine containing acid anhydride should be suitable because it is low-cost, readily available, and applicable for a range of organic syntheses.[4-9] In this context, we developed allylic fluoroalkylations of simple alkenes by using the acid anhydrides as fluoroalkyl sources (Scheme 1).[7,8] Specifically, diacyl peroxide is generated *in situ* from acid anhydride and urea-hydrogen peroxide and used as a reagent for allylic fluoroalkylation reaction of alkenes. The key to success is to control the selectivity with the aid of a copper catalyst, because a complex mixture is obtained in its absence. For perfluoroalkylation (R_f = CF_3, C_2F_5, and C_3F_7), [Cu(CH$_3$CN)]$_4$PF$_6$ is the optimal catalyst, affording the desired product in up to 95% yield (Scheme 1a).[7] The combination of Cu(O$_2$CCF$_3$)$_2$ as the catalyst and pyridine as an additive enables efficient allylic chlorodifluromethylation using chlorodifluoroacetic anhydride (Scheme 1b).[8]

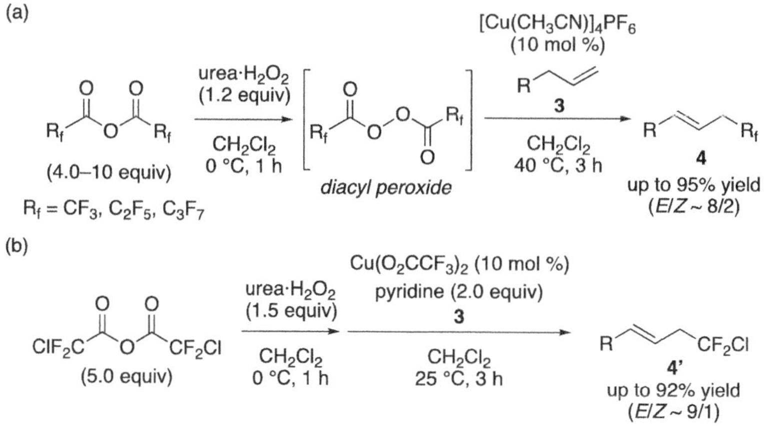

(a)

Scheme 1. Cu-catalyzed allylic fluoroalkylations using acid anhydrides

Notably, as an exceptional case of successful fluoroalkylation of simple alkenes using acid anhydrides in the absence of a Cu catalyst, an alkene bearing a pendant aromatic ring **5** provides carbo-fluoroalkylation product **6** *via* intramolecular carbocycle formation (Scheme 2).[7–9] Catalyst-controlled switching between allylic and carbo-fluoroalkylations is possible and greatly expands the scope of this method, as well as increasing the availability of fluoroalkyl group-containing molecules. The present procedure is a powerful synthetic tool to prepare fluoroalkyl group-containing molecules as a candidate for pharmaceuticals and functional materials from commonly used starting materials.

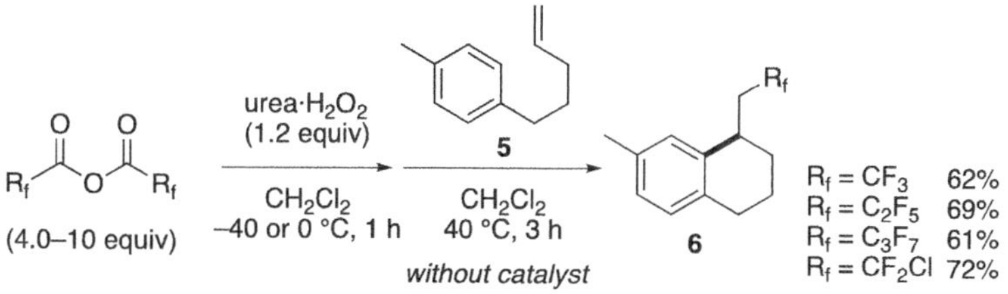

Scheme 2. Carbo-fluoroalkylation of alkene without any catalyst

References

1. Center for Sustainable Resource Science, RIKEN, 2-1 Hirosawa, Wako, Saitama 351-0198, Japan, and Synthetic Organic Chemistry Laboratory,

RIKEN, 2-1 Hirosawa, Wako, Saitama 351-0198, Japan. E-mail: sodeoka@riken.jp. ORCID: https://orcid.org/0000-0002-1344-364X. Acknowledgements: This work was supported by JSPS KAKENHI (No. 15K17860), and Project Funding from RIKEN.

2. Kopitzky, R.; Willner, H.; Hermann, A.; Oberhammer, H. *Inorg. Chem.* **2001**, *40*, 2693–2698. See also, Miller, W. T.; Dittman, A. L.; Reed, S. K.; US Patent **1951**, 2,580,0358.
3. Selected books: (a) Gouverneur, V.; Müller, K. *Fluorine in Pharmaceutical and Medicinal Chemistry: From Biophysical Aspects to Clinical Applications;* Imperial College Press: London, 2012. (b) Ojima, I. *Fluorine in Medicinal Chemistry and Chemical Biology;* Wiley-Blackwell: Chichester, 2009.
4. Selected reports on perfluoroalkylation reactions using perfluoro acid anhydrides: (a) Yoshida, M.; Yoshida, T.; Kobayashi, M.; Kamigata, N. *J. Chem. Soc., Perkin Trans. 1.* **1989**, 909–914. (b) Zhong, S.; Hafner, A.; Hussal, C.; Nieger, M.; Bräse, S. *RSC Adv.* **2015**, *5*, 6255–6258. (c) Beatty, J. W.; Douglas, J. J.; Cole, K. P.; Stephenson, C. R. J. *Nat. Commun.* **2015**, *6*, 7919. (d) Valverde, E.; Kawamura, S.; Sekine, D.; Sodeoka, M. *Chem. Sci.* **2018**, *9*, 7115–7121.
5. Selected reports on chlorodifluoromethylation reactions using chlorodifluorodifluoroacetic anhydrides: (a) Yoshida, M.; Morinaga, Y.; Iyoda, M. *J. Fluorine Chem.* **1994**, *68*, 33–38. (b) McAtree, R. C.; Betty, J. W.; McAtee, C. C.; Stephenson, C. R. J. *Org. Lett.* **2018**, *20*, 3491–3495.
6. Kawamura, S.; Sodeoka, M. *Bull. Chem. Soc. Jpn.* **2019**, *92*, 1245–1262.
7. Kawamura, S.; Sodeoka, M. *Angew. Chem., Int. Ed.* **2016**, *55*, 8740–8743.
8. Kawamura, S.; Henderson, C. J.; Aoki, Y.; Sekine, D.; Kobayashi, S.; Sodeoka, M. *Chem. Commun.* **2018**, *54*, 11276–11279.
9. Kawamura, S.; Dosei, K.; Valverde, E. Ushida, K.; Sodeoka, M. *J. Org. Chem.* **2017**, *82*, 12539–12553.

Appendix
Chemical Abstracts Nomenclature (Registry Number)

Urea hydrogen peroxide (urea·H_2O_2); (124-43-6)
Trifluoroacetic anhydride; (407-25-0)
N-Allyl-*p*-toluenesulfonamide; (50487-71-3)
Tetrakis(acetonitrile)copper(I) hexafluorophosphate; ([Cu(CH_3CN)$_4$]PF_6); (64443-05-6)

Yuma Aoki is received his B.E. M.E., and Ph. D. degrees from Kyoto University under the supervision of Professor Masaharu Nakamura. Since 2018, he has worked in Professor Sodeoka's group at RIKEN.

Shintaro Kawamura received his B.E. degree from Doshisha University in 2007. He received his M.E. and Ph. D. degrees from Kyoto University under the supervision of Professor Masaharu Nakamura. Since 2012, he has worked in Professor Sodeoka's group at RIKEN. In 2017, he was promoted from a postdoctoral researcher to a Research Scientist at RIKEN.

Mikiko Sodeoka received her B.S., M.S., and Ph. D. degrees from Chiba University. After working at the Sagami Chemical Research Center, Hokkaido University, Harvard University, and the University of Tokyo, she became a Group Leader at the Sagami Chemical Research Center in 1996. She moved to the University of Tokyo as an Associate Professor and then to Tohoku University as a Full Professor in 2000. Since 2004, she has been a Chief Scientist at RIKEN.

Martina Drescher is the lead technician of the Maulide group at the University of Vienna, where she has worked with several group leaders over the course of 38 years.

Daniel Kaiser received his Ph.D. at the University of Vienna in 2018, completing his studies under the supervision of Prof. Nuno Maulide. After a postdoctoral stay with Prof. Varinder K. Aggarwal at the University of Bristol, he returned to Vienna in 2020 to assume a position as senior scientist in the Maulide group. His current research focusses on the chemistry of destabilized carbocations and related high-energy intermediates.

Synthesis of a Phosphorous Sulfur Incorporating Reagent for the Enantioselective Synthesis of Thiophosphates

Prantik Maity*, Amitha S. Anandamurthy, Vijaykumar Shekarappa, Rajappa Vaidyanathan, Bin Zheng, Jason Zhu, Michael A Schmidt*, Richard J. Fox, Kyle W. Knouse, Julien C. Vantourout, Phil S. Baran, and Martin D. Eastgate

Chemical Development and API Supply, Biocon Bristol Myers Squibb Research and Development Center. Chemical Process Development, Bristol Myers Squibb. Department of Chemistry, The Scripps Research Institute

Checked by Álvaro Gutiérrez-Bonet and Kevin Campos

Procedure (Note 1)

A. *Triethylammonium bis-(pentafluorothiophenol)phosphorousdithioate (3)*. A dry 1 L, four-necked [middle neck size = 29/32 for overhead stirrer; size of other three necks = 24/29], round-bottomed flask is equipped with a 4 cm Teflon-coated overhead mechanical stirrer, a temperature sensor (thermocouple), an addition funnel (size = 24/29) connected with a nitrogen inlet and bubbler, and an outlet connected to an empty gas tube which is vented into a 500 mL round-bottomed flask half-filled (250 mL) with 10 wt% aqueous bleach solution (Figure 1).

Figure 1. Reaction set up (photo provided by submitters)

Toluene (150 mL) (Note 2) is charged into the flask followed by phosphorus pentasulfide **2** (28 g, 0.13 mol, 1.0 equiv) (Note 3) under a nitrogen atmosphere. The resulting pale-yellow slurry is stirred, then 2,3,4,5,6-pentafluorothiophenol **1** (50 g, 0.25 mol, 2.0 equiv) (Note 4) is added through an addition funnel, followed by another charge of toluene (100 mL) to rinse down the vessel wall. The slurry is cooled to an internal temperature of 10–15 °C using an ice-water bath, and then triethylamine (35.5 mL, 0.26 mol, 2.05 equiv) (Note 5) is charged to the addition funnel and added dropwise to the reaction mixture over a period of 30 min. During this addition, the internal temperature of the reaction mixture increased to ~30 °C, and the solution becomes homogeneous, turning to a golden yellow color. After completion of the addition, the ice-water bath is removed, and the solution is allowed to reach ambient temperature (22–26 °C). After stirring for 4–5 h, the reaction mixture becomes a pale-yellow heterogeneous slurry (Figure 2).

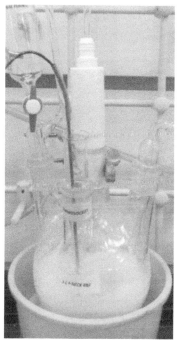

Figure 2. Reaction appearance after 4 h (photo provided by submitters)

The slurry is stirred at ambient temperature (22–26 °C) for a total of ~30 h under a nitrogen atmosphere and then sampled for reaction completion by UPLC (Note 6). The reaction mass is filtered over a Celite bed (20 g) prepared on top of a Büchner funnel and subsequently washed using toluene (100 mL) (Figure 3A). The combined filtrate is taken into a 1 L separatory funnel and washed with water (2 × 250 mL), followed by a 20 wt% aqueous brine solution (100 mL) (Figure 3B) (Note 7). The organic layer is transferred into a 1 L single-necked round-bottomed flask and concentrated to a final volume of ~100 to 150 mL on a rotary evaporator (bath temperature: 45–50 °C, 38 mmHg) (Figure 3C).

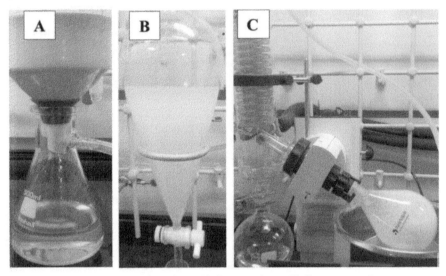

Figure 3. A) Filtration, B) Separation, C) Concentration
(photos provided by submitters)

The solution becomes a white suspension during concentration. The suspension is cooled to ambient temperature (25–35 °C), then methanol (100 mL) (Note 8) is added, forming a pale-yellow solution, which is transferred into a 1 L four-necked, round-bottomed flask equipped with a 4 cm Teflon-coated overhead mechanical stirrer, a temperature sensor (thermocouple), and a 250 mL addition funnel, and a glass stopper. The top of the addition funnel is fitted with a nitrogen inlet. An additional charge of methanol (75 mL) is used to completely transfer the material from the 1 L single-necked, round-bottomed flask. *n*-Heptane (175 mL) (Note 9) is added directly to the solution, resulting in a biphasic mixture. After stirring the biphasic mixture for ~5 min, a hazy suspension forms (Figure 4A). Water (150 mL) is charged to the addition funnel and is added over a period of 20 min to the white, hazy mixture, while maintaining an internal temperature of 25–35 °C. After the addition of water is complete, crystals form (Note 10) and the slurry thickens. Stirring is continued for 16 h at the same temperature. The crystals are filtered through a Büchner funnel fitted with a 30 μm filter cloth. The round-bottomed flask is washed with an aqueous methanol solution (3:2 v/v, water/methanol, 75 mL) and the rinse is passed through the filter cake (Figure 4B). The filter cake is washed with water (2 × 90 mL) then with *n*-heptane (2 × 45 mL). The wet cake is deliquored on the filter for 3 h. The white cake is placed into a vacuum oven (50 °C, 600 mmHg) and dried for 18 h. The oven-dried material is unloaded to give 66.07 g (89%) of compound **3** as a white crystalline solid with 99.7% purity (Notes 11, 12, and 13) (Figure 4C).

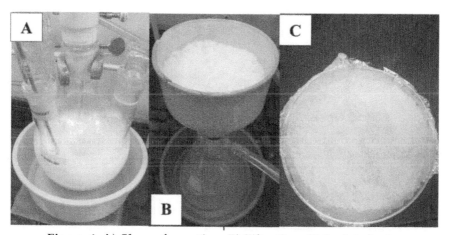

Figure 4. A) Slurry formation, B) Filtration, C) Isolated cake
(photos provided by submitters)

B.   (2R,3aR,6S,7aR)-3a-methyl-2-((perfluorophenyl)thio)-6-(prop-1-en-2-yl) *hexahydrobenzo[d][1,3,2]oxathiaphosphole 2-sulfide* (**5**). A 1 L, four-necked, round-bottomed flask is equipped with a 4 cm Teflon-coated overhead mechanical stirrer, a temperature sensor (thermocouple), rubber septum and a 50 mL addition funnel connected with a nitrogen inlet and bubbler. 1,2-Dichloroethane (300 mL) (Note 14) is charged into the flask followed by compound **3** (50 g, 83.80 mmol, 99.8 mass%) under a nitrogen atmosphere affording a yellow, homogeneous solution. Di-*n*-butyl phosphate (20.8 mL, 111 mmol, 1.32 equiv) (Note 15) is added followed by *cis*-(–)-limonene oxide (19.2 g, 126.2 mmol, 1.5 equiv) (Note 16). An additional charge of 1,2-dichloroethane (75 mL) is used to rinse the vessel wall. Dichloroacetic acid (17.5 mL, 212 mmol, 2.53 equiv) (Note 17) is charged into the 50 mL addition funnel and added dropwise over a period of 30 min. with a concomitant internal temperature rise to ~30 °C. The reaction mixture is stirred for 2 h at 50–55 °C under a nitrogen atmosphere (Figure 5A). After 2 h a homogenous, dark yellow solution is formed. The reaction is monitored by UPLC (Note 18). Upon completion of the reaction after 2 h, the mixture is cooled to 25–30 °C

(Figure 5B) and transferred to a 2 L single-necked, round-bottomed flask and concentrated on a rotary evaporator (bath temperature: 40–45 °C, 30 mmHg) until an approximate final volume of 250–300 mL is reached.

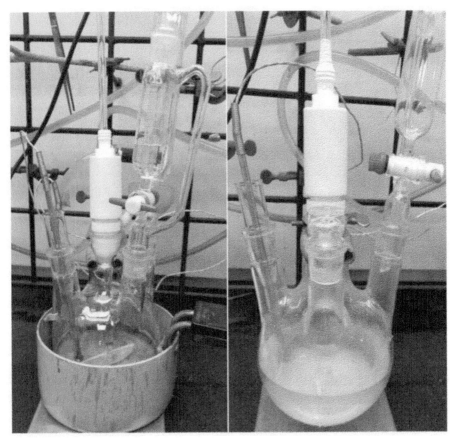

Figure 5. A) Initial reaction set up, B) Final reaction appearance (photos provided by submitters)

Methanol (500 mL) is charged into the flask and the solution is concentrated to an approximate final volume of 250–300 mL. This process is repeated a total of five times (Figure 6) to remove 1,2-dichloroethane (Note 19) and the thin slurry is transferred to a 1 L four-necked, round-bottomed flask equipped with a 4 cm Teflon-coated overhead mechanical stirrer, a temperature sensor (thermocouple), and a 100 mL addition funnel.

Additional methanol (50 mL) is used to rinse and assist the transfer to the 1 L four-necked, round-bottomed flask containing the slurry.

Figure 6. Appearance after final methanol addition (photos provided by submitters)

Water (25 mL) is added through the addition funnel over a period of 10 min into the reaction mixture while maintaining an internal temperature of 25–35 °C. A white precipitate forms, and the slurry is stirred for 2 h at the same temperature (Figure 7A). The solids are collected by filtration over a Büchner funnel fitted with a 30 μm filter cloth (Figure 7B). The filter cake is washed using 1:1 mixture of methanol/water (250 mL). The wet cake is deliquored on the filter for 1 h. The filtered solid is unloaded to afford 22.91 g of crude compound **5** as a white solid (Figure 8) (Notes 20 and 21). This solid is subjected to a further recrystallization to improve the purity.

Figure 7. A) Slurry formation, B) Filtration
(photos provided by submitters)

Figure 8. Isolated product 5
(photo provided by submitters)

Compound **5** (21 g) is added to a 1 L, single-necked, round-bottomed flask. Dichloromethane (21 mL) (Note 22) is charged into the flask and the solid dissolves immediately to give a clear solution. *n*-Heptane (42 mL) is added and the mixture is concentrated on a rotary evaporator under reduced pressure (bath temperature: 45–50 °C, 550 mmHg) to a final volume of ~ 40–45 mL. The addition of *n*-heptane and the concentration are repeated twice more, whereupon a slurry forms. The mass is cooled to room temperature and transferred to a 1 L, three-necked, round-bottomed flask, fitted with an overhead mechanical stirrer and temperature sensor (thermocouple), and a glass stopper. *n*-Heptane (42 mL) is charged slowly into the round-bottomed flask, and the mass is stirred for 2 h at room temperature. The crystals of **5** that formed are collected by filtration over a Büchner funnel fitted with a 30 µm filter cloth, followed by washing with *n*-heptane (21 mL), and the crystals are deliquored for 2 h. The wet crystals are dried in a vacuum oven (45 °C, 600 mmHg) for 6 h. The material is unloaded from the oven to give 19.8 g (51%) of compound **5** as a white crystalline solid (Notes 23, 24, and 25).

Notes

1. Prior to performing each reaction, a thorough hazard analysis and risk assessment should be carried out with regard to each chemical substance and experimental operation on the scale planned in the context of the laboratory where the procedures will be carried out. Guidelines for carrying out risk assessments and for analyzing the hazards associated with chemicals can be found in references such as Chapter 4 of "Prudent Practices in the Laboratory" (The National Academies Press, Washington, D.C., 2011; the full text can be accessed free of charge at https://www.nap.edu/catalog/12654/prudent-practices-in-the-laboratory-handling-and-management-of-chemical. See also "Identifying and Evaluating Hazards in Research Laboratories" (American Chemical Society, 2015) which is available via the associated website "Hazard Assessment in Research Laboratories" at https://www.acs.org/content/acs/en/about/governance/committees/chemicalsafety/hazard-assessment.html. In the case of this procedure, the risk assessment should include (but not necessarily be limited to) an evaluation of the potential hazards associated with phosphorous pentasulfide, 2,3,4,5,6-pentafluorophenol, toluene, *n*-heptane, methanol, dichloroethane, dichloromethane, triethylamine, (+) or (-)-limonene oxide, di-*n*-butyl phosphate, dichloroacetic acid and Celite, as well as the proper procedures for handling malodorous chemistry.

2. Anhydrous toluene (99.8%) was purchased by the checkers from Sigma-Aldrich and used as received. Dry toluene (100 mass%) was purchased by the submitters from Sonia Industries, India and used as received.
3. Phosphorous pentasulfide (98+%) was purchased by the checkers from ACROS Organics and used as received. Phosphorous pentasulfide (100 mass%) was purchased by the submitters from Leonid Chemicals Private Ltd., India and used as received. Fresh lots of high quality phosphorous pentasulfide should be used. Lower quality and/or older lots of this material can lead to variable conversion and isolated yields.
4. 2,3,4,5,6-Pentafluorothiophenol (100 mass%) was purchased by the checkers from Apollo Scientific and used as received. 2,3,4,5,6-Pentafluorothiophenol (100 mass%) was purchased from Spectrochem, China and used as received.
5. Triethylamine (99%) was purchased by the checkers from Fisher Scientific and used as received. Triethylamine (99.8 mass%) was purchased by the submitters from Sonia Industries, India and used as received.
6. A supernatant aliquot (0.5 mL) was withdrawn from the reaction mixture. Approximately 100 mg was diluted to 10 mL with acetonitrile and submitted as such for UPLC analysis. The UPLC analysis indicated that relative area percent of pentafluorothiophenol versus product <5%. UPLC Conditions are as follows. Column: Agilent Poroshell 120, EC-C18(50 x 2.1 mm, 1.9 µm particle size). Flow rate: 0.8 mL/min. Injection Volume: 1.0 µL. Detector Wavelength: 210 nm. Column temperature: 40 °C. Sample Temperature: 25 °C. Run Length: 5 min. Mobile Phase "A": 0.1% H_3PO_4 in water. Mobile Phase "B": Acetonitrile. Gradient (Time, %B): (0.00 min, 5%), (4.00 min, 95%), (5.00 min, 95%). Pentafluorothiophenol retention time is 2.441 min, product **3** retention time is 2.814 min. UPLC Conditions are as follows. Column: Waters Acquity CSH C18 (100 x 2.1 mm, 1.7 µm particle size).
7. The filtrate was washed with water to remove excess Et_3N. If this step is omitted and the filtrate is directly concentrated, a sharp exotherm is observed in the crystallization. This is due to residual Et_3N.
8. Methanol (Optima, 99.9%) was purchased by the checkers from Fisher Chemical and used as received. Methanol (Laboratory Grade) was purchased by the submitters from Sonia Industries, India and used as received.
9. *n*-Heptane (ReagentPlus® 99%) was purchased by the checkers from Sigma Aldrich and used as received. *n*-Heptane (Laboratory Grade) was

purchased by the submitters from Sonia Industries, India and used as received.
10. Occasionally, unseeded crystallization, as written, does not happen instantaneously. Upon prolonged, vigorous stirring (~30 min) it will begin to self-nucleate and crystallize.
11. Triethylammonium bis(pentafluorothiophenol)phosphorousdithioate (3) has the following characterization properties: mp 95.0 °C. ^{1}H NMR (500 MHz, CDCl$_3$) δ: 1.42 (t, J = 7.3 Hz, 9H), 3.26 (qd, J = 7.3, 5.3 Hz, 6H), 8.89 (s, 1H); ^{13}C NMR (126 MHz, CDCl$_3$) δ: 8.7, 46.7, 109.4 (t, J = 18.0 Hz), 128.7 (d, J = 101.8 Hz), 136.7 (t, J = 15.6 Hz), 138.7 (t, J = 15.2 Hz), 141.2 (t, J = 15.5 Hz), 143.2 (t, J = 15.5 Hz), 147.1 (d, J = 10.1 Hz), 149.1 (d, J = 9.8 Hz). ^{31}P NMR (202 MHz, CDCl$_3$) δ: 99.42 (s, br). ^{19}F NMR (470 MHz, CDCl$_3$) δ –128.65 (d, J = 17.8 Hz), –150.96 (td, J = 21.1, 7.2 Hz), –161.68 (t, J = 20.4 Hz); FTIR (film): 2997, 1636, 1469, 1389, 1090, 979 cm^{-1}. HRMS: ESI [M + H + TFA + Na] calcd for $C_{14}H_2F_{13}NaO_2PS_4$: 630.8360. Found: 630.8403. qNMR: 24.9 mg of **3** and 28.9 mg of acenaphthene (99%, Sigma Aldrich) were dissolved in 3.0 mL of CDCl$_3$.
12. Triethylammonium was assayed by UPLC for purity (sample concentration: 0.3 mg/mL in acetonitrile) (see Note 6 for UPLC conditions), which was determined to be >99.9% by area. Quantitative ^{1}H NMR using 24.9 mg of **3** and 28.9 mg of acenaphthene (99%, Sigma Aldrich) as an internal standard provided a purity assessment of >99.9% by weight.
13. A second reaction performed on similar scale provided 66.09 g (88%) of compound **3** in 99.6% purity as determined by qNMR.
14. Anhydrous 1,2-dichloroethane (99.8%) was purchased by the checkers from Sigma-Aldrich and used as received. Dry 1,2-dichloroethane (100 mass%) was purchased by the submitters from Sonia Industries, India and used as received.
15. Dibutyl phosphate (97%) was purchased by the checkers from Oakwood Chemical and used as received. Dibutyl phosphate (100 mass%) was purchased by the submitters from Changzhou Sinowa Chemicals, China and used as received.
16. *Cis*-Limonene oxide (100 mass%) was purchased from Keminntek Laboratories, India and used as received. *Cis*-Limonene oxide can also be prepared as in Reference 6. The reaction profile and isolated yield is highly dependent on the quality of limonene oxide used.

17. Dichloroacetic acid (99%) was purchased by the checkers from Oakwood Chemical and used as received. Dichloroacetic acid (100 mass%) was purchased by the submitters from Sigma-Aldrich and used as received.
18. An aliquot (~0.5 mL) was withdrawn from reaction mass. Approximately 100 mg was diluted to 10 mL with acetonitrile and submitted as such for UPLC analysis. For UPLC conditions see Note 6.
19. The submitters reported that GC could be used to monitor 1,2-dichloroethane content, which should be no more than 0.5 ppm.
20. Compound **5** was assayed by UPLC for purity (sample concentration: 0.2 mg/mL in acetonitrile) (please see note 6 for UPLC conditions). Purity, as determined by area, was assessed at 98.3%, with 1.7% of the isomer.
21. A second run on similar scale provided 23.04 g of the same material before recrystallization. UPLC was used to assess purity of 97.5%, with 2.5% of the isomer.
22. Anhydrous dichloromethane (99.8%) was purchased by the checkers from Sigma-Aldrich and used as received. Dichloromethane (100 mass%) was purchased by the submitters from Sonia industries and used as received.
23. Characterization of **5**: mp 108.6 °C; ^{1}H NMR (500 MHz, CDCl$_3$) δ: 1.67 (s, 3H), 1.71–1.79 (m, 1H), 1.81 (s, 3H), 1.86–2.07 (m, 4H), 2.31–2.38 (m, 1H), 2.60 (s, 1H), 4.28 (ddd, J = 12.7, 4.8, 3.7 Hz, 1H), 4.86 (s, 1H), 5.02 (d, J = 0.5 Hz, 1H); ^{13}C NMR (126 MHz, CDCl$_3$) δ: 22.1, 22.6, 23.5, 27.8 (d, J = 14.7 Hz), 33.8 (d, J = 9.0 Hz), 39.0, 65.7, 86.7 (d, J = 3.3 Hz), 105.3–104.7 (m), 111.8, 137.2–136.7 (m), 139.2–138.7 (m), 142.3–141.8 (m), 144.4–143.9 (m), 145.1, 147.0 (dt, J = 11.0, 4.2 Hz), 149.0 (dt, J = 10.9, 4.2 Hz); ^{31}P NMR (202 MHz, CDCl$_3$) δ: 99.6–99.7 (m); ^{19}F NMR (470 MHz, CDCl$_3$) δ: –129.27 – –130.30 (m), –148.02 (tdt, J = 21.0, 7.5, 3.8 Hz), –158.32 – –160.28 (m). FTIR (film): 2967, 1636, 1511, 1482, 1090, 973 cm^{-1}. HRMS: ESI [M + H] calcd for C$_{16}$H$_{17}$F$_5$OPS$_3$: 447.0094. Found: 447.0074.
24. Compound **5** was assayed by UPLC for purity (sample concentration: 0.3 mg/mL in acetonitrile) (see Note 6 for UPLC conditions), which was determined to be 99.9% by area. Quantitative ^{1}H NMR using acenaphthene (99%, Sigma Aldrich) as an internal standard provided a purity assessment of 95.5% by weight.

25. Recrystallization of the reaction product from a second reaction (described in Notes 13 and 21) provided 20.49 g (54%) of compound **5** in 99.9% UPLC area percent purity and 98.8 wt% purity as assessed by qNMR.

Working with Hazardous Chemicals

The procedures in *Organic Syntheses* are intended for use only by persons with proper training in experimental organic chemistry. All hazardous materials should be handled using the standard procedures for work with chemicals described in references such as "Prudent Practices in the Laboratory" (The National Academies Press, Washington, D.C., 2011; the full text can be accessed free of charge at http://www.nap.edu/catalog.php?record_id=12654). All chemical waste should be disposed of in accordance with local regulations. For general guidelines for the management of chemical waste, see Chapter 8 of Prudent Practices.

In some articles in *Organic Syntheses*, chemical-specific hazards are highlighted in red "Caution Notes" within a procedure. It is important to recognize that the absence of a caution note does not imply that no significant hazards are associated with the chemicals involved in that procedure. Prior to performing a reaction, a thorough risk assessment should be carried out that includes a review of the potential hazards associated with each chemical and experimental operation on the scale that is planned for the procedure. Guidelines for carrying out a risk assessment and for analyzing the hazards associated with chemicals can be found in Chapter 4 of Prudent Practices.

The procedures described in *Organic Syntheses* are provided as published and are conducted at one's own risk. *Organic Syntheses, Inc.*, its Editors, and its Board of Directors do not warrant or guarantee the safety of individuals using these procedures and hereby disclaim any liability for any injuries or damages claimed to have resulted from or related in any way to the procedures herein.

Discussion

The replacement of a single oxygen atom with a sulfur atom is a common modification found in antisense oligonucleoside research. This modification can affect the properties of the molecule, for example by improving its stability to metabolic cleavage.[2] This single atom modification now creates a chiral center at phosphorous, the consequences of which are currently under investigation by practitioners in the field. Currently, the method of choice by which this functionality is constructed is through the use of P(III) phosphoramidites which are loaded with chiral auxiliaries based on, for example, proline. This phosphoramidite is coupled with another nucleoside with a specific activator, then the phosphorous is oxidized with a sulfur source.[3] While typically a high yielding methodology, there are some challenges that can be encountered. This chemistry is highly water sensitive owing to the reactivity of the phosphoramidite and strict care must be taken to limit moisture. Additionally, the chiral auxiliary must be prepared often times in multiple steps. To circumvent these issues, we sought to devise a system that would be more tolerant to water, be more readily accessible and be isohypsic[4] with respect to the final phosphorous oxidation state. Drawing inspiration from the oxathiaphospholane work originating with Stec,[5] we developed the phosphorous-sulfur incorporation reagents (stylized as psi or Ψ) as a stable, and easily prepared reagent that links two entities (e.g. nucleosides) affording a chiral thiophosphate with excellent levels of stereocontrol.[6] This is operationally carried out by loading the reagent on the first nucleoside with DBU (Scheme 1). This loaded nucleoside can be purified by chromatography, crystallization, or precipitation.

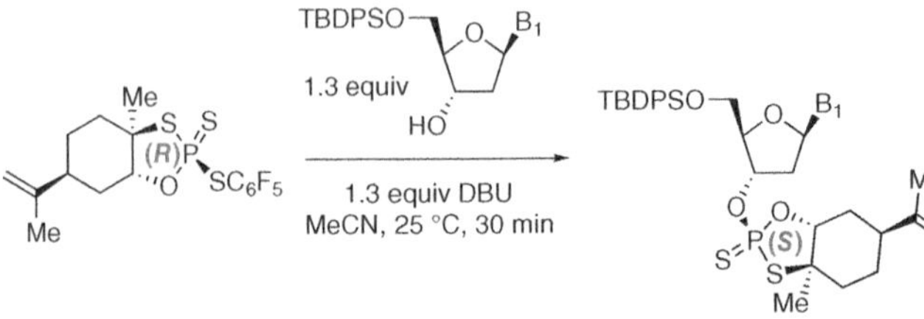

B_1	Yield[a]	Yield[b]
A	96%	97%
T	90%	90%
G^{dmf}	91%	76%
C	87%	81%

[a]Isolated yields with the P(*R*) reagent (shown). [b]Isolated yields with the enatiomeric (*S*) reagent (not shown).

Scheme 1. Loading a nucleoside

The loaded nucleoside is then coupled with the next nucleoside, again with DBU (Scheme 2).

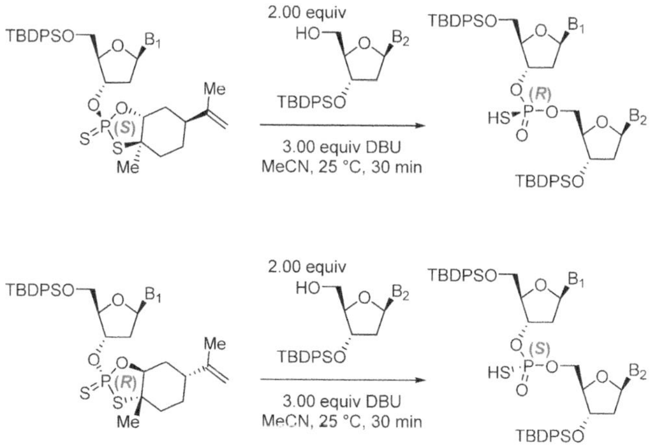

B1	B2	Yield (d.r.)
A	A^{Bz}	65% (>99:1)
C	C^{Bz}	73% (>99:1)
C	$G^{i\text{-}Bu}$	82% (>99:1)
G^{dmf}	$G^{i\text{-}Bu}$	86% (>99:1)

B1	B2	Yield (d.r.)
A	C^{Bz}	91% (>99:1)
A	T	61% (>99:1)
A	$G^{i\text{-}Bu}$	79% (>99:1)
C	T	76% (>99:1)
T	T	72% (>99:1)
T	$G^{i\text{-}Bu}$	67% (>99:1)

Scheme 2. Stereoselective coupling of nucleosides

Both enantiomers of the reagents are commercially available through Sigma-Aldrich (the P(*R*) PSI reagent, i.e., **5** catalog number ALD00604 and the enatiomeric P(*S*) PSI reagent, i.e. **ent-5** catalog number ALD00602). The availability of the reagents and ease of use compare favorably to the current art.

References

1. Chemical Development and API Supply, Biocon Bristol-Myers Squibb Research and Development Center, Biocon Park, Jigani Link Road, Bommasandra IV, Bangalore-560099, India. Chemical and Synthetic Development, Bristol-Myers Squibb, 1 Squibb Dr. New Brunswick, NJ 08903. Department of Chemistry, The Scripps Research Institute, 10550 North Torrey Pines Road, La Jolla, CA 92037, USA.
2. Chlebowski, J. F.; Coleman, J. E. *J. Biol. Chem.* **1974**, *249*, 7192–7202.
3. (a) Oka, N.; Yamamoto, M.; Sato, T.; Wada, T. *J. Am. Chem. Soc.* **2008**, *130*, 16031–16037. (b) Oka, N.; Kondo, T.; Fujiwara, S.; Maizuru, Y.; Wada, T. *Org. Lett.* **2009**, *11*, 967–970. (c) Nukaga, Y.; Yamada, K.; Ogata, T.; Oka, N.; Wada, T. *J. Org. Chem.* **2012**, *77*, 7913–7922.
4. Newhouse, T.; Baran, P. S.; Hoffmann, R. W. *Chem. Soc. Rev.* **2009**, *38*, 3010–3021.
5. Guga, P.; Stec, W. J. *Curr. Protoc. Nucleic Acid Chem.* **2003**, *14*, 4.17.1–4.17.28.

6. Knouse, K.; deGruyter, J. N.; Schmidt, M. A.; Zheng, B.; Vantourout, J. C.; Kingston, C.; Mercer, S. E.; McDonald, I. M.; Olson, R. E.; Zhu, Y.; Hang, C.; Zhu, J.; Yuan, C.; Wang, Q.; Park, P.; Eastgate, M. D.; Baran, P. S. *Science*, **2018**, *361*, 1234–1238.

Appendix
Chemical Abstracts Nomenclature (Registry Number)

Phosphorous Pentasulfide: Phosphorous Sulfide; (1314-80-3)
2,3,4,5,6-Pentafluorothiophenol 2,3,4,5,6-Pentafluorobenzenethiol; (771-62-0)
Triethylamine: *N*,*N*-Diethylethanamine; (121-44-8)
Cis-(-)-Limonene oxide: (1*S*,4*S*,6*R*)-1-Methyl-4-(1-methylethenyl)-7-oxabicyclo[4.1.0]heptane; (32543-51-4)
Dibutyl Phosphate: Phosphoric Acid, Dibutyl Ester; (107-66-4)
Dichloroacetic acid: 2,2-Dichloroacetic Acid; (79-43-6)
Celite: Diatomaceous earth; (68855-54-9)
(2*R*,3a*R*,6*S*,7a*R*)-3a-Methyl-2-((perfluorophenyl)thio)-6-(prop-1-en-2-yl)hexahydrobenzo[d][1,3,2]oxathiaphosphole 2-sulfide (**5**): (2*R*,3a*R*,6*S*,7a*R*)-Hexahydro-3a-methyl-6-(1-methylethenyl)-2-[(2,3,4,5,6-pentafluorophenyl)thio]-1,3,2-Benzoxathiaphosphole; (2245335-71-9)

Prantik Maity obtained his M.Sc. degree from IIT Madras, Chennai. After acquiring his Ph.D. at University of Regensburg, Germany in the laboratory of Prof. Burkhard König, 2008, he commenced two postdoctoral research at University of Freiburg and University of Delaware under Prof. Bernhard Breit and Prof. Mary P Watson respectively. In 2014, he joined at Biocon Bristol Myers Squibb Research and Development Center (BBRC), where currently he holds the position of Principal Investigator in the department of Chemical Development and API Supply.

Amitha Anandamurthy was born in Shivamogga Karnataka, India in 1992 and acquired her M.Sc. in Chemistry in 2014. She had previous work experience with Rentokil PCI as Research Associate. She is currently working with Biocon Bristol Myers Squibb Research and Development Center (BBRC) as a senior Research Associate.

Vijaykumar Shekarappa was born in Karnataka, India in 1985 and acquired his M.Sc. in Organic Chemistry in 2008 from Kuvempu University, India. He worked as a Research Associate in R&D centre, The Himalaya Drug Company, Bangalore. Currently he is working with Biocon Bristol Myers Squibb Research and Development Center (BBRC) as a Senior Associate Scientist.

Rajappa Vaidyanathan was born in Madras, India. He completed his Ph.D. in 1998 from the University of California, Irvine working in the laboratories of Prof. Scott Rychnovsky. After a post-doctoral appointment at Eli Lilly and Company, he joined the Chemical Process R&D group of Pharmacia Corporation, Kalamazoo, MI, and subsequently Pfizer in Groton, CT. During this period, he led several inter-disciplinary teams in the discovery and development of practical, environmentally responsible processes for New Chemical Entities, three of which were commercialized as approved drugs. He is currently Group Director and Head of Chemical Development and API Supply at Bristol Myers Squibb in Bangalore, India.

Bin Zheng received his Ph.D. from the University of Toledo. After postdoctoral studies at Caltech under the supervision of Professor Andrew Myers, he joined the Process R&D group of Bristol Myers Squibb in New Brunswick, NJ, where he is currently a Principal Scientist in Chemical Process Development.

Jason Zhu obtained his Master's Degree in 1993 from Boston University. After working at a CRO for 3 years, he then joined the Process R&D group of Bristol Myers Squibb in New Brunswick, NJ, where he is currently a Research Scientist in Chemical Process Development.

Michael Schmidt obtained his Ph.D. in 2008 from the Massachusetts Institute of Technology working in the group of Professor Mohammad Movassaghi in the area of alkaloid total synthesis. Thereafter, he joined Bristol Myers Squibb where he is currently a Principal Scientist in Chemical Process Development.

Richard (Rich) Fox completed his Ph.D. in 2006 from the University of Pennsylvania working in the group of Professor Amos B Smith, III. After a post-doctoral appointment at the University of California, Berkeley with Professor Robert Bergman, he joined the Process R&D group of Bristol Myers Squibb in New Brunswick, NJ, where he is currently a Principal Scientist in Chemical Process Development.

Kyle Knouse was born in Pennsylvania and received his B.S in Chemistry from Temple University in Philadelphia (2011). He is currently undergoing his graduate studies at the Scripps Research Institute under the guidance of Professor Phil S. Baran.

Julien Vantourout completed his Ph.D. in 2018 from the collaborative program between the University of Strathclyde and GlaxoSmithKline, UK, where he studied in the laboratories of Dr. Allan Watson. Thereafter, he started a post-doctoral appointment at Scripps Research with Prof. Phil Baran. He is currently a staff scientist in the Baran lab.

Phil S. Baran was born in New Jersey in 1977 and received his undergraduate education from New York University in 1997. After earning his Ph.D. at TSRI in 2001, he pursued postdoctoral studies at Harvard University until 2003, at which point he returned to TSRI to begin his independent career. He was promoted to the rank of Professor in 2008 and is currently the Darlene Shiley Professor of chemistry. The mission of his laboratory is to educate students at the intersection of fundamental organic chemistry and translational science.

Martin Eastgate completed his Ph.D. in 2002 from the University of Cambridge, UK, where he studied in the laboratories of Dr. Stuart Warren. After a post-doctoral appointment at the University of Illinois Urbana-Champaign, with Prof. Scott Denmark, he joined the process chemistry team at Bristol Myers Squibb, where he is currently Head of Chemical Research in Chemical Process Development

Álvaro Gutiérrez-Bonet was born in Madrid (Spain) in 1989. He received his Ph.D. from the ICIQ (Institute of Chemical Research of Catalonia) working in the laboratories of Professor Ruben Martin. After postdoctoral studies at the University of Pennsylvania under the supervision of Professor Gary Molander, he joined the Process R&D group of Merck, in West Point, PA, where he is currently a Senior Scientist in Discovery Process Chemistry.

Discussion Addendum for:

Preparation of (S)-tert-ButylPyOx and Palladium-Catalyzed Asymmetric Conjugate Addition of Arylboronic Acids

Stephen R. Sardini and Brian M. Stoltz*[1]

Warren and Katharine Schlinger Laboratory for Chemistry and Chemical Engineering, Division of Chemistry and Chemical Engineering, California Institute of Technology, Pasadena, California 91125, United States

Original Article: Holder, J.; Shockley, S.; Wiesenfeldt, M.; Shimizu, H.; Stoltz, B. M. Org. Synth. **2015**, 92, 247–266.

The use of asymmetric catalysis in the enantioselective construction of all-carbon quaternary stereocenters is a contemporary challenge in organic synthesis.[2-7] Conjugate addition reactions of carbon nucleophiles to β,β-disubstituted α,β-unsaturated carbonyl compounds has emerged as an effective strategy to address this challenge.[8] Copper-catalyzed asymmetric conjugate addition reactions generally employ highly reactive nucleophiles such as dialkylzinc,[9-17] triorganoaluminum,[18-27] organozirconium,[28-30] and organomagnesium reagents.[31-34] Organoaluminum nucleophiles have also been reported to participate in asymmetric conjugate addition reactions to β,β-disubstituted α,β-unsaturated carbonyl compounds under rhodium catalysis.[35] Unfortunately, use of such highly reactive organometallic species require rigorous exclusion of moisture and oxygen, and can limit the functional group tolerance of these transformations.

Efforts by Hayashi and coworkers have led to the development of rhodium complexes to catalyze the addition of organoboron nucleophiles to α,β-unsaturated carbonyl compounds with excellent levels of enantioselectivity.[36] In contrast to the previously mentioned organometallic nucleophiles, organoboron nucleophiles are easily handled and stored on the benchtop, and the mild nature of these reagents allow for rhodium-catalyzed conjugate addition reactions to tolerate a wide variety of functional groups. While reports of rhodium-catalyzed conjugate addition with β,β-disubstituted α,β-unsaturated carbonyl compounds are rare, chiral diene ligated rhodium complexes have proven to be effective catalysts for the formation of stereogenic all-carbon quaternary centers via this reaction manifold. Unfortunately, commercially available boronic acids are not suitable for this process, which often requires the use of tetraaryl borates[37] or boroxines.[38] Despite the associated advantages of rhodium-catalyzed conjugate addition, the high cost of rhodium, and oxygen sensitivity of these processes are undesirable.

Palladium catalysis has recently emerged as a more robust and cost-effective alternative to rhodium-catalyzed conjugate addition processes. The asymmetric construction of tertiary β-substituted ketones utilizing a cationic palladium(II)-DuPHOS complex was described by Minnaard and coworkers in 2011, and displays a remarkable tolerance to air and moisture.[39] Unfortunately, the conditions outlined in this report are not suitable for the asymmetric construction of quaternary stereocenters. A subsequent report from Lu and colleagues disclosed that a cationic palladium(II)-bipyridine complex is a competent catalyst for the reaction of β,β-disubstituted α,β-unsaturated ketones, although it is worth noting that these conditions do not allow for the construction of enantioenriched all-carbon quaternary centers.[40] Given our interests in the asymmetric construction of all-carbon quaternary stereocenters, we focused our attention toward an enantioselective variant of this transformation. Described herein are recent developments in the field of palladium-catalyzed conjugate addition toward the asymmetric construction of all-carbon quaternary centers.

Reaction Development and Substrate Scope

In 2011 our laboratory disclosed the first report of an asymmetric palladium-catalyzed conjugate addition to generate all-carbon quaternary stereocenters.[41] Our studies revealed that chiral pyridineoxazoline (PyOx) ligands (**L4-L9**) complexed to Pd(OCOCF$_3$)$_2$ were particularly adept in providing the conjugate addition products in high yield. Other chiral ligands such as bisoxazoline (**L1**), pyridinebisoxazoline (**L2**), and (–)-spartine (**L3**) did not give rise to catalytically active palladium complexes. The palladium catalyst derived from *t*-BuPyOx (**L7**) provided the highest levels of enantioselectivity, whereas ligands derived from phenylalanine (**L4**), leucine (**L5**), or valine (**L6**) provided significantly lower levels of enantioselectivity. Finally, electronic rich (**L8**), and electron poor (**L9**) *t*-BuPyOx derivatives were both inferior to the unsubstituted variant (Table 1). While *t*-BuPyOx can be synthesized in one step from methyl picolinimidate and *tert*-leucinol, the two-step procedure outlined in the original *Organic Syntheses* article was more efficient to produce larger quantities of *t*-BuPyOx in our hands.

Table 1. Investigation of Chiral Ligands

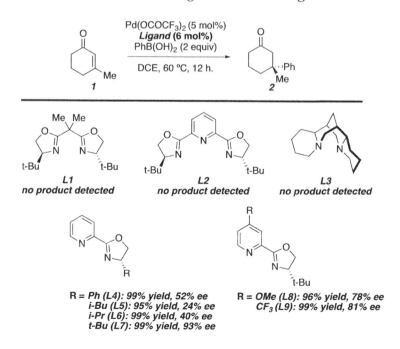

Our initial report details the addition of a wide variety of boronic acids including electron rich (product **3**), electron poor (products **4** and **5**), and boronic acids containing potentially reactive functional groups such as ketones (product **5**). Notably, electron rich boronic acids tend to exhibit lower levels of enantioselectivity. With respect to the enone component, the reaction of five-, six-, and seven-membered rings were well tolerated (products **6**, **7**, and **8** respectively). Moreover, substrates with increased steric hindrance about the enone react smoothly under the standard reaction conditions to generate highly congested all-carbon quaternary centers (products **9** and **10**). After our initial report, Stanley and coworkers disclosed a similar study that details the conjugate addition of arylboronic acids to β-aryl α,β-unsaturated ketones to yield enantioenriched bis-benzylic quaternary stereocenters.[42]

Table 2. Selected substrate scope

3
96% yield
74% ee

4
99% yield
96% ee

5
99% yield
96% ee

6
84% yield
91% ee

7
97% yield
91% ee

8
85% yield
93% ee

9
86% yield
85% ee

10
68% yield
88% ee

Despite the broad substrate scope and efficiency of the process, attempts to scale up the reaction with gram quantities of **1** were initially unsuccessful (Scheme 1A, Eq. 1). To this end, we set out to study the mechanism computationally in collaboration with Ken Houk's group at UCLA and devise an improved set of reaction conditions to address the problem of scalability.[43] An important feature of our proposed mechanism is the enantiodetermining carbopalladation to provide the palladium(II) enolate (**IV**), which must then be protonated to close the catalytic cycle (Scheme 1B). We initially hypothesized that the boronic acid could act as a proton source to turn over the catalytic cycle; however, the lack of scalability of the reaction caused us to reconsider this hypothesis. We posited that addition of an external proton source would be beneficial for the reaction. To our delight, addition of 5 equivalents of water to the previously employed reaction conditions allowed for facile scale-up of this process with no change in yield or enantioselectivity (Scheme 1A, Eq. 2).

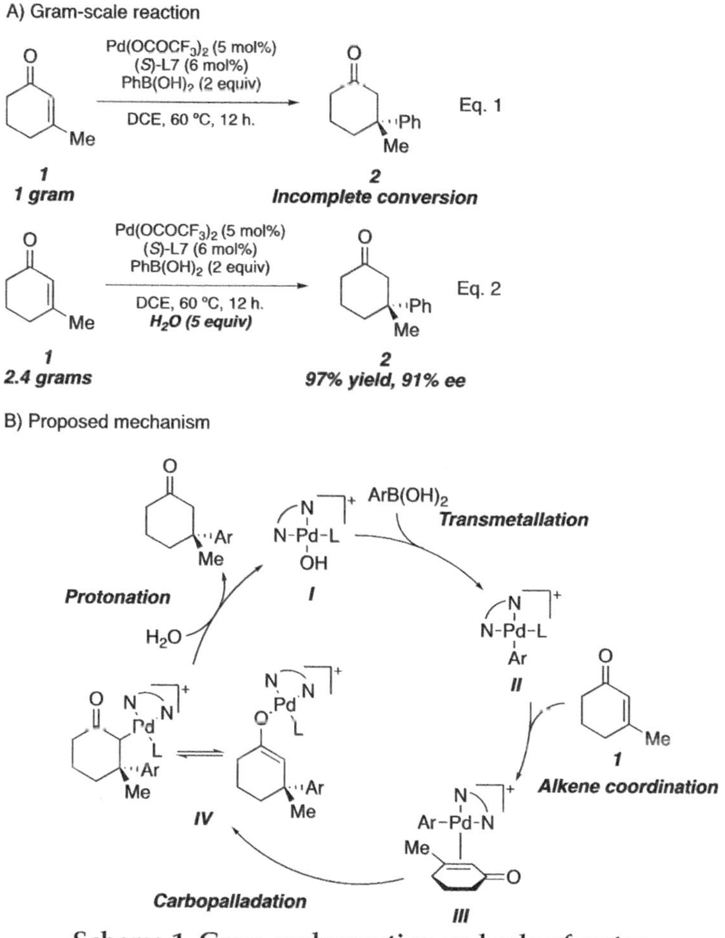

Scheme 1. Gram-scale reaction and role of water

We then turned our attention to the addition of metal salts containing weakly coordinating counter ions (PF_6^-, SbF_6^-, BF_4^-, etc) in an effort to improve the activity of the palladium catalyst. This would allow for lower catalyst loadings and lower reaction temperatures. We hypothesize that weakly coordinating anions could undergo a salt metathesis with the palladium(II) trifluoroacetate, resulting in a more reactive, cationic palladium(II) species. In line with our hypothesis, addition of NaCl (Table 3, entry 1) inhibited reactivity, presumably due to the strongly coordinating nature of chloride anion. Conversely, when sodium salts bearing weakly coordinating counterions were employed, enhanced reactivity was observed (Table 3, entries 2–4); however, the enantioselectivity of the process was diminished. Enantioselectivity could be restored at the expense of the reaction rate when tetrabutylammonium salts were employed (Table 3, entries 5–6). Sodium tetraphenylborate (Table 3, entry 7) failed to promote the desired reactivity due to concomitant formation of biphenyl, presumably via an oxidative coupling of the tetraphenylborate nucleophile. Finally, ammonium salts were effective in providing an optimal balance of reactivity and enantioselectivity (Table 3, entries 8–9), with the hexafluorophosphate anion yielding the best result (Table 3, entry 9).

Table 3. Effect of Salt Additives

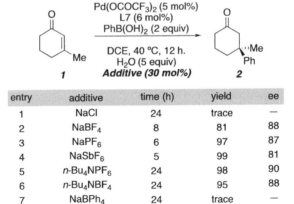

entry	additive	time (h)	yield	ee
1	NaCl	24	trace	—
2	$NaBF_4$	8	81	88
3	$NaPF_6$	6	97	87
4	$NaSbF_6$	5	99	81
5	$n\text{-}Bu_4NPF_6$	24	98	90
6	$n\text{-}Bu_4NBF_4$	24	95	88
7	$NaBPh_4$	24	trace	—
8	NH_4BF_4	15	93	89
9	NH_4PF_6	12	96	91

The combination of 5 equivalents of water, and NH_4PF_6 allowed for the reaction to be conducted at lower temperatures, and with lower catalyst loadings to provide product **2** in essentially the same yield and enantioselectivity as the standard reaction conditions (Scheme 2A). Moreover, the addition of water, and NH_4PF_6 allows for more efficient reaction of substrates that proved difficult under the initial set of conditions (Scheme 2B).

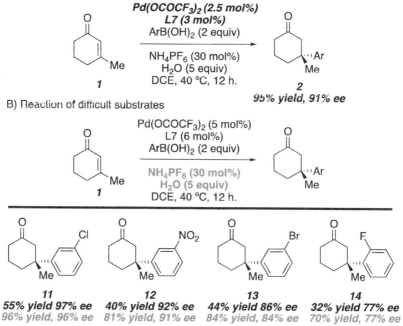

Scheme 2. Low Catalyst Load and Reactivity of Difficult Substrates

Subsequent efforts from our group have led to the expansion of this method to include chromenones as electrophiles, providing access to tertiary stereocenter containing flavanones (Scheme 3A).[44] Unfortunately, even the improved conditions outlined above were unable to promote the reaction of C(2)-substituted chromenones. Interestingly, Stanley and coworkers reported the reaction of C(2)-substituted chromenones with a palladium(II)-phenanthroline complex in the presence of aqueous $Na(OCOCF_3)$; however, these reaction conditions generate a racemic mixture of the products bearing tetrasubstituted stereocenters.[45] Subsequent investigation of different ligand scaffolds in collaboration with Sukwon Hong's group of GIST, revealed that pyridine-dihydroisoquinoline (PyDHIQ) ligands, and an oxygen atmosphere allowed for smooth reaction of C(2)-substituted chromenone electrophiles with excellent levels of enantioselectivity (Scheme 3B).[46] Finally, we became interested in the reaction of 3-acyl cyclohexenones, a challenging substrate class due to the possibility of the formation of constitutional isomers. Employing palladium(II) trifluoroacetate and (S)-t-BuPyOx allows for selective formation of compound **20**, which bears a quaternary stereocenter at the exclusion of forming a tertiary center via addition to the other end of the olefin (Scheme 3C).[47]

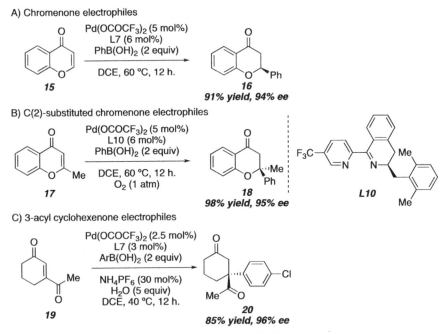

Scheme 3. Selected Electrophile Examples
Applications in Synthesis

The robust nature and scalability of this reaction has led to its use in the synthesis natural products and active pharmaceutical ingredients. In 2019, AbbVie reported the synthesis of ABBV-2222 (**24**), which is a preclinical candidate for the treatment of cystic fibrosis.[48] Synthesis of ABBV-2222 was accomplished via a key conjugate addition of boronic acid **22** to chromenone **21** to produce 13 grams of conjugate addition adduct **23** with excellent enantioselectivity, in a single pass. Enantioenriched flavanone **23** could then be carried forward through 5 more steps to generate >130 g of the active pharmaceutical ingredient **24** (Scheme 4).[49]

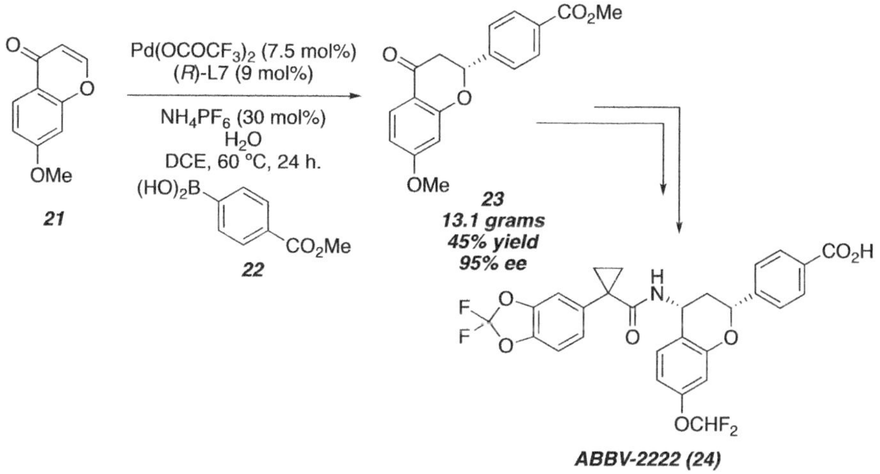

Scheme 4. Synthesis of ABBV-2222

Additionally, our group has utilized a palladium-catalyzed conjugate addition to accomplish a formal synthesis of (+)-dichroanone and (+)-taiwaniaquinone H.[50] Plotting the enantiomeric ratio versus the Hammett (σ_P) value for a variety of arylboronic acids revealed a strong positive linear correlation. This suggests that the difference in the diastereomeric transition states in the enantiodetermining step is larger with more electron withdrawing *para* substituents on the arylboronic acid. To this end, *para*-halogenated boronic acid **25** was selected to undergo conjugate addition with enone **1** to provide enantioenriched β-quaternary ketone **26** in excellent yield and enantioselectivity. This material was then advanced to intermediate **27**, which is a known intermediate in the synthesis of (+)-dichroanone and (+)-taiwaniaquinone H.[51]

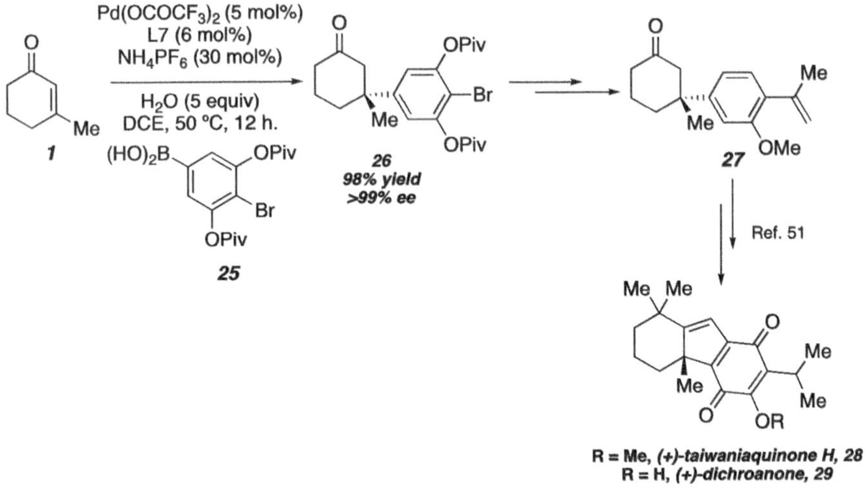

Scheme 5. Formal Synthesis of (+)-Dichroanone and (+)-Taiwaniaquinone H

More recently, Wood and coworkers utilized an enantioselective palladium-catalyzed conjugate addition toward the total synthesis of (–)-caesalpinflavans A and B in 2019.[52] Addition of phenyl boronic acid to chromenone **30** allowed for enantioselective synthesis of flavanone **31**. A divergent approach from this key intermediate can then grant access to (–)-caesalpinflavans A and B (Scheme 6). This strategy highlights the mild nature of boronic acid nucleophiles, which allowed for the use of **30** in its unprotected form.

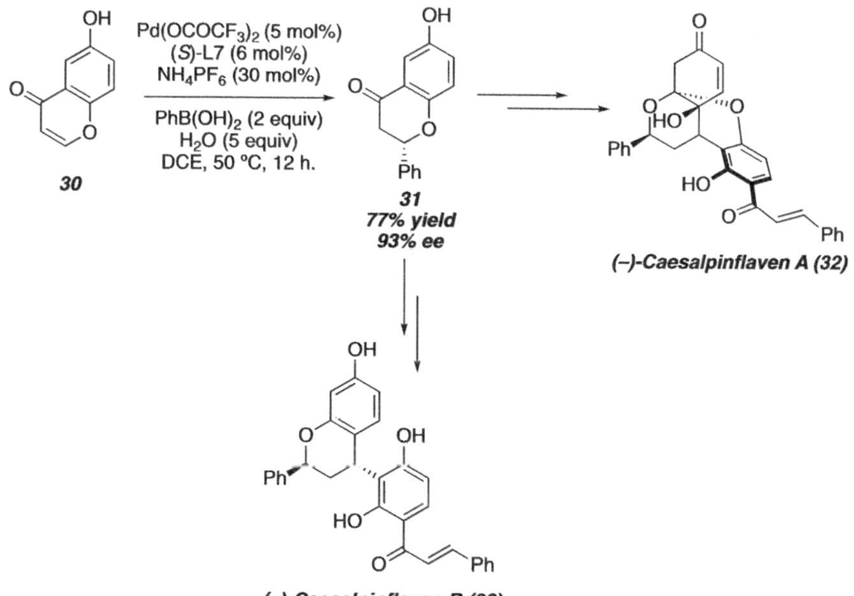

Scheme 6. Synthesis of (–)-Caesalpinflavens A and B

Conclusion

In summary, the asymmetric palladium-catalyzed conjugate addition of arylboronic acids to β,β-disubstituted α,β-unsaturated carbonyl compounds has emerged as a useful strategy for the construction of enantioenriched quaternary centers. The *t*-BuPyOx ligand required for the catalytically active cationic palladium(II) complex is easily synthesized in two steps from commercially available starting materials, as outlined in the original article. Furthermore, this process is advantageous due to the use of commercially available boronic acid nucleophiles, which are mild and easily handled on the benchtop. Finally, the reaction is easily scaled and extremely tolerant to air and water. The aforementioned attributes make this strategy attractive for the facile construction of enantioenriched building blocks toward the synthesis of complex molecules.

References

1. Division of Chemistry and Chemical Engineering, California Institute of Technology, Pasadena, California, 91125, United States. Email: stoltz@caltech.edu. We thank the NIH (R01GM080269) and Caltech for financial support. S.R.S thanks the NIH for a Ruth L. Kirschstein NRSA Postdoctoral fellowship (F32GM139300-02).
2. Denissova, I.; Barriault, L. *Tetrahedron* **2003**, *59*, 10105–10146.
3. Douglas, C. J.; Overman, L. E. *Proc. Natl. Acad. Sci. U.S.A.* **2004**, *101*, 5363–5367.
4. Christoffers, J.; Baro, A. *Adv. Synth. Catal.* **2005**, *347*, 1473–1482.
5. Trost, B. M.; Jiang, C. *Synthesis* **2006**, 369–396.
6. Mohr, J. T. Stoltz, B. M. *Chem. –Asian J.* **2007**, *2*, 1476–1491.
7. Cozzi, P. G.; Hilgraf, R.; Zimmermann, N. *Eur. J. Org. Chem.* **2007**, *36*, 5969–5994.
8. Hawner, C.; Alexakis, A. *Chem. Commun.* **2010**, *46*, 7295–7306.
9. Feringa, B. L. *Acc. Chem. Res.* **2000**, *33*, 346–353.
10. Wu, J.; Mampreian, D. M.; Hoveyda, A. H. *J. Am. Chem. Soc.* **2005**, *127*, 4584–4585.
11. Hird, A. W.; Hoveyda, A. H. *J. Am. Chem. Soc.* **2005**, *127*, 14988–14989.
12. Wilsily, A.; Fillion, E. *J. Am. Chem. Soc.* **2006**, *128*, 2774–2775.
13. Lee, K. -S.; Brown, M. K.; Hird, A. W.; Hoveyda, A. H. *J. Am. Chem. Soc.* **2006**, *128*, 7182–7184.
14. Brown, M. K.; May, T. L.; Baxter, C. A.; Hoveyda, A. H. *Angew. Chem., Int. Ed.* **2007**, *46*, 1097–1100.
15. Wilsey, A.; Fillion, E. *Org. Lett.* **2008**, *10*, 2801–2804.
16. Wilsey, A.; Fillion, E. *J. Org. Chem.* **2009**, *74*, 8583–8594.
17. Dumas, A. M.; Fillion E. *Acc. Chem. Res.* **2010**, *43*, 440–454.
18. D'Augustin, M.; Palais, L.; Alexakis, A. *Angew. Chem., Int. Ed.* **2005**, *44*, 1376–1378.
19. Fuchs, N.; d'Augustin, M.; Humam, M.; Alexakis, A.; Taras, R.; Gladiali, S. *Tetrahedron: Assym.* **2005**, *16*, 3143–3146.
20. Vuagnoux-d'Augustin, M.; Alexakis, A. *Chem. –Eur. J.* **2007**, *13*, 9647–9662.
21. Vuagnoux-d'Augustin, M.; Kehrli, M.; Alexakis, A. *Synlett* **2007**, 2057–2060.
22. Palais, L.; Mikhel, I. S.; Bournaud, C.; Micouin, L.; Falciola, C. A.; Vuagnoux-d'Augustin, M.; Rosset, S.; Bernardinelli, G.; Alexakis, A. *Angew. Chem., Int. Ed.* **2007**, *46*, 7462–7465.

23. May, T. L.; Brown, M. K.; Hoveyda, A. H. *Angew. Chem., Int. Ed.* **2008**, *47*, 7358–7362.
24. Hawner, C.; Li, K.; Cirriez, V.; Alexakis, A. *Angew. Chem., Int. Ed.* **2008**, *47*, 8211–8214.
25. Ladjel, C.; Fuchs, N.; Zhao, J.; Bernardinelli, G. Alexakis, A. *Eur. J. Org. Chem.* **2009** 4949–4955.
26. Palais, L.; Alexakis, A. *Chem. -Eur. J.* **2009**, *15*, 10473–10485.
27. Müller, D.; Hawner, C.; Tissot, M.; Palais, L.; Alexakis, A. *Synlett* **2010**, 1694–1698.
28. Roth, P. M. C.; Sidera, M.; Maksymowicz, R. M.; Fletcher, S. P. *Nat. Protoc.* **2014**, *9*, 104–111.
29. Gao, Z.; Fletcher, S. P. *Chem. Sci.* **2016**, *8*, 641–646.
30. Ardkhean, R.; Mortimore, M.; Paton, R. S.; Fletcher, S. P. *Chem Sci.* **2018**, *9*, 2628–2632.
31. Martin, D.; Kehrli, S.; Vuagnoux-d'Augustin, M.; Clavier, H.; Mauduit, M.; Alexakis, A. *J. Am. Chem. Soc.* **2006**, *128*, 8416–8417.
32. Hénon, H.; Mauduit, M.; Alexakis, A. *Angew. Chem., Int. Ed.* **2008**, *47*, 9122–9124.
33. Matsumoto, Y.; Yamada, K.-I.; Tomioka, K. *J. Org. Chem.* **2008**, *73*, 4578–4581.
34. Kehrli, S.; Martin, D.; Rix, D.; Mauduit, M.; Alexakis, A. *Chem. -Eur. J.* **2010** *16*, 9890–9904.
35. Hawner, C.; Müller, D.; Gremaud, L.; Felouat, A.; Woodward, S.; Alexakis, A. *Angew. Chem., Int. Ed.* **2010**, *49*, 7769–7772.
36. Hayashi, T.; Yamasaki, K. *Chem. Rev.* **2003**, *103*, 2829–2844.
37. Shintani, R.; Tsutsumi, Y.; Nagaosa, M.; Nishimura, T.; Hayashi, T. *J. Am. Chem. Soc.* **2009**, *131*, 13588–13589.
38. Shintani, R.; Takeda, M.; Nishimura, T.; Hayashi, T. *Angew. Chem., Int. Ed.* **2010**, *49*, 3969–3971.
39. Gini, F.; Hessen, B.; Minnaard, A. J. *Org. Lett.*, **2005**, *7*, 5309–5312.
40. Lin, S.; Lu, X. *Org. Lett.* **2010**, *12*, 2536–2539.
41. Kikushima, K.; Holder, J.; Gattie, M.; Stoltz, B. M. *J. Am. Chem. Soc.* **2011**, *133*, 6902–6905.
42. Kadam, A.; Ellern, A.; Stanley, L. *Org. Lett.* **2017**, *19*, 4062–4065.
43. Holder, J.; Zou, L.; Marziale, A.; Liu, P.; Lan, Y.; Gatti, M.; Kikushima, K.; Houk, K. N.; Stoltz, B. M. *J. Am. Chem. Soc.* **2013**, *135*, 14996–15007.
44. Holder, J.; Marziale, A.; Gatti, M.; Mao, B.; Stoltz, B. M. *Chem. Eur. J.* **2013**, *19*, 74–77.
45. Gerten, A.; Stanley, L. *Tetrahedron Lett.* **2016**, *57*, 5460–5463.

46. Baek, D.; Ryu, H.; Ryu, J. Lee, J.; Stoltz, B. M.; Hong, S. *Chem. Sci.* **2020**, *11*, 4602–4607.
47. Holder, J.; Goodman, E.; Kikushima, K.; Gatti, M.; Marziale, Stoltz, B. M. *Tetrahedron* **2015**, *71*, 5781–5792.
48. Wang, X.; Liu, B.; Searle, X.; Yeung, C.; Bogdan, A.; Greszler, S.; Singh, A.; Fan, Y.; Swensen, A.; Vortherms, T.; Balut, C.; Jia, Y.; Desino, K.; Gao, W.; Yong, H.; Tse, C.; Kym, P. *J. Med. Chem.* **2018**, *61*, 1436–1449
49. Greszler, S.; Shelat, B.; Voight, E. *Org. Lett.* **2019**, *21*, 5725–5727.
50. Shockley, S.; Holder, J.; Stoltz, B. M. *Org. Lett.* **2014**, *16*, 6362–6365.
51. Li, L.-Q.; Li, M.-M.; Chen, D.; Liu, H.-M.; Geng, H.-C.; Jin, J.; Qin, H.-B. *Tetrahedron Lett.* **2014**, *55*, 5960–5962.
52. Timmerman, J.; Sims, N.; Wood, J. L. *J. Am. Chem. Soc.* **2019**, *141*, 10082–10090.

Stephen R. Sardini was born in Fairfax, VA in 1990 and earned his B.S. in chemistry from The Ohio State University in 2012. He then earned his Ph.D. in 2020 under the direction of Prof. M. Kevin Brown at Indiana University. Stephen is currently an NIH Ruth L. Kirschstein NRSA postdoctoral fellow conducting research in the laboratory of Prof. Brian Stoltz at the California Institute of Technology.

Brian M. Stoltz was born in Philadelphia, PA in 1970 and obtained his B.S. degree from the Indiana University of Pennsylvania in Indiana, PA. After acquiring his Ph.D. at Yale University in the laboratory of John L. Wood, he commenced an NIH postdoctoral fellowship at Harvard under E. J. Corey. In 2000, he accepted a position at Caltech, where he is currently a Professor of Chemistry. His research interests lie in the development of new methodologies with general applications in synthetic chemistry.

Palladium-Catalyzed Hydrodefluorination of Fluoroarenes

Joseph J. Gair, Ronald L. Grey*, Simon Giroux, and Michael A. Brodney

Vertex Pharmaceuticals Inc, 50 Northern Avenue, Boston, Massachusetts 02210, United States

Checked by Vincent Porte, Daniel Kaiser, and Nuno Maulide

Procedure (Note 1)

3-Phenyl-1H-pyrazole. An oven dried 250-ml, three-necked, round-bottomed flask equipped with a Teflon-coated magnetic stir bar (3 cm) is charged with 3-(4-fluorophenyl)-1H-pyrazole (6.49 g, 40.0 mmol, 1.00 equiv) (Note 2) and RuPhos Palladacycle Gen. 4 (1.02 g, 1.20 mmol, 3.00 mol%) (Note 3). The vessel is equipped with an oven-dried reflux condenser and capped with two rubber septa. The reflux condenser is equipped with an argon inlet attached to a Schlenk line. The apparatus is evacuated, refilled with argon (3x). Toluene (55 mL, 0.73M) (Note 4) is added in two portions at room temperature via syringe through the side arm of the three-necked flask under flow of argon. The flask is charged with 2-propanol (15.3 mL, 200 mmol, 5.00 equiv) (Note 5) at room temperature via syringe through the side arm of the three-necked flask. The yellow suspension is stirred for five min under positive argon pressure. The resulting yellow solution (Figure 1A) is treated with a solution of sodium *tert*-pentoxide in toluene (30 mL, 3.3M, 100 mmol, 2.5 equiv) (Note 6) under flow of argon (exothermic reaction). The resulting black suspension (Figure 1B) is transferred to an aluminum heating block equilibrated to 80 °C. Water is circulated through the reflux condenser. A thermometer is inserted through one neck, replacing one of the septa, and the reaction is stirred at 80 °C (internal temperature = 70 °C) under argon atmosphere (Figure 1C). After 18 h, ^{1}H and ^{19}F NMR showed no evidence of starting material (Note 7).

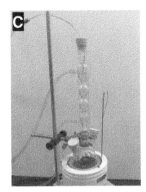

Figure 1. (A) Reaction mixture prior to addition of sodium *tert*-pentoxide; (B) Reaction mixture immediately after adding sodium tert-pentoxide; and (C) Reaction setup during heating

The black heterogeneous reaction mixture is removed from the heating block and allowed to cool to room temperature (Figure 2A). The reaction mixture is filtered through a pad of sand (1 cm) layered with Celite (20 g) on a fritted funnel (porosity 2) into a 500-mL round-bottomed flask (Figure 2B). The filter cake is carefully triturated with a spatula to insure a better filtration. The reaction vessel and Celite cake are then washed with ethyl acetate until the eluent is colorless (9 × 20 mL). The filtrate is concentrated by rotary evaporation (50 °C, 70 mmHg) to afford a thick brown paste (Figure 2C).

Figure 2. (A) Heterogeneous reaction mixture after heating for 18 hours; (B) Reaction mixture after filtration through Celite; and (C) Filtrate after concentration by rotary evaporation

The concentrated crude material is quenched with saturated aqueous sodium bicarbonate (40 mL) at room temperature. The resulting slurry is suspended with swirling and transferred via plastic funnel to a 125-mL separatory funnel. The round-bottomed flask containing quenched material is washed with a biphasic mixture of deionized water (5 mL) and dichloromethane (5 mL) (5 × biphasic wash) to transfer any residual product to the separatory funnel. The separatory funnel is shaken and the layers (Figure 3A) are allowed to separate over five min. The dichloromethane layer is collected in a 250 mL Erlenmeyer flask, and the aqueous layer is re-extracted with dichloromethane (4 × 20 mL) until the organic phase is colorless (Figure 3B,C). The organic fractions are treated with sodium sulfate (20 g), left to dry for 10 min and clarified by vacuum filtration on a fritted funnel into a 500-mL round-bottomed flask. The drying agent is washed with ethyl acetate (3 × 40 mL) to extract any residual product and the ethyl acetate washes are filtered into the dichloromethane filtrate. The combined filtrates are charged with Celite (17 g) and concentrated by rotary evaporation (50 °C, 70 mmHg).

  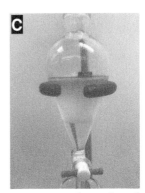

Figure 3. (A) Biphasic mixture after transferring to separatory funnel and shaking; (B) Biphasic mixture after second extraction with dichloromethane; and (C) Biphasic mixture after fourth extraction with dichloromethane

A glass chromatography column (inner diameter 5 cm) is charged with silica gel (200g) (Note 8) loaded as a slurry with 40 vol% ethyl acetate in *n*-heptane. The silica gel slurry is allowed to settle, and excess mobile phase is eluted by gravity. Atop the silica gel slurry is loaded the crude solid adsorbed on Celite followed by a layer of sand (4 cm). The packed column is then eluted with 40 vol% ethyl acetate in *n*-heptane (1L) followed by ethyl acetate (1L) (Note 9). The eluent is collected in 25 × 150 mm test tubes (50 mL fractions) for 40 fractions. Fractions 13–28 contain the desired product (Note 10) and are concentrated by rotary evaporation (50 °C, 50 mmHg) in a 500-mL round-bottomed flask. Thick oil is dissolved with EtOAc and transferred to a 50-mL round-bottomed flask and concentrated under reduced pressure to yield the desired product (5.70 g) as a tan solid in >90% purity as assayed by ^{1}H NMR (Figure 4A).

The chromatographed material is further purified by crystallization (Note 11): The solid is transferred to a tared 30-mL vial as a solid, and any residual material in the 500-mL round-bottomed flask is transferred by washing with hot benzene (4 × 2 mL, 70 °C) followed by pipette transfer to the 30-mL vial. The tan solid suspended in benzene is dissolved by heating the vial (aluminum heating block, 90 °C) until the mixture is gently refluxing (5 min reflux with intermittent swirling). The resulting, hazy orange mixture (Figure 4B) is layered with *n*-heptane (8 mL) (Figure 4C) and allowed to cool to room temperature over 4 h (Note 12) (Figure 4D). The recrystallization setup is then gently swirled to facilitate mixing of the benzene and *n*-heptane layers. The recrystallization setup is stored for 24 h at 4 °C, which affords large colorless crystals (Figure 4E). The supernatant is removed by pipette. The colorless crystals are washed with cold *n*-heptane (2 × 6 mL, 4 °C) and dried first by rotary evaporation (40 °C, 10 mmHg) then in a vacuum oven (30 °C, 0.1 mmHg, 4 h) to afford 3-phenyl-1H-pyrazole (5.06 g, 88% yield (4.97 g, 86% based on purity)) as a colorless solid (Figure 4F) (Notes 13 and 14).

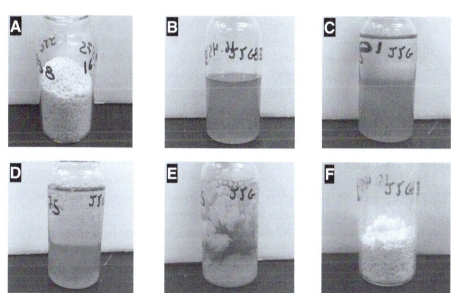

Figure 4. (A) Tan material obtained from column chromatography; (B) Concentrated solution of desired product in hot benzene; (C) After layering with hot *n*-heptane; (D) Small crystals formed after cooling to room temperature; (E) Large crystals after cooling for 24 h at 4 °C; and (F) Isolated colorless crystalline solid

Notes

1. Prior to performing each reaction, a thorough hazard analysis and risk assessment should be carried out with regard to each chemical substance and experimental operation on the scale planned and in the context of the laboratory where the procedures will be carried out. Guidelines for carrying out risk assessments and for analyzing the hazards associated with chemicals can be found in references such as

Chapter 4 of "Prudent Practices in the Laboratory" (The National Academies Press, Washington, D.C., 2011; the full text can be accessed free of charge at https://www.nap.edu/catalog/12654/prudent-practices-in-the-laboratory-handling-and-management-of-chemical. See also "Identifying and Evaluating Hazards in Research Laboratories" (American Chemical Society, 2015) which is available via the associated website "Hazard Assessment in Research Laboratories" at https://www.acs.org/content/acs/en/about/governance/committees/chemicalsafety/hazard-assessment.html. In the case of this procedure, the risk assessment should include (but not necessarily be limited to) an evaluation of the potential hazards associated with 3-(4-fluorophenyl)-1*H*-pyrazole, RuPhos Palladacycle Gen. 4, 2-propanol, sodium *tert*-pentoxide, toluene, ethyl acetate, sodium bicarbonate, dichloromethane, sodium sulfate, *n*-heptane, celite, silica gel, benzene, acetonitrile, trifluoroacetic acid, and deuterated chloroform.

2. 3-(4-Fluorophenyl)-1*H*-pyrazole (>97%) was obtained from Alfa Aesar (H34355) as an off-white powder and used as received.
3. RuPhos Palladacycle Gen. 4 (>98%) was obtained from Strem Chemicals (46-0395) as a beige powder and used as received.
4. Toluene (anhydrous, 99.8%) was purchased from Sigma Aldrich (244511) in a Sure/Seal bottle and used as received. Toluene was taken up from the Sure/Seal bottle via syringe under positive argon pressure.
5. 2-Propanol (>99.5%) was purchased from Sigma Aldrich (I9516) and used as received. 2-Propanol was measured and handled under air with no precautions to omit oxygen or moisture. Reactions conducted with lower concentrations of *i*-PrOH gave slower and sometimes incomplete conversion.
6. Sodium *tert*-pentoxide solution (40%) in toluene was purchased from Sigma Aldrich (752096) in a Sure/Seal bottle and used as received. Sodium *tert*-pentoxide was taken up from the Sure/Seal bottle via syringe under positive argon pressure. Reactions conducted with lower concentrations of sodium *tert*-pentoxide gave slower and sometimes incomplete conversion.
7. The submitters monitored reaction progress using LCMS: Reactions were monitored by liquid chromatography mass spectrometry (LCMS) on a Waters Acquity UPLC with Waters Acquity HSS T3 C18 column (2.1 mm × 50 mm) and eluting with a gradient from 10–60% acetonitrile in water (0.1% trifluoroacetic acid modifier) over 1.6 min (product R_t = 0.80 min, starting material R_t = 0.84 min).

8. Silica gel (pore size 60Å, 230-400 mesh particle size) was purchased from Sigma-Aldrich (227196). The checkers used silica gel purchased from Macherey-Nagel (particle size 0.040–0.063 mm), which was used as received.
9. Thin layer chromatography (TLC) of starting material (left lane, R_f 0.32), reaction mixture (center lane), and product (right line R_f, 0.34) on silica gel 60 (Merck KGaA) eluting with 40 vol% ethyl acetate in n-heptane (Figure 5). The reaction mixture is monitored by ^{1}H and ^{19}F NMR or LCMS (Note 7). Reaction monitoring by TLC is not recommended due to poor separation of starting material and product.

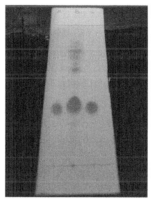

Figure 5. TLC of starting material (left lane), reaction mixture (center lane), and product (right lane)

10. Thin layer chromatography (TLC) on Silicagel 60 (Merck KGaA) eluting with 40vol% ethyl acetate in n-heptane showing fractions 1-30 from column chromatography (Figure 6).

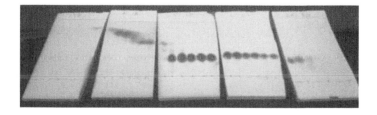

Figure 6. TLC of fractions 1-30 from column chromatography

11. An alternative crystallization procedure is reported here: To the round-bottomed flask, fitted with a condenser containing the chromatographed material was added 8 mL of benzene, and the resulting slurry was heated with intermitting swirling until all material had dissolved. To the orange mixture was added 8 mL of heptane and mixture was heated to reflux for 5 min. The mixture was allowed to return to room temperature, left standing at this temperature for 15 h, before being cooled with an ice bath (0 °C) to induce crystallization. The supernatant was removed, the crystals were washed with heptane (2 × 6 mL) and dried first by rotary evaporation (50 °C, 10 mmHg) then with under high vacuum for 7 h.
12. The submitters observed the formation of crystals at this point, while the checkers only observed crystal formation upon subsequent cooling in the fridge.
13. Analytical data for 3-phenyl-1H-pyrazole as follows: mp 77–79 °C; IR (neat) 3166, 2962, 1456, 1353, 1071, 1046, 954, 931 cm^{-1}. ^{1}H NMR (400 MHz, CDCl$_3$, 0.7M) δ: 6.63 (d, J = 2.2 Hz, 1H), 7.32–7.44 (m, 3H), 7.61 (d, J = 2.2 Hz, 1H), 7.80 (d, br, J = 7.3 Hz, 2H), 13.01 (s, br, 1H). ^{13}C NMR (100 MHz, CDCl$_3$, 0.7M) δ: 102.7, 126.0 (2C), 128.1, 128.9 (2C), 132.4, 133.4 (br), 149.3 (br). HRMS (*m/z*) calcd for C$_9$H$_9$N$_2$ [M+H]$^+$: 145.0760, found: 145.0757. Purity was assessed at 98.3% by qNMR using 11.1 mg of the product and 1,3,5-trimethoxybenzene (9.1 mg) as an internal standard.
14. The checkers performed a second reaction on half scale, yielding 2.56 g (89%) of the product, while a second reaction performed by the submitters on full scale afforded 5.20 g (90% yield) of 3-phenyl-1H-pyrazole.

Working with Hazardous Chemicals

The procedures in *Organic Syntheses* are intended for use only by persons with proper training in experimental organic chemistry. All hazardous materials should be handled using the standard procedures for work with chemicals described in references such as "Prudent Practices in the Laboratory" (The National Academies Press, Washington, D.C., 2011; the full text can be accessed free of charge at http://www.nap.edu/catalog.php?record_id=12654). All chemical waste should be disposed of in accordance with local regulations. For general

guidelines for the management of chemical waste, see Chapter 8 of Prudent Practices.

In some articles in *Organic Syntheses*, chemical-specific hazards are highlighted in red "Caution Notes" within a procedure. It is important to recognize that the absence of a caution note does not imply that no significant hazards are associated with the chemicals involved in that procedure. Prior to performing a reaction, a thorough risk assessment should be carried out that includes a review of the potential hazards associated with each chemical and experimental operation on the scale that is planned for the procedure. Guidelines for carrying out a risk assessment and for analyzing the hazards associated with chemicals can be found in Chapter 4 of Prudent Practices.

The procedures described in *Organic Syntheses* are provided as published and are conducted at one's own risk. *Organic Syntheses, Inc.*, its Editors, and its Board of Directors do not warrant or guarantee the safety of individuals using these procedures and hereby disclaim any liability for any injuries or damages claimed to have resulted from or related in any way to the procedures herein.

Discussion

Fluorine plays an important role in drug discovery due to its ability to fine tune the physiochemical and metabolic properties of biologically active compounds.[2] Given the important impacts that a single fluorine substituent can have on key properties (both beneficial and detrimental), drug discovery teams frequently encounter situations in which direct comparison of fluoro- and des-fluoro-analogues is insightful. Preparation of des-fluoro-analogues typically requires lengthy resynthesis from des-fluoro building blocks. A method for direct conversion of fluorinated compounds to des-fluoro-analogues by hydrodefluorination would circumvent this resynthesis and thereby provide a useful tool to accelerate drug discovery.

Transition metal-catalyzed hydrodefluorination methodology has undergone significant growth in recent years; however, these advances have not yet translated to reliable tools for drug discovery.[3] This gap is likely a consequence of the fact that many of the reported methods are optimized for substrates with activated polyfluoro-arenes[4] or lacking functional groups (especially heterocycles) encountered in drug discovery.[5]

To close this gap, we recently reported a general method for hydrodefluorination of (hetero)arenes that is suitable for immediate implementation in drug discovery settings by virtue of the fact that the method requires only commercially available catalysts and reagents, can be set up under air and affords good yields on a variety of functional groups and heterocycles encountered in drug discovery.[6]

The substrate scope for a set of general reaction conditions suitable for hydrodefluorination of a variety of heterocyclic scaffolds is provided in Table 1. The reaction is compatible with ortho, meta, and para-substituted fluoroarenes (**1–4**). The reaction proceeds efficiently for substrates bearing acidic and phenolic functionality (**5**, **16**, **24**, and **26**) and NH fragments in amides (**1** and **7**), pyrazole (**3**), indazoles (**4** and **10**), imidazoles (**12–14**), azaindole (**17**), and secondary amines (**21** and **25**). The reaction also tolerates the presence of Lewis basic functional that might coordinate to transition metal catalysts including tertiary anilines (**2**), benzimidazoles (**8** and **9**), benzotriazole (**15**), (iso)quinolines (**18** and **19**), and pyridines (**20–23**). Due to its expanded scope and compatibility with a variety of functional groups encountered in medicinal chemistry, this method is well suited for immediate implementation in drug discovery settings.

Table 1. Substrate scope in palladium-catalyzed hydrodefluorination

1a 2-F 94%
1b 3-F 71%
1c 4-F 81%

2a 2-F 13%[a]
2b 3-F 24%[a]
2c 4-F 41%[a]

3a 4-F 93%[b]
3b 2-F 92%

4 77%

5 55%[b]

7 29%

8 73%

9 97%

10a 73%

10b 67%

11 50%

12 88%

13 73%

14 41%[b]

15 89%

16 42%

17 68%

18 71%

19a 31%

19b 76%

20a 85%

20b 72%[a,c]

21 37%

22 79%

23 70%

24 55%

25 29%[a]
(Paroxetine, anti-depressant)

26 61%[a]
(Flurbiprofen, anti-inflammatory)

[a]10 mol% Pd, [b]3 mol% Pd, [c]60 °C

References

1. Vertex Pharmaceuticals Inc, 50 Northern Avenue, Boston, MA 02210, USA. Email: Ron_Grey@vrtx.com ORCID 0000-0001-8948-5056.
2. (a) Böhm, H.-J.; Banner, D.; Bendels, S.; Kansy, M.; Kuhn, B.; Müller, K.; Obst-Sander, U.; Stahl, M. *ChemBioChem* **2004**, *5*, 637–643. (b) Kirk, K. L. *J. Fluor. Chem.* **2006**, *127*, 1013–1029. (c) Müller, K.; Faeh, C.; Diederich, F. *Science* **2007**, *317*, 1881–1886. (d) Ricci, P.; Khotavivattana, T.; Pfeifer, L.; Médebielle, M.; Morphy, J. R.; Gouverneur, V. *Chem. Sci.* **2017**, *8*, 1195–1199. (e) Hagmann, W. K. *J. Med. Chem.* **2008**, *51*, 4359–4369. (f) Wang, J.; Sánchez-Roselló, M.; Aceña, J. L.; del Pozo, C.; Sorochinsky, A. E.; Fustero, S.; Soloshonok, V. A.; Liu, H. *Chem. Rev.* **2013**, *114*, 2432–2506. (g) Gillis, E. P.; Eastman, K. J.; Hill, M. D.; Donnelly, D. J.; Meanwell, N. A. *J. Med. Chem.* **2015**, *58*, 8315–8359. (h) Meanwell, N. A. *J. Med. Chem.* **2018**, *61*, 5822–5880. (i) Dimagno, S. G.; Sun, H. *Curr. Top. Med. Chem.* **2006**, *6*, 1473–1482. (j) Smart, B. E. *J. Fluor. Chem.* **2001**, *109*, 3–11.
3. (a) Whittlesey, M. K.; Peris, E. *ACS Catal.* **2014**, *4*, 3152–3159. (b) Amii, H.; Uneyama, K. *Chem. Rev.* **2009**, *109*, 2119–2183. (c) Alonso, F.; Beletskaya, I. P.; Yus, M. *Chem. Rev.* **2002**, *102*, 4009–4092.
4. (a) Tsuzuki, H.; Kamio, K.; Fujimoto, H.; Mimura, K.; Matsumoto, S.; Tsukinoki, T.; Mataka, S.; Yonemitsu, T.; Tashiro, M. *J. Label. Compd. Radiopharm.* **1993**, *33*, 205–212. (b) Maron, L.; Werkema, E. L.; Perrin, L.; Eisenstein, O.; Andersen, R. A. *J. Am. Chem. Soc.* **2005**, *127*, 279–292. (c) Beltrán, T. F.; Feliz, M.; Llusar, R.; Mata, J. A.; Safont, V. S. *Organometallics* **2011**, *30*, 290–297. (d) Chen, Z.; He, C.-Y.; Yin, Z.; Chen, L.; He, Y.; Zhang, X. *Angew. Chem. Int. Ed.* **2013**, *52*, 5813–5817. (e) Senaweera, S. M.; Singh, A.; Weaver, J. D. *J. Am. Chem. Soc.* **2014**, *136*, 3002–3005. (f) Schwartsburd, L.; Mahon, M. F.; Poulten, R. C.; Warren, M. R.; Whittlesey, M. K. **2014**, *33*, 6165–6170. (g) Ekkert, O.; Strudley, S. D. A.; Rozenfeld, A.; White, A. J. P.; Crimmin, M. R. *Organometallics* **2014**, *33*, 7027–7030. (h) Cybulski, M. K.; Riddlestone, I. M.; Mahon, M. F.; Woodman, T. J.; Whittlesey, M. K. *Dalton Trans.* **2015**, *44*, 19597–19605. (i) McKay, D.; Riddlestone, I. M.; Macgregor, S. A.; Mahon, M. F.; Whittlesey, M. K. *ACS Catal.* **2015**, *5*, 776–787. (j) Podolan, G.; Jungk, P.; Lentz, D.; Zimmer, R.; Reissig, H.-U. *Adv. Synth. Catal.* **2015**, *357*, 3215–3228. (k) Liu, X.; Wang, Z.; Zhao, X.; Fu, X. *Inorg. Chem. Front.* **2016**, *3*, 861–865. (l) Matsunami, A.; Kuwata, S.; Kayaki, Y. *ACS Catal.* **2016**, *6*, 5181–5185. (m) Krüger, J.; Leppkes, J.; Ehm, C.; Lentz, D.

Chem. Asian J. **2016**, *11*, 3062–3071. (n) Mai, V. H.; Nikonov, G. I. *ACS Catal.* **2016**, *6*, 7956–7961. (o) Cybulski, M. K.; McKay, D.; Macgregor, S. A.; Mahon, M. F.; Whittlesey, M. K. *Angew. Chem. Int. Ed.* **2017**, *129*, 1537–1541. (p) Chen, J.; Huang, D.; Ding, Y. *ChemistrySelect* **2017**, *2*, 1219–1224. (q) Cybulski, M. K.; Nicholls, J. E.; Lowe, J. P.; Mahon, M. F.; Whittlesey, M. K. **2017**, *36*, 2308–2316. (r) Kikushima, K.; Grellier, M.; Ohashi, M.; Ogoshi, S. *Angew. Chem. Int. Ed.* **2017**, *56*, 16191–16196. (s) Matsunami, A.; Kayaki, Y.; Kuwata, S.; Ikariya, T. *Organometallics* **2018**, *37*,
1958–1969. (t) Jaeger, A. D.; Ehm, C.; Lentz, D. *Chem. Eur. J.* **2018**, *24*, 6769–6777. (u) Cybulski, M. K.; Davies, C. J. E.; Lowe, J. P.; Mahon, M. F.; Whittlesey, M. K. *Inorg. Chem.* **2018**, *57*, 13749–13760.

5. (a) Ukisu, Y.; Miyadera, T. *J. Mol. Catal. A Chem.* **1997**, *125*, 135–142. (b) Young, R. J.; Grushin, V. V. *Organometallics* **1999**, *18*, 294–296. (c) Aramendía, M. A.; Borau, V.; García, I. M.; Jiménez, C.; Marinas, A.; Marinas, J. M.; Urbano, F. J. *Comptes Rendus de l'Académie des Sciences - Series IIC - Chemistry* **2000**, *3*, 465–470. (d) Desmarets, C.; Kuhl, S.; Schneider, R.; Fort, Y. *Organometallics* **2002**, *21*, 1554–1559. (e) Kuhl, S.; Schneider, R.; Fort, Y. *Adv. Synth. Catal.* **2003**, *345*, 341–344. (f) Cellier, P. P.; Spindler, J.-F.; Taillefer, M.; Cristau, H.-J. *Tetrahedron Lett.* **2003**, *44*, 7191–7195. (g) Davies, C. J. E.; Page, M. J.; Ellul, C. E.; Mahon, M. F.; Whittlesey, M. K. *Chem. Commun.* **2010**, *46*, 5151–5153. (h) Wu, J.; Cao, S. *ChemCatChem* **2011**, *3*, 1582–1586. (i) Sawama, Y.; Yabe, Y.; Shigetsura, M.; Yamada, T.; Nagata, S.; Fujiwara, Y.; Maegawa, T.; Monguchi, Y.; Sajiki, H. *Adv. Synth. Catal.* **2012**, *354*, 777–782. (j) Xiao, J.; Wu, J.; Zhao, W.; Cao, S. *J. Fluor. Chem.* **2013**, *146*, 76–79. (k) Sabater, S.; Mata, J. A.; Peris, E. *Nat. Commun.* **2013**, *4*, 1–7. (l) Wu, W.-B.; Li, M.-L.; Huang, J.-M. *Tetrahedron Lett.* **2015**, *56*, 1520–1523. (m) Sabater, S.; Mata, J. A.; Peris, E. *Organometallics* **2015**, *34*, 1186–1190. (n) Xu, Y.; Ma, H.; Ge, T.; Chu, Y.; Ma, C.-A. *Electrochem. Commun.* **2016**, *66*, 16–20. (o) Hokamp, T.; Dewanji, A.; Lübbesmeyer, M.; Mück-Lichtenfeld, C.; Würthwein, E.-U.; Studer, A. *Angew. Chem. Int. Ed.* **2017**, *56*, 13275–13278. (p) Tashiro, M.; Nakamura, H.; Nakayama, K. *Org. Prep. Proced. Int.* **1987**, *19*, 442–446.
6. Gair, J. J.; Grey, R. L.; Giroux, S.; Brodney, M. A. *Org. Lett.* **2019**, *21* (7), 2482–2487.

Appendix
Chemical Abstracts Nomenclature (Registry Number)

Toluene; (108-88-3)
2-Propanol: propan-2-ol; (67-63-0)
Sodium *tert*-pentoxide; (14593-46-5)
3-(4-Fluorophenyl)-1*H*-pyrazole; (154258-82-9)
3-Phenyl-1*H*-pyrazole; (2458-26-6)
RuPhos palladacycle generation 4: Methanesulfonato(2-dicyclohexylphosphino-2',6'-di-i-propoxy-1,1'-biphenyl)(2'-dimethylamino-1,1'-biphenyl-2-yl)palladium(II); (1599466-85-9)

Joseph Gair studied organometallic synthesis with Chris Schaller at Saint John's University, where he earned a B.A. in 2012. Following additional training with Jeffrey Johnson at Hope College, he spent one year teaching secondary school chemistry in Hanga Tanzania with the Benedictine Volunteer Corps. He obtained his Ph.D. from the University of Chicago in 2018 as an NSF research fellow with Jared Lewis. He spent one year as a Co-Op in medicinal chemistry at Vertex Pharmaceuticals before starting his current position as an NIH postdoctoral fellow in the lab of Eric Jacobsen at Harvard University.

Ron Grey received his B.S. in Biochemistry from the University of Delaware in 1997. He subsequently spent 4 years at Dupont's Experimental Station in Wilmington, DE focusing on parallel synthesis. Since 2001, he has served in the Medicinal Chemistry Department at Vertex Pharmaceuticals.

Simon Giroux received his Ph.D. with Prof. Stephen Hanessian from the Université de Montréal in 2006. He subsequently spent 2 years in the laboratory of Prof. E. J. Corey at Harvard University, as an NSERC postdoctoral fellow. He is the author of more than 30 patents and publications and currently serves as a Research Fellow in the Medicinal Chemistry Department at Vertex Pharmaceuticals.

Michael Brodney is currently the Head of Medicinal Chemistry at Vertex Pharmaceuticals. Prior to joining Vertex, Michael was a senior director at Pfizer working on a range of programs in the neuroscience area. He received his Ph.D. with Prof. Al Padwa from Emory University and then moved to Stanford University for an NIH postdoctoral fellowship with Prof. Paul Wender. He is the author of more than 75 patents and publications.

Vincent Porte received his M.Sc in organic chemistry and diploma in Chemical Engineering from the Ecole Nationale Supérieure de Chimie de Montpellier in 2019. During his undergraduate studies, he performed internships in medicinal chemistry and process chemistry at Hoffmann-La Roche and Firmenich. He is currently a second-year graduate student in the group of Prof. Nuno Maulide. His research focuses on the minimization of the flexibility of drug molecules using fused ring systems and the synthesis of natural-product-like structures.

Daniel Kaiser received his Ph.D. at the University of Vienna in 2018, completing his studies under the supervision of Prof. Nuno Maulide. After a postdoctoral stay with Prof. Varinder K. Aggarwal at the University of Bristol, he returned to Vienna in 2020 to assume a position as senior scientist in the Maulide group. His current research focusses on the chemistry of destabilized carbocations and related high-energy intermediates.

Synthesis of the Isocyanide Building Block Asmic, anisylsulfanylmethylisocyanide

Embarek Alwedi, Bilal Altundas, Allen Chao, Zachary L. Ziminsky, Maanasa Natrayan, and Fraser F. Fleming[*1]

Department of Chemistry, Drexel University, 32 South 32nd St., Philadelphia, PA 19104

Checked by Names Anthony J. Fernandes, Martina Drescher, and Nuno Maulide

Procedure (Note 1)

A. *N-((2-Methoxyphenylthio)methyl)formamide (1)*. A 3-necked, 500 mL round-bottomed flask (Note 2), equipped with a magnetic stir bar (PTFE-coated, cylindrical, 3 cm) (Figure 1a) is charged with paraformaldehyde (17.7 g, 584 mmol, 4.1 equiv), formamide (43 mL, 48.6 g, 1.08 mol, 7.6 equiv), formic acid (27 mL, 33 g, 713 mmol, 5 equiv), and 2-methoxybenzenethiol (17.4 mL, 20 g, 142.6 mmol, 1 equiv) (Note 3) (Figure 1b). The left and right necks are stoppered with a rubber septum and the middle neck is connected to a reflux condenser stoppered with a rubber septum that is pierced with a 20-gauge disposable needle open to the atmosphere (Figure 2).

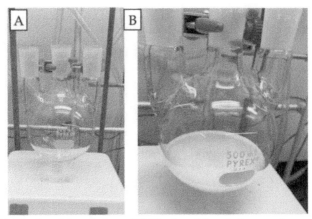

Figure 1. (A) Flask with stir bar; (B) Flask with stir bar and reagents (photos provided by submitters)

Figure 2. Reaction set-up prior to heating (photo provided by submitter)

The flask is immersed in an oil bath that is gradually heated to 100 °C over 1 h; the temperature is then maintained at 100 °C for an additional 3 h (Note 4) (Figure 3a-c).

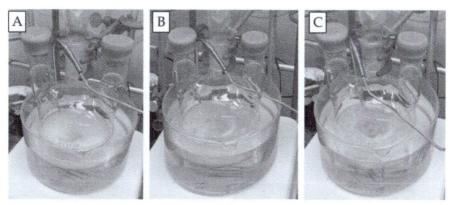

Figure 3. (A) Reaction mixture at 91 °C; (B) Reaction mixture at 100 °C; (C) Reaction mixture after 3h at 100 °C (photos provided by submitters)

The reaction is monitored by silica gel thin layer chromatography with both 5% (5:95) EtOAc-hexanes as the eluent to check for the presence of 2-methoxybenzenethiol and 75% (3:1) EtOAc-hexanes as eluent to monitor the formation of the formamide **1**. The TLC plate is visualized with UV light (Note 5) (Figure 4a). The reaction mixture is then allowed to cool to room temperature (Figure 4b).

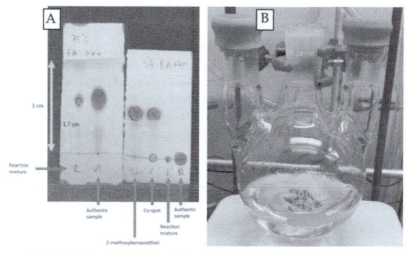

Figure 4. (A) TLC analysis after 3 h at 100 °C; (B) Reaction mixture after cooling to rt (photos provided by submitters)

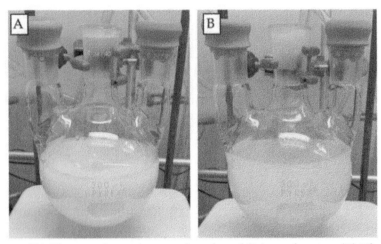

Figure 5. (A) The reaction mixture after the addition of water; (B) The reaction mixture after the addition of dichloromethane (photos provided by submitters)

Distilled water (200 mL) and dichloromethane (80 mL) are added to the crude reaction mixture (Note 6) (Figure 5). The contents of the flask are transferred to a 1 L separatory funnel (Figure 6a). The phases are separated, and then the aqueous phase is extracted with dichloromethane (3 × 100 mL). The combined organic extract is washed with brine (200 mL) (Figure 6b), dried for 10 min over anhydrous sodium sulfate (Na_2SO_4, 80 g) and concentrated to afford the crude formamide as a clear, pale yellow viscous oil (Figure 6c). The formamide is purified by precipitation from a dichloromethane solution through the addition of hexane. The crude formamide is dissolved in dichloromethane (95 mL) to which hexanes is added portion-wise until no further precipitation is observed (Note 7) (Figure 7). The resultant suspension is placed in a –20 °C freezer.

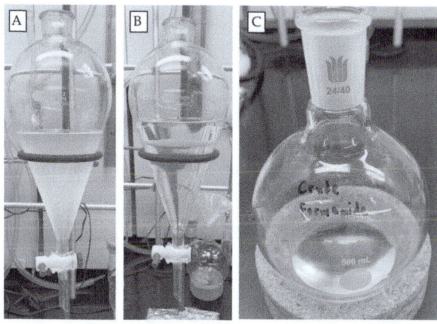

Figure 6. (A) Crude reaction mixture; (B) The crude reaction mixture after aqueous washing with brine; (C) Crude formamide (photos provided by submitters)

Figure 7. (A) Dissolution of the crude formamide in CH_2Cl_2; (B) Addition of 100 mL of hexanes; (C) Addition of a further 50 mL of hexanes – 150 mL total; (D) Addition of a further 50 mL of hexanes – 200 mL total; (E) Addition of a further 50 mL of hexanes – 250 mL total (photos provided by submitters)

After 19 h, the white solid (Figure 8a) is isolated by filtration through a 500 mL medium porosity, sintered Büchner funnel (Figure 8b). The solid on the frit is washed with cold (0 °C) hexanes (2 × 25 mL). The solid is collected, dissolved in CH_2Cl_2 (50 mL) and then precipitated with hexanes (200 mL) (Note 8). The white solid is filtered and then washed with cold (0 °C) hexanes (2 × 25 mL). The solid is again collected, dissolved in CH_2Cl_2 (50 mL), precipitated with hexanes (200 mL), filtered, and washed with cold (0 °C) hexanes (2 × 25 mL). The white solid is then dried under vacuum (1 mmHg) for 1 h at room temperature to afford 19.0 g (68% yield) of formamide **1** as a white solid (Figure 8c) (Notes 10, 11, and 12).

Figure 8. (A) Formamide 1 after 19 h at –20 °C (B) Formamide 1 after filtration (C) Formamide after third precipitation (submitters); (D) Formamide after third precipitation (checkers)

B. *Anisylsulfanylmethylisocyanide (Asmic (2)*. A 3-necked, 500 mL round-bottomed flask, equipped with a magnetic stir bar (PTFE-coated, cylindrical, 3 cm) (Note 2) is charged with *N-((2-methoxyphenylthio)methyl)formamide* (18.5 g, 93.8 mmol, 1 equiv) (Note 9) (Note 13). The left and right necks of the flask are stoppered with rubber septa (Figure 9a). Dichloromethane (108 mL) (Note 14) is added at room temperature (20 °C) (Figure 9b). After the formamide is completely dissolved, diisopropylethylamine (81.7 mL, 60.6 g, 469.0 mmol, 5 equiv) (Note 15) is added from a graduated cylinder through the middle neck (Note 16).

Figure 9. (A) Reaction set-up with formamide 1; (B) Dissolution of formamide 1 in CH$_2$Cl$_2$; (C) Reaction mixture after addition of diisopropylethylamine (photos provided by submitters)

After 10 min, the middle neck is equipped with a 100 mL, pressure-equalizing addition funnel capped with a septum pierced with a 20-gauge needle open to the atmosphere (Figure 10). The two side necks are capped with septa that are each pierced with a 20-gauge needle open to the atmosphere. The reaction mixture is immersed in an ice-salt bath that is cooled to –10 °C (Note 17) (Figure 10a). The addition funnel is charged with POCl$_3$ (16.4 mL, 25.8 g, 188.0 mmol, 2 equiv) (Note 18) that is added dropwise, over 20 min; the reaction temperature is maintained between –12 and –9 °C during the addition (Note 19) (Figures 10b and 10c).

Figure 10. (A) Reaction mixture at –12 °C with POCl$_3$ charged in the addition funnel; (B) Reaction mixture 20 min after the addition of POCl$_3$; (C) Reaction mixture 90 min after the addition of POCl$_3$ (photos provided by submitters)

The reaction is monitored by thin layer chromatography (TLC) on silica gel with both 10% (1:9) EtOAc-hexanes and 75% (3:1) EtOAc-hexanes as eluent and visualized under UV light (Note 20) (Figure 11). After 1.5 h, TLC analysis indicated that the reaction is complete.

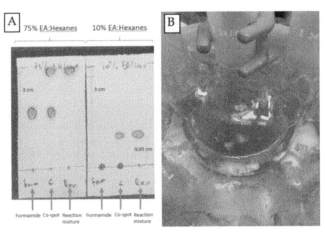

Figure 11. (A) TLC analysis of the reaction after 1.5 h; (B) The crude reaction mixture after 1.5 h (photos provided by submitters)

The light brown reaction mixture is diluted with dichloromethane (95 mL), transferred to a 2-L Erlenmeyer flask equipped with a magnetic stir bar (PTFE-coated, cylindrical, 3 cm) and then cooled in an ice bath (Figure 12a). A cold (0 °C), saturated, aqueous solution (500 mL) of NaHCO$_3$ is slowly added down the side of the flask via pipette into the stirred reaction mixture until the initially vigorous reaction subsided; the remaining NaHCO$_3$ solution is added by slowly pouring the remaining NaHCO$_3$ solution into the reaction flask (Note 21) (Figure 12b) until the solution reached a pH = 8 (Note 22) (Figure 12c).

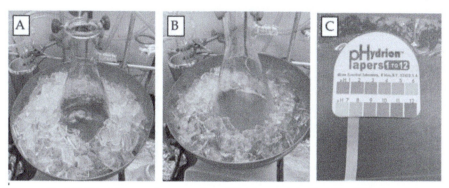

Figure 12. (A) Reaction mixture after dilution with CH_2Cl_2; (B) Reaction mixture after neutralization; (C) The pH after addition of saturated, aqueous $NaHCO_3$ (photos provided by submitters)

The crude reaction mixture is allowed to warm to room temperature (Figure 13a) and then transferred to a 1-L separatory funnel (Figure 13b). The phases are separated (Note 23), the organic layer is collected, and the aqueous phase further extracted with dichloromethane (95 mL) (Figure 13c).

Figure 13. (A) Crude reaction mixture at room temperature; (B) Reaction mixture after transfer to the separatory funnel; (C) CH_2Cl_2 extract of the aqueous phase (photos provided by submitters)

The combined organic phase is concentrated by rotary evaporation (30 °C, 20 mm Hg) to give an amber oil (Figure 14a). The crude oil is dissolved in ethyl acetate (370 mL) and then washed with brine (4 × 230 mL) (Figures 14b and c).

Figure 14. (A) Appearance of crude reaction mixture after concentration; (B) Dissolution of crude Asmic in EtOAc followed by washing with brine; (C) Crude Asmic after the fourth wash with 250 mL brine (photos provided by submitters)

The phases are separated and then the organic phase is combined (Figure 15a). The crude solution of Asmic (**2**) is dried for 10 min over anhydrous Na_2SO_4 (37 g, Figure 15b), filtered (Figure 15c) and then concentrated by rotary evaporation (30 °C, 20 mmHg). Crude Asmic is subjected to a high vacuum (20 °C, 0.1 mmHg) for 15 min, which gives a viscous amber oil in 17.1 g, 101% yield (Figure 15d, Note 24). Asmic (**2**) obtained at this stage is sufficiently pure for many applications without requiring further purification.

Figure 15. (A) The combined organic phase after extraction; (B) The dried organic phase over Na$_2$SO$_4$; (C) The dry, filtered organic phase; (D) Crude Asmic after removal of the solvent (photos provided by submitters)

Asmic (**2**) (17.1 g) is dissolved in CH$_2$Cl$_2$ (95 mL) and 21 g Celite (21 g) is added (Note 25). The solvent is removed by rotary evaporation (30 °C, 20 mmHg), followed by high vacuum (20 °C, 0.1 mmHg) for 30 min. The adsorbed material (Figure 16) is charged on a column (6 × 12 cm) of 118 g of silica gel (Note 26) pre-equilibrated with heptane (Figure 17), and the column is eluted with 2.2 L of 10% ethyl acetate in *n*-heptane (1:9 Ethyl acetate:*n*-heptane). After 450 mL of solvent elutes, 25 mL fractions are collected, and the elution continues for 56 fractions. Pure Asmic (**2**) is obtained in fractions 16-77 (Figure 18), which are concentrated by rotary evaporation (30 °C, 20 mmHg) and then under high vacuum (20 °C, 0.1 mmHg) for 15 min to afford a clear oil. The oil is stored in a –20 °C freezer for 12 h resulting in 14.2 g (85% yield) of an off-white crystalline solid (Notes 27, 28, 29, and 30) (Figure 19). Asmic (**2**) is stable for at least 9 months when stored at –20 °C.

Figure 16. Asmic (2) adsorbed on Celite (photo provided by submitters)

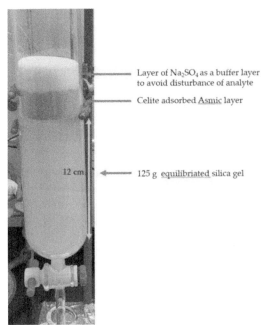

Figure 17. Silica gel column setup with adsorbed Asmic (2) (photo provided by submitters)

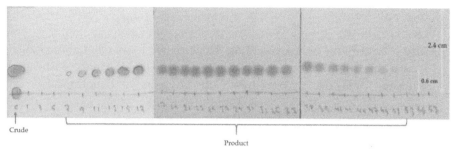

Figure 18. TLC of fractions 7–53 containing product (2) (photo provided by submitters)

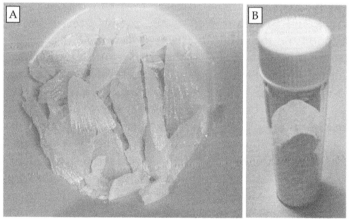

Figure 19. (A) Off-white crystalline Asmic (2) after rotary evaporation, high vacuum, and storage in a –20 °C freezer for 12 h (submitters); (B) crystalline Asmic (2) (checkers)

Notes

1. Prior to performing each reaction, a thorough hazard analysis and risk assessment should be carried out with regard to each chemical substance and experimental operation on the scale planned and in the context of the laboratory where the procedures will be carried out. Guidelines for carrying out risk assessments and for analyzing the hazards associated with chemicals can be found in references such as Chapter 4 of "Prudent Practices in the Laboratory" (The National Academies Press, Washington, D.C., 2011; the full text can be accessed

free of charge at https://www.nap.edu/catalog/12654/prudent-practices-in-the-laboratory-handling-and-management-of-chemical. See also "Identifying and Evaluating Hazards in Research Laboratories" (American Chemical Society, 2015) which is available via the associated website "Hazard Assessment in Research Laboratories" at https://www.acs.org/content/acs/en/about/governance/committees/chemicalsafety/hazard-assessment.html. In the case of this procedure, the risk assessment should include (but not necessarily be limited to) an evaluation of the potential hazards associated with paraformaldehyde, formic acid, formamide, 2-methoxybenzenethiol, N,N-diisopropylethylamine, phosphorous oxychloride, sodium hydrogen carbonate, hexane, dichloromethane, ethyl acetate, heptane, silica gel, and Celite, as well as the proper procedures for solvent evaporation and the application of high vacuum.
2. All glassware was dried in an oven (120 °C) for 1 h prior to use.
3. Paraformaldehyde (96%), formic acid (97%), formamide (high purity) and 2-methoxybenzenethiol (95%) were purchased from Acros, Alfa Aesar, VWR, and Oakwood Chemicals, respectively, and used as is without purification. Formamide and formic acid are used in excess as they serve as the solvent for the reaction. The Process Mass Intensity, PMI, is 71, considerably less than the industry standard of 200. The checkers purchased formamide (≥99%) and 2-methoxybenzenethiol (97%) from Sigma Aldrich.
4. Upon heating, the reaction mixture changed from an initial white heterogeneous suspension of paraformaldehyde (Figure 3a) to a yellow suspension (Figure 3b) and then to a clear yellow solution after 3 h at 100 °C (Figure 3c). Upon cooling to rt, a clear, very pale yellow solution of formamide **1** is formed (Figure 4d).
5. An aliquot (0.1 mL) is removed and diluted with CH_2Cl_2 (3 mL) to analyze the reaction mixture via TLC.
6. The reaction mixture may appear white and cloudy due to some precipitation of the desired formamide. If significant precipitation is observed, the solid can be filtered and the remaining solution subjected to the precipitation procedure described. The best result is obtained when the reaction mixture is diluted with dichloromethane and washed as described.
7. Portion-wise addition of hexane is recommended as rapid addition of hexane sometimes results in the formation of an oil that is unable to be precipitated.

8. The first recrystallization results in a white powder that may require iterative recrystallizations to obtain white crystalline solid formamide.
9. The material was dried for 1 h under high vacuum (0.1 mmHg) before performing the dehydration.
10. A second run by the checkers was performed at half scale and provided 12.3 g (87%) of the product.
11. N-((2-Methoxyphenylthio)methyl)formamide was obtained as an off-white, crystalline solid. The sample presented the following analytical data: mp 82–84 °C; IR (neat) cm^{-1} 3285, 3059, 2939, 2837, 1664, 1477, 1243, 751. At 25 °C, two rotamers, one major and one minor, of the formamide are observed in the NMR spectra: ^{1}H NMR (700 MHz, CDCl$_3$) δ: 3.89 (s, 1H), 3.91 (s, 4H), 4.55 (d, J = 7.0 Hz, 1H), 4.69 (d, J = 6.1 Hz, 2H), 5.94 (s, 1H), 6.04 (s, 1H), 6.95 – 6.89 (m, 3H), 7.32 – 7.27 (m, 1H), 7.35 (dd, J = 11.0, 4.5 Hz, 1H), 7.44 – 7.38 (m, 2H), 7.74 (d, J = 11.7 Hz, 1H), 8.10 (s, 1H). ^{13}C NMR (176 MHz, CDCl$_3$) δ: 41.0, 46.3, 56.0, 56.1, 111.3, 111.5, 119.1, 121.1, 121.4, 121.5, 129.7, 131.0, 133.5, 136.4, 158.7, 159.6, 160.6, 163.8. HRMS (m/z) calcd for C$_9$H$_{11}$NNaO$_2$S [M+Na]$^+$: 220.0403, found: 220.0399.
12. The precipitated formamide was used directly for the synthesis of Asmic. The submitters reported the following crystallization procedure: Formamide (24 g) is dissolved in 50 mL of hot (60 °C) isopropyl alcohol, which is then allowed to cool to room temperature (22 °C) over 12 h to afford 14.1 g (50% yield) of an off-white, crystalline solid. The checkers assessed purity before and after recrystallization by qNMR using 1,3,5-trimethoxybenzene as an internal standard. Purity was assessed as 78% prior to crystallization and 83% post-crystallization.
13. The submitters started with 20 g, (101 mmol) of *N-((2-methoxyphenylthio)methyl)formamide*, owing to their larger yield (75%) for the previous step.
14. Dichloromethane (ACS grade) was purchased from Fisher Scientific and used as is without purification.
15. *N,N*-Diisopropylethylamine (99%) was purchased from Oakwood Chemicals and used as is. Excess *N,N*-diisopropylethylamine was employed to minimize hydrolysis by maintaining a basic medium throughout.
16. Precipitation may be observed after addition of *N,N*-diisopropylethylamine, but this is not detrimental to the overall reaction.

17. The temperature is maintained below –9 °C during the course of the reaction.
18. Phosphorus oxychloride (POCl₃, 99%) was purchased from ACROS Organics and used as is without purification.
19. The addition commenced slowly at –12 °C to modulate a rapid exotherm that results in an initial, slight elevation in the reaction temperature. The temperature was maintained between –12 °C and –9 °C throughout the addition.
20. An aliquot (0.2 mL) was removed from the reaction mixture, diluted with CH₂Cl₂ (3 mL) and carefully quenched with saturated aqueous NaHCO₃ solution. The mini-workup provided a clear visualization of the reaction mixture which avoids streaking due to the excess amine.
21. The saturated aqueous NaHCO₃ was added dropwise to control a rapid exotherm that occurred upon the initial addition. Once the initial exotherm subsides, the saturated, aqueous NaHCO₃ solution can be slowly added by pouring the solution down the side of the flask. The pH was determined by testing with pH paper.
22. The pH of the mixture should not be higher than pH = 9 as this can sometimes result in the detrimental basic hydrolysis of the isocyanide to the formamide.
23. Care should be taken during the extraction; the separatory funnel should be gently shaken to avoid emulsions and the risk of a pressure build up from the evolution of CO_2.
24. The submitters reported 18.05 g (99%, based on their larger scale) of the product prior to column chromatography.
25. Celite™ 545 (CAS # 68855-54-9) was purchased from Millipore Sigma and used as is.
26. Silica gel (SiliaFlash P60, 40–63 μm (230–400 mesh), 60 Å) was purchased from Silicycle and used as is.
27. A second run on half-scale provided 7.65 g (84%) of Asmic.
28. Analytical data for Asmic (2): mp 27–29 °C; IR (neat) cm⁻¹ 3065, 2969, 2940, 2837, 2137, 1581, 1477, 1243, 1021, 749. ¹H NMR (600 MHz, CDCl₃) δ: 3.92 (s, 3H), 4.58 (s, 2H), 6.94 (d, J = 8.3 Hz, 1H), 6.99 (td, J = 7.5, 1.0 Hz, 1H), 7.38 (td, J = 8.2, 1.7 Hz, 1H), 7.53 (dd, J = 7.6, 1.6 Hz, 1H). ¹³C NMR (151 MHz, CDCl₃) δ: 42.9, 42.9, 43.0, 56.0, 111.3, 118.8, 121.5, 131.0, 134.9, 158.9 (2C, broad). HRMS (m/z) calcd for C₉H₉NNaOS [M+Na]⁺: 202.0297. Found: 202.0295.

29. Purity of Asmic (**2**) (17.9 mg) was assessed as 99.5 wt% using qNMR with 1,3,5-trimethoxybenzene (16.8 mg) as an internal standard. No recrystallization was performed.
30. The submitters report that Asmic (**2**) can be recrystallized from CCl_4 and pentane: Asmic (5.76 g) was dissolved in CCl_4 (8 mL) at rt and then was diluted with small volume of pentane and then cooled, initially to 0 °C and then to –20 °C. Additional pentane was added to initiate crystallization and the solution left at –20 °C for 3 days. The crystals were filtered, washed three times 5 mL portions of CCl_4/pentane (1:5), and then dried under high vacuum at rt to afford 4.04 g (73%) of pure Asmic as a colorless crystalline solid. Concentration and recrystallization of the filtrate afforded additional pure Asmic (**2**).

Working with Hazardous Chemicals

The procedures in *Organic Syntheses* are intended for use only by persons with proper training in experimental organic chemistry. All hazardous materials should be handled using the standard procedures for work with chemicals described in references such as "Prudent Practices in the Laboratory" (The National Academies Press, Washington, D.C., 2011; the full text can be accessed free of charge at http://www.nap.edu/catalog.php?record_id=12654). All chemical waste should be disposed of in accordance with local regulations. For general guidelines for the management of chemical waste, see Chapter 8 of Prudent Practices.

In some articles in *Organic Syntheses*, chemical-specific hazards are highlighted in red "Caution Notes" within a procedure. It is important to recognize that the absence of a caution note does not imply that no significant hazards are associated with the chemicals involved in that procedure. Prior to performing a reaction, a thorough risk assessment should be carried out that includes a review of the potential hazards associated with each chemical and experimental operation on the scale that is planned for the procedure. Guidelines for carrying out a risk assessment and for analyzing the hazards associated with chemicals can be found in Chapter 4 of Prudent Practices.

The procedures described in *Organic Syntheses* are provided as published and are conducted at one's own risk. *Organic Syntheses, Inc.*, its Editors, and its Board of Directors do not warrant or guarantee the safety of individuals using these procedures and hereby disclaim any liability for any injuries or damages claimed to have resulted from or related in any way to the procedures herein.

Discussion

Isocyanides are extremely important precursors for a diverse suit of bond forming reactions: multi-component reactions,[2] transition metal insertions,[3] and radical reactions.[4] The promiscuous reactivity, driven by conversion of the terminal, divalent carbon to a more stable tetravalent state, obscures the limited availability of isocyanides; only 24 isocyanides are commercially available at prices less than $50/g of which almost half are derivatives of Tosmic.[5] Structurally complex isocyanides therefore require assembly from simpler precursors, most often via sequential formylation and dehydration of primary amines.[6]

Asmic (**2**), Anisylsulfanylmethylisocyanide,[7] fills the void by providing an efficient isocyanide building block that allows rapid access to structurally diverse substituted isocyanides[8] and heterocycles.[9] Formed by a sequential Mannich formylation-dehydration sequence, Asmic is a bench-stable, crystalline solid that stores well with minimal odor. Asmic is readily deprotonated by a variety of bases (NaH, LDA, BuLi) to afford a nucleophilic organometallic that efficiently intercepts electrophiles (Scheme 1, **2** → **3**). The deprotonation-alkylation can be performed in separate steps or efficiently telescoped into a single synthetic operation to convert Asmic into dialkylAsmic **3**; the telescoped alkylation with α, ω -dihalides affords cyclic dialkylAsmic derivatives **3**.

Treatment of dialkylAsmic **3** with BuLi triggers the rapid formation of a highly nucleophilic organolithium whose trapping with a variety of electrophiles efficiently generates tri-substituted isocyanides **4** (Scheme 1). The conversion of disubstituted Asmic derivatives **3** to reactive intermediates is extremely fast; the reaction with BuLi is essentially complete within 5 min at –78 °C. Subsequent trapping with an array of electrophiles ranging from alkyl halides to carbonyls to heteroatom species affords the corresponding trisubstituted isocyanide (Scheme 1 **4a** – **4b**, **4e** – **4k**, and **4n**, respectively); ketones react to afford oxazolines formed by attack of the intermediate alkoxide on the electrophilic isocyanide (Scheme 1, **4l** and **4m**). Disubstituted isocyanides are accessed through two sequential alkylations of Asmic followed by the addition of BuLi and NH₄Cl (see **4c** and **4d**, Scheme 1). The Asmic-based syntheses allows an efficient route to substituted isocyanides with functionality that is otherwise difficult to access, such as the ketoisocyanides **4g** and **4h**.

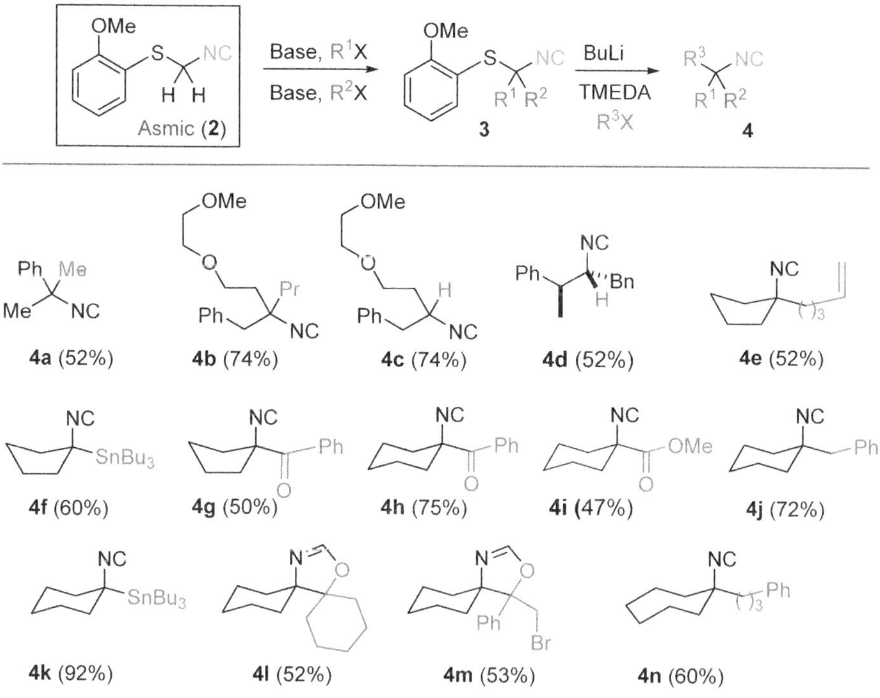

Scheme 1. Asmic-Based Synthesis of Di- and Tri-Substituted Isocyanides

DialkylAsmic derivatives **3** provide an efficient route to quaternary nitriles **6** (Scheme 2).[10] Omitting TMEDA during the addition of BuLi triggers a facile sulfur-lithium exchange-isomerization to afford the corresponding lithiated nitrile **5**, excellent nucleophiles that efficiently intercept a range of electrophiles.[11] Trapping the lithiated nitriles **5** with carbon electrophiles installs quaternary centers. The BuLi-initiated isocyanide-nitrile isomerization is equally efficient in forming cyclic nitriles (**6a** and **6b**) as for acyclic nitriles (**6c** and **6d**).

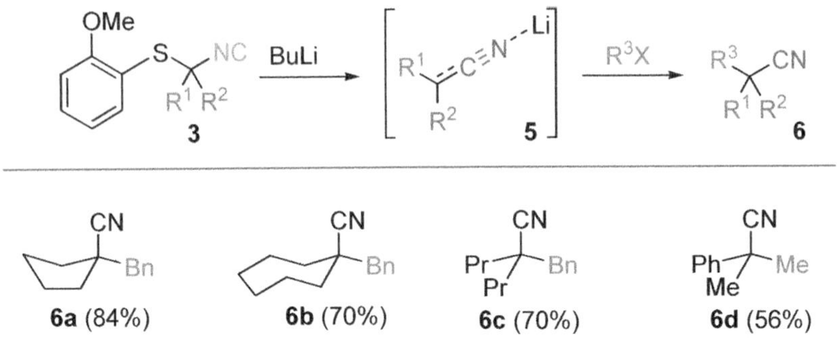

Scheme 2. Asmic-Based Synthesis of Quaternary Nitriles

Asmic is a versatile isocyanide building block whose sequential alkylations afford di- and tri-substituted isocyanides. The sequenced double alkylation, BuLi-exchange-alkylation provides a rapid route to isocyanides or nitriles that are otherwise difficult to access. The Asmic-based isocyanide synthesis is ideally suited to accessing homologous isocyanides for structure activity assays because the strategy is modular in nature and highly efficient.

References

1. Fraser Fleming, Department of Chemistry, Drexel University, 32 South 32nd St., Philadelphia, PA 19104. ORCID 0002-9637-0246. Financial support from NSF (1953128) is gratefully acknowledged.
2. Dömling, A. *Chem. Rev.* **2006**, *106*, 17–89.
3. (a) Qiu, G.; Ding, Q.; Wu J. *Chem. Soc. Rev.* 2013, 42, 5257–5269. (b) Chakrabarty, S.; Choudhary, S.; Doshi, A.; Liu, F.-Q.; Mohan, R.; Ravindra, M. P.; Shah, D. Yang, X.; Fleming, F. F. *Adv. Synth. Cat.* 2014, 356, 2135–2196.
4. Minozzi, M.; Nanni, D.; Spagnolo, P. *Curr. Org. Chem.* **2007**, *11*, 1366–1384.
5. Determined by SciFinder structure searches (11/25/2020).
6. Suginome, M.; Ito, Y. in *Science of Synthesis*, 2012, ch. 7, pp. 445–531.
7. Alwedi, E.; Lujan-Montelongo, J. A.; Pitta, B. R.; Chao, A.; Cortés-Mejía, R.; del Campo, J. M.; Fleming, F. F. *Org. Lett.* **2018**, *20*, 5910–5913.
8. Fleming, F. F.; Altundas, B. eEROS, 2020.

9. Kumar, C. V. S.; Holyoke, C. W. Jr., Keller, T.; Fleming, F. F. *J. Org. Chem.* **2020**, *85*, 9153–9160.
10. Alwedi, E.; Lujan-Montelongo, J. A.; Cortés-Mejía, R.; del Campo, J. M.; Altundas, B.; Fleming, F. F. *Eur. J. Org. Chem.* **2019**, 4644–4648.
11. Yang, X.; Fleming, F. F. *Acc. Chem. Res.* **2017**, *50*, 2556–2568.

Appendix
Chemical Abstracts Nomenclature (Registry Number)

N-((2-Methoxyphenylthio)methyl)formamide: Formamide, *N*-[[(2-methoxyphenyl)thio]methyl]-; (118617-57-5)
Asmic, anisylsulfanylmethylisocyanide: Benzene, 1-[(isocyanomethyl)thio]-2-methoxy-; (1803329-89-6)
Paraformaldehyde; (30525-89-4)
Formamide; (75-12-7)
Methoxybenzenethiol; (7217-59-6)
Diisopropylethylamine: *N*-Ethyl-*N*-(propan-2-yl)propan-2-amine; (7087-68-5)

Fraser Fleming obtained a BS (Hons.) at Massey University, New Zealand, in 1986 and a Ph. D. in 1990 from the University of British Columbia, Canada under the direction of Edward Piers. He completed postdoctoral research with James D. White at Oregon State University before taking his first faculty position at Duquesne University, Pittsburgh in 1992. In 2013 he began a two-year appointment as a Program Director at the National Science Foundation working in the Synthesis and the Catalysis Programs. In 2015 he moved to Drexel University where his research is focused on the reactions of isocyanides and nitriles.

Embarek Alwedi received his B.S in 2001 and M.S in 2006 from Alfatah University, Libya. He remained at his alma mater for a further two years working as a lecturer before relocating to the US to pursue Ph.D. with Prof. Paul Blakemore at Oregon State University. He completed his postdoctoral studies under the supervision of Prof. Fraser Fleming before moving to Merck, Rahway where he is working as research chemist in process R&D. His research interests lie in the development of new synthetic methodology and processes.

Bilal Altundas obtained his BS with honors in chemistry from Middle East Technical University in Ankara, Turkey in 2013 before moving to Miami University to pursue an MS in organic chemistry under the direction of Professor Hong Wang. Upon completing his MS in 2016, he started his Ph. D. under the guidance of Professor Fraser F. Fleming in the Department of Chemistry at Drexel University. His research interests are in synthetic methodology as applied to the reactions of ketenimines and in uncovering new reactions of metalated isocyanides.

Allen Chao obtained a B.S. in chemistry from the University of Pittsburgh in 2007 followed by industrial experience before embarking on a Ph. D. He began his graduate career at Duquesne University, then moved with Prof. Fleming to complete his Ph. D. at Drexel University in 2018. He is currently working as a post-doctoral fellow in the Wistar Institute, Philadelphia. His research interests include development of PROTACs as therapeutic agents in treating cancer. He is invested in helping bridge the communication gap between the chemistry and biology disciplines.

Zachary L. Ziminsky began his undergraduate career at Drexel University. He worked on the large-scale synthesis of Asmic as a summer research student with Prof. Fleming.

Maanasa Natrajan completed her high school at Jnanodaya school, Bengaluru in India. In 2016, she started her undergraduate at Drexel University where she is currently a senior pursuing an individualized course of interdisciplinary study drawing upon neuroscience, chemistry, biophysics, and computation. She started research in organic chemistry with Prof. Fleming in her first year and then continued as a research assistant. She has performed systems biology research at Thomas Jefferson University, bioinformatics and phage biology at the Genome Institute of Singapore, and theoretical and computational neuroscience at Howard Hughes Medical Institute, Janelia. She plans to pursue a Ph. D. in neuroscience.

Martina Drescher is the lead technician of the Maulide group at the University of Vienna, where she has worked with several group leaders over the course of 38 years.

Anthony J. Fernandes received his Ph.D. in 2018 at the University of Bordeaux, under the supervision of Prof. Yannick Landais and Dr. Frédéric Robert. He moved to the University of Strasbourg and the University of Poitiers, starting a post-doctoral position within a collaboration with Dr. Frédéric Leroux and Prof. Sébastien Thibaudeau. He then moved to the University of Vienna, where he is currently working as a postdoctoral researcher in the group of Prof. Nuno Maulide. Research interests involve highly reactive intermediates, rearrangements and the synthesis of biologically active compounds.

Mild *mono*-Acylation of 4,5-Diiodoimidazole: Preparation of 1-(5-Iodo-1*H*-imidazole-4-yl)pent-4-en-1-one

Michael Morgen,[1] Jasmin Lohbeck,[1] and Aubry K. Miller[1]*

Cancer Drug Development Group, German Cancer Research Center (DKFZ), 69126 Heidelberg, Germany

Checked by Anji Chen, Hari P. R. Mangunuru, Nathaniel D. Kaetzel, Gopal Sirasani, and Chris Senanayake

Procedure (Note 1)

A. *N-Methoxy-N-methylpent-4-enamide* (**1**). A 1 L round-bottomed flask equipped with a 3.5 cm Teflon-coated barbell-shaped stir bar is loaded with 4-pentenoic acid (25.5 mL, 249.9 mmol, 1.0 equiv) (Note 2) and CH$_2$Cl$_2$ (500 mL) (Note 3). The resulting solution is magnetically stirred (500 rpm) and oxalyl chloride (22.2 mL, 262.3 mmol, 1.05 equiv) (Note 4) is added dropwise at 24 °C via a 50 mL dropping funnel that is capped with a rubber septum (Figure 1A). The addition is complete within 5 min, at which time the dropping funnel is removed and DMF (0.1 mL) (Note 5) is added to the reaction mixture in one portion with a syringe. The solution begins to vigorously bubble.

The reaction flask is quickly connected in series, first to an empty wash bottle and then to a wash bottle filled with a 5 M NaOH solution (200 mL), which itself is vented to the back of the hood (Figure 1B) (Note 6). The colorless reaction mixture is stirred at 24 °C for 2 h, during which bubbling ceases.

Figure 1. (A) Reaction setup consisting of a 1 L flask equipped with a stir bar that is further connected to a dropping funnel – wash bottles are arranged to be quickly connected to the reaction flask. Oxalyl chloride is slowly dropping into a stirring solution of 4-pentenoic acid; (B) Reaction mixture after addition of DMF. The reaction flask is now connected to a series of wash bottles. (Photos provided by submitters)

The reaction mixture is concentrated on a rotary evaporator (40 °C, 375 then 110 mmHg) leaving the stir bar inside the flask (Note 7). The remaining orange oily residue is dissolved in CH$_2$Cl$_2$ (250 mL) and again concentrated on a rotary evaporator (40 °C, 375 then 110 mmHg) (Note 8). The residue is then dissolved in CH$_2$Cl$_2$ (400 mL), cooled to 0 °C in an ice bath, and N,O-dimethylhydroxylamine hydrochloride (24.35 g, 249.7 mmol, 1.0 equiv) (Note 9) is added in one portion to make a suspension. The flask is equipped with a 100 mL dropping funnel filled with triethylamine

(72.7 mL, 524.4 mmol, 2.10 equiv) (Note 10). Triethylamine is added dropwise within 5 to 10 min to the magnetically stirred (500 rpm) reaction mixture at 0 °C (Figure 2A).

After the addition is complete, the reaction mixture is stirred at 0 °C for 2 h and a white suspension forms. The suspension is poured into a 1 L separatory funnel containing a saturated solution of $NaHCO_3$ (400 mL) at 24 °C (Figure 2B). The reaction flask is rinsed with CH_2Cl_2 (2 × 25 mL) and the resulting solutions are added to the separatory funnel. After shaking, the phases are separated and the yellow-orange organic layer is further washed with water (2 × 300 mL) (Note 11) and a saturated aqueous NaCl solution (300 mL), dried over $MgSO_4$ (50 g) (Note 12), filtered through fluted filter paper in a glass funnel (Note 13), and concentrated on a rotary evaporator (40 °C, 500 then 15 mmHg). The residual orange-colored oil (contaminated with solid $Et_3N•HCl$) is transferred into a 100 mL round-bottomed flask (NS24/40) equipped with a 2.5 cm long oval Teflon coated stir bar (Note 14). This is connected to a short distillation bridge (Liebig condenser) equipped with a thermometer and a distillation cow bearing four 100 mL flasks. The distillation flask is submerged into an oil bath on a magnetic hot plate and is covered with aluminum foil from above the oil level up to the thermometer (Figure 2C). The apparatus is connected via high vacuum-grade silicone tubing to a Schlenk-line and evacuated. By using a needle valve, a pressure of 7–10 mmHg is maintained. After waiting 2–3 min until residual solvent is

Figure 2. (A) Reaction mixture while adding Et_3N – a white suspension forms during the addition; (B) Washing of the organic phase with saturated aqueous NaCl; (C) Distillation of the product using a Liebig condenser equipped with a thermometer and a distillation cow bearing four flasks; (D) Appearance of pure 1. (Photos provided by submitters)

removed (indicated by ceased bubble formation) the oil bath temperature is gradually increased from 24 °C to 95 °C. The pure product distills at a temperature of 85 °C (oil bath temperature: 95 °C) and a pressure of 7–10 mmHg (Note 15). Weinreb amide **1** is obtained as a colorless oil in a yield of 69% (24.69 g, 172.4 mmol) (Figure 2D) (Note 16 and 17).

B. *4,5-Diiodo-1H-imidazole (2)*. A 500 mL Erlenmeyer flask equipped with a 4.0 cm long Teflon-coated flat stir bar is charged with potassium iodide (97.53 g, 587.6 mmol, 4.0 equiv) (Note 18). Water (300 mL) (Note 11) at 24 °C is then added to make a colorless solution. Iodine (74.56 g, 293.8 mmol, 2.0 equiv) (Note 19) is then added and the resulting mixture is magnetically stirred (700 rpm) until complete dissolution of iodine (~15–20 min) (Note 20) provides a dark brown solution.

In parallel, a 1 L round-bottomed flask equipped with a 3.2 cm long oval Teflon-coated stir bar is charged with a 2 M NaOH solution (NaOH (35.25 g, 881.3 mmol, 6.0 equiv) in H_2O (440 mL)) (Note 21). To this is added imidazole (10.0 g, 146.9 mmol, 1.0 equiv) (Note 22) at 24 °C to make a colorless solution. This flask is connected to a 500 mL dropping funnel capped with a rubber septum, which is filled with the before mentioned KI/I_2 solution (Figure 3) (Note 23).

Figure 3. Reaction setup before the addition of the KI/I_2 solution. (Photo provided by submitters)

The KI/I₂ solution is added dropwise to the imidazole solution at 24 °C within 60–90 min. Upon addition, a brownish color with concomitant formation of a white precipitate is observed, both of which rapidly disappear (Figure 4A). After complete addition, the resulting yellow solution (Figure 4B) is stirred at 24 °C for 3 h, during which the solution becomes pale-yellow to colorless (Figure 4C). The reaction mixture is cooled to 0 °C using an ice bath and acidified. This is done by adding concentrated HCl (~48–49 mL, ~4 equiv) (Note 24) portion wise (5 × 9 mL) and then dropwise to the reaction mixture, during which a number of visual changes are observed. After each of the five 9 mL additions, a dark red/brown precipitate forms, which slowly re-dissolves with vigorous stirring (Figure 4D). The solution remains essentially colorless. As the endpoint nears, dropwise addition of HCl produces a white precipitate and a yellow/orange color in the solution, both of which disappear with stirring. Closer to the endpoint, the white precipitate persists (Figure 4E). The endpoint can be determined by color change when a yellow/orange color of the solution no longer returns to colorless. Temporarily discontinuing stirring can be helpful for making this determination. The resulting mixture is pH 8, which can be checked by pH-indicator paper. The resulting white solid is collected by vacuum filtration using a round filter paper (ø = 9 cm) in a Büchner funnel and washed with cold (4 °C) water (2 × 250 mL). The solid material is air dried in the funnel for 1 h with house vacuum (Figure 4F) and transferred into a 1 L round-bottomed flask, using EtOH (Note 25) to quantitate the transfer. The resulting suspension is heated to reflux on a boiling water bath (Note 26) with EtOH (350 mL) to fully dissolve the solids (Note 27). The orange-colored solution is allowed to cool to 24 °C on the benchtop (Figure 4G) and afterwards in the freezer at –20 °C for 18 h. The resulting solid is collected by vacuum filtration on filter paper (ø = 9 cm) in a Büchner funnel and washed with cold (–20 °C) ethanol (2 × 50 mL) and air dried on the filter for 1 h using house vacuum. The product is obtained as an off-white solid. The mother liquor is concentrated to dryness on a rotary evaporator (40 °C, 35 mmHg), suspended in 250 mL of EtOH and recrystallized as described above to obtain a second crop of **2**. The two crops are combined, finely mortared, transferred to a 250 mL round-bottomed flask (NS 24/40), and dried under high vacuum (24 °C, 2 × 10^{-2} mmHg) for 18 h to give 4,5-diiodoimidazole (**2**) as an off-white solid in a yield of 59% (27.90 g, 87.22 mmol) (Figure 4H) (Note 28 and 29).

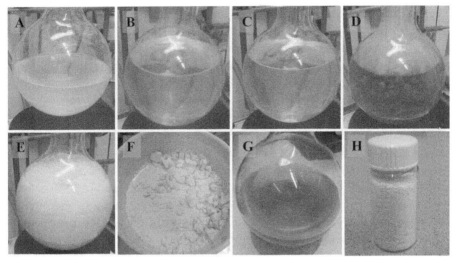

Figure 4. (A) Reaction mixture while adding the KI/I$_2$ solution; (B) Immediately after complete addition of the KI/I$_2$ solution; (C) After stirring at 24 °C for 3 h; (D) While acidifying with conc. HCl; (E) After adjusting to pH 8 and continued stirring for 2 min; (F) Material obtained after vacuum filtration of the precipitated product; (G) Dissolved in hot ethanol for recrystallization; (H) Final product 2. (Photos provided by submitters)

C. *1-(5-Iodo-1H-imidazole-4-yl)pent-4-en-1-one (3)*. An oven-dried 1 L three-necked, round-bottomed flask (3 × NS 24/40) is filled with anhydrous THF (500 mL) (Note 30 and 31), equipped with two glass stoppers, and connected to a Schlenk line with a glass adapter. Under positive nitrogen pressure, the middle glass stopper is exchanged with a glass rod equipped with a 5 cm long oval-shaped Teflon vane that is connected to an overhead stirrer (Figure 5A). Via the remaining side neck, lithium chloride (3.31 g, 78.2 mmol, 1.0 equiv) (Note 32) is added and the mixture is stirred (160 rpm) (Note 33) at 24 °C under an atmosphere of nitrogen until the lithium chloride is fully dissolved (~18 h). Then finely crushed 4,5-diiodoimidazole **2** (25.0 g, 78.2 mmol, 1.0 equiv) (Note 34) is added. The resulting yellow solution is cooled to 0 °C using an ice bath, whereupon a white precipitate forms (Note 35). Then the glass stopper is replaced with a silicone septum and methylmagnesium chloride (3 M solution in THF; 27.4 mL, 82.1 mmol, 1.05 equiv) (Note 36 and 37) is added dropwise with a syringe to the reaction mixture within 5–10 min

(Note 38). During addition of methylmagensium chloride the reaction mixture clarifies, and the resulting orange solution is stirred (160 rpm) at 0 °C. After 5 min the ice bath is removed and stirring is continued for another 5 min (Note 39) (Figure 5D). Then isopropylmagnesium chloride lithium chloride complex (1.3 M solution in THF; 66.1 mL, 86.0 mmol, 1.10 equiv) (Note 40 and 41) is added with a syringe to the reaction mixture within 10–15 min. The resulting off-white suspension is stirred (160 rpm) (Note 33) at 24 °C for 2.0–2.5 h until the iodine/magnesium exchange is considered complete by TLC monitoring (Note 42) (Figure 5E). Afterwards, neat Weinreb amide **1** (12.31 g, 86.0 mmol, 1.10 equiv) is added to the suspension dropwise with a syringe within 2–3 min (Note 43) and stirring (160 rpm) is continued for 2.0–2.5 h at 24 °C, during which the reaction mixture clarifies again to an orange colored solution (Figure 5F). The reaction is monitored by TLC (Note 44). After completion, the reaction mixture is transferred into a 2 L separatory funnel and quenched at 24 °C by adding a saturated aqueous solution of NH_4Cl (400 mL). The reaction flask is rinsed with saturated aqueous NH_4Cl (25 mL) and EtOAc (25 mL) and the resulting mixture is added to the separatory funnel. The phases are separated and the aqueous layer is extracted with EtOAc (2 × 200 mL) (Note 45). The combined organic phases are washed with brine (400 mL), dried with $MgSO_4$ (50 g) (Note 12), filtered through a fluted filter paper in a glass funnel (Note 46), and concentrated on a rotary evaporator (40 °C, 35 mmHg). The obtained off-white solid is crystallized as follows: on a gently boiling water bath, the solid is dissolved in 100 mL of EtOAc (Note 26). To the boiling hot solution is added in a portionwise fashion enough hexanes (~50 mL) (Note 47) so that the white precipitate that forms upon addition slowly disappears. The flask is then covered with a plastic yellow cap and allowed to cool to 24 °C on the bench and then at –20 °C in a freezer for 18 h. The resulting solid is collected by vacuum filtration through a round filter paper (ø = 9 cm) in a Büchner funnel, successively washed with a 1:1-mixture of cold (4 °C) EtOAc/hexanes (100 mL), cold (4 °C) hexanes (100 mL), and afterwards air dried with suction for 30 min (Note 48). The material is further dried under high vacuum (24 °C, 2 × 10^{-2} mmHg) for 18 h. The desired product **3** is obtained in a yield of 78% (16.91 g, 61.25 mmol) (Figure 5G) (Note 49 and 29).

Figure 5. (A) Reaction apparatus consisting of a 1 L three-necked flask, which is connected to a Schlenk line (positive nitrogen pressure) via one side neck, to an overhear stirrer via the middle neck, with the remaining side neck equipped with a glass stopper; (B) Appearance of the reaction mixture containing LiCl in THF immediately after adding 4,5-diiodoimidazole (2); (C) Upon cooling to 0 °C in an ice bath – an off-white suspension forms; (D) Appearance of the reaction mixture after complete addition of MeMgCl – the suspension clarifies; (E) Appearance of the reaction mixture upon addition of *i*-PrMgCl•LiCl – again a white suspension is obtained; (F) Appearance of the reaction mixture upon complete addition of the Weinreb amide 1 – an orange-colored solution is obtained; (G) Appearance of the pure product 3 after work-up, recrystallization from hot EtOAc and hexanes and drying under high vacuum. (Photos provided by submitters)

Notes

1. Prior to performing each reaction, a thorough hazard analysis and risk assessment should be carried out with regard to each chemical substance and experimental operation on the scale planned and in the context of the laboratory where the procedures will be carried out. Guidelines for carrying out risk assessments and for analyzing the hazards associated with chemicals can be found in references such as Chapter 4 of "Prudent Practices in the Laboratory" (The National Academies Press, Washington, D.C., 2011; the full text can be accessed free of charge at https://www.nap.edu/catalog/12654/prudent-practices-in-the-laboratory-handling-and-management-of-chemical. See also "Identifying and Evaluating Hazards in Research Laboratories" (American Chemical Society, 2015) which is available via the associated website "Hazard Assessment in Research Laboratories" at https://www.acs.org/content/acs/en/about/governance/committees/chemicalsafety/hazard-assessment.html. In the case of this procedure, the risk assessment should include (but not necessarily be limited to) an evaluation of the potential hazards associated with 4-pentenoic acid, dichloromethane, oxalyl chloride, dimethylformamide, N,O-dimethylhydroxylamine hydrochloride, triethylamine, potassium iodide, iodine, imidazole, sodium hydroxide, hydrochloric acid, ethanol, tetrahydrofuran, lithium chloride, methylmagnesium chloride, isopropylmagnesium chloride, ethyl acetate and hexanes, as well as the proper procedures for vacuum distillations.
2. 4-Pentenoic acid (97%) was obtained from Sigma-Aldrich and was used as received.
3. Dichloromethane (≥99.8%, stabilized with 40–150 ppm amylene, anhydrous) was obtained from Sigma-Aldrich and was used as received.
4. Oxalyl chloride (≥99%) was obtained from Sigma-Aldrich and was used as received.
5. Dimethyl formamide (99.8%, anhydrous) was obtained from Sigma-Aldrich and was used as received.
6. This reduces the copious amount of evolving HCl from being released into the hood. CO_2 and CO are also produced as gaseous byproducts in this reaction.

7. The product (4-pentenoyl chloride) has a reported boiling point of 125 °C. Care must be taken not to co-distill the product on the rotary evaporator. The product also has an unpleasant odor and it is advisable to use a rotary evaporator that is housed in a ventilated hood, and transport the flask through the lab only while capped.
8. Redissolving the residue and subsequent concentration ensures the removal of any unreacted oxalyl chloride and residual HCl.
9. N,O-Dimethylhydroxylamine hydrochloride (98%) was obtained from Sigma-Aldrich and was used as received.
10. Triethylamine (≥99.5%) was obtained from Sigma-Aldrich and was used as received.
11. Deionized water was used from in-house supply.
12. Magnesium sulfate (>99.5%, anhydrous) was obtained from Sigma Aldrich and was used as received.
13. The remaining $MgSO_4$ was washed once with 20 mL of CH_2Cl_2.
14. The reaction flask is rinsed with CH_2Cl_2 (2 × 2 mL) to quantitate the transfer. The residual solvent was removed on a rotary evaporator (40 °C, 15 mmHg) prior to distillation.
15. A small forerun of a few drops was taken before collecting the pure product **1**.
16. A second run on the same scale provided 24.37 g (68%) of the product. The following analytical data were obtained for **1**: R_f = 0.5 (EtOAc/hexanes 1:1); bp 85 °C (7–10 mmHg); ^{1}H NMR (600 MHz, CDCl$_3$) δ: 2.35 – 2.41 (m, 2H), 2.48 – 2.55 (m, 2H), 3.17 (s, 3H), 3.67 (s, 3H), 4.98 (ddt, J = 10.2, 2.0, 1.4 Hz, 1H), 5.06 (dq, J = 17.1, 1.6 Hz, 1H), 5.85 (ddt, J = 16.8, 10.2, 6.5 Hz, 1H). ^{13}C NMR (150 MHz, CDCl$_3$) δ: 28.5, 31.2, 32.1, 61.2, 115.2, 137.5, 173.9. LCMS–ESI (*m/z*): [M + H]$^+$ calcd for $C_7H_{14}NO_2^+$: 144; Found: 144. The purity was determined to be 98.6% by qNMR spectroscopy in CDCl$_3$ using 33.3 mg of compound (**1**) and 5.3 mg of dimethyl fumarate (99%) as an internal standard.
17. After one month storage at 4 °C in a refrigerator, the submitters observed no decomposition of **1**.
18. Potassium iodide (99%, ACS reagent) was obtained from Sigma-Aldrich and was used as received.
19. Iodine (99.8%, ACS grade) was obtained from VWR and was used as received.

20. Occasionally the beaker was placed into an ultrasonication bath and small pieces of iodine had to be broken up with a spatula to facilitate dissolution.
21. NaOH (97%, pellets) was obtained from Oakwood Chemical and was used as received.
22. Imidazole (99%) was obtained from Sigma-Aldrich and was used as received.
23. Potassium iodide was used in this reaction to increase the solubility of iodine in water by forming potassium triiodide (KI_3).
24. Hydrochloric acid (36.5–38%, ACS grade) was obtained from Oakwood Chemical and was used as received.
25. Ethanol absolute (200 proof) was obtained from KOPTEC (analytical reagent grade) and was used as received.
26. A wooden stick was placed in the flask to prevent bumping.
27. In the case that small residual particles did not easily go into solution the solid was gently crushed with a spatula and the flask was put into an ultrasonic bath for half a minute.
28. A second run on the same scale provided 27.73 g (59%) of the product. The following analytical data were obtained for **2**: R_f = 0.46 (CH_2Cl_2/MeOH 9:1 + 0.5% NH_4OH). mp (uncorrected): 201–202 °C (provided by submitters). ^{1}H NMR (600 MHz, d_6-DMSO) δ: 7.78 (s, 1H), 12.91 (br s, 1H). ^{13}C NMR (150 MHz, d_6-DMSO) δ: 78.5, 95.8, 141.6. LCMS–ESI (m/z): $[M + H]^+$ calcd for $C_3H_3I_2N_2^+$: 321, Found: 321. The purity was determined to be 99.6% by qNMR spectroscopy in d_6-DMSO using 29.2 mg of compound (**2**) and 13.3 mg of dimethyl fumarate (99%) as an internal standard.
29. The compound was stored at 4 °C in the dark (with aluminum foil covering the storage bottle).
30. Tetrahydrofuran (THF, anhydrous, ≥99.9%, inhibitor-free) was obtained from Sigma-Aldrich.
31. The submitters obtained tetrahydrofuran (THF, Chromosolv Plus for HPLC, inhibitor-free, ≥99.9%) from Honeywell Riedel-de Haën and dispensed it from a solvent purification system (MB SPS-800, MBraun) under an atmosphere of nitrogen. The submitters proceeded as follows to obtain the exact amount of solvent: The desired reaction flask was filled with 500 mL of THF, which had been accurately measured with a graduated cylinder. The solvent level in the flask was labelled with a permanent marker. This filling mark was then used to measure the

correct amount of solvent (after oven drying the flask) from the solvent purification system.
32. Lithium chloride was obtained from Sigma-Aldrich (≥99%, ACS reagent) and was used as received.
33. The checkers used a IKA EUROSTAR 60 digital overhead stirrer, which was set on 160 rpm during the reaction mixture stirring.
34. 4,5-Diiodoimidazole (**2**) is now commercially available from a number of suppliers. An attempt to use commercial **2** (ABCR, Karlsruhe, Germany) in this reaction resulted in no conversion. Recrystallization of purchased **2** as described above resulted in material that performed as well as the "home-made" **2**. The submitters hypothesize the commercial material was wet.
35. Solutions of 4,5-diiodoimidazole **2** and LiCl in dry THF, regardless of temperature or concentration, form thick white precipitates within minutes (see discussion section). At larger reaction scales, the mixture cannot be stirred magnetically, and overhead stirring is required.
36. Methylmagnesium chloride (3 M solution in THF) was obtained from Sigma-Aldrich and was used as received.
37. The checkers used a 50 mL plastic syringe with a 4 inch long, gauge 22 needle to measure the amount of methylmagnesium chloride solution used.
38. Significant bubbling was observed, resulting from methane which was formed by deprotonation of 4,5-diiodoimidazole (**2**) with methylmagnesium chloride.
39. In case of remaining material sticking on the wall of the flask, the stirring rate was temporarily increased in order to wash that material down into the reaction mixture.
40. Isopropylmagnesium chloride lithium chloride complex solution (1.3 M in THF) was obtained from Sigma-Aldrich and was used as received.
41. The checker used two 50 mL plastic syringe with a 4 inches long gauge 22 needle to measure the amount of isopropylmagnesium chloride lithium chloride solution used.
42. For reaction monitoring thin-layer chromatography with silica gel 60 F_{254}-plates from Merck, Darmstadt, Germany and a mixture of CH_2Cl_2/MeOH 9:1 (+0.5% NH_4OH) as eluent was used (Figure 6). A mini-work-up of the reaction mixture was performed by removing the septum, pulling ~0.1 mL of reaction mixture into a glass pipette, and quenching this into ~1.0 mL of saturated aqueous ammonium chloride. EtOAc (~0.2 mL) was added, and after agitation and separation of the

two layers, the top organic layer was used for TLC spotting. During this work-up, the desired Grignard intermediate is converted to 4-iodoimidazole, which is what is observed as "product" on the TLC plate, showing an R_f value of 0.40 in CH_2Cl_2/MeOH 9:1 (+0.5% NH_4OH) under UV light and with iodine-staining. The starting material is visible under UV light and shows an R_f value of 0.46 in CH_2Cl_2/MeOH 9:1 (+0.5% NH_4OH). The magnesium/iodine exchange reaction does not always go to completion, but the next step can be conducted with an observed outcome as shown below. The submitters found that prolonged stirring or a larger excess of i-PrMgCl•LiCl led to undesired byproducts (see discussion section).

Figure 6. Reaction monitoring of the magnesium/halogen exchange reaction of 4,5-diiodoimidazole (2) with i-PrMgCl•LiCl (left lane: 4,5-diiodomidazole (2); right lane: reaction mixture quenched with conc. NH_4Cl solution and extracted with EtOAc; middle lane: co-spot). (Photo provided by submitters)

43. The checkers weighed **1** directly into a tared 12 mL plastic syringe with a 4 inches long gauge 22 needle. After complete addition of **1**, the transfer was quantitated by two times pulling 1 mL of reaction mixture into the syringe and dispensing it back into the reaction mixture.

44. For reaction monitoring thin-layer chromatography with silica gel 60 F_{254}-plates from Merck, Darmstadt, Germany and a mixture of CH_2Cl_2/MeOH 9:1 (+0.5% NH_4OH) as eluent was used (Figure 7). A mini work-up was performed as in Note 42. This was spotted in comparison to the previous quench (4-iodoimidazole) as well as Weinreb amide **1**. The desired product was visible under UV light and stained with iodine showing an R_f value of 0.50.

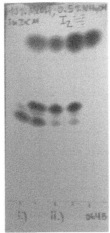

Figure 7. Reaction monitoring of the Grignard reaction with Weinreb amide 1 using iodine staining (first lane from left: "4-iodoimidazole" (see above); third lane: reaction mixture; fifth lane: Weinreb amide 1; second and fourth lanes: co-spots). Traces of starting material (2) show-up after long staining time; the reaction can be worked-up with an observed outcome as shown above. (Photo provided by submitters)

45. Ethyl acetate was obtained from Oakwood Chemical (99.9%, ACS grade) and was used as received.
46. The remaining $MgSO_4$ was washed once with 20 mL of ethyl acetate.
47. Hexanes was obtained from Oakwood Chemical (45% *n*-hexane ACS grade) and was used as received.
48. A second and third crop of material was obtained by concentrating the mother liquor and recrystallizing the residue from a smaller volume of ethyl acetate and hexanes as mentioned above.
49. A second run on the same scale provided 16.77 g (78%) of the product. The following analytical data were obtained for compound **3**: R_f = 0.50

(CH$_2$Cl$_2$/MeOH 9:1 + 0.5% NH$_4$OH). ^{1}H NMR (600 MHz, d_6-DMSO) δ: 2.30 – 2.37 (m, 2H), 3.01 (t, J = 7.4 Hz, 2H), 4.97 (d, J = 10.1, 1H), 5.05 (d, J = 17.2, 1H), 5.80 – 5.91 (m, 1H), 7.88 (s, 1H), 13.25 (br s, 1H). ^{13}C NMR (150 MHz, d_6-DMSO) δ: 27.6, 38.1, 115.1, 137.6, 140.6 (see printed spectra for more detail). LCMS–ESI (m/z): [M + H]$^+$ calcd for C$_8$H$_{10}$IN$_2$O$^+$: 277, Found: 277. The purity was determined to be 98.6% by qNMR spectroscopy in CDCl$_3$ using 10.4 mg of compound (3) and 7.4 mg of dimethyl fumarate (99%) as an internal standard.

Working with Hazardous Chemicals

The procedures in *Organic Syntheses* are intended for use only by persons with proper training in experimental organic chemistry. All hazardous materials should be handled using the standard procedures for work with chemicals described in references such as "Prudent Practices in the Laboratory" (The National Academies Press, Washington, D.C., 2011; the full text can be accessed free of charge at http://www.nap.edu/catalog.php?record_id=12654). All chemical waste should be disposed of in accordance with local regulations. For general guidelines for the management of chemical waste, see Chapter 8 of Prudent Practices.

In some articles in *Organic Syntheses*, chemical-specific hazards are highlighted in red "Caution Notes" within a procedure. It is important to recognize that the absence of a caution note does not imply that no significant hazards are associated with the chemicals involved in that procedure. Prior to performing a reaction, a thorough risk assessment should be carried out that includes a review of the potential hazards associated with each chemical and experimental operation on the scale that is planned for the procedure. Guidelines for carrying out a risk assessment and for analyzing the hazards associated with chemicals can be found in Chapter 4 of Prudent Practices.

The procedures described in *Organic Syntheses* are provided as published and are conducted at one's own risk. *Organic Syntheses, Inc.*, its Editors, and its Board of Directors do not warrant or guarantee the safety of individuals using these procedures and hereby disclaim any liability for any injuries or damages claimed to have resulted from or related in any way to the procedures herein.

Discussion

Imidazole is an important and commonly used heterocycle, finding uses in a wide range of chemistry-related disciplines. Imidazoles are precursors for 1,3-dialkyl imidazolium salts,[2] which are used as ionic liquids and as precursors for the synthesis of N-heterocyclic carbenes (NHCs) and their metal complexes.[3] In addition, imidazoles are found in many drugs and pesticides as exemplified by the H_2-receptor antagonist Cimetidine and the AT_1-receptor antagonist Losartan as well as Cyazofamid, all of which are 4,5-substituted imidazoles (Figure 8). Thus, the development of reliable and scalable procedures for the regioselective synthesis of sophisticated 4,5-disubstituted imidazoles remains a highly valuable task.

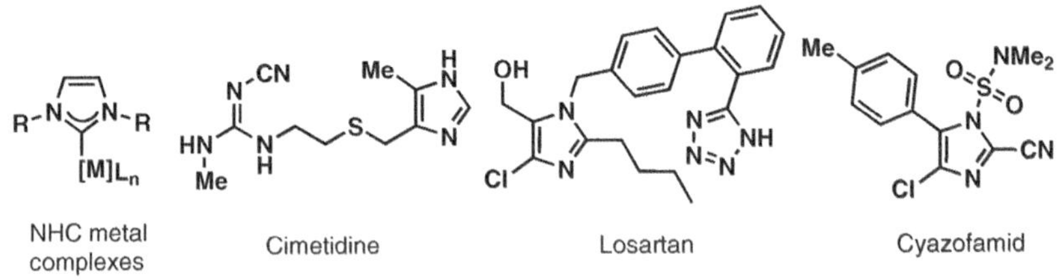

Figure 8. Examples of imidazole-containing compounds.

Herein, we report a multigram-scale synthesis of 1-(5-iodo-1H-imidazole-4-yl)pent-4-en-1-one (3) that arose as an important intermediate in a drug discovery project focusing on irreversible methionine aminopeptidase 2 (MetAP2) inhibitors. Starting from 3, N-alkylation, Heck reaction, and epoxidation gave a library of epoxyimidazoles as our first-generation MetAP2 inhibitors (Figure 9).[4] These substances were designed to mimic the natural product ovalicin.

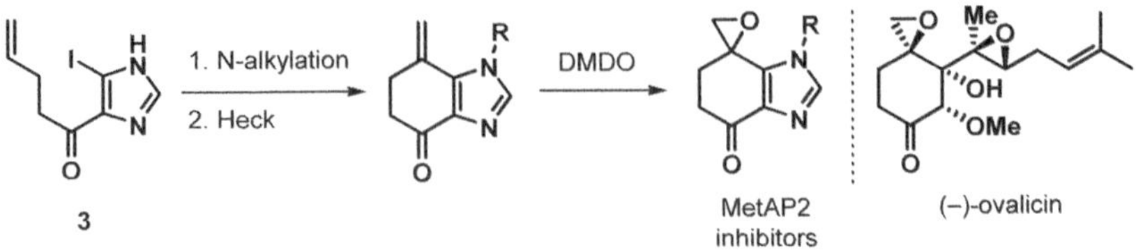

Figure 9. Synthetic route from 3 to putative MetAP2 inhibitors.

Our first reported synthesis of **3**[4] also called for magnesiation of 4,5-diiodoimidazole (**2**) using conditions initially described by Kopp and Knochel.[5] In that case, the resulting Grignard intermediate was treated with 4-pentenal to give an alcohol, which was oxidized to **3** with IBX (Scheme 1). Although this sequence reliably gave good yields on scale, it had drawbacks. First, 4-pentenal is low-boiling (bp 96 °C) and has an extremely noxious odor, making its large-scale laboratory preparation unpleasant.[6] Second, using a reagent at the aldehyde oxidation state necessitated an oxidation to arrive at **3**. We therefore developed the current route which gives **3** directly from Weinreb amide **1**, which itself is prepared in a two-step one pot reaction from inexpensive 4-pentenoic acid.[7]

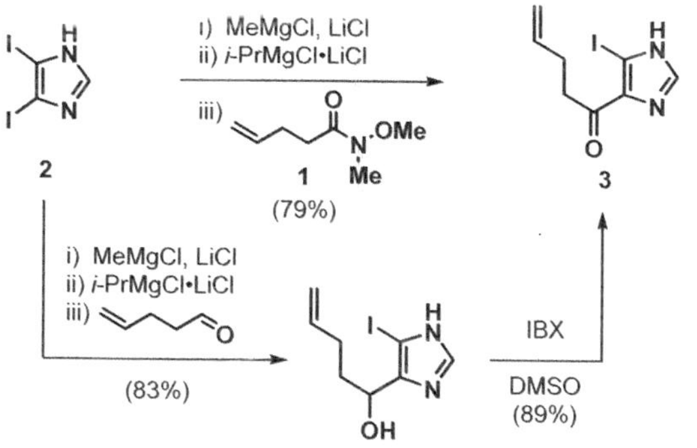

Scheme 1. Original (lower) and current (upper) approaches to building block 3.

It is of interest to note that at the outset of our MetAP2 inhibitor project (2009) multiple publications reported that reaction of imidazole with ≥3 equivalents of I_2 produced triiodoimidazole.[8] This apparent error can be traced back to 1908 when Pauly *et al.* first described the iodination of different N-heterocycles, including imidazole.[9] Due to the limited analytical techniques available at the time, only melting point and elemental analysis were used to characterize the product. The authors wrote of extreme difficulties in obtaining an accurate elemental analysis, which is assuredly the source of the error. Interestingly, this error apparently was not uncovered, or at least not reported in the primary literature, for nearly 100 years. While our preparation only calls for 2 equivalents of I_2, we have found that additional equivalents do not promote further iodination. This has been corroborated in the meantime by others.[10]

As mentioned above, THF solutions of **2** and LiCl produce a white precipitate, which on large scale can cause problems with stirring. Curious to determine what adduct was formed, we prepared a dilute THF solution of equimolar **2** and LiCl and stored it at 4 °C. Thin colorless needles formed, which were amenable to X-ray crystallographic analysis. The experimental structure shows a tetrahedral complex of lithium, which is coordinated by two THF solvent molecules, chlorine and N3 of 4,5-diiodoimidazole (**2**) (Figure 10, left). In the solid state, this complex forms a network through intermolecular H-bonds of N^1H and chlorine (Figure 10, right).[11]

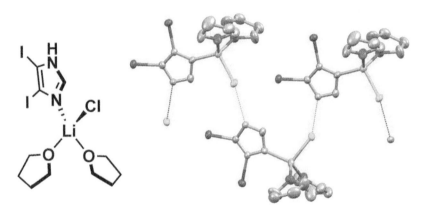

Figure 10. (Left) Depiction of the precipitated adduct of 4,5-diiodoimidazole (**2**) and LiCl in THF; (Right) the experimental structure of the tetrahedral complex. Ellipsoids are shown at 50% probability level; color code: carbon, grey; oxygen, red; nitrogen, blue; iodine, purple; lithium, pink; chlorine, green; hydrogen omitted for clarity; dashed lines indicate the network formed by this complex via H-bonds (imidazole-N^1H--Cl).

As described above, we often observed that magnesium/iodine exchange of **2** with 1.1 equiv of *i*-PrMgCl•LiCl did not go to completion (TLC monitoring of a mini work-up showed small amounts of remaining 4,5-diiodoimidazole (**2**)). An attempt to drive the reaction to completion with additional *i*-PrMgCl followed by a prolonged reaction time with **1** (over night; approximately 16 h) led to a byproduct via side reaction with THF which we identified as 5-iodo-4-(tetrahydrofuran-2-yl)-1*H*-imidazole

(Scheme 2). The compound is highly crystalline, difficult to remove from the desired product, and has an R_f value (in 9:1 MeOH/CH$_2$Cl$_2$ + 0.5% NH$_4$OH) that is identical to 4-iodoimidazole. The reaction of aryl Grignards with THF, mediated by alkyl Grignards and iodoalkanes, at room temperature has been reported.[12] In our system, there is an aryl Grignard (the desired magnesium/iodine exchange product), an alkyl Grignard (excess *i*-PrMgCl), an iodoalkane (*i*-PrI – the other magnesium/iodine exchange product), and THF as solvent.

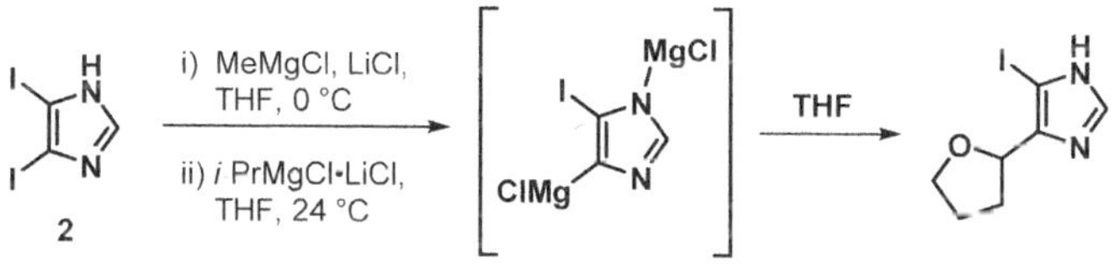

Scheme 2. Observed side-product of the iodine-magnesium exchange reaction in THF as solvent (at a prolonged reaction time).

References

1. Cancer Drug Development Group, German Cancer Research Center (DKFZ), Im Neuenheimer Feld 280, D-69120 Heidelberg, Germany. E-mail: aubry.miller@dkfz.de. The Helmholtz Drug Initiative and the German Cancer Consortium (DKTK) are gratefully acknowledged for financial support. We thank Frank Rominger (Organic Chemistry Institute, Heidelberg University) for X-ray data of the coordination complex of LiCl with 4,5-diiodoimidazole 2 and THF.
2. Dupont, J.; Consorti, C. S.; Suarez, P. A. Z.; de Souza, R. F. *Org. Synth.* **2004**, *79*, 236–243.
3. Hopkins, M. N.; Richter, C.; Schedler, M.; Glorius, F. *Nature* **2014**, *510*, 485–496.
4. Miller, A.; Jöst, C.; Klein, C. "Preparation of spiroepoxide tetrahydrobenzotriazoles and -imidazoles useful as MetAP-II-inhibitors", Eur. Pat. Appl. (2014), EP 2711367 A1 20140326.
5. Kopp, F.; Knochel, P. *Synlett* **2007**, *6*, 980–982.
6. Crimmins, M. T.; Ellis, J. M. *J. Am. Chem. Soc.* **2005**, *127*, 17200–17201.
7. (a) Sherry, B. D.; Fürstner, A. *Chem. Commun.* **2009**, 7116–7118. (b) Doyle, K. J.; Kerrigan, F.; Watts, J. P. "Dihydroimidazol[2,1-B]thiazole

and Dihydro-5H-Thiazolo[3,2-A]pyrimidines as Antidepressant Agents", Intl. Pat. Appl. (2001), WO 01/68653 A1.
8. We have found one manuscript from the time describing the direct formation of "only" 4,5-dioodoimidazole (2) with excess I_2: Knochel, P.; Yang, X. *Chem. Commun.* **2006**, *20*, 2170–2172.
9. Pauly, H.; Gundermann, K. *Ber. Dtsch. Chem. Ges.* **1908**, *41*, 3999–4012.
10. (a) Zou, Y.; Wang, F.; Wang, Y.; Sun, Q.; Hu, Y.; Li, Y.; Liu, W.; Guo, W.; Huang, Z.; Zhang, Y.; Xu, Q.; Lai, Y. *Eur. J. Med. Chem.* **2017**, *140*, 293–304. (b) Sandtorv, A. H.; Bjørsvik, H.-R. *Eur. J. Org. Chem.* **2015**, 4658–4666. (c) Jin, H.; Tan, T. T. Y.; Hahn, F. E. *Angewandte Chem. Int. Ed.* **2015**, *54*, 13811–13815.
11. The crystallographic data of the complex 4,5-diiodoimidazole **2** coordinated to lithium chloride in THF can be obtained free of charge form the Cambridge Crystallographic Data Center (cif-file reference code CCDC 1978199).
12. Inoue, A.; Shinokubo, H.; Oshima, K. *Synlett* **1999**, *10*, 1582–1584.

Appendix
Chemical Abstracts Nomenclature (Registry Number)

4-Pentenoic acid; (591-80-0)
Oxalyl chloride: Ethanedioyl dichloride; (79-37-8)
N, O-Dimethylhydroxylamine hydrochloride: Methanamine, N-methoxy-, hydrochloride (1:1); (6638-79-5)
Triethyl amine: Ethanamine, N,N-diethyl-; (121-44-8)
4,5-Diiodoimidazole: 1H-Imidazole, 4,5-diiodo-; (15813-09-9)
Potassium iodide; (7681-11-0)
Iodine; (7553-56-2)
Imidazole: 1H-Imidazole; (288-32-4)
Sodium hydroxide; (1310-73-2)
Hydrochloric acid; (7647-01-0)
Lithium chloride; (7447-41-8)
Methylmagnesium chloride (3M solution in THF); Magnesium, chloromethyl- (676-58-4)
iso-Propylmagnesium chloride lithium chloride complex: Magnesate(1-), dichloro(1-methylethyl)-, lithium (1:1); (745038-86-2)
1-(5-Iodo-1H-imidazol-4-yl)pent-4-en-1-one: 4-Penten-1-one, 1-(5-iodo-1H-imidazol-4-yl)-; (1585257-59-5)

Michael Morgen received his Ph.D. in inorganic chemistry from the Ruprecht-Karls-University, Heidelberg in 2013 while working with Prof. Dr. Peter Comba on the design of novel bispidine-based multimodal imaging agents. Afterwards he joined the Cancer Drug Development Group of Dr. Aubry Miller and Dr. Nikolas Gunkel at the German Cancer Research Center (DKFZ). His current research interests lie in the design and synthesis of organic and metal-organic inhibitors for cancer-related drug targets.

Jasmin Lohbeck was trained as chemistry laboratory assistant at BASF in Ludwigshafen, Germany, from 2006 – 2009. She subsequently joined the chemistry lab of Dr. Aubry Miller at the German Cancer Research Center and specialized in preparative organic synthesis.

Aubry Miller leads the Cancer Drug Development Group at the German Cancer Research Center (DKFZ) in Heidelberg, Germany. His laboratory focuses on the discovery and biological characterization of chemical probes for use as tools and drug leads in cancer research.

Dr. Anji Chen received his Ph.D. in organic chemistry at Old Dominion University, VA under the mentorship of Prof. Guijun Wang in 2018, where he made contributions on two distinct areas of research: 1) synthesis of new classes of glycoclusters and glycomimetics and their applications as soft biomaterials and 2) synthesis of novel bis-triazole linked carbohydrate-based macrocycles and their application for accelerating copper sulfate mediated click reaction. After completing his graduate studies in 2019, Anji joined TCG GreenChem, Inc. as a post-doctoral researcher. His research interests include the design and development of green and robust large-scale production of active pharmaceutical ingredients (APIs) and synthetic methodology of novel organic transformations.

Hari P. R. Mangunuru earned his Ph.D. in organic chemistry from the Old Dominion University, 2012 under the guidance of Prof. Guijun Wang, where his research focused on developing carbohydrate based hydrogelators as drug delivery carriers. He joined the Department of Process Research & Development at Boehringer Ingelheim, Ridgefield, CT, USA, as a postdoctoral fellow, where he developed a new class of phosphine ligands for asymmetric catalysis. In 2016 he moved to Virginia Commonwealth University as a postdoctoral fellow to work with Prof. B. Frank Gupton, where he was responsible for the discovery and development of green, robust, and safe synthetic routes to active pharmaceutical ingredients (APIs) using continuous flow technology. Currently working as a Senior scientist in Process Chemistry at TCG GreenChem, Inc., his research interests include the development, study, applications of novel organic transformations and continuous flow asymmetric catalysis.

Nathaniel Kaetzel received his B.S. in chemistry from Virginia Commonwealth University while under the mentorship of Prof. Joshua Sieber, where he studied novel total synthetic applications of copper-catalyzed reductive coupling reactions, and Prof. Christopher Kelly, where he studied the oxidative formation of nitriles utilizing an oxoammonium salt, as well as the reactivity of silylated bis-boronic ester species. After completing his undergraduate studies in 2020, Nathaniel joined TCG GreenChem Inc. as a research scientist. His research interests include the instrumental analysis of in-development manufacturing processes for active pharmaceutical ingredients (APIs), as well as the design and synthesis of new chiral phosphine ligands.

Dr. Gopal Sirasani received his Bachelor's and Master's degrees in Hyderabad, India. He obtained his Ph.D. in synthetic organic chemistry in 2011 from Temple University, Philadelphia under the guidance of Prof. Rodrigo B. Andrade. His doctoral research was focused on developing novel methodologies, total syntheses of natural products and their analogs thereof. He got his post-doctoral training in the laboratory of Prof. Emily Balskus at Harvard University, where he developed biocompatible organic reactions utilizing microbially generated reagents to realize transition metal catalysis in the presence of microbes. In 2013, Gopal began his industrial career at Melinta Therapeutics, NewHaven, CT. He is currently working at TCG GreenChem, Inc. as a Director in the department of process research and development.

Enantioselective Michael-Proton Transfer-Lactamization for Pyroglutamic Acid Derivatives: Synthesis of Dimethyl-(S,E)-5-oxo-3-styryl-1-tosylpyrrolidine-2,2-dicarboxylate

Christian M. Chaheine, Conner J. Song, Paul T. Gladen, and Daniel Romo[*1]

Department of Chemistry & Biochemistry, Baylor University, Waco, TX 76710

Checked by Matthew J. Genzink and Tehshik P. Yoon

Procedure (Note 1)

A. *Dimethyl 2-amino((4-methylphenyl)sulfonamido)malonate* (**2**). To a single-necked, 500 mL round-bottomed flask containing a football-shaped, teflon-coated stir bar (5 cm) and fitted with a 24/40 glass, threaded gas-inlet adapter with a silicone/PTFE septa (Figure 1A) (Note 2), in-turn connected via chemically resistant tubing to a vacuum/nitrogen manifold is added commercially available dimethyl aminomalonate hydrochloride (**1**, 12.5 g, 68.2 mmol, 1.00 equiv) and *p*-toluenesulfonic anhydride (27.2 g of 90 wt. % purity, 74.9 mmol, 1.1 equiv) (Note 3) as solids. The atmosphere in the flask is replaced with nitrogen by three cycles of vacuum (Note 4) and nitrogen back-filling via the vacuum/nitrogen manifold. The glass adapter is briefly removed, and tetrahydrofuran (270 mL) is then added from a polyethylene graduated cylinder. The resulting beige suspension is stirred vigorously and cooled to 0 °C in an ice/water bath (*e.g.* recrystallizing dish, Figure 1B) for 15 min before adding freshly distilled triethylamine (28.5 mL, 204 mmol, 3.0 equiv; Note 5) by rapid dropwise addition via a 60 mL, plastic, luer-lock syringe fitted with a stainless steel 18-gauge needle over ~5 min. A yellow-orange suspension forms within 1 h (Figure 1C) and gradually becomes a light-yellow color as the reaction mixture is stirred over 24 h (Note 6) (Figure 1D) and allowed to slowly warm to ambient temperature (23 °C).

Figure 1. N-Tosylation of dimethylaminomalonate hydrochloride: A) Threaded gas-inlet adapter, 24/40 joint size, fitted with a red/white PTFE/silicone septa in an open-top screw cap. B) Prior to addition of triethylamine. C) 1 hour after addition of triethylamine. D) 16 h after addition of triethylamine (photos provided by submitters).

The yellow-orange suspension is then vacuum-filtered through a pad of celite (Note 7) (Figure 2A) into a 1 L, heavy-walled vacuum Erlenmeyer flask to remove the white precipitate by-product, which is triethylammonium *p*-toluenesulfonate. Quantitative transfer of the crude material to the filter funnel is carried out by rinsing with ethyl acetate from a polyethylene squirt bottle (~25 mL). The filter cake is rinsed with ethyl acetate (250 mL) and the filtrate is transferred to a 1 L recovery flask and concentrated by rotary evaporation (~150 to 75 mmHg, 30 °C bath temperature) to a yellow oil that is dry-loaded onto 70 g of silica gel (Notes 8 and 9). The resulting fine yellow powder is loaded onto a pre-packed flash chromatography column for purification (Figures 2B-C) (Note 10) and eluted to yield 13.3 g (65%) of dimethyl 2-((4-methylphenyl)sulfonamido)malonate (**2**) as an off-white grainy powder (Note 11) (Figure 2D).

Figure 2. N-Tosylation reaction work-up and column chromatography: A) Crude reaction solution after filtering through Celite. B) Flash column prior to fraction collection with crude mixture pre-adsorbed onto silica gel. C) Flash column following product collection; yellow color that progresses through column is not collected. D) Appearance of N-tosyl-2-aminomalonate (photos provided by submitters).

B. *(2E,4E)-5-Phenylpenta-2,4-dienoyl chloride (**4**)*. To an oven-dried (Note 12), 500 mL, single-necked, round-bottomed flask containing a football-shaped, teflon-coated stir bar (4 cm) and fitted with a 24/40 glass, threaded gas-inlet adapter with a silicone/PTFE septa (Figure 1A, Note 2), in-turn attached via chemically resistant tubing to a vacuum/nitrogen manifold, is added commercially available 5-phenyl-2,4-pentadienoic acid (**3**) (10.0 g, 57.5 mmol, 1.00 equiv) (Note 13). The atmosphere in the vessel is replaced with nitrogen by three cycles of evacuation (Note 4) and back-filling with nitrogen via the vacuum/nitrogen manifold. The hose connection to the manifold is then quickly replaced with tygon tubing connected in sequence to a mineral oil bubbler followed by a solution of saturated, aqueous sodium bicarbonate (150 mL) in a 250 mL Erlenmeyer flask (Figure 3A). Dichloromethane (230 mL) is then added in four portions via a 60 mL plastic syringe fitted with an 18-gauge stainless steel needle, forming a beige-colored suspension that is stirred at ambient temperature (23 °C). N,N-Dimethylformamide (0.67 mL, 8.6 mmol, 15 mol%) is added via a plastic 1 mL syringe in one portion, followed by dropwise addition of oxalyl chloride (14.6 mL, 173 mmol, 3.0 equiv) over 15 min via a 30 mL plastic luer-lock syringe fitted with a 20-gauge stainless-steel needle connected to a syringe pump (59 mL/h flow rate). The reaction mixture is stirred at ambient temperature (23 °C) while venting gaseous by-products through the mineral oil bubbler/sodium bicarbonate solution (Note 14) until it became a homogenous, gold-colored solution and had ceased gas evolution (~3–4 h).

Figure 3. Acid chloride synthesis: A) Reaction set up with mineral oil bubbler and sodium bicarbonate solution in sequence to control and quench exhaust gases. B) Crude acid chloride used in the next reaction after drying on high vacuum with an attached acid scrubber (KOH pellets) (photos provided by submitters).

The stir bar is removed and the crude reaction mixture is concentrated by rotary evaporation (150 to 75 mmHg, 30 °C bath temperature, ~1 h) to a light-brown, amorphous solid that is further dried under high vacuum (~3 mmHg) (Note 15) (Figure 3B) with an intervening acid scrubber column for at least 3 h to provide (2E,4E)-5-phenylpenta-2,4-dienoyl chloride (4). The atmosphere in the flask is charged with nitrogen via the Schlenk manifold yielding 11.2 g (101%) of crude acid chloride as a yellow solid which is used directly in the next reaction without purification (Note 16).

C. *Dimethyl-(S,E)-5-oxo-3-styryl-1-tosylpyrrolidine-2,2-dicarboxylate* (**5**): To an oven-dried (Note 12), 1 L, single-necked, round-bottomed flask containing a football-shaped, teflon-coated stir bar (5 cm) is added dimethyl sulfonamidomalonate (**2**) (12.4 g, 41.3 mmol, 1.00 equiv). The reaction vessel is fitted with a T-bore Schlenk adapter attached to both vacuum and an argon balloon (Notes 17 and 18) (Figure 4A). The atmosphere in the flask is replaced with argon by three cycles of vacuum/back-filling via the Schlenk adapter, which is then quickly replaced with a rubber septum and an argon balloon. Tetrahydrofuran is added (210 mL, 0.20 M initial concentration of sulfonamidomalonate) (Notes 19 and 20) in four portions via a plastic 60 mL syringe fitted with an 18-gauge stainless steel needle. The resulting clear solution is stirred vigorously and cooled to 0 °C in an ice bath over 15 min. A newly opened 100 mL bottle of lithium hexamethyldisilazide (LiHMDS, 1.0 M in tetrahydrofuran, 47.5 mL, 47.5 mmol, 1.15 equiv) (Note 21) is then added dropwise from a 60 mL plastic luer-lock syringe fitted with an 18-gauge stainless-steel needle connected to a syringe pump over 15 min (189 mL/h flow rate, Figure 4B).

Figure 4. Nucleophile-catalyzed Michael-proton transfer-lactamization (NCMPL); Deprotonation of sulfonamidomalonate: A) T-bore Schlenk adapter attached to 1 L, single-necked round-bottomed flask. B) Addition of LiHMDS solution via syringe pump. C) Following complete addition of LiHMDS (photos provided by submitters).

The resulting yellow solution (Figure 4C) is further cooled for 15 min in a cryobath set to –10 °C (Note 22). 1,8-Diazobyciclo[5.4.0]undec-7-ene (DBU, 6.85 mL, 45.4 mmol, 1.10 equiv) is then added as a very viscous liquid in a single portion via a 12 mL plastic syringe equipped with an 18 gauge needle followed by the addition of O-trimethylsilylquinine (TMSQN) solution via a 12 mL plastic syringe via a 20 gauge needle in one portion (Note 23) within

5 minutes of DBU addition. A solution of crude, (2E,4E)-5-phenylpenta-2,4-dienoyl chloride (4) (1.0 M in tetrahydrofuran, 57.5 mmol, 1.39 equiv in 57.4 mL) (Note 24) is then added dropwise over 3 h from a 60 mL luer-lock plastic syringe fitted with a 16-gauge stainless-steel needle connected to a syringe pump (19 mL/h flow rate; Figure 5B-C).

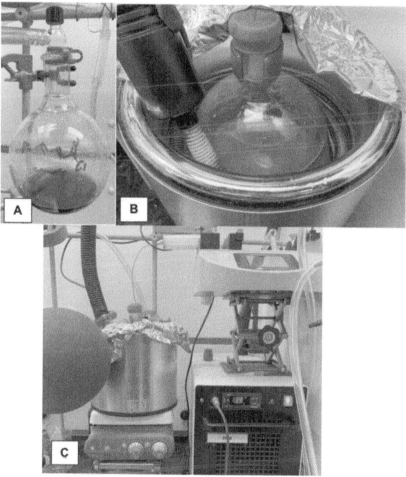

Figure 5. NCMPL Organocascade: A) Appearance of crude unsaturated acid chloride solution in THF. B) Reaction solution following addition of DBU and TMSQN. C) Addition of crude unsaturated acid chloride solution via syringe pump with cooling from a cryobath system (photos provided by submitters).

The reaction mixture is stirred at –10 to –14 °C for an additional 3 h (total reaction time = 6 h) (Note 25) and is quenched by carefully pouring the cold orange-red solution into a 1 L separatory funnel containing aqueous hydrochloric acid (1M, 150 mL). Quantitative transfer of the crude reaction mixture to the separatory funnel is carried out by rinsing the reaction vessel with ethyl acetate (150 mL). The layers are separated (Figure 6A) and the aqueous layer is extracted with ethyl acetate (4 × 125 mL) (Figure 6B). The combined organic layers are washed with saturated, aqueous sodium chloride solution (150 mL), dried over magnesium sulfate (5 g), and filtered through Celite (Note 7) (Figure 6C) into a 1 L Erlenmeyer flask. The filter cake is rinsed with ethyl acetate (75 mL).

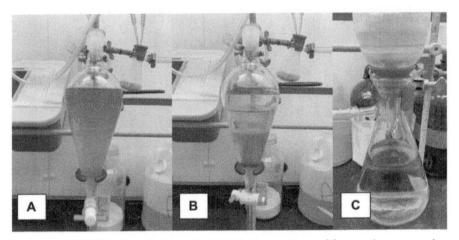

Figure 6. NCMPL reaction work up: A) Formation of layers in separation funnel upon first extraction with ethyl acetate. B) Appearance of layers in separation funnel following final extraction with ethyl acetate. C) Combined organic layers after drying and filtering through Celite (photos provided by submitters).

The filtrate is transferred to a 2 L recovery flask and concentrated by rotary evaporation (~200 mm Hg, 30 °C bath temperature) until a yellow, amorphous solid forms (Figure 7A). This crude, yellow solid is transferred to a Büchner funnel (7 cm diameter) by use of a stainless-steel spatula. The solid is agitated with a spatula and washed with a minimal amount of methanol (~200 mL) from a polyethylene squirt bottle while under vacuum filtration (Note 26) (Figure 7B). After transferring the off-white solid to

6-dram vials and removing the residual solvent under high vacuum for at least 4 h (Notes 4 and 27), 13.5 g of dimethyl-(S,E)-5-oxo-3-styryl-1-tosylpyrrolidine-2,2-dicarboxylate (**5**) (71% isolated yield, >99:1 er) is obtained as an off-white crystalline powder that is of sufficient purity (97% by q^{1}H-NMR) for subsequent reactions (Notes 28 and 29).

Figure 7. NCMPL product isolation: A) Crude solid following thorough removal of volatiles by rotary evaporation. B) Product after washing with methanol under gentle vacuum filtration in a Büchner funnel. C) Appearance of γ-lactam (**5**) as a powdery crystalline solid (photos provided by submitters).

Notes

1. Prior to performing each reaction, a thorough hazard analysis and risk assessment should be carried out with regard to each chemical substance and experimental operation on the scale planned and in the context of the laboratory where the procedures will be carried out. Guidelines for carrying out risk assessments and for analyzing the hazards associated with chemicals can be found in references such as Chapter 4 of "Prudent Practices in the Laboratory" (The National Academies Press, Washington, D.C., 2011; the full text can be accessed free of charge at https://www.nap.edu/catalog/12654/prudent-practices-in-thelaboratory-handling-and-management-of-chemical. See also "Identifying and Evaluating Hazards in Research Laboratories" (American Chemical Society, 2015) which is available via the associated website "Hazard Assessment in Research Laboratories" at https://www.acs.org/content/acs/en/about/governance/committees

/chemicalsafety/hazard-assessment.html. In the case of this procedure, the risk assessment should include (but not necessarily be limited to) an evaluation of the potential hazards associated with dimethyl aminomalonate hydrochloride, *p*-toluenesulfonic anhydride, triethylamine, 5-phenyl-2,4-pentadienoic acid, oxalyl chloride, methylene chloride, *N,N*-dimethylformamide, mineral oil, sodium bicarbonate, dihydrogen monoxide, carbon monoxide, carbon dioxide, hydrogen chloride, lithium hexamethyldisilazide, 1,8-diazobyciclo[5.4.0]undec-7-ene, methanol, tetrahydrofuran, ethyl acetate, hexanes, Celite, magnesium sulfate, and silica gel.

2. A rubber septum and needle connected to a nitrogen/vacuum manifold can also be used if the glass adapter shown in Figure 1A is unavailable.

3. Dimethyl aminomalonate hydrochloride (97%) was purchased from Sigma Aldrich, *p*-toluenesulfonic anhydride (90%) was purchased from Oakwood, triethylamine (99%), tetrahydrofuran (HPLC grade, unstabilized) ethyl acetate (ACS grade), and hexanes (ACS grade) were purchased from Fisher Scientific. All reagents and solvents, excluding triethylamine, were used as received. *p*-Toluenesulfonyl chloride (TsCl) was also studied for this tosylation but was ineffective even in the presence of DMAP.

4. A Welch Duo Seal 1400 rotary vane vacuum pump (~3 mm Hg) connected to the vacuum/nitrogen manifold through a cold-trap was used for high-vacuum applications throughout this procedure.

5. Triethylamine was distilled from calcium hydride (95%, purchased from Oakwood) under an atmosphere of dry nitrogen prior to use.

6. The formation of product can be monitored by thin-layer chromatography (TLC, 30% ethyl acetate/hexanes, sulfonamidomalonate product R_f = 0.40, visualized under a 254 nm UV-lamp). Glass-backed, 250 μm-thickness TLC plates were purchased from Silicycle inc.

Figure 8. TLC analysis of Step A (photo provided by submitters)

7. Celite 545 filter aid was purchased from Fisher Scientific. A pad of Celite (1.5 cm tall) was used in a 150 mL funnel (6.5 cm height × 7 cm inner diameter) with a glass frit of porosity M.
8. Silica Gel (Silicycle Ultrapure SilicaFlash silica gel, 60 Å pore size, particle size distribution 40-63 microns) was purchased from Fisher Scientific.
9. Silica gel (70g) was added to the crude yellow oil following rotary evaporation. Ethyl acetate was used to rinse the walls of the 1 L recovery flask to ensure complete and uniform absorption onto the silica. Solvent that could interfere with chromatographic separation was thoroughly removed by rotary evaporation until the yellow powder thus formed was dry and free-flowing.
10. A heavy-walled, glass flash chromatography column 10 cm in diameter was used. The free-flowing fine yellow powder silica gel with the absorbed crude material was carefully loaded into the column, which, beforehand, was wet-packed with 250 g of fresh silica gel using hexanes and topped with a layer of sand. A step-wise gradient was used (hexanes (500 mL), then 20% ethyl acetate/hexanes (500 mL), then 40% ethyl acetate/hexanes (500 mL), then 60% ethyl acetate/hexanes (2500 mL). The eluent received when eluting with hexanes was collected in a 1 L Erlenmeyer flask. The eluent from 20% ethyl acetate in hexanes was collected in 500 mL Erlenmeyer flasks. The product was then collected in 250 mL fractions using Erlenmeyer flasks starting at 40% and ending at 60% ethyl acetate/hexanes. Product elution was monitored via thin-layer chromatography (50% ethyl acetate/hexanes, R_f = 0.60 visualized under a 254 nm UV-lamp). Glass-backed 250 μm-thickness TLC plates were purchased from Silicycle, Inc.

Figure 9. TLC analysis of column fractions for Step A (photo provided by submitters)

11. Characterization data for dimethyl 2-((4-methylphenyl)sulfonamido)-malonate (2): IR (thin film) 3235, 1749, 1160 cm^{-1}; ^{1}H NMR (500 MHz, CDCl$_3$) δ: 2.43 (s, 2H), 3.68 (s, 3H), 4.69 (d, J = 8.5 Hz, 1H), 5.56 (d, J = 8.5 Hz, 1H), 7.30 (d, J = 8.3 Hz, 1H), 7.74 (d, J = 8.3 Hz, 1H); ^{13}C NMR (125 MHz, CDCl$_3$) δ: 21.6, 53.6, 58.5, 127.3, 129.7, 136.4, 144.1, 166.0; mp = 118–122 °C; HRMS (ESI) [M + H]$^+$ calculated for C$_{12}$H$_{15}$NO$_6$S: 302.06928, found: 302.0686; This compound can be stored at ambient temperature (23 °C) indefinitely without decomposition. The purity was assessed at 82% by qNMR using 14.2 mg of the product and dimethyl terephthalate (20.2 mg) as an internal standard. The corrected yield for the reaction would be 10.9 g (53%). The checkers performed a second reaction yielding 13.9 g (69%) of the product.
12. Reaction flasks were dried in an oven at 125 °C for at least 4 h prior to use.
13. 5-Phenyl-2,4-pentadienoic acid (98%), purchased from Combi-Blocks, oxalyl chloride (99% reagent plus), purchased from Sigma Aldrich, and N,N-dimethylformamide (99.8% extra dry, AcroSeal), purchased from Acros, were used as received. Methylene chloride (HPLC grade, cyclohexane preservative) was purchased from Fisher Scientific and passed through a column of activated alumina under an atmosphere of ultra-high purity argon (JC Meyer Solvent Purification System) prior to use. It was empirically determined that 3.0 equiv of oxalyl chloride was required for complete consumption of the starting carboxylic acid.
14. *Caution*! This reaction rapidly expels gaseous by-products: carbon dioxide, caustic hydrogen chloride, and toxic carbon monoxide. Thus, careful dropwise addition of neat oxalyl chloride is performed in a uniform, controlled fashion using a syringe pump over ~15 min.
15. A column of potassium hydroxide pellets (Fisher Scientific) was connected in-line between the Schlenk manifold and the round-bottomed flask with a fritted gas outlet adapter and thick-walled Tygon tubing to act as an acid scrubber and prevent HCl corrosion of the vacuum pump.
16. This compound (4) could be stored for up to a week under inert atmosphere in a sealed vessel at –20 °C, but both the submitters and checkers used it immediately. The checkers performed a second reaction on half scale yielding 5.64 g (102%) of the crude product.
17. The submitters found the use of an argon balloon was more convenient in providing an inert atmosphere for the nucleophile-catalyzed

Michael-proton transfer lactamization, but a typical nitrogen line with a needle and rubber septum can also be used.

18. Thick-walled, natural latex rubber balloons were purchased from Sigma Aldrich (SKU- Z154997).
19. Tetrahydrofuran (HPLC grade, unstabilized) was purchased from Fisher Scientific and passed through a column of activated alumina under an atmosphere of ultra-high purity argon (JC Meyer Solvent Purification System). Methanol (ACS grade) was purchased from Fisher Scientific and used as received. Lithium hexamethyldisilazide (1.0 M in tetrahydrofuran) and 1,8-diazobyciclo[5.4.0]undec-7-ene (DBU, 98%) were purchased from Sigma Aldrich, and both were used as received. Sodium chloride was purchased from Fisher Scientific and used as received. Magnesium sulfate (99%, anhydrous powder) was purchased from Oakwood and used as received by the submitters. Magnesium sulfate (97%, anhydrous powder) was purchased from Sigma Aldrich and used as received by the checkers.
20. The initial concentration of the sulfonamidomalonate starting material is essential to the yield of this reaction as variations led to formation of insoluble salts as the base is added and greatly reduced yields.
21. The quality of the LiHMDS used is very important and a newly opened 100 mL bottle for this procedure is required to ensure titer of base is as close as possible to the nominal 1.0 M indicated. The appearance of the LiHMDS solution in the plastic syringe should be a very light yellow/orange in color and clear (not dark orange or cloudy). Complete formation of the enolate from the starting sulfonamidomalonate is essential to the yield of the product lactam.
22. Maintaining the reaction temperature between −10 and −14 °C is critical to obtain high enantioselectivity in the initial Michael-addition step.
23. O-Trimethylsilylquinine (TMSQN), an off-white, amorphous solid, was prepared according to the *Organic Syntheses* procedure previously reported by the submitters.[2] TMSQN is prepared as a solution in tetrahydrofuran under an atmosphere of nitrogen (1.0 M, 3.29 g, 8.30 mmol, 0.20 equiv in 8 mL THF) in an oven-dried (Note 12) 25 mL pear-shaped flask fitted with a glass, threaded gas-inlet adapter with a silicone/PTFE septa (Note 2).
24. Both the submitters and checkers observed a light-yellow, undissolved solid which remained undissolved when forming the THF solution of the crude unsaturated acyl chloride. While some solid may be pulled into the syringe prior to dropwise addition to the reaction mixture, it does not

create an issue with the reaction. However, the syringe/syringe pump apparatus should be checked periodically to ensure the needle is not clogged and regular dropwise addition is proceeding.
25. The consumption of starting material and the formation of the product could be monitored by thin-layer chromatography (TLC, 30% ethyl acetate/hexanes, sulfonamidomalonate substrate R_f = 0.40, lactam product R_f = 0.50, visualized under a 254 nm UV-lamp) Glass-backed, 250 µm-thickness TLC plates were purchased from Silicycle, Inc.

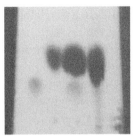

Figure 10. TLC analysis of Step C (photo provided by submitters)

26. The methanol rinse under vacuum filtration removes the yellow color of the crude solid and should be done with caution to prevent the resulting fine crystalline powder from passing underneath the filter paper. The product is sparingly soluble in MeOH. Up to 200 mL of methanol can be required to completely remove the yellow impurity. A stainless steel spatula is used to break apart clumps of yellow solid to expose product surface for thorough rinsing.
27. Thorough removal of residual methanol from the product by high vacuum is necessary in order to avoid decomposition upon storage at ambient temperature.
28. Characterization data for dimethyl-(S,E)-5-oxo-3-styryl-1-tosylpyrrolidine-2,2-dicarboxylate (5): Absolute stereochemistry was assigned by the submitters based on X-ray analysis using anomalous dispersion; $[\alpha]^{20}_D$ +0.78 (c 10.0, CHCl$_3$); IR (thin film) 1736, 1162 cm^{-1}; mp = 164–165 °C; ^{1}H NMR (500 MHz, CDCl$_3$) δ: 2.45 (s, 3H), 2.65 (dd, J = 16.9, 9.4 Hz, 1H), 2.70 (dd, J = 17.0, 8.3 Hz, 1H), 3.54 (app q, J = 8.5 Hz, 1H), 3.81 (s, 3H), 3.94 (s, 3H), 6.07 (dd, J = 15.8, 8.0 Hz, 1H), 6.47 (d, J = 15.8 Hz, 1H), 7.36–7.27 (m, 7H), 8.06 (d, J = 8.4 Hz, 2H); ^{13}C NMR (125 MHz, CDCl$_3$) δ: 21.7, 35.8, 45.1, 53.4, 53.8, 76.2, 123.3, 126.6, 128.4, 128.7, 129.0, 129.9, 134.5, 135.1, 135.8, 145.4, 165.9, 167.5, 171.6; HRMS

(ESI) [M + Na]⁺ m/z calculated for $C_{23}H_{23}NO_7S$: 480.1087, found: 480.1077; This compound can be stored at ambient temperature (23 °C) on the benchtop indefinitely without decomposition. The purity was assessed at 97% by qNMR using 12.1 mg of the product and dimethyl terephthalate (22.7 mg) as an internal standard. The corrected yield as per qNMR would be 13.1 g (69%). The checkers performed a second reaction on half scale yielding 6.4 g (67%) of the product.

29. The submitters determined the enantiomeric ratio by chiral-HPLC analysis in comparison with an authentic racemic sample (synthesized using racemic catalyst (*i.e.* 1:1 TMSQN/TMSQD)) of the product using a Chiralcel AD-H column: hexanes/ⁱPrOH = 80:20, flow rate of 0.5 mL/min, λ = 254 nm: t_{minor} = 75.3 min, t_{major} = 80.1 min. The checkers determined the enantiomeric ratio by chiral-SFC analysis in comparison with an authentic racemic sample (synthesized using racemic catalyst (*i.e.* 1:1 TMSQN/TMSQD)) of the product using a Chiralcel OJ-H column: 10% ⁱPrOH, flow rate of 3 mL /min, t_{major} = 11.0 min, t_{minor} = 14.8 min.

Working with Hazardous Chemicals

The procedures in *Organic Syntheses* are intended for use only by persons with proper training in experimental organic chemistry. All hazardous materials should be handled using the standard procedures for work with chemicals described in references such as "Prudent Practices in the Laboratory" (The National Academies Press, Washington, D.C., 2011; the full text can be accessed free of charge at http://www.nap.edu/catalog.php?record_id=12654). All chemical waste should be disposed of in accordance with local regulations. For general guidelines for the management of chemical waste, see Chapter 8 of Prudent Practices.

In some articles in *Organic Syntheses*, chemical-specific hazards are highlighted in red "Caution Notes" within a procedure. It is important to recognize that the absence of a caution note does not imply that no significant hazards are associated with the chemicals involved in that procedure. Prior to performing a reaction, a thorough risk assessment should be carried out that includes a review of the potential hazards associated with each chemical and experimental operation on the scale that is planned for the procedure. Guidelines for carrying out a risk assessment and for analyzing the hazards associated with chemicals can be found in Chapter 4 of Prudent Practices.

The procedures described in *Organic Syntheses* are provided as published and are conducted at one's own risk. *Organic Syntheses, Inc.*, its Editors, and its Board of Directors do not warrant or guarantee the safety of individuals using these procedures and hereby disclaim any liability for any injuries or damages claimed to have resulted from or related in any way to the procedures herein.

Discussion

The development of synthetic methods for enantioselective access to functionalized pyroglutamate,[3,4] and, more generally, γ-lactam (aka pyrrolidine-2-one, γ-butyrolactam, azolidine-2-one, 2-oxopyrrolidine) containing compounds remains a highly active topic in the field of organic synthesis. This is due to their versatility as synthetic intermediates and their presence in many natural products, metabolites, and pharmaceuticals with broad-ranging and potent biological activities (Figure 11).[5–22]

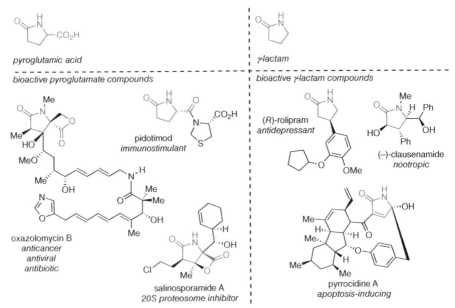

Figure 11. Enantioselective transition metal-catalyzed γ-lactam synthesis

There are an abundance of chemical methods for the synthesis of γ-lactams in racemic form, primarily through intramolecular cyclizations of functionalized linear substrates via C-N as well as C-C bond formation.[7, 21,23-27] In only the last three years, several methods for the construction of racemic γ-lactams have been reported– tandem addition/cyclization,[28-31] multi-component,[32-35] intramolecular C-H activation[36-40], photoredox-catalyzed,[41-43] transition metal-catalyzed cyclizations,[44, 45] rearrangement,[46] direct α-alkylation of primary amines with acrylates,[47] ring contraction/expansion,[48,49] and flow chemistry.[50] Chemoenzymatic,[51] enzymatic resolution,[52] and chiral pool synthesis[53, 54] have also been reported recently as asymmetric strategies. While these myriad, powerful methods are now available for the synthesis of γ-lactams, fewer examples illustrate a robust means of *enantioselective construction* of this key *N*-heterocycle from achiral starting materials. Even fewer methods have demonstrated scalability while maintaining a high degree of enantiopurity in the product.

Enantioselective synthesis of γ-lactams:

The following will briefly highlight recent (within the last three years) advances in synthetic methods to access chiral γ-lactams in an enantioselective fashion from achiral starting materials. This will be followed by a discussion of our group's research activity in the area of unsaturated acylammonium catalysis, centering on the nucleophile-catalyzed Michael-proton transfer lactamization (NCMPL) used in the described procedure.

Asymmetric organocatalytic routes to γ-lactams primarily employ chiral *N*-heterocyclic carbenes (NHCs),[55] which is a complementary strategy to the NCMPL methodology in its use of α,β-unsaturated carbonyl derivatives as substrates in a cascade process. In addition to multiple NHC-based organocatalytic methods, chiral phosphoric acids,[56] phase-transfer,[57] and hydrogen bond donor catalysts,[58] and multi-component reactions[59, 60] have also appeared recently as alternate avenues to optically active γ-lactams. Huang disclosed a Michael-addition/cyclization reaction using chiral NHC **A3** as a precatalyst, α-bromo-α,β-enals and α-sulfonamidoketones as substrates, and an excess of alkylamine base (Scheme 1A).[61] High-enantioselectivities and diastereoselectivites are obtained, and the reaction was performed on gram scale without significant loss of yield or enantiopurity. Notably, heteroaryl and cycloalkyl substituted ketones are efficient reactants in this process, delivering functionalized *trans*-γ-lactams that are aptly suited to the synthesis of clausenamide analogues.

A chiral NHC-catalyzed homoenolate addition/cyclization was reported by Li.[62] Utilizing the same precatalyst **A3** and various α,β-enals as homoenolate precursors, the Li group demonstrated a formal (3+2) annulation occurs with α-sulfonamidoacrylates (Scheme 1B). Mechanistic investigations revealed that, rather than the anticipated homoenolate Michael-addition/cyclization sequence, a tautomerization of the sulfonamidoacrylate to an α-iminoester precedes homoenolate addition and cyclization, yielding pyroglutamate derivatives that bear an *aza*-quaternary center in good yield and enantioselectivity.

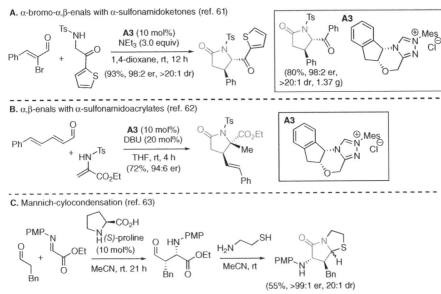

Scheme 1. Organocatalytic enantioselective γ-lactam synthesis

In contrast, Córdova developed an asymmetric cascade processes to construct bicyclic γ- and δ-lactam *N,S*-acetals through chiral amine catalysis (Scheme 1C).[63] The bicyclic products are intriguing scaffolds that occur in 5-membered ring analogs of penicillin antibiotics. In this reaction, Mannich-addition of an enolizable aldehyde onto an α-iminoester is catalyzed by

(S)-proline, forming a Mannich-base intermediate that is then treated with an aminothiol to afford bicyclic α-amino as well as α-hydroxy-γ-lactam N,S-acetals with excellent enantioselectivity.

More prevalent in the literature are methods based on transition metal catalysis. The asymmetric reductive amination of α-ketoesters for the synthesis of enantioenriched N-unprotected-γ-lactams was realized by the Yin group.[64] A chiral bidentate phosphine ligand **L1** in combination with Ru(OAc)$_2$ under an atmosphere of H$_2$ was identified as an efficient catalytic system for the reductive amination to NH-γ-lactams (Scheme 2A). Only aryl-substituted ketone substrates led to high enantioselectivities, however. This method was used on up to a gram-scale to build a pyrrolidine intermediate toward an anti-cancer therapeutic, Larotrectinib, as well as a benzolactam intermediate toward a quinolone antibiotic drug, Garenoxacin.

An enantioselective cobalt-catalyzed hydroboration cyclization of linear N-allyl-propiolamides was recently reported by Ge.[65] Borylated γ-lactams bearing stereogenic all-carbon quaternary centers were constructed in high yield and enantioselectivity with the application of chiral bidentate phosphine ligand **L2** and Co(acac)$_2$ (Scheme 2B). This reaction was conducted on a gram-scale and the versatility of the boryl functionality was demonstrated through a variety of known transformations.

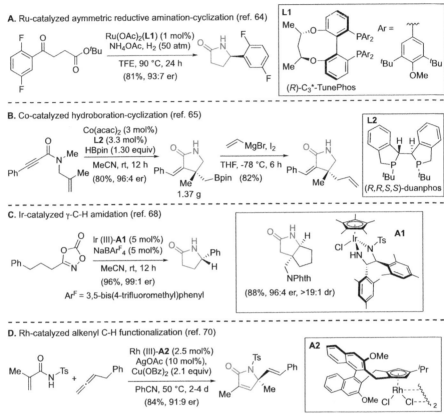

Scheme 2. Enantioselective transition metal-catalyzed γ-lactam synthesis

Activation of sp³ C–H bonds by transition metal catalysis is an important and fertile sub-field in organometallic chemistry and recently has enabled the enantioselective synthesis of NH-γ-lactams from readily-available carboxylic acid derivatives like 1,4,2-dioxazol-5-ones.[66–68] Chang optimized the intramolecular, Ir (III)-catalyzed γ-C–H amidation using a chiral diamine ligand and 1,4,2-dioxazol-5-ones as acylnitrenoid precursors (Scheme 2C).[69] The Curtius rearrangement pathway that commonly impedes the desired C–H-amidation through the formation of isocyanate side-products is completely suppressed with Ir catalyst **A1** under their established conditions. An impressive substrate scope is demonstrated using a variety of aliphatic sp³ C–H containing dioxazolones, leading to functionalized γ-stereogenic-γ-lactams in good yields and high enantioselectivities. The reaction can be carried out in air and has been demonstrated on half-gram scale without erosion of product enantiopurity.

Alkenyl C-H-activation has also been employed recently in the enantioselective synthesis of γ-lactams by Cramer.[70] Acrylamides and allenes are reacted in a (4+1) annulation to γ-stereogenic-α,β-unsaturated-γ-lactams via chiral cyclopentadienyl Rh (III) catalyst **A2** (Scheme 2D). AgOAc is used to generate the active Rh (III) catalyst by ligand substitution and Cu(OBz)$_2$ was found to be the optimal source of terminal oxidant to regenerate the catalyst. Mild heating and extended reaction times (2-4 days) lead to chiral α,β-unsaturated-γ-lactams in good yields and high enantioselectivities, although the scalability of this method is unclear. Other methods of note for the enantioselective construction and reactions of γ-lactams include hydrogenation,[71,72] C-H functionalization,[73,74] desaturation,[75] and cyclopropanations.[76]

Synthetic utility of α,β-unsaturated acylammonium salts:

Initial reports of acylammonium salts as reactive intermediates emerged as early as the 1930s with Wegler's description of an asymmetric acyl-transfer process.[77] The synthetic potential of acylammonium salts has since expanded to include a multitude of asymmetric organocascade processes initiated by different types of chiral acylammonium intermediates (Figure 12), leading to mono- and polycyclic frameworks of varying complexity. Developments in this area have been recently reviewed.[78, 79] In the last four years additional examples of asymmetric organocascade reactions using α,β-unsaturated acylammonium salts have been reported.

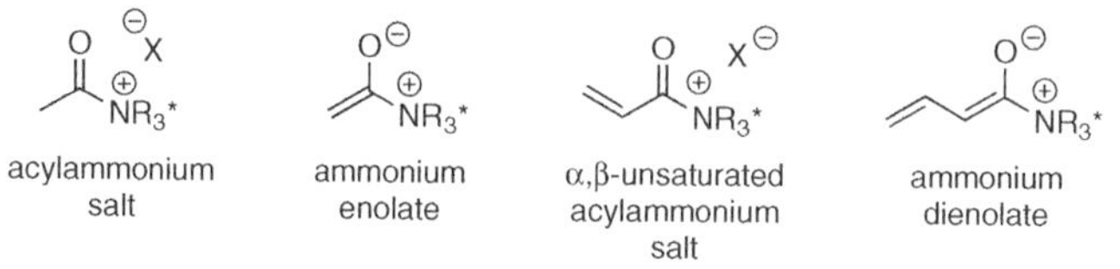

Figure 12. Chiral acylammonium salt intermediates

In 2016, Birman disclosed an asymmetric thio-Michael addition-aldol-β-lactonization-decarboxylation cascade process toward the synthesis of thiochromenes and in a related process accessed thiochromanes (Scheme 3).[80] Catalyzed by HBTM-2.0[81] or H-PIP[82], highly enantioselective generation of these heterocycles starting from readily available α,β unsaturated thioesters was demonstrated.

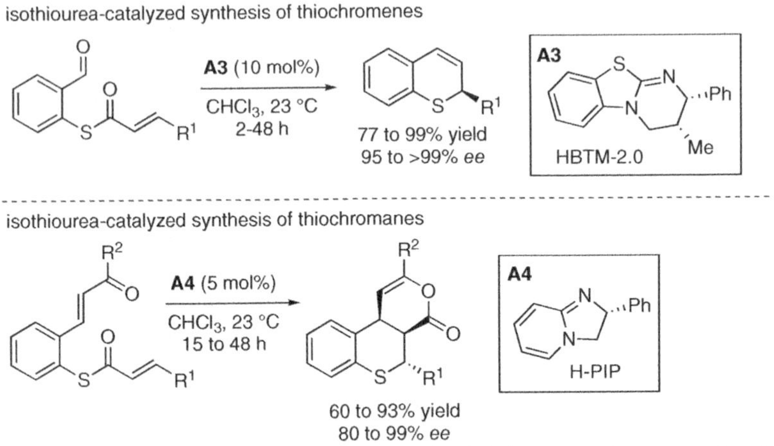

isothiourea-catalyzed synthesis of thiochromenes

Scheme 3. Asymmetric thio-Michael addition-aldol-γ-lactonization-decarboxylation cascade and thio-Michael-Michael-enol lactonization

In the same year, Smith reported chemo- and enantioselective, benzazole annulations catalyzed by HBTM-2.1 (**A5**) that employ α,β-unsaturated homoanhydride electrophiles and afford benzazole δ-lactams and lactones from acylbenzothiazole and acylbenzoxazole nucleophiles (Scheme 4).[83] Computational studies suggested that the presence of two 1,5-S–O n→σ* interactions following the initial Michael-addition step, dictate the chemoselectivity of δ-lactone vs δ-lactam formation.

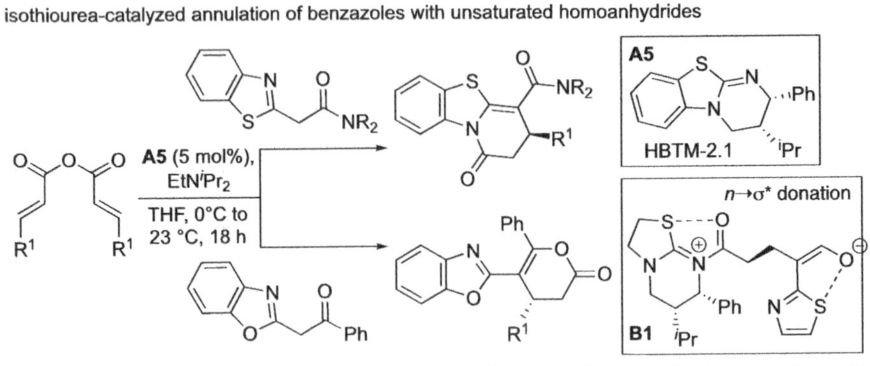

isothiourea-catalyzed annulation of benzazoles with unsaturated homoanhydrides

Scheme 4. Chemo-and enantioselective benzazole annulation catalyzed by HBTM-2.1

Until recently, catalytic turnover of the chiral Lewis-base in acylammonium organocascades was driven by acyl substitution with a pendant, *in situ* generated nucleophile. In 2017, Smith showed that catalyst regeneration can be facilitated by an aryloxide counterion that is produced following the formation of the acylammonium intermediate from α,β-unsaturated aryl esters, expanding the scope of potential nucleophiles applicable in these types of organocascades (Scheme 5).[84] The Smith group demonstrated the utility of this approach with an enantioselective isothiourea-catalyzed Michael-addition of nitroalkanes to the α,β-unsaturated aryl esters in up to 79% yield and 99:1 *er*. The branched nitroalkene products could be elaborated to optically active γ-lactams following chemoselective reduction of the nitro moiety. Smith quite recently described the use of aryloxide-facilitated turnover toward the enantioselective Michael-addition of N-heterocyclic pronucleophiles[85] and silylnitronates[86] to α,β-unsaturated aryl esters.

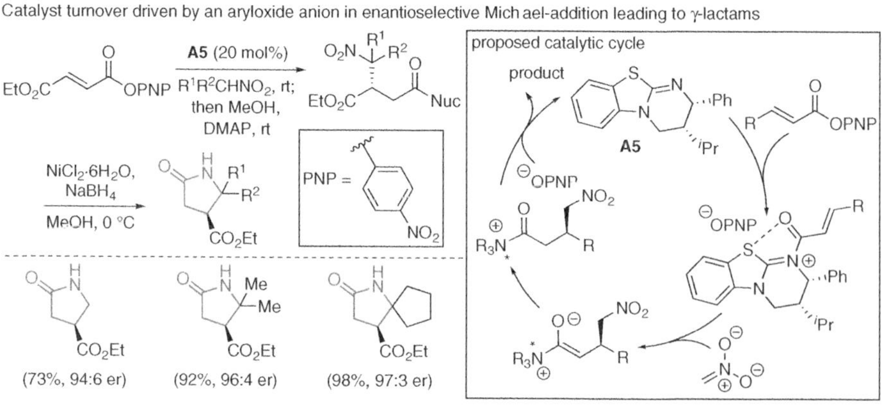

Scheme 5. Catalyst regeneration facilitated by an aryloxide counterion produced following acylammonium formation

A long-standing research focus of our group has been the application of chiral acylammonium salts in the organocatalytic, asymmetric synthesis of heterocycles. Since 2001, we developed a nucleophile-catalyzed/aldol/lactonization (NCAL),[87,88] nucleophile-catalyzed Michael/aldol/β-lactonization (NCMAL),[89] Diels-Alder/lactonization (DAL),[90,91] and a nucleophile-catalyzed Michael/proton transfer/lactamization (NCMPL),[92] the basis of the current procedure.

Nucleophile-catalyzed Michael/proton transfer/lactamization (NCMPL)

In 2013, we disclosed an enantioselective, organocatalytic nucleophile-catalyzed Michael/proton transfer/lactamization (NCMPL) cascade. Using α,β-unsaturated acylammonium salts generated from commercially available α,β-unsaturated acid chlorides, readily accessible O-trimethylsilylquinidine (TMSQD) or O-trimethylsilylquinine catalysts (TMSQN, **A6**), and N-protected aminomalonates as bis-nucleophilic reaction partners. The substrate scope from the original publication as well as the product of the current procedure (a new entry in the table) is illustrated below (Table 1).

During optimization of this method, the importance of 1,8-diazabicyclo[5.4.0]undec-7-ene (DBU) as an acid scavenger was noted, omitting this amine base resulting in <5% yield. The use of cinchona alkaloid derived, as opposed to isothiourea Lewis-bases, led to superior enantioselectivities. Additionally, substitution of LiHMDS with sodium bis(trimethylsilyl)amide (NaHMDS) also drastically reduced enantioselectivity, implicating the lithium cation as a crucial coordinating Lewis acid in the transition-state arrangement. This method was also applicable to the enantioselective synthesis of enol lactones when β-ketoesters are used as bis-nucleophilic reactants. The current procedure represents another example of the robustness of the NCMPL wherein a functionalized lactam is produced as virtually a single enantiomer on a decagram scale. The submitters have been able to scale the reaction up to 20 g of sulfonamidomalonate starting material without detriment to yield or enantiopurity. The above procedure can also be conducted using *diethyl*aminomalonate hydrochloride, which is significantly less costly, however, the submitters were unable to find conditions for the recrystallization of the product γ-lactam from the crude reaction mixture. The exclusion of column chromatography following the NCMPL to obtain the desired γ-lactam product, as is possible in the case of *dimethyl*aminomalonate, significantly simplifies the purification and was chosen for this reason.

Table 1. Substrate scope of NCMPL organocascade

Further expanding on the utility of chiral, α,β-unsaturated acylammonium salts, we adapted the NCMPL to the synthesis of medium-sized lactams (Table 2).[93] The resulting azepanones, benzazepinones, azocanones, and benzazocinones were generated in high enantiopurity and synthetically useful yield.

In summary, the NCMPL organocascade provides a convenient route to functionalized, small to medium-sized lactams, including pyroglutamic acid derivatives, in optically active form typically with high enantiopurity. The utility of the derived lactams, including applications to natural product synthesis, and established manipulations can be found in the published work from our group in this area.[92, 93, 94]

References

1. C.M.C, C.J.S. & D.R., Department of Chemistry & Biochemistry, Baylor University, 101 Bagby Ave., Waco, TX, 76710; daniel_romo@baylor.edu; ORCID: 0000-0003-3805-092X; P.T.J., Department of Chemistry, Lake Forest College, 555 North Sheridan Rd., Lake Forest, Ill, 60045. Support from NIH (GM052964, GM134910) and NSF (CHE-1800411) is gratefully acknowledged.
2. Nguyen, H.; Oh, S.; Henry-Riyad, H.; Sepulveda, D.; Romo, D. *Org. Synth.* **2011**; *88*, 121–137.
3. Panday, S. K.; Prasad, J.; Dikshit, D. K. *Tetrahedron Asymm.* **2009**, *20* (14), 1581–1632.
4. Nájera, C.; Yus, M. *Tetrahedron Asymm* **1999**, *10* (12), 2245–2303.
5. Saldívar González, F. I.; Lenci, E.; Trabocchi, A.; Medina-Franco, J. L. *RSC Adv.* **2019**, *9* (46), 27105–27116.
6. Khan, M. K.; Wang, D.; Moloney, M. G. *Synthesis* **2020**, *52* (11), 1602–1616.
7. Nay, B.; Riache, N.; Evanno, L. *Nat. Prod. Rep.* **2009**, *26* (8), 1044–1062.
8. Moloney, M. G.; Trippier, P., C.; Yaqoob, M.; Wang, Z. *Curr. Drug Disc. Technol.* **2004**, *1* (3), 181–199.
9. Nishimaru, T.; Eto, K.; Komine, K.; Ishihara, J.; Hatakeyama, S. *Chem. Eur. J.* **2019**, *25* (33), 7927–7934.
10. Kim, J. H.; Kim, I.; Song, Y.; Kim, M. J.; Kim, S. *Angew. Chem., Int. Ed.* **2019**, *58* (32), 11018–11022.
11. Eto, K.; Yoshino, M.; Takahashi, K.; Ishihara, J.; Hatakeyama, S. *Org. Lett.* **2011**, *13* (19), 5398–5401.
12. Onyango, E. O.; Tsurumoto, J.; Imai, N.; Takahashi, K.; Ishihara, J.; Hatakeyama, S. *Angew. Chem., Int. Ed.* **2007**, *46* (35), 6703–6705.
13. Kende, A. S.; Kawamura, K.; Devita, R. J. *J. Am. Chem. Soc.* **1990**, *112* (10), 4070–4072.
14. Mahashur, A.; Thomas, P. K.; Mehta, P.; Nivangune, K.; Muchhala, S.; Jain, R. *Lung India* **2019**, *36* (5), 422–433.
15. Gulder, T. A. M.; Moore, B. S. *Angew. Chem., Int. Ed.* **2010**, *49* (49), 9346–9367.
16. Shibasaki, M.; Kanai, M.; Fukuda, N. *Chem – Asian J.* **2007**, *2* (1), 20–38.
17. Uesugi, S.; Fujisawa, N.; Yoshida, J.; Watanabe, M.; Dan, S.; Yamori, T.; Shiono, Y.; Kimura, K.-i. *J. Antibiot.* **2016**, *69* (3), 133–140.

18. Chu, S.; Liu, S.; Duan, W.; Cheng, Y.; Jiang, X.; Zhu, C.; Tang, K.; Wang, R.; Xu, L.; Wang, X.; Yu, X.; Wu, K.; Wang, Y.; Wang, M.; Huang, H.; Zhang, J. *Pharmacol. Ther.* **2016**, *162*, 179–187.
19. Chu, S.-f.; Zhang, J.-t. *Acta Pharm. Sin. B* **2014**, *4* (6), 417–423.
20. He, H.; Yang, H. Y.; Bigelis, R.; Solum, E. H.; Greenstein, M.; Carter, G. T. *Tetrahedron Lett.* **2002**, *43* (9), 1633–1636.
21. Caruano, J.; Muccioli, G. G.; Robiette, R. *Org. Biomol. Chem.* **2016**, *14* (43), 10134–10156.
22. Shorvon, S. *Lancet* **2001**, *358* (9296), 1885–1892.
23. Varvounis, G.; Gerontitis, I. E.; Gkalpinos, V. *Chem. Heterocycl. Compd.* **2018**, *54* (3), 249–268.
24. Rivas, F.; Ling, T. *Org. Prep. Proced. Int.* **2016**, *48* (3), 254–295.
25. Ye, L.-W.; Shu, C.; Gagosz, F. *Org. Biomol. Chem.* **2014**, *12* (12), 1833–1845.
26. Ordóñez, M.; Cativiela, C. *Tetrahedron: Asymm.* **2007**, *18* (1), 3–99.
27. Soleimani-Amiri, S.; Vessally, E.; Babazadeh, M.; Hosseinian, A.; Edjlali, L. *RSC Adv.* **2017**, *7* (45), 28407–28418.
28. Deng, B.; Rao, C. B.; Zhang, R.; Li, J.; Liang, Y.; Zhao, Y.; Gao, M.; Dong, D. *Adv. Synth. Catal.* **2019**, *361* (19), 4549–4557.
29. Zhu, X. Q.; Yuan, H.; Sun, Q.; Zhou, B.; Han, X. Q.; Zhang, Z. X.; Lu, X.; Ye, L. W. *Green Chem.* **2018**, *20* (18), 4287–4291.
30. Zhmurov, P. A.; Ushakov, P. Y.; Novikov, R. A.; Sukhorukov, A. Y.; Ioffe, S. L. *Synlett* **2018**, *29* (14), 1871–1874.
31. Çinar, S.; Ünaleroglu, C. *Turk. J. Chem.* **2018**, *42* (1), 29–35.
32. Mardjan, M. I. D.; Mayooufi, A.; Parrain, J.-L.; Thibonnet, J.; Commeiras, L. *Org. Process Res. Dev.* **2020**, *24* (5), 606–614.
33. Borja-Miranda, A.; Sánchez-Chávez, A. C.; Polindara-García, L. A. *Eur. J. Org. Chem.* **2019**, *2019* (14), 2453–2471.
34. Gockel, S. N.; Buchanan, T. L.; Hull, K. L. *J. Am. Chem. Soc.* **2018**, *140* (1), 58–61.
35. De Marigorta, E. M.; De Los Santos, J. M.; De Retana, A. M. O.; Vicario, J.; Palacios, F. *Synthesis* **2018**, *50* (23), 4539–4554.
36. Audic, B.; Cramer, N. *Org. Lett.* **2020**, *22*, 5030-5034.
37. Jung, H.; Schrader, M.; Kim, D.; Baik, M. H.; Park, Y.; Chang, S. *J. Am. Chem. Soc.* **2019**, *141* (38), 15356–15366.
38. Huh, S.; Hong, S. Y.; Chang, S. *Org. Lett.* **2019**, *21* (8), 2808–2812.
39. Zhou, D.; Wang, C.; Li, M.; Long, Z.; Lan, J. *Chin. Chem. Lett.* **2018**, *29* (1), 191–193.

40. Png, Z. M.; Cabrera-Pardo, J. R.; Peiró Cadahía, J.; Gaunt, M. J. *Chem. Sci.* **2018**, *9* (39), 7628–7633.
41. Zheng, S.; Gutiérrez-Bonet, Á.; Molander, G. A. *Chem* **2019**, *5* (2), 339–352.
42. Koleoso, O. K.; Elsegood, M. R. J.; Teat, S. J.; Kimber, M. C. *Org. Lett.* **2018**, *20* (4), 1003–1006.
43. Jia, J.; Ho, Y. A.; Bülow, R. F.; Rueping, M. *Chem. Eur. J.* **2018**, *24* (53), 14054–14058.
44. Rao, W. H.; Jiang, L. L.; Liu, X. M.; Chen, M. J.; Chen, F. Y.; Jiang, X.; Zhao, J. X.; Zou, G. D.; Zhou, Y. Q.; Tang, L. *Org. Lett.* **2019**, *21* (8), 2890–2893.
45. Fukuyama, T.; Okada, T.; Nakashima, N.; Ryu, I. *Helv. Chim. Acta* **2019**, *102* (10), e1900186.
46. Ganesh Kumar, M.; Veeresh, K.; Nalawade, S. A.; Nithun, R. V.; Gopi, H. N. *J. Org. Chem.* **2019**, *84* (23), 15145–15153.
47. Ye, J.; Kalvet, I.; Schoenebeck, F.; Rovis, T. *Nat. Chem.* **2018**, *10* (10), 1037–1041.
48. Wang, F.; Zhang, X.; He, Y.; Fan, X. *Org. Biomol. Chem.* **2019**, *17* (1), 156–164.
49. Dražić, T.; Roje, M. *Chem. Heterocycl. Compd.* **2017**, *53* (9), 953–962.
50. Schröder, F.; Erdmann, N.; Noël, T.; Luque, R.; Van der Eycken, E. V., *Adv. Synth. Catal.* **2015**, *357* (14-15), 3141–3147.
51. Mourelle-Insua, Á.; Zampieri, L. A.; Lavandera, I.; Gotor-Fernández, V. *Adv. Synth. Catal.* **2018**, *360* (4), 686–695.
52. Su, Y.; Gao, S.; Li, H.; Zheng, G. *Process Biochem* **2018**, *72*, 96–104.
53. Bagum, H.; Christensen, K. E.; Genov, M.; Pretsch, A.; Pretsch, D.; Moloney, M. G. *Tetrahedron* **2019**, *75* (40), 130561.
54. Verho, O.; Maetani, M.; Melillo, B.; Zoller, J.; Schreiber, S. L. *Org. Lett.* **2017**, *19* (17), 4424–4427.
55. Enders, D.; Niemeier, O.; Henseler, A. *Chem. Rev.* **2007**, *107* (12), 5606–5655.
56. del Corte, X.; Maestro, A.; Vicario, J.; Martinez de Marigorta, E.; Palacios, F. *Org. Lett.* **2018**, *20* (2), 317–320.
57. Hu, B.; Deng, L. *Angew. Chem. Int. Ed.* **2018**, *57* (8), 2233–2237.
58. Collar, A. G.; Trujillo, C.; Lockett-Walters, B.; Twamley, B.; Connon, S. J. *Chem. Eur. J.* **2019**, *25* (30), 7275–7279.
59. Sućs Sajko, J.; Ljoljić Bilić, V.; Kosalec, I.; Jerić, I. *ACS Comb. Chem.* **2019**, *21* (1), 28–34.

60. de Gracia Retamosa, M.; Ruiz-Olalla, A.; Bello, T.; de Cozar, A.; Cossio, F. P. *Angew. Chem. Int. Ed.* **2018**, *57* (3), 668–672.
61. Hu, Z.; Zhu, Y.; Fu, Z.; Huang, W. *J. Org. Chem.* **2019**, *84* (16), 10328–10337.
62. Li, X. S.; Zhao, L. L.; Wang, X. K.; Cao, L. L.; Shi, X. Q.; Zhang, R.; Qi, J. *Org. Lett.* **2017**, *19* (14), 3943–3946.
63. Zhang, K.; Deiana, L.; Grape, E. S.; Inge, A. K.; Córdova, A. *Eur. J. Org. Chem.* **2019**, *2019* (29), 4649–4657.
64. Shi, Y.; Tan, X.; Gao, S.; Zhang, Y.; Wang, J.; Zhang, X.; Yin, Q. *Org. Lett.* **2020**, *22* (7), 2707–2713.
65. Wang, C.; Ge, S. *J. Am. Chem. Soc.* **2018**, *140* (34), 10687–10690.
66. Zhou, Z.; Chen, S.; Hong, Y.; Winterling, E.; Tan, Y.; Hemming, M.; Harms, K.; Houk, K. N.; Meggers, E. *J. Am. Chem. Soc.* **2019**, *141* (48), 19048–19057.
67. Xing, Q.; Chan, C. M.; Yeung, Y. W.; Yu, W. Y. *J. Am. Chem. Soc.* **2019**, *141* (9), 3849–3853.
68. Wang, H.; Park, Y.; Bai, Z.; Chang, S.; He, G.; Chen, G. *J. Am. Chem. Soc.* **2019**, *141* (17), 7194–7201.
69. Park, Y.; Chang, S. *Nat. Catal.* **2019**, *2* (3), 219–227.
70. Wang, S. G.; Liu, Y.; Cramer, N. *Angew. Chem. Int. Ed.* **2019**, *58* (50), 18136–18140.
71. Lang, Q.; Gu, G.; Cheng, Y.; Yin, Q.; Zhang, X. *ACS Catal.* **2018**, *8* (6), 4824–4828.
72. Yuan, Q.; Liu, D.; Zhang, W. *Org. Lett.* **2017**, *19* (5), 1144–1147.
73. Jette, C. I.; Geibel, I.; Bachman, S.; Hayashi, M.; Sakurai, S.; Shimizu, H.; Morgan, J. B.; Stoltz, B. M. *Angew. Chem. Int. Ed.* **2019**, *58* (13), 4297–4301.
74. Nanjo, T.; De Lucca, E. C.; White, M. C. *J. Am. Chem. Soc.* **2017**, *139* (41), 14586–14591.
75. Chen, M.; Dong, G. *J. Am. Chem. Soc.* **2017**, *139* (23), 7757–7760.
76. Harris, L.; Gilpin, M.; Thompson, A. L.; Cowley, A. R.; Moloney, M. G. *Org. Biomol. Chem.* **2015**, *13* (23), 6522–6550.
77. Wegler, R. *Justus Liebigs Ann. Chem.* **1932**, *498* (1), 62–76.
78. Vellalath, S.; Romo, D. *Angew. Chem. Int. Ed.* **2016**, *55* (45), 13934–13943.
79. Biswas, A.; Mondal, H.; Maji, M. S. *J. Heterocycl. Chem.* **2020**, *57*, 3818–3844.

80. Ahlemeyer, N. A.; Streff, E. V.; Muthupandi, P.; Birman, V. B. *Org. Lett.* **2017,** *19* (24), 6486–6489.
81. Birman, V. B.; Li, X. *Org. Lett.* **2008,** *10* (6), 1115–1118.
82. Birman, V. B.; Uffman, E. W.; Jiang, H.; Li, X.; Kilbane, C. J. *J. Am. Chem. Soc.* **2004,** *126* (39), 12226–12227.
83. Robinson, E. R. T.; Walden, D. M.; Fallan, C.; Greenhalgh, M. D.; Cheong, P. H.-Y.; Smith, A. D. *Chem. Sci.* **2016,** *7* (12), 6919–6927.
84. Matviitsuk, A.; Greenhalgh, M. D.; Antúnez, D.-J. B.; Slawin, A. M. Z.; Smith, A. D. *Angew. Chem., Int. Ed.* **2017,** *56* (40), 12282–12287.
85. Shu, C.; Liu, H.; Slawin, A. M. Z.; Carpenter-Warren, C.; Smith, A. D. *Chem. Sci.* **2020,** *11* (1), 241–247.
86. Matviitsuk, A.; Greenhalgh, M. D.; Taylor, J. E.; Nguyen, X. B.; Cordes, D. B.; Slawin, A. M. Z.; Lupton, D. W.; Smith, A. D. *Org. Lett.* **2020,** *22* (1), 335–339.
87. Morris, K. A.; Arendt, K. M.; Oh, S. H.; Romo, D. *Org. Lett.* **2010,** *12* (17), 3764–3767.
88. Cortez, G. S.; Tennyson, R. L.; Romo, D. *J. Am. Chem. Soc.* **2001,** *123* (32), 7945–7946.
89. Liu, G.; Shirley, M. E.; Van, K. N.; McFarlin, R. L.; Romo, D. *Nat. Chem.* **2013,** *5* (12), 1049–1057.
90. Abbasov, M. E.; Hudson, B. M.; Tantillo, D. J.; Romo, D. *Chem. Sci.* **2017,** *8* (2), 1511–1524.
91. Abbasov, M. E.; Hudson, B. M.; Tantillo, D. J.; Romo, D. *J. Am. Chem. Soc.* **2014,** *136* (12), 4492–4495.
92. Vellalath, S.; Van, K. N.; Romo, D. *Angew. Chem., Int. Ed.* **2013,** *52* (51), 13688–13693.
93. Kang, G.; Yamagami, M.; Vellalath, S.; Romo, D. *Angew. Chem. Int. Ed.* **2018,** *57* (22), 6527–6531.
94. Chaheine, C. M.; Gladen, P. T.; Abbasov, M. E.; Romo, D. *Org. Lett.* **2020,** *22* (23), 9282–9286.

Appendix
Chemical Abstracts Nomenclature (Registry Number)

Dimethyl aminomalonate hydrochloride; (16115-80-3)
p-Toluenesulfonic anhydride; (4124-41-8)
THF: Tetrahydrofuran; (109-99-9)
Triethylamine: N,N-Diethylethanamine, (121-44-8)
5-Phenyl-2,4-pentadienoic acid; (1552-94-9)
Dichloromethane; Methylene chloride; (75-09-2)
N,N-Dimethylformamide; (68-12-2)
Oxalyl chloride: Ethanedioyl dichloride; (79-37-8)
LiHMDS: Lithium hexamethyldisilazide; (79-43-6)
DBU: 1,8-Diazobyciclo[5.4.0]undec-7-ene; (6674-22-2)

Christian Michael Chaheine was born in El Paso, Texas. He obtained his bachelor's degree in biochemistry from the University of Texas at Arlington in 2014 and began graduate studies in synthetic organic chemistry under the mentorship of Prof Daniel Romo at Texas A&M University in the same year. He moved with the Romo group to Baylor University in the fall of 2015 where his research efforts centered around the total synthesis of the oxazolomycins and simplified derivatives toward the elucidation of new bioactivities and protein target identification. He began post-doctoral studies with Prof. Christopher Parker (Scripps Florida) in 2021.

Conner J. Song received his B.S. in Biochemistry & Molecular Biology and his B.A. in Art History from Wake Forest University in 2019 where he worked on characterizing prostate cancer stem cells under the mentorship of Dr. Bethany Kerr. He is currently working toward his Ph.D. under the guidance of Dr. Daniel Romo at Baylor University. His research is focused on mechanism of action studies of the curromycins and related natural products.

Paul T. Gladen obtained a B.A. in Chemistry from St. Olaf College and a PhD in Organic Chemistry from Indiana University where he studied the synthesis of complex terpenoid natural products with Professor David Williams. He continued his study of natural products as a postdoctoral research associate at Texas A&M and Baylor University with Professor Daniel Romo before joining the faculty at Lake Forest College as an Assistant Professor in 2016. His independent research is in the area of natural product synthesis with a current focus on the synthesis of antimicrobial heterocycles.

Daniel Romo received his B.A. in Chemistry/Biology from Texas A&M University and a Ph.D. in Chemistry from Colorado State University as a NSF Minority Graduate Fellow under the tutelage of the late Professor Albert I. Meyers. Following postdoctoral studies at Harvard as an American Cancer Society Fellow with Professor Stuart L. Schreiber, he began his independent career at Texas A&M in 1993 and then moved to Baylor University in 2015 where he is the Schotts Professor of Chemistry. Research interests in the Romo Group are at the interface of chemistry and biology focused on application of pharmacophore-directed retrosynthesis to assist with cellular mechanism of action studies of bioactive natural products, synthetic methodology focused on new organocascade processes including those directed toward the synthesis and application of β-lactones as intermediates for organic synthesis and exploring their utility as cellular probes and drug leads.

Matthew Genzink completed a B.Sc. in Chemistry from Grove City College in 2018 where he performed research with Charles Kriley. During this time, he performed a research internship for three months at KU Leuven (Belgium) where he studied fluorescent probes with Wim Dehaen. In 2018 Matthew started at UW–Madison, joining the group of Tehshik Yoon where he was an NSF Graduate Research Fellow. Matthew currently studies Brønsted acid-catalyzed asymmetric photochemistry in the Yoon group.

Preparation of 1-Hydrosilatrane, and Its Use in the Highly Practical Synthesis of Secondary and Tertiary Amines from Aldehydes and Ketones via Direct Reductive Amination

Fawwaz Azam and Marc J. Adler[1*]

Department of Chemistry & Biology, Ryerson University, 350 Victoria St., Toronto, ON, M5B2K3, Canada

Checked by Jin Su Ham and Richmond Sarpong

Procedure (Note 1)

A. *1-Hydrosilatrane* (**2**). A 1-L, three-necked (24/40 joints), round-bottomed flask containing a 5.0 cm × 0.5 cm Teflon-coated magnetic stir bar, is fitted with a reflux condenser. A rubber septum is attached to the top of the reflux condenser with a nitrogen inlet and an outlet needle, and a 24/40 glass stopper is attached to the side necks of the 1-L round-bottomed flask. The entire apparatus is flame-dried and, after removing the outlet needle, is allowed to cool to 23 °C under a flow of N_2. The glass stoppers are switched to rubber septa at this stage. The round-bottomed flask is charged in one portion with boratrane (6.28 g, 40 mmol, 1.0 equiv) (Note 2) through the side neck under a nitrogen flow. Then, using a 500-mL graduated cylinder, xylenes (400 mL) (Note 2) is added in one portion through the side neck. Then, using a 20-mL plastic syringe, triethoxysilane (11.1 mL, 60 mmol, 1.5 equiv) (Notes 2 and 3) is added in one portion through the septum. Finally, aluminum ethoxide (0.12 g, 0.76 mmol, 0.019 equiv) (Note 2) is added in one portion through the side neck. The aluminum ethoxide turns into a black suspension, but the reaction mixture homogenizes within an hour. Cold water is run through the condenser and the apparatus is lowered into a silicone oil bath (Note 4) pre-heated to 177 °C on a hot plate equipped with a temperature probe (Figure 1).

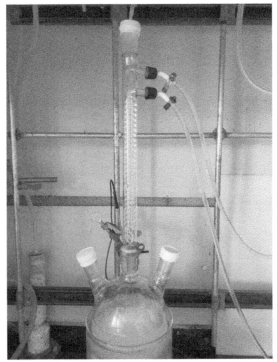

Figure 1. Reaction set-up (provided by checker)

The reaction mixture is stirred over 4 h under reflux while maintaining the oil bath at 177 °C. The clear, colorless reaction mixture is removed from the oil to cool to 23 °C under nitrogen (Figure 2), followed shortly by crashing out of the product (Figure 3).

Figure 2. Reaction mixture cooling (provided by checker)

Figure 3. Product crashing out (provided by checker)

The round-bottomed flask, disassembled from the condenser, is cooled in an ice/water bath for an additional 15 min, followed by vacuum filtration using a Büchner funnel, a 1-L filter flask, and a 185-mm diameter Whatman filter paper. An additional 40 mL of cold (0 °C) xylenes is used to transfer residual product from the round-bottomed flask and to rinse the filter cake. White, fibrous solid remains and is dried on the funnel under vacuum for an additional 10 min (Figure 4).

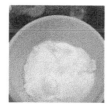

Figure 4. Filter cake

The solid is transferred to a 100-mL, single-necked, round-bottomed flask (24/40 joint) containing a 2.5 cm × 0.6 cm rod-shaped stir bar. CH_2Cl_2 (60 mL) (Note 5) is added using a 100-mL graduated cylinder and the content is stirred for 5 min (Figure 5). The entire suspension is passed through a celite plug (Note 6) with the aid of air pressure and collected in a 250-mL, single-necked (24/40 joint), round-bottomed flask. The clear, colorless eluent is concentrated on a rotary evaporator (Note 7) for 30 min to remove the solvent to yield a fine, white, crystalline powder. Before use, the product is dried overnight under high vacuum (0.30 mmHg) for 12 h to yield 1-hydrosilatrane (**2**) (Figure 6) (5.20 g, 74% yield, >99.5% purity) (Note 8 and 9).

Figure 5. After dissolving in CH_2Cl_2 (provided by checker)

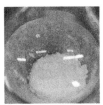

Figure 6. Final product (2)

B. *N-(4-Methylbenzyl)aniline* (**4**). A 50-mL, single-necked, round-bottomed flask (24/40 joint) containing a 1.2 cm × 0.8 cm egg-shaped stir bar is charged sequentially and in one portion with *p*-tolualdehyde (1.6 mL, 13.6 mmol, 1.0 equiv) (Note 10) and aniline (1.5 mL, 16.4 mmol, 1.2 equiv) (Note 10) using a 3-mL plastic syringe through the septum with a vent needle, resulting in a cloudy, opaque solution (Figure 7A). Following this, acetic acid (14 mL) (Note 10) is added over 15 sec via a 20-mL plastic syringe through the septum, resulting in a homogenous bright-yellow solution (Figure 7B). The reaction mixture is stirred and then 1-

hydrosilatrane (4.80 g, 27.4 mmol, 2.0 equiv) (**2**) (Note 10) is added in one portion through the neck. The resulting opaque reaction mixture homogenizes as more silatrane dissolves (Figure 7C). After removing the vent needle, the reaction mixture is stirred over 3 h at 23 °C under ambient atmosphere.

A. After addition of aniline

B. After addition of acetic acid

C. After addition of hydrosilatrane

D. Reaction mixture at 3 h

Figure 7. Reaction mixture

The reaction mixture, now a pale-yellow solution (Figure 7D), is transferred to a 500-mL separatory funnel and diluted with CH_2Cl_2 (140 mL). (Note 5). Aqueous NaOH (1 M, 300 mL) (Note 11), which is added to the separatory funnel in 50 mL portions to minimize an exotherm, is used to wash the solution through inversion of the separatory funnel, with frequent venting (Figure 8) (Note 12). The organic layer is collected, and the aqueous layer is then extracted with CH_2Cl_2 (2 × 140 mL). The combined organic layer is washed with H_2O (100 mL) and dried over $MgSO_4$ (4 g) (Note 13) for 5 min with occasional swirling by hand and then filtered by gravity, using a plastic funnel and a 185-mm diameter Whatman filter paper, into a 1-L, single-necked, round-bottomed flask (24/40 joint).

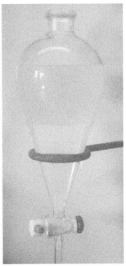

Figure 8. After 1M NaOH addition (provided by checker)

Removal of the solvent by rotary evaporation is performed until the volume is small enough to transfer to a 100-mL single-necked, round-bottomed flask (24/40) (Note 14). After transferring the pale-yellow liquid to 100-mL single-necked, round-bottomed flask, an additional 10 mL of CH_2Cl_2 is used to transfer residual product from the 1-L round-bottomed flask. Final removal of the solvent by rotary evaporation (Note 15) yields a pale-yellow liquid (Figure 9), which is then placed in an oil bath preheated to 60 °C and further dried under high vacuum (0.30 mmHg) for 12 h. After cooling to 23 °C, the resulting solid (Figure 10) is washed with cold (0 °C) petroleum ether (1 mL) and, after the liquid is removed by trituration, is dried under high vacuum for 12 h (0.30 mmHg) (Note 16). The solid product is heated to 50 °C on a water bath until melted (~5 mins) and then cooled to 23 °C while under high vacuum (0.30 mmHg) overnight to yield the product *N*-(4-methylbenzyl)aniline (**4**) as a pale-brown crystalline solid (Figure 11) (2.22 g, 83% yield, 98.9% purity) (Notes 17 and 18).

Figure 9. Before hi-vacuum

Figure 10. After hi-vacuum

Figure 11. Final product (4)

Notes

1. Prior to performing each reaction, a thorough hazard analysis and risk assessment should be carried out with regard to each chemical substance and experimental operation on the scale planned and in the context of the laboratory where the procedures will be carried out. Guidelines for carrying out risk assessments and for analyzing the hazards associated with chemicals can be found in references such as Chapter 4 of "Prudent Practices in the Laboratory" (The National Academies Press, Washington, D.C., 2011; the full text can be accessed free of charge at https://www.nap.edu/catalog/12654/prudent-practices-in-the-laboratory-handling-and-management-of-chemical. See also "Identifying and Evaluating Hazards in Research Laboratories" (American Chemical Society, 2015) which is available via the

associated website "Hazard Assessment in Research Laboratories" at https://www.acs.org/content/acs/en/about/governance/committees/chemicalsafety/hazard-assessment.html. In the case of this procedure, the risk assessment should include (but not necessarily be limited to) an evaluation of the potential hazards associated with boratrane, triethoxysilane, xylenes, aluminum ethoxide, p-tolualdehyde, aniline, acetic acid, 1 M NaOH, and dichloromethane. Appropriate care must be taken when performing reactions at high temperatures. Triethoxysilane, in particular, must be handled in a fume hood.

2. Boratrane (97%) was obtained from Sigma-Aldrich. It can also be synthesized according to published procedure.[2] Xylenes (ACS reagent grade) was obtained from ACP chemicals. Triethoxysilane (96%) was obtained from Alfa Aesar. Aluminum ethoxide (97%) was obtained from Sigma-Aldrich. All chemicals were used as received.

3. An excess of triethoxysilane was used to ensure full conversion and faster reaction time. Using less than 1.5 equivalents may result in incomplete conversion and slower reaction time.

4. Silicone oil bath was purchased from Alfa Aesar. Alternatively, a heating mantle works equally well for this protocol.

5. Dichloromethane (Reagent grade) was purchased from ACP chemicals and used as received.

6. A column was packed with glass wool and sand. Celite was added on top of the sand layer (1 inch). Glass wool was purchased from Thermo Scientific. Sand was purchased from Fisher Chemical. Celite was purchased from ACP chemicals.

7. The following rotary evaporator settings were used: 500 mmHg, 40 °C water bath, for 15 minutes. Then 40 mmHg, 40 °C water bath for 15 min.

8. Characterization data for compound (**2**): mp 250–256 °C; ^{1}H NMR (700 MHz, CDCl$_3$) δ: 2.87 (t, J = 5.9 Hz, 6H), 3.79 (t, J = 6.0 Hz, 6H), 3.88 (s, 1H); ^{13}C NMR (176 MHz, CDCl$_3$) δ: 51.0 (s), 57.1 (s); ^{29}Si NMR (119 MHz, CDCl$_3$) δ: –83.1 (s); FTIR (powder): 2974, 2934, 2885, 2085, 1860, 1456, 1348, 1268, 1166, 1082, 1019, 930, 910, 862, 752 cm^{-1}. The product is air- and moisture-stable and can be stored under air for several months. The purity of the product was determined by qNMR by dissolving 18.8 mg of pyrazine (>99%, purchased from Sigma-Aldrich and used as received) and 40.8 mg of compound (**2**) in 0.9 mL CDCl$_3$. The purity was determined to be >99.5%.

9. A second reaction on the same scale afforded the product (2) with 73% yield and 98.5% purity.
10. *p*-Tolualdehyde was purchased from Sigma-Aldrich (97%), aniline was purchased from Alfa Aesar (99+%), glacial acetic acid (ACS reagent grade) was purchased from ACP chemicals. All chemicals were used as received.
11. 1 M NaOH was made using NaOH pellets purchased from EMPLURA and used as received. A single washing of the organic solvent was performed using all 300 mL of aq NaOH, but the NaOH solution was added 50 mL portions to minimize exotherm concerns.
12. A white insoluble suspension formed as the additional volume of aq. NaOH increased. The solution became clear after the addition of 300 mL NaOH.
13. MgSO$_4$ was purchased from Alfa Aesar (powder, anhydrous, 99.5%).
14. The following rotary evaporator settings were used: 520 mmHg, 40 °C.
15. The following rotary evaporator settings were used: 412 mmHg, 50 °C water bath for 20 min.
16. Petroleum ether was purchased from Sigma-Aldrich (ACS grade) and used as received.
17. Characterization data for compound (4): mp 42–45 °C; ^{1}H NMR (700 MHz, CDCl$_3$) δ: 2.38 (s, 3H), 3.96 (br, 1H), 4.29 (s, 2H), 6.65 (d, *J* = 7.9 Hz, 2H), 6.74 (t, *J* = 7.3 Hz, 1H), 7.22–7.15 (m, 4H), 7.28 (d, *J* = 7.8 Hz, 2H); ^{13}C NMR (176 MHz, CDCl$_3$): 21.2, 48.2, 112.9, 117.6, 127.7, 129.4, 129.4, 136.5, 137.0, 148.3; FTIR (neat): 3417, 3015, 2914, 1600, 1510, 1438, 1323, 1269, 1178, 1097, 1042, 983 $^{-1}$. HRMS: (ESI) calc'd for C$_{14}$H$_{16}$N [M + H]$^+$: 198.1277. Found: 198.1278. The product is stable and can be stored under air for several months. The purity of the product was determined by qNMR by dissolving 13.9 mg of pyrazine (>99%, purchased from Sigma-Aldrich and used as received) and 32.0 mg of compound (4) in 0.9 mL CDCl$_3$. The purity was determined to be 98.9%. Based on this % purity determination, the mass of product is 2.19 g (82%).
18. A second reaction on the half-scale afforded the product (4) with 90% yield and >99.5% purity.

Working with Hazardous Chemicals

The procedures in *Organic Syntheses* are intended for use only by persons with proper training in experimental organic chemistry. All hazardous materials should be handled using the standard procedures for work with chemicals described in references such as "Prudent Practices in the Laboratory" (The National Academies Press, Washington, D.C., 2011; the full text can be accessed free of charge at http://www.nap.edu/catalog.php?record_id=12654). All chemical waste should be disposed of in accordance with local regulations. For general guidelines for the management of chemical waste, see Chapter 8 of Prudent Practices.

In some articles in *Organic Syntheses*, chemical-specific hazards are highlighted in red "Caution Notes" within a procedure. It is important to recognize that the absence of a caution note does not imply that no significant hazards are associated with the chemicals involved in that procedure. Prior to performing a reaction, a thorough risk assessment should be carried out that includes a review of the potential hazards associated with each chemical and experimental operation on the scale that is planned for the procedure. Guidelines for carrying out a risk assessment and for analyzing the hazards associated with chemicals can be found in Chapter 4 of Prudent Practices.

The procedures described in *Organic Syntheses* are provided as published and are conducted at one's own risk. *Organic Syntheses, Inc.*, its Editors, and its Board of Directors do not warrant or guarantee the safety of individuals using these procedures and hereby disclaim any liability for any injuries or damages claimed to have resulted from or related in any way to the procedures herein.

Discussion

The synthesis of amines is highly desired for their wide range of applications from pharmaceuticals to fine chemicals.[3] By far the most practical method to prepare amines is through direct reduction amination[4] (DRA), which involves reacting an amine with an aldehyde/ketone to generate an imine or an iminium, which is then reduced to the corresponding amine. The most common reducing agents used are $NaBH_3CN$ and $NaBH(OAc)_3$ due to their widespread availability and the mild reaction conditions required.[5,6] However, there are certain drawbacks associated with their usage: NaBH$_3$CN is toxic and forms toxic by-products upon workup, NaBH(OAc)$_3$ is not compatible with some aromatic ketones or amines.[6,7]

Organosilanes can be used as alternatives to borohydrides as reducing agents for DRA reactions.[8] Polymethylhydrosiloxane (PMHS) has been shown to carry out these reactions and presents certain advantages over the borohydride reducing agents mentioned such as its non-toxicity, low price, and inertness in the presence of an activator.[9] However, they require activating catalysts, such as TFA, Bu$_2$SnCl, or InCl$_3$ to promote reduction.[10,11,12] The use of these catalysts introduces issues such as incompatibility with acid-labile functional groups, toxicity, and cost, which are undesirable.

Our lab recently demonstrated a DRA method using aldehydes and ketones to access secondary and tertiary amines using 1-hydrosilatrane (**2**) as the reducing agent.[13] (We have also used hydrosilatrane to carry out reduction of aryl aldehydes to alcohols,[2] reduction of ketones to alcohols,[14] and reductive acetylation of aldehydes;[15] we have also published asymmetric versions of the ketone reduction[16] to secondary alcohols and the DRA[17] of ketones.) Hydrosilatrane is easily prepared and is air- and moisture-stable making it easy to handle. The preparation of secondary amines requires AcOH as the solvent while tertiary amines can be accessed under neat conditions. This method is applicable to alkyl- and aryl- amines and shows broad functional group tolerance. The factors mentioned make this an attractive alternative method to access secondary and tertiary amines via DRA. A selection of the substrate scope for the preparation of secondary amines and tertiary amines from aldehydes and ketones is shown in Tables 1-4 below.

Table 1. Scope for aldehydes and primary amines for DRA to form secondary amines

Table 2. Scope for aldehydes and secondary amines for DRA to form tertiary amines

[Reaction scheme: R¹CHO + HNR²R³ → (2 (2 eq.), 70 °C, overnight) → R¹CH₂NR²R³]

Products:
- Bn-N(Et)₂ from benzyl: 80%
- (i-Pr)₃Si-O-C₆H₄-CH₂-N(Ph)₂ : 87%
- 4-MeO-C₆H₄-CH₂-NBn₂ : 97%
- i-Bu-CH₂-NBn₂ : 90%

Table 3. Scope for ketones and primary amines for DRA to form secondary amines

[Reaction scheme: R¹C(O)R² + H₂N-R³ → (2 (2 eq.), 70 °C, overnight) → R¹R²CH-NHR³]

Products:
- Ph-CH(Me)-NHPh : 98%
- Cyclopentyl-NHBn : 76%
- Bu-CH(Me)-NHBn : 92%
- Bu-CH(Me)-NHPh : 72%

Table 4. Scope for ketones and secondary amines for DRA to form tertiary amines

[Reaction scheme: R¹C(O)R² + HNR³R⁴ → (2 (2 eq.), CHCl₃, 60 °C, overnight) → R¹R²CH-NR³R⁴]

Products:
- 4-O₂N-C₆H₄-CH(Me)-NBu₂ : 91%
- Ph-CH(Me)-pyrrolidine : 85%
- Cyclohexyl-pyrrolidine : 75%
- Bu-CH(Me)-pyrrolidine : 80%

References

1. Prof. Marc J. Adler, Department of Chemistry & Biology, Ryerson U Department of Chemistry & Biology, Ryerson University, 350 Victoria St., Toronto, ON, M5B2K3, Canada, marcjadler@ryerson.ca, ORCID ID 0000-0002-1049-509X. The authors acknowledge Ryerson University, Northern Illinois University and the Natural Sciences and Engineering Research Council of Canada (NSERC) for financial support.
2. Skrypai, V.; Hurley, J. J. M.; Adler, M. J. *Eur. J. Org. Chem.* **2016**, *2016*, 2207–2211.
3. Abdel-Magid, A. F.; Mehrman, S. J. *Org. Process Res. Dev.* **2006**, *10*, 971–1031.
4. (a) Baxter, E. W.; Reitz, A. B. *Org. React.* **2002**, *59*, 1–714; (b) Alinezhad, H.; Yavari, H.; Salehian, F. *Curr. Org. Chem.* **2015**, *19*, 1021–1049.
5. Borch, R. F.; Bernstein, M. D.; Durst, H. P. *J. Am. Chem. Soc.* **1971**, *93*, 2897–2904.
6. Abdel-Magid, A. F.; Carson, K. G.; Haris, B. D.; Maryanoff, C. A.; Shah, R. D., *J. Org. Chem.* **1996**, *61*, 3849–3862.
7. Miura, K.; Ootsuka, K.; Suda, S.; Nishikori, H.; Hosomi, A. *Synlett.* **2001**, *10*, 1617–1619.
8. Kangasmetsa, J. J.; Johnson, T. *Org. Lett.* **2005**, *7*, 5653–5655.
9. Lawrence, N. J.; Drew, M. D.; Bushell, S. M. *J. Chem. Soc. Perkin Trans. 1* **1999**, 3381–3391.
10. Dube, D.; Scholte, A. A. *Tetrahedron Lett.* **1999**, *40*, 2295–2298.
11. Kato, H.; Shibata, I.; Yasaka, Y.; Tsunoi, S.; Yasuda, M.; Baba, A. *Chem. Commun.* **2006**, *40*, 4189–4191.
12. Nakajima, M.; Takahashi, H.; Sasaki, M.; Kobayashi, Y.; Awano, T.; Irie, D.; Sakemi, K.; Ohno, Y.; Usami, M. *Teratog. Carcinog. Mutagen.* **1998**, *18*, 231–238.
13. Varjosaari, S. E.; Skrypai, V.; Suating, P.; Hurley, J. J. M.; De Lio, A. M.; Gilbert, T. M.; Adler, M. J. *Adv. Synth. Catal.* **2017**, *359*, 1872–1878.
14. Varjosaari, S.E.; Skrypai, V.; Suating, P.; Hurley, J.J.M.; Gilbert, T.M.; Adler, M.J. *Eur. J. Org. Chem.* **2017**, *2017*, 229–232.
15. James, R.R.; Herlugson, S.M.; Varjosaari, S.E.; Skrypai, V.; Gilbert, T.M.; Adler, M.J. *SynOpen.* **2019**, *3*, 1–3.
16. Varjosaari, S.E.; Skrypai, V.; Herlugson, S.M.; Gilbert, T.M.; Adler, M.J. *Tetrahedron Lett.* **2018**, *59*, 2839–2843.
17. Skrypai, V.; Varjosaari, S.E.; Gilbert, T.M.; Adler, M.J. *J. Org. Chem.* **2019**, *84*, 5021–5026.

Appendix
Chemical Abstracts Nomenclature (Registry Number)

Boratrane: 2,8,9-Trioxa-5-aza-1-borabicyclo[3.3.3]undecane; (283-56-7)
Triethoxysilane: Silane, triethoxy-; (998-30-1)
Aluminum ethoxide: Ethanol, aluminum salt (3:1); (555-75-9)
Xylenes: Benzene, dimethyl-; (1330-20-7)
1-Hydrosilatrane: 2,8,9-Trioxa-5-aza-1-silabicyclo[3.3.3]undecane; (283-60-3)
p-Tolualdehyde: Benzaldehyde, 4-methyl-; (104-87-0)
Aniline: Benzenamine; (62-53-3)
N-(4-Methylbenzyl)aniline: Benzenemethanamine, 4-methyl-*N*-phenyl-; (15818-64-1)

Fawwaz Azam was born in tropical Penang, Malaysia and raised in Dubai, United Arab Emirates. He moved to Toronto, Canada in 2014 and obtained his BSc in Chemistry from the University of Toronto in 2018. He is currently in the MSc program in Molecular Science at Ryerson University, where his research focuses on developing new synthetic organic methods using silanes.

Marc J. Adler is an Assistant Professor in the Department of Chemistry & Biology at Ryerson University (Canada). He was born in San Diego, CA (USA), received degrees in chemistry from University of California, Berkeley (USA, BSc) and Duke University (USA, Ph.D.), and further trained as a postdoctoral researcher at Yale University (USA) and University of Oxford (UK). He began his independent academic career at Northern Illinois University (USA). He loves family and friends, and lives by the motto "stay far from timid, only make moves when your heart's in it, and live the phrase 'sky's the limit'" (Christopher Wallace).

Jin Su Ham received his B. S. and M. S. in chemistry from KAIST where he worked in the lab of Prof. Hee-Yoon Lee, researching total synthesis of natural products. After 9-years of industrial experience at SK Innovation, he is currently a Ph.D. student in the laboratories of Prof. Richmond Sarpong at UC Berkeley. His research focuses on developing new synthetic methods using transition metal catalysis.

Synthesis of Tetraaryl-, Pentaaryl-, and Hexaaryl-1,4-dihydropyrrolo[3,2-b]pyrroles

Maciej Krzeszewski, Mariusz Tasior, Marek Grzybowski, and Daniel T. Gryko*[1]

Institute of Organic Chemistry, Polish Academy of Sciences, Kasprzaka 44-52, 01-224 Warsaw, Poland

Checked by Shu Nakamura, Koichi Hagiwara, and Masayuki Inoue

Procedure (Note 1)

A. **1,4-Bis(4-(tert-butyl)phenyl)-2,5-bis(4-cyanophenyl)-1,4-dihydropyrrolo[3,2-b]pyrrole (1).** A three-necked, 200 mL round-bottomed flask (main neck 29/32 joint, side neck 15/25) is equipped with a 5 cm Teflon-coated football shaped magnetic stir bar. All three necks are left open to the atmosphere. Acetic acid (50 mL) (Note 2) and toluene (50 mL) (Note 3) are poured into the flask *via* a glass funnel. 4-Cyanobenzaldehyde (7.87 g, 60.0 mmol, 2 equiv) (Note 4) is dissolved in the solvent mixture, then 4-*tert*-butylaniline (9.0 g, 9.6 mL, 60 mmol, 2 equiv) (Note 5) is added *via* a plastic syringe (Figure 1A). The flask is immersed in a silicone oil bath preheated to 50 °C (Notes 6 and 7). The clear yellow solution is stirred for 30 min under air

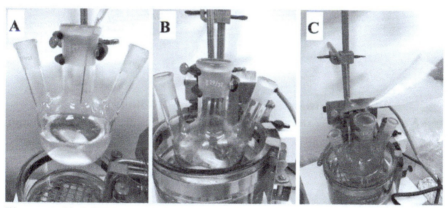

Figure 1. A) Addition of 4-*tert*-butylaniline *via* a syringe into dissolved 4-cyanobenzaldehyde. B) Solution stirred for 30 min at 50 °C. C) Solution color change after addition of Fe(OTs)$_3$· 6H$_2$O.

atmosphere (Figure 1B). Iron(III) *p*-toluenesulfonate hexahydrate (1.22 g, 1.80 mmol, 6 mol%) (Note 8) is added, causing immediate reddening of the solution (Figure 1C). Immediately after addition of iron(III) *p*-toluenesulfonate hexahydrate, butane-2,3-dione (2.6 g, 2.6 mL, 30 mmol, 1 equiv) is added dropwise over 3 min *via* a plastic syringe, causing blackening of the solution (Note 9) (Figure 2A). The mixture is heated at 50 °C. After ca. 10 min, the reaction mixture turns orange and a yellow precipitate is formed (Figure 2B). The suspension is vigorously stirred for 16 h at 50 °C with all three necks uncapped to provide access of the oxygen from the air (Note 10) (Figure 2C). The oil bath is removed to allow the reaction mixture to cool to room temperature over 15 min (Note 11). The resulting dark mixture with yellow precipitate (Figure 3A) is treated with methanol (150 mL) (Note 12) and vacuum filtered through a Kiriyama-funnel (S-60, filter paper: No.5B, 60 mmφ). The filtered solid is rinsed with methanol (150 mL) (Figure 3B), followed by diethyl ether (50 mL) (Note 13) (Figure 3C). The resulting solid is dried under vacuum (4.8 mmHg) at 95 °C for 3 h (Figure 4A) to give the tetraarylpyrrolo[3,2-*b*]pyrrole (TAPP, compound **1**) as a yellow solid (10.21 g, 17.82 mmol, 59%) (Notes 14, 15, and 16) (Figure 4B).

Figure 2. A) Dropwise addition of butane-2,3-dione via a syringe. B) Precipitate formed after ca. 10 min after dione addition. C) Suspension after 1 hour stirring.

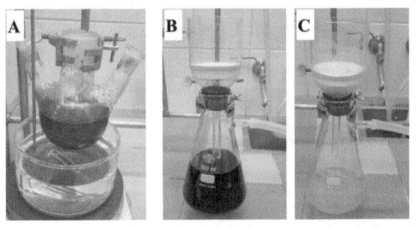

Figure 3. A) Reaction mixture after 16 h. B) Appearance after rinsing with methanol. C) Appearance after rinsing with diethyl ether (photos provided by submitters).

Figure 4. A) Drying set-up. B) Pure and dry TAPP (compound 1) (photos provided by submitters).

B. *[1,4-Bis(diphenylphosphino)butane](η³-allyl)palladium (II) chloride (PdCl(C₃H₅) (dppb))* (Note 17). A 30 mL Schlenk flask (15/25 joint) is equipped with a 2 cm Teflon-coated football shaped magnetic stir bar. The flask is connected to an argon/vacuum line and the reaction setup is dried under vacuum (4.8 mmHg) by heating with a heat gun and allowed to cool to room temperature. The reaction set-up is backfilled with argon and the argon atmosphere is maintained throughout the reaction using an argon manifold system. The flask is charged with allylpalladium chloride dimer ([Pd(C₃H₅)Cl]₂, 200 mg, 0.547 mmol) (Note 18) and 1,4-bis(diphenylphosphino)butane (dppb, 460 mg, 1.08 mmol) (Note 19). The reaction set-up is evacuated and backfilled with argon three times. Anhydrous dichloromethane (10 mL) (Note 20) is added and the flask is immersed in the silicone oil bath preheated to 25 °C. The solution is stirred at 25 °C for 30 min. The initial yellow solution (Figure 5A) changes color to red in ca. 15 min (Figure 5B). The solvent is removed in vacuo (4.8 mmHg at room temperature) and dried under vacuum (4.8 mmHg at room temperature for 3 h) to give a pinkish powder, which is used without further purification (Note 21) (Figure 5C-D).

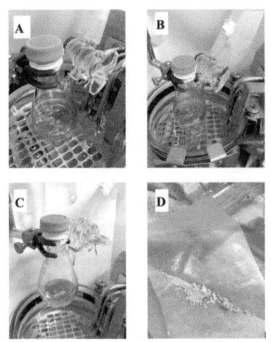

Figure 5. A) Appearance of reaction mixture immediately after addition of dichloromethane. B) Solution stirred for 15 min at 25 °C C) Drying set-up. D) Appearance of the crude PdCl(C₃H₅)(dppb) catalyst.

C. *1,4-Bis(4-(tert-butyl)phenyl)-2,5-bis(4-cyanophenyl)-3,6-bis(4-nitrophenyl)-1,4-dihydropyrrolo[3,2-b]pyrrole (2)*. A three-necked, 500 mL round-bottomed flask (main neck 29/32 joint, side neck 15/25 joint) (flask A) is equipped with a 5 cm Teflon-coated football shaped magnetic stir bar. A 29/32 three-way stop cock is attached to the main neck of the flask and sealed with Teflon tape. The flask is connected to an argon/vacuum line and the reaction setup is dried under vacuum (4.8 mmHg) by heating with a heat gun and allowed to cool to room temperature. Next, the reaction set-up is backfilled with argon and the flask is charged with **1** (5.73 g, 10.0 mmol, 1 equiv), 1-bromo-4-nitrobenzene (12.1 g, 60.0 mmol, 6 equiv) (Note 22), PdCl(C₃H₅)(dppb) (0.60 g, 0.98 mmol, 10 mol%) and potassium acetate (5.30 g, 54.0 mmol, 5.4 equiv) (Note 23) while purging the flask with argon. The reaction set-up is evacuated and backfilled with argon three times and sealed with a rubber septum. Dry *N,N*-dimethylacetamide (DMA, 250 mL) (Note 24) is added to flask A *via* a cannula from a three-necked, 500 mL round-bottomed flask equipped with a three-way stopcock and sealed with rubber septa and Teflon tape) (flask B) with the constant argon flow (Figure 6A). The three-way stopcock of the flask A is closed after solvent addition. The cannula is removed and the flask A is immersed in the silicone oil bath preheated to 150 °C (Note 25).

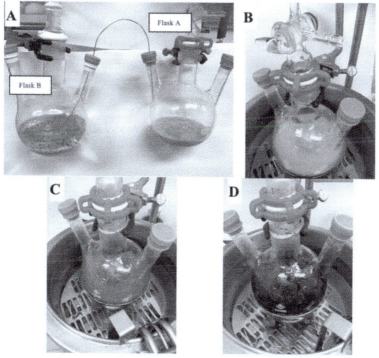

Figure 6. A) Cannulation of DMA from flask B to reaction flask A. B) Initial color of the solution. C) Appearance after 30 min. D) Appearance after 90 min.

Gradual color change of the solution from yellow through green to black is apparent (Figures 6B-D). The solution turns black completely in ca. 90 min. The mixture is stirred at 150 °C for 24 h. After cooling to room temperature over 30 min, the stir bar is removed and the resulting reaction mixture is transferred to 1 L round-bottomed flask, which is rinsed with ethyl acetate (60 mL) that is added to the reaction mixture (Note 26). The reaction mixture is concentrated on a rotary evaporator (70 °C, 10 mmHg) and dried under vacuum (4.8 mmHg, 30 min, room temperature). After evaporation of the reaction mixture, the tarry-black residue (Figure 7A) is dissolved in dichloromethane (250 mL) and transferred to the 1L separatory funnel. Organic layer is washed with distilled water (2 × 250 mL), brine (250 mL) and dried over anhydrous $MgSO_4$ (30 g) (Note 27) for 2 h. The drying agent is filtered off through a cotton plug and rinsed with dichloromethane (150 mL). The filtrate is concentrated on a rotary evaporator (40 °C, 600 to 70 mmHg) to dryness (Figure 7B) and acetonitrile (30 mL) (Note 28) is poured onto the residue. After stirring for 30 min, yellow precipitate formed is vacuum filtered through a Kiriyama-funnel (S-60, filter paper: No.5B, 60 mmφ) and rinsed with acetonitrile (20 mL) (Figures 7C and 10) (Notes 29 and 30). Collected solid is dissolved in dichloromethane (125 mL), and silica gel (50 g)

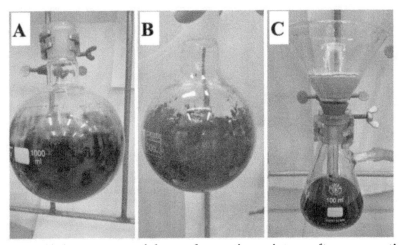

Figure 7. A) Appearance of the crude reaction mixture after evaporation. B) Appearance of the concentrated organic phase after liquid-liquid extraction. C) Filtration set-up after rinsing with acetonitrile (photos provided by submitters).

(Note 31) is added to the solution. The suspension is concentrated on a rotary evaporator (40 °C, 540 to 70 mmHg) to dryness. The resulting solid is purified by a gradient column chromatography on silica gel using hexane/dichloromethane mixture as an eluent (Figures 8A and 11) (Notes 32, 33, and 34). Fractions containing solely the desired product are collected, combined, evaporated and dried under vacuum (4.8 mmHg, 30 min, room temperature) (Figure 8B). The solid material is treated with diethyl ether (25 mL) and vacuum filtered through a Kiriyama-funnel (S-60, filter paper:

No.5B, 60 mmφ) and washed with diethyl ether (25 mL) (Figure 8C). The yellow crystals are collected, dried under vacuum (4.8 mmHg) at 85 °C for 1 hour to give desired hexaarylpyrrolopyrrole (compound **2**) as a yellow solid (4.98 g, 6.11 mmol, 61%) (Figure 8D) (Note 35).

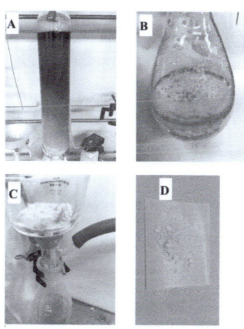

Figure 8. A) Appearance of a column chromatography in progress. B) Drying set-up for combined fractions 13-24. C) Filtration set-up after rinsing with diethyl ether. D) Appearance of the final hexaarylpyrrolo[3,2-*b*]pyrrole (compound **2**).

Notes

1. Prior to performing each reaction, a thorough hazard analysis and risk assessment should be carried out with regard to each chemical substance and experimental operation on the scale planned and in the context of the laboratory where the procedures will be carried out. Guidelines for carrying out risk assessments and for analyzing the hazards associated with chemicals can be found in references such as Chapter 4 of "Prudent Practices in the Laboratory" (The National

Academies Press, Washington, D.C., 2011; the full text can be accessed free of charge at https://www.nap.edu/catalog/12654/prudent-practices-in-the-laboratory-handling-and-management-of-chemical. See also "Identifying and Evaluating Hazards in Research Laboratories" (American Chemical Society, 2015) which is available via the associated website "Hazard Assessment in Research Laboratories" at https://www.acs.org/content/acs/en/about/governance/committees/chemicalsafety/hazard-assessment.html. In the case of this procedure, the risk assessment should include (but not necessarily be limited to) an evaluation of the potential hazards associated with acetic acid, toluene, 4-cyanobenzaldehyde, 4-*tert*-butylaniline, iron(III) *p*-toluenesulfonate hexahydrate, butane-2,3-dione, methanol, diethyl ether, allylpalladium chloride dimer, 1,4-bis(diphenylphosphino)butane, dichloromethane, 1-bromo-4-nitrobenzene, potassium acetate, *N,N*-dimethylacetamide, ethyl acetate, magnesium sulfate, acetonitrile, silica gel, hexanes, and dibromomethane.
2. Acetic acid (>99.5%) was purchased from Chempur (http://en.chempur.pl/) and was used as received (submitters). Acetic acid (>99.7%) was purchased from Nacalai Tesque, Inc. and was used as received (checkers).
3. Toluene (>99.5%) was purchased from POCh (http://www.english.poch.com.pl/) and was used as received (submitters). Toluene (>99.5%) was purchased from Nacalai Tesque, Inc. and was used as received (checkers).
4. 4-Cyanobenzaldehyde (99%) was purchased from Fluorochem and was used as received.
5. 4-*tert*-Butylaniline (98%) was purchased from Fluorochem and was used as received (submitters). 4-*tert*-Butylaniline (98%) was purchased from Tokyo Chemical Industry Co., Ltd. and was used as received (checkers).
6. A stir plate was purchased from Heidolph Instruments GmbH & Co. KG. It has an input power of 230–240 V (50–60 Hz, 825 W), and the stirring range is from 100 rpm to 1400 rpm. Unless specified differently, 300 rpm was used for stirring (submitters). A stir plate was purchased from Yazawa-Kagaku. It has an input power of 100 V (50–60 Hz, 10 W), and the stirring range is from 100 rpm to 1500 rpm. Unless specified differently, 900 rpm was used for stirring (checkers).
7. The silicone oil for the oil bath was purchased from Carl Roth. The boiling point is over 205 °C. Unless specified differently, the oil in the oil bath should cover the reaction mixture in the reaction flask while

heating. Unless reported, the temperatures throughout this manuscript refer to temperatures of oil in the oil baths (submitters). The silicone oil for the oil bath was purchased from Shin-Etsu Chemical Co., Ltd. The flash point is over 300 °C (checkers).

8. Fe(OTs)$_3 \cdot$ 6H$_2$O (technical grade) was purchased from Sigma Aldrich and was used as received.
9. Butane-2,3-dione (99%) was purchased from Alfa Aesar and was used as received.
10. If the precipitate prevented stirring within 15 min of heating, it needed to be crushed manually to facilitate the stirring. Efficient stirring providing access of an oxygen in the whole reaction's volume is crucial for the large-scale syntheses of tetraarylpyrrolo[3,2-*b*]pyrroles (see ref 8).
11. The room temperature throughout this manuscript refers to temperature between 22 °C and 25 °C.
12. Methanol (>99.0%) was purchased from POCh and was used as received (submitters). Methanol (>99.0%) was purchased from Nacalai Tesque, Inc. and was used as received (checkers).
13. Diethyl ether (>99.5%) was purchased from POCh and was used as received (submitters). Diethyl ether (>99.5%) was purchased from Nacalai Tesque, Inc. and was used as received (checkers).
14. When the reaction was carried out on a 15 mmol scale, 5.56 g (65%) of the product was obtained.
15. TLC analysis of the product **1** is shown below. The R$_f$ value of the product **1** in hexane/dichloromethane (1/4, v/v) was 0.74. The spot of the product can be viewed on silica gel 60 F$_{254}$ plates (TLC Silica gel 60 F$_{254}$, purchased from Merck KGaA) with UV light (254 nm).

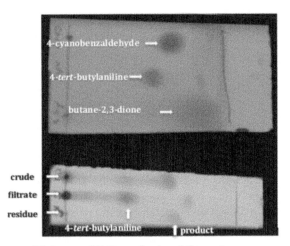

Figure 9. TLC analysis of Step A

16. 1,4-Bis(4-(tert-butyl)phenyl)-2,5-bis(4-cyanophenyl)-1,4-dihydropyrrolo[3,2-b]pyrrole (**1**) has the following physical properties: mp 343.5–354.5 °C (dec.); IR (film): 2958, 2868, 2224, 1599, 1518, 1459, 1380, 1362, 1171, 766 cm^{-1}; ^{1}H NMR (500 MHz, CDCl$_3$) δ: 1.36 (s, 18H), 6.49 (s, 2H); 7.18 (AA'XX', d, J = 8.6 Hz, 4H), 7.28 (AA'XX', d, J = 8.6 Hz, 4H), 7.42 (AA'XX', d, J = 8.6 Hz, 4H), 7.47 (AA'XX', d, J = 8.6 Hz, 4H); ^{13}C NMR (125 MHz, CDCl$_3$) δ: 31.4, 34.6, 96.1, 109.0, 119.2, 124.8, 126.4, 127.8, 131.9, 133.4, 135.0, 136.6, 137.7, 149.7; HRMS (ESI) m/z calcd for C$_{40}$H$_{36}$N$_4$Na$^+$ [M + Na]$^+$ 595.2832, found 595.2822; Anal. calcd for C$_{40}$H$_{36}$N$_4$: C, 83.88; H, 6.34; N, 9.78. Found: C, 83.77; H, 6.49; N, 9.70.
17. PdCl(C$_3$H$_5$)(dppb) catalyst was prepared according to the procedure reported by Doucet et. al. in *ChemCatChem* **2013**, *5*, 255–262 (Ref. 12c).
18. Allylpalladium chloride dimer (>97%) was purchased from Tokyo Chemical Industry Co., Ltd. and was used as received.
19. 1,4-Bis(diphenylphosphino)butane (>98%) was purchased from Tokyo Chemical Industry Co., Ltd. and was used as received.
20. Dichloromethane (99%) was purchased from Linegal Chemicals (http://www.linegal.com.pl/en/) and was used as received. For water-sensitive reactions it was dried using Solvent Purification System from MBraun (https://www.mbraun.com/us/) (submitters). Dichloromethane (>99.5%) was purchased from FUJIFILM Wako Pure Chemical Corporation and purified by Glass Contour solvent dispensing system (Nikko Hansen & Co., Ltd.) when it was used for

Step B. Otherwise, dichloromethane (>99%) was purchased from Kanto Chemical Co., Inc. and was used as received (checkers).

21. NMR analysis of the crude palladium catalyst (PdCl(C$_3$H$_5$)(dppb)): ^{1}H NMR (500 MHz, CDCl$_3$) δ: 1.71–1.88 (m, 4H), 2.54–3.11 (m, 4H), 3.74 (bs, 4H), 5.60 (quint, J = 10.3 Hz,1H), 7.35-7.60 (m, 20H); ^{13}C NMR (125 MHz, CDCl$_3$) δ: 23.3 (m), 27.2 (t), 74.6 (t), 120.4, 128.8 (t), 130.5, 132.7; ^{31}P NMR (202 MHz, CDCl$_3$): 18.5 (s).
22. 1-Bromo-4-nitrobenzene (99%) was purchased from Sigma Aldrich and was used as received.
23. Anhydrous KOAc was purchased from Chempur and was additionally dried under vacuum (0.80 mmHg) at 150 °C for 2 h prior to use (submitters). Anhydrous KOAc (>99%) was purchased from Sigma Aldrich and was additionally dried under vacuum (4.8 mmHg) at 150 °C for 11 h prior to use (checkers).
24. N,N-Dimethylacetamide (DMA), (>99%) was purchased from Sigma Aldrich and was additionally dried with molecular sieves 4Å (50 g) for 16 h, which were heated at 300 °C for 6 h in a furnace (submitters). N,N-Dimethylacetamide (DMA), (>99%) was purchased from Sigma Aldrich and was additionally dried in flask B with molecular sieves 4Å (50 g) for 12 h, which were heated by microwave oven (500W, 3 × 90 seconds) (checkers).
25. The reaction was performed under argon atmosphere in the flask sealed with septum. To prevent over-pressurization, after stirring for 1 h at 150 °C the rubber septum was equipped with a needle adapter and the three-way stop cock connected to the argon/vacuum line was opened to faciliate constant argon flow. The flask was flushed with argon for 30 seconds, after which the three-way stopcock was closed and the needle adapter removed.
26. Ethyl acetate (>99%) was purchased from Kanto Chemical Co., Inc. and was used as received (checkers).
27. Anhydrous MgSO$_4$ was purchased from Chempur and was used as received (submitters). Anhydrous MgSO$_4$ was purchased from Tokyo Chemical Industry Co., Ltd. and was used as received (checkers).
28. Acetonitrile (99.9%) was purchased from POCh and was used as received (submitters). Acetonitrile (99%) was purchased from Nacalai Tesque, Inc. and was used as received (checkers).

29. In order to get better separation of the products on the column chromatography by preventing overloading, the crude reaction mixture was treated with acetonitrile.
30. TLC analysis of the crude product before column chromatography is shown below. The R_f value of the product **2** in hexane/dichloromethane (1/4, v/v) was 0.31. The spot of the product can be viewed on silica gel 60 F_{254} plates (TLC Silica gel 60 F_{254}, purchased from Merck KGaA) with UV light (254 nm).

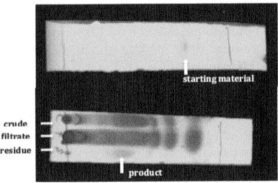

Figure 10. TLC analysis of Step C before column chromatography

31. Silica gel (Silica gel 60, 0.040–0.063 mm) was purchased from Merck KGaA and was used as received.
32. Hexanes (95%) were purchased from Linegal Chemicals and distilled prior to use, collecting fraction boiling at 68 °C (submitters). Hexane (>95%) was purchased from Kanto Chemical Co., Inc. and was used as received (checkers).
33. Flash column chromatography was performed using 950 g of silica gel (Silica gel 60, 0.040–0.063 mm, purchased from Merck KGaA). It was wet packed (9.5 cm in diameter, bed height = 32 cm) using hexane/dichloromethane (2/3, v/v) and the silica gel with the crude material adsorbed was loaded onto the column. Sand with 2 cm minimum height was added to the bottom and to the top of the column (sand was used to assist packing). Chromatography reservoir (500 mL, 29/32 joint) was attached. The column was eluted with 4 L of hexane/dichloromethane (2/3, v/v), then 1.5 L of hexane/dichloromethane (1/2, v/v), then 11 L of hexane/dichloromethane (1/4, v/v). Fraction 0 containing 4 L of hexane/dichloromethane (2/3, v/v) was collected in four 1 L glass bottles and fractions 1-25 were collected in 500 mL glass bottles. TLC analyses of

the fractions were performed on silica gel TLC plates coated with fluorescent indicator F_{254} (purchased from Merck) with hexane/dichloromethane (1/9, v/v) as eluent and visualized with 254 nm UV light (The R_f value of **2** was 0.77). Fractions 13-24 contained the product and were combined and concentrated on a rotary evaporator (40 °C, 640 to 60 mmHg).

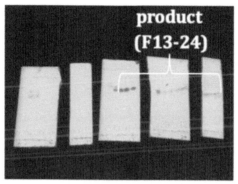

Figure 11. TLC analysis of column fractions

34. When the reaction was carried out on a 5 mmol scale, 2.62 g (64%) of the product was obtained.
35. 1,4-Bis(4-(tert-butyl)phenyl)-2,5-bis(4-cyanophenyl)-3,6-bis(4-nitrophenyl)-1,4-dihydropyrrolo[3,2-b]pyrrole (**2**) has the following physical properties: mp 353.0–392.8 °C (dec.); IR (film): 2962, 2869, 2226, 1599, 1515, 1341, 1115, 1587, 1017 cm^{-1}; ^{1}H NMR (500 MHz, CDCl$_3$) δ: 1.28 (s, 18H), 6.78–6.82 (m, 8H), 7.01 (AA'XX', d, J = 8.6 Hz, 4H), 7.11 (AA'XX', d, J = 8.6 Hz, 4H), 7.36 (AA'XX', d, J = 8.0 Hz, 4H), 7.75 (AA'XX', d, J = 9.2 Hz, 4H); ^{13}C NMR (125 MHz, CDCl$_3$) δ: 31.3, 34.7, 107.9, 110.6, 118.5, 122.7, 125.8, 127.3, 129.2, 130.9, 131.3, 131.8, 132.9, 135.0, 135.7, 140.1, 145.6, 151.6; HRMS (EI) m/z calcd for C$_{52}$H$_{42}$N$_6$O$_4$Na$^+$ [M + Na]$^+$ 837.3160, found 837.3159; Anal. calcd for C$_{52}$H$_{42}$N$_6$O$_4$: C, 76.64; H, 5.19; N, 10.31. Found: C, 76.19; H, 5.43; N, 10.22. Purity was determined to be 98.3% by qNMR using 35.7 mg of **2** and 11.2 mg of dibromomethane (Nacalai Tesque, Inc., >99.0%). The corrected weight and yield based on qNMR purity is 4.89 g (60%).

Working with Hazardous Chemicals

The procedures in *Organic Syntheses* are intended for use only by persons with proper training in experimental organic chemistry. All hazardous materials should be handled using the standard procedures for work with chemicals described in references such as "Prudent Practices in the Laboratory" (The National Academies Press, Washington, D.C., 2011; the full text can be accessed free of charge at http://www.nap.edu/catalog.php?record_id=12654). All chemical waste should be disposed of in accordance with local regulations. For general guidelines for the management of chemical waste, see Chapter 8 of Prudent Practices.

In some articles in *Organic Syntheses*, chemical-specific hazards are highlighted in red "Caution Notes" within a procedure. It is important to recognize that the absence of a caution note does not imply that no significant hazards are associated with the chemicals involved in that procedure. Prior to performing a reaction, a thorough risk assessment should be carried out that includes a review of the potential hazards associated with each chemical and experimental operation on the scale that is planned for the procedure. Guidelines for carrying out a risk assessment and for analyzing the hazards associated with chemicals can be found in Chapter 4 of Prudent Practices.

The procedures described in *Organic Syntheses* are provided as published and are conducted at one's own risk. *Organic Syntheses, Inc.*, its Editors, and its Board of Directors do not warrant or guarantee the safety of individuals using these procedures and hereby disclaim any liability for any injuries or damages claimed to have resulted from or related in any way to the procedures herein.

Discussion

The on-going progress in the field of organic electronics requires compounds possessing a suitable combination of photophysical and electronic properties.[2] Heteropentalenes are a class of 10 π-electron aromatic compounds. The most widely studied members of this family are thieno[3,2-*b*]thiophenes and thieno[3,2-*b*]pyrroles.[3] Their optoelectronic properties makes them suitable candidates in various applications.[3] Pyrrolopyrroles offer the most electron rich core among the 10π-electron heteropentalene family.[4] In 2013 our group reported a serendipitous discovery of a multicomponent reaction between aromatic aldehydes, aromatic amines, and butane-2,3-dione giving straightforward route to previously inaccessible 1,2,4,5-tetraarylpyrrolo[3,2-b]pyrroles.[5] Subsequently, various catalysts have been employed and the methodology has been improved several times[6-8] resulting in improving the yields from ~10–20% to 50–77%. At the same time scope of this reaction has been broadened to embrace variety of primary aromatic amines and aromatic aldehydes including the derivatives of furan, benzofuran, thiophene, pyrrole, pyridine and quinoline.[7,8] Since then, we have intensively explored the chemistry of pyrrolopyrroles and their π-extended analogs.[9] Despite their young age, tetraarylpyrrolo[3,2-b]pyrroles have already been applied in research related to studying symmetry breaking in the excited state, organic light-emitting diodes, MOFs, solvatofluorochromism, resistive memory devices, bulk heterojunction organic solar cells, aggregation-induced emission photochromic analysis of halocarbons, and dye-sensitized solar cells.[10] Compared to thieno[3,2-b]thiophenes, pyrrolo[3,2-b]pyrroles offer instalment of total 6 instead of 4 substituents giving opportunity for fine tuning of the physicochemical properties. Strongly electron-rich positions 3 and 6 of pyrrolopyrroles are ideal entry points for further transformation of the central core *via* electrophilic aromatic substitution, the Scholl reaction[11] or by means of direct arylation. The latter one is especially attractive given that attempts to introduce additional arene rings to positions 3 and 6 of the pyrrolopyrrole core *via* traditional Suzuki-Miyaura coupling procedures turned out to be futile. After careful optimization we have discovered that the most feasible conditions for mono- and bis-arylation of tetrarylpyrrolopyrroles are the ones reported by Doucet and co-workers for the direct palladium-catalyzed polyarylation of pyrroles (Table 1).[12] Both electron-deficient and electron-rich aryl bromides proved to be good substrates under these conditions.[6] In the case of electron-donating bromoarenes, monoarylation products were almost exclusively formed (Table 1, entries 1, 7 and 8). The opposite situation emerged when haloarenes with electron-withdrawing substituents were used (Table 1, entries 2, 3, 5 and 6).

Table 1.[a] Palladium-catalyzed direct arylation reaction of parent tetraaryl-pyrrolo[3,2-b]pyrrole (TAPP) with various aryl bromides

entry	Ar-Br	PAPP[b] Yield	HAPP Yield
1	2-(9,9-dioctylfluorenyl)	30%	-
2	4-SF$_5$-C$_6$H$_4$	-	35%
3	4-O$_2$N-C$_6$H$_4$	24%	53%
4	4-NC-C$_6$H$_4$	37%	-
5	3-pyridyl	21%	47%
6	4-pyridyl	-	56%
7	4-MeO-C$_6$H$_4$	34%	11%
8	3,5-(MeO)$_2$-C$_6$H$_3$	29%	12%

[a] reaction conditions: parent TAPP (0.5 mmol), aryl bromide (2.0 mmol), PdCl(C$_3$H$_5$)(dppb) (4 mol%), KOAc (2.0 mmol), DMA (8 mL) in a sealed tube. [b] PAPP- pentaaryl-pyrrolopyrrole

References

1. Institute of Organic Chemistry, Polish Academy of Sciences, Kasprzaka 44–52, 01-224 Warsaw, Poland. E-mail: dtgryko@icho.edu.pl. We thank the Foundation for Polish Science (TEAM POIR.04.04.00-00-3CF4/16-00 and Mistrz programme) and the National Science Centre, Poland (HARMONIA 2016/22/M/ST5/00431) for the financial support.
2. (a) Koezuka, H.; Tsumura, A.; Ando, T. *Synth. Met.* **1987**, *18*, 699–704. (b) Tang, C. W.; Vanslyke, S. A. *Appl. Phys. Lett.* **1987**, *51*, 913–915. (c) Hu, W. Organic Optoelectronics; Wiley-VCH: Weinheim, Germany, **2013**.
3. (a) Fuller, L. S.; Iddon, B.; Smith, K. A. *J. Chem. Soc. Perkin Trans. 1* **1997**, *22*, 3465–3470. (b) Behringer, H.; Meinetsberger E. *Tetrahedron Lett.* **1973**, *14*, 1915–1918. (c) Kuhn, M.; Falk, F. C.; Paradies, J. *Org. Lett.* **2011**, *13*, 4100–4103. (d) Nenajdenko, V. G.; Sumerin, V. V.; Chernichenko, K. Y.; Balenkova, E. S. *Org. Lett.* **2004**, *6*, 3437–3439. (e) Koh, K.; Wong-Foy, A. G.; Matzger, A. J. *J. Am. Chem. Soc.* **2009**, *131*, 4184–4185. (f) Wong-Foy, A. G.; Matzger, A. J.; Yaghi, O. M. *J. Am. Chem. Soc.* **2006**, *128*, 3494–3495. (g) Roncali, J. *Chem. Rev.* **1992**, *92*, 711–738.
4. Janiga, A.; Gryko, D. T. *Chem.–Asian J.* **2014**, *9*, 3036–3045 and references therein.
5. Janiga, A.; Glodkowska-Mrowka, E.; Stoklosa, T.; Gryko, D. T. *Asian J. Org. Chem.* **2013**, *2*, 411–415.
6. Krzeszewski, M.; Thorsted, B.; Brewer, J.; Gryko, D. T. *J. Org. Chem.* **2014**, *79*, 3119–3128.
7. Tasior, M.; Koszarna, B.; Young, D. C.; Bernard, B.; Jacquemin, D.; Gryko, D.; Gryko D. T. *Org. Chem. Front.* **2019**, *6*, 2939–2948.
8. Tasior, M.; Vakuliuk, O.; Koga, D.; Koszarna, B.; Górski, K.; Grzybowski, M.; Kielesiński, Ł.; Krzeszewski, M.; Gryko, D. T. *J. Org. Chem.* **2020**, *85*, 13529–13543.
9. (a) Friese, D. H.; Mikhaylov, A.; Krzeszewski, M.; Poronik, Y. M.; Rebane, A.; Ruud, K.; Gryko, D. T. *Chem. Eur. J.* **2015**, *21*, 18364–18374. (b) Łukasiewicz, Ł. G.; Ryu, H. G.; Mikhaylov, A.; Azarias, C.; Banasiewicz, M.; Kozankiewicz, B.; Ahn, K. H.; Jacquemin, D.; Rebane, A.; Gryko, D. T. *Chem.–Asian J.* **2017**, *12*, 1736–1748. (c) Krzeszewski, M.; Kodama, T.; Espinoza, E. M.; Vullev, V. I.; Kubo, T.; Gryko, D. T. *Chem. Eur. J.* **2016**, *22*, 16478–16488. (d) Krzeszewski, M.; Gryko, D.; Gryko, D. T. *Acc. Chem. Res.* **2017**, *50*, 2334–2345.

10. (a) Domínguez, R.; Montcada, N. F.; de la Cruz, P.; Palomares, E.; Langa, F. *ChemPlusChem* **2017**, *82*, 1096–1104. (b) Balasubramanyam, R. K. C.; Kumar, R.; Ippolito, S. J.; Bhargava, S. K.; Periasamy, S. R.; Narayan, R.; Basak, P. *J. Phys. Chem. C* **2016**, *120*, 11313–11323. (c) Dereka, B.; Rosspeintner, A.; Krzeszewski, M.; Gryko, D. T.; Vauthey, E. *Angew. Chem. Int. Ed.* **2016**, *55*, 15624–15628. (d) Wu, J.-Y.; Yu, C.-H.; Wen, J.-J.; Chang, C.-L.; Leung, M.-k. *Anal. Chem.* **2016**, *88*, 1195–1201. (e) Zhou, Y.; Zhang, M.; Ye, J.; Liu, H.; Wang, K.; Yuan, Y.; Du, Y.-Q.; Zhang, C.; Zheng, C.-J.; Zhang, X.-H. *Org. Electron.* **2019**, *65*, 110–115. (f) Li, K.; Liu, Y.; Li, Y.; Feng, Q.; Hou, H.; Tang, B. Z. *Chem. Sci.* **2017**, *8*, 7258–7267. (g) Ma, Y.; Zhang, Y.; Kong, L.; Yang, J. *Molecules* **2018**, *23*, 3255. (h) Hawes, C. S.; Máille, G. M. Ó.; Byrne, K.; Schmitt W.; Gunnlaugsson, T. *Dalton Trans.* **2018**, *47*, 10080–10092. (i) Wang, J.; Chai, Z.; Liu, S.; Fang, M.; Chang, K.; Han, M.; Hong, L.; Han, H.; Li Q.; Li, Z. *Chem. Eur. J.* **2018**, *24*, 18032–18042.
11. (a) Krzeszewski, M.; Gryko, D. T. *J. Org. Chem.* **2015**, *80*, 2893–2899. (b) Krzeszewski, M.; Świder, P.; Dobrzycki, Ł.; Cyrański, M. K.; Danikiewicz, W.; Gryko, D. T. *Chem. Commun.* **2016**, *52*, 11539–11542. (c) Krzeszewski, M.; Sahara, K.; Poronik, Y. M.; Kubo, T.; Gryko, D. T. *Org. Lett.*, **2018**, *20*, 1517–1520. (d) Mishra, S.; Krzeszewski, M.; Pignedoli, C. A.; Ruffieux, P.; Fasel, R.; Gryko, D. T. *Nature Commun.* **2018**, *9*, 1714.
12. (a) Bensaid, S.; Doucet, H. *Tetrahedron* **2012**, *68*, 7655–7662. (b) Xu, Y.; Zhao, L.; Li, Y.; Doucet, H. *Adv. Synth. Catal.* **2013**, *355*, 1423–1432. (c) Zhao, L.; Bruneau, C.; Doucet, H. *ChemCatChem* **2013**, *5*, 255–262.

Appendix
Chemical Abstracts Nomenclature (Registry Number)

4-Cyanobenzaldehyde; (105-07-7)
4-*tert*-Butylaniline; (769-92-6)
Fe(OTs)$_3$· 6H$_2$O: Iron(III) *p*-toluenesulfonate hexahydrate; (312619-41-3)
2,3-Butanedione; (431-03-8)
Allylpalladium(II) chloride dimer; (12012-95-2)
1-Bromo-4-nitrobenzene; (586-78-7)
DMA: *N,N*-Dimethylacetamide; (127-19-5)
Dppb: 1,4-Bis(diphenylphosphino)butane; (7688-25-7)

Maciej Krzeszewski obtained his Ph.D. from the Institute of Organic Chemistry of the Polish Academy of Sciences in 2017 under the supervision of Professor Daniel Gryko for the research on pyrrolo[3,2-*b*]pyrroles and their nonplanar π-extended analogs. During his Ph.D. studies, he joined Professor Valentine Vullev's group at University of California Riverside (USA) as a visiting researcher studying influence of dipole effects on electron transfer. Since 2017 to 2019, he has been working with Professor Kenichiro Itami at Nagoya University (Japan) as a JSPS postdoctoral fellow on nonplanar molecular nanocarbons. He is currently working as an assistant in Professor Daniel Gryko's group.

Mariusz Tasior received his M.Sc. from the Warsaw University while conducting research under the guidance of Professor Daniel Gryko. He continued his research in the same group and obtained his Ph.D. from the Institute of Organic Chemistry of the Polish Academy of Sciences, in 2008. He worked as a postdoctoral research fellow under the supervision of Professor Donal F. O'Shea in University College Dublin (2008-2010), before obtaining a postdoctoral position at IOC PAS in Professor Gryko's group. His current research interests are focused on the synthesis and spectroscopy of advanced functional dyes with a special emphasis given to pyrrole derivatives.

Marek Grzybowski obtained his Ph.D. from the Institute of Organic Chemistry of the Polish Academy of Sciences in 2014 under the supervision of Prof. D. T. Gryko. From 2015 to 2018 he carried out postdoctoral research under a JSPS fellowship in the group of Prof. S. Yamaguchi at Nagoya University aiming at the development of photostable near-infrared fluorophores for bioimaging. Currently he is an assistant professor at the Institute of Organic Chemistry PAS. His research interests are focused on the synthesis of extended π-systems and strained polycyclic arenes.

Daniel T. Gryko obtained his Ph.D. from the Institute of Organic Chemistry of the Polish Academy of Sciences in 1997, under the supervision of Prof. J. Jurczak. After a postdoctoral stay with Prof. J. Lindsey at North Carolina State University, he started his independent career in Poland. He received the Society of Porphyrins and Phthalocyanines Young Investigator Award in 2008 and Foundation for Polish Science Award in 2017. His current research interests are focused on the synthesis of functional dyes as well as on two-photon absorption, excited-state intramolecular proton transfer and fluorescence probes.

Shu Nakamura was born in Fukuoka, Japan. He graduated from the University of Tokyo in 2020 with B.Sc. in Pharmaceutical Sciences. He is continuing his graduate studies at the University of Tokyo under the supervision of Prof. Masayuki Inoue. His research interests are in the area of the total synthesis of complex natural products.

Koichi Hagiwara was born in Kanagawa, Japan., in 1989. He received his B.Sc. degree in 2013 from the University of Tokyo, and his Ph.D. degree in Pharmaceutical Sciences in 2019 under the supervision of Prof. Masayuki Inoue. He was appointed as an assistant professor in the Graduate School of Pharmaceutical Sciences at the University of Tokyo in 2017. His research interests include the total synthesis of bioactive and highly complex natural products.

Preparation of Hindered Aniline CyanH and Application in the Allyl-Ni-Catalyzed α,β-Dehydrogenation of Carbonyls

Alexandra K. Bodnar, Aneta Turlik, David Huang, Will Butcher, Joanna K. Lew, and Timothy R. Newhouse[*1]

Department of Chemistry, Yale University, 225 Prospect Street, New Haven, Connecticut 06520-8107, United States

Checked by Aoi Takeuchi, Hiroaki Itoh, and Masayuki Inoue

Procedure (Note 1)

A. CyanH (N-cyclohexyl-2,6-diisopropylaniline) (1). An oven-dried, 1-L 3-necked round-bottomed flask (central neck 29/32 joint, side necks 15/25 joint) is equipped with a 4.5 cm × 2 cm Teflon-coated stir bar. One side neck is fitted with a 15/25 rubber septum with argon inlet and outlet needles, and the other is fitted with a 100-mL addition funnel (15/25 joint). The central neck is fitted with a 29/42 rubber septum (Figure 1). After the substitution of the gas by an argon flow, the rubber septum is removed from the central neck of the flask. $NaBH_4$ (17.5 g, 463 mmol, 3.0 equiv) (Note 2) and 1,2- dichloroethane (300 mL, 0.5 M) (Note 3) are added through the central neck (Figure 2A). Then, the central neck is fitted with the 29/42 rubber septum again. The white cloudy suspension is cooled to 0 °C in an ice-water bath for 10 min. Acetic acid (80 mL, 1400 mmol, 9.0 equiv) (Note 4) is added to the mixture via addition funnel over 25 min with vigorous H_2 gas evolution (Figure 2B). After the H_2 evolution ceases, the resulting white cloudy suspension is stirred for an additional 30 min at 0 °C. 2,6-Diisopropylaniline (29 mL, 154 mmol, 1.0 equiv) (Note 5) is added dropwise via addition funnel over 5 min, followed by the addition of cyclohexanone (32 mL, 308 mmol, 2.0 equiv) (Note 6) dropwise via addition funnel over 5 min. The addition funnel is replaced with a 15/25 rubber septum (Figure 2C). The resulting white cloudy solution is placed in a pre-heated 35 °C oil bath and is stirred for 40 h (500 rpm) (Note 7).

Figure 1. Reaction set-up photo (photo provided by checkers)

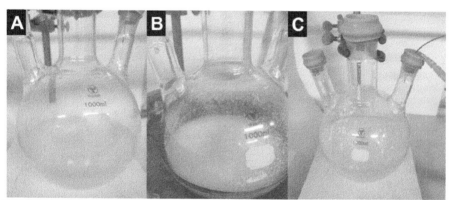

Figure 2. Reaction after addition of A) sodium borohydride, B) acetic acid, and C) 2,6-diisopropylaniline and cyclohexanone and removal of the addition funnel (photos provided by checkers)

A 2.5 M aqueous NaOH solution (400 mL) (Note 8) is slowly added to the reaction flask, which warms upon quenching over 10 min (Figure 3). The mixture is stirred for an additional 10 min. The warm mixture is then cooled in an ice-water bath for 30 min and subsequently transferred to a 1-L separatory funnel, and the reaction vessel is rinsed with EtOAc and water. The organic layer is separated, and the aqueous layer is extracted with EtOAc (3 × 100 mL). The 1,2-dichloroethane and EtOAc layers are combined, washed with brine (250 mL), dried over anhydrous Na_2SO_4 (70 g) (Note 9), filtered through a glass funnel equipped with cotton and concentrated under reduced pressure (100 mmHg, 36 °C) by rotary evaporation to provide 49.2 g of crude material as a viscous light pink oil (Figure 4). The product is purified by flash column chromatography on silica gel (Note 10) using 5% EtOAc in hexane as eluent (Note 11). The fractions containing the product are combined and concentrated under reduced pressure (100 mmHg, 36 °C) by rotary evaporation to provide CyanH (**1**) as white crystals (21.8 g, 55%) (Note 12) (Figure 5).

Figure 3. Reaction mixture after quenching with 2.5 M NaOH (photo provided by checkers)

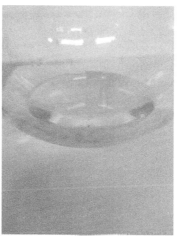

Figure 4. Crude material after concentration under reduced pressure, as a viscous, light pink oil (photo provided by checkers)

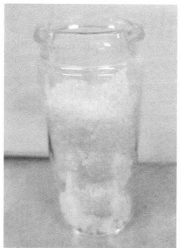

Figure 5. Purified Product 1 (photo provided by checkers)

B. *N,N-Dibenzyl-3-methylbutanamide (2)*. To an oven-dried 300-mL round-bottomed flask (29/42 joint) equipped with a 4.5 cm × 2 cm Teflon-coated stir bar and a 29/42 rubber septum with argon inlet and outlet needles (Figure 6) are added isovaleryl chloride (2.6 mL, 21 mmol, 1.0 equiv) (Note 13) and anhydrous diethyl ether (100 mL, 0.2 M) (Note 14) by a syringe through the rubber septum. The reaction vessel is moved to a 0 °C ice-water bath and stirred for 10 min. Triethylamine (8.9 mL, 64 mmol, 3.0

equiv) (Note 15) is added to the mixture by a syringe through the rubber septum over 3 min, resulting in a cloudy white suspension (Figure 7A). To the mixture is added dibenzylamine (Note 16) (4.5 mL, 23 mmol, 1.1 equiv) dropwise by a syringe through the rubber septum over 5 min, and the reaction is stirred for an additional hour at 0 °C (Note 17) (Figure 7B).

Figure 6. Reaction set-up (photo provided by checkers)

Figure 7. Reaction after addition of A) triethylamine and B) dibenzylamine (photos provided by checkers)

The reaction is quenched at 0 °C by the addition of a saturated aqueous solution of NH_4Cl (100 mL) over 5 min (Note 18). The reaction mixture is transferred to a 500 mL separatory funnel and the organic layer is separated.

The aqueous phase is extracted with diethyl ether (2 × 100 mL), and the combined organic extracts are washed with brine (100 mL), dried over anhydrous Na_2SO_4 (30 g), filtered through a glass funnel fitted with a cotton plug and concentrated under reduced pressure (100 mmHg, 36 °C) by rotary evaporation to provide 6.5 g of a light yellow cloudy oil (Figure 8A). The product is purified by flash column chromatography on silica gel (Note 10) using 13% EtOAc in hexane as eluent (Note 19). The fractions containing the product are combined and concentrated under reduced pressure (100 mmHg, 36 °C) by rotary evaporation to afford N,N-dibenzyl-3-methylbutanamide (2) as a light yellow oil (5.6 g, 94%) (Note 20) (Figure 8B).

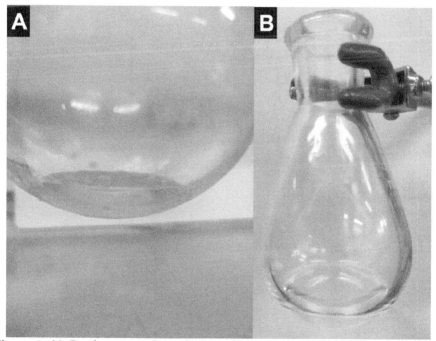

Figure 8. A) Crude material as a light yellow cloudy oil. B) Dried product 2 (photos provided by checkers)

C. *N,N-Dibenzyl-3-methylbut-2-enamide (3).* An oven-dried, 1-L 3-necked round-bottomed flask (central neck 29/32 joint, side necks 15/25 joint) is equipped with a 4.5 cm × 2 cm Teflon-coated stir bar. One side neck is fitted with a 15/25 rubber septum, and the other is fitted with a 50-mL addition funnel (15/25 joint), which is topped with a 15/25 rubber septum.

The central neck is equipped with a connecting adapter (upper outer joint 15/25, lower inner joint 29/42) and a water-cooled reflux Dimroth condenser (15/25 joint), which is topped with a 15/25 three-way stopcock connected to an argon inlet and a vacuum line. All junctures of the glassware and the rubber septa are sealed with silicone grease and Teflon tape. The rubber septum of the side neck of the flask is removed, and CyanH (1) (6.5 g, 25 mmol, 1.3 equiv) (Note 21) is added through the side neck. The rubber septum is reattached and sealed with Teflon tape. The flask is connected to the vacuum line for 2 min, after which the flask is backfilled with argon. This substitution of the gas is repeated three times (Figure 9).

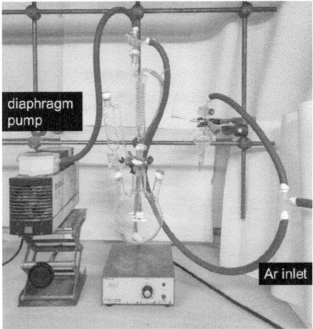

Figure 9. Reaction set up after addition of CyanH (1) (photo provided by checkers)

Tetrahydrofuran (THF, 250 mL, 0.1 M) (Note 22) is added to the flask by a syringe through a rubber septum on the side neck. The clear colorless solution is cooled to −45 °C (dry ice in acetonitrile) (Figure 10A) for 1 h. To the solution is added *n*-butyllithium (9.2 mL of 2.6 M in hexane (checker), 24 mmol, 1.25 equiv) (Note 23) dropwise by a syringe through a rubber septum over 5 min, resulting in a light yellow heterogeneous mixture (Figure 10B). The mixture is stirred for an additional 1.5 h at –45 °C. To the mixture is added a solution of N,N-dibenzyl-3-methylbutanamide (**2**) (5.4 g, 19 mmol, 1.0 equiv) (Note 24) in THF (50 mL, 0.4 M) dropwise via addition funnel over 15 min. The reaction vessel is moved to a 0 °C ice-water bath and stirred for 30 min, resulting in an orange homogeneous solution (Figure 10C).

To an oven-dried 100-mL round-bottomed flask (15/25 joint), equipped with a 2.0 × 0.9 cm Teflon-coated stir bar, is added $ZnBr_2$ (8.6 g, 38 mmol, 2.0 equiv) (Note 25). The neck of flask is fitted with a 15/25 rubber septum with argon inlet and outlet needles. THF (40 mL, 1.0 M) is added to the flask by a syringe through the rubber septum over 1 minute. The mixture is stirred at room temperature until complete dissolution of $ZnBr_2$ resulting in a clear, colorless solution. This $ZnBr_2$ solution is then transferred to the addition funnel by a syringe through the rubber septum and added to the reaction mixture via addition funnel over 5 min. The reaction is stirred for an additional 30 min at 0 °C, resulting in a light yellow homogeneous solution (Figure 10D).

To an oven-dried 100-mL round-bottomed flask (15/25 joint), equipped with a 2.0 × 0.9 cm Teflon-coated stir bar, is added $NiBr_2(dme)$ (590 mg, 1.9 mmol, 0.1 equiv) (Note 26). The neck of the flask is fitted with a 15/25 rubber septum with argon inlet and outlet needles. THF (40 mL, 0.6 M) and diethyl allyl phosphate (4.1 mL, 23 mmol, 1.2 equiv) (Note 27) are added to the flask via syringe over 1 min, resulting in a deep purple homogeneous solution (Figure 10E). This solution is stirred at room temperature for 10 min and then transferred to the addition funnel by a syringe through the rubber septum (Note 28). The solution is subsequently added to the reaction mixture via addition funnel over 5 min. The reaction vessel is then placed in a pre-heated 80 °C oil bath and stirred at reflux for 4 h (Figure 10F) (Note 29).

Figure 10. A) Reaction after addition of CyanH (1) and THF and cooling to -45 °C, B) after LiCyan formation, C) after addition of 2, D) after ZnBr$_2$ addition. E) Premixed solution of NiBr$_2$(dme) and diethyl allyl phosphate. F) Reaction mixture during reflux (photos provided by checkers)

The reaction vessel is removed from the oil bath and allowed to return to room temperature over 30 min. The reaction is then quenched by the addition of saturated aqueous NH$_4$Cl (500 mL) over 3 min. The quenched reaction mixture is transferred to a 2-L separatory funnel, and the reaction vessel is rinsed with EtOAc. The organic layer is separated, and the aqueous layer is extracted with EtOAc (3 × 250 mL). The combined organic layers are washed with brine (250 mL), dried over anhydrous Na$_2$SO$_4$ (50 g), filtered through a glass funnel equipped with cotton, and concentrated under reduced pressure (100 mmHg, 36 °C) by rotary evaporation to provide 13 g of crude material as a viscous orange oil. The product is purified by flash column chromatography on silica gel (Note 10) using a gradient of 7% EtOAc in hexane to 15% EtOAc in hexane as eluent (Note 30). The fractions containing the product are combined and concentrated under reduced pressure (100 mmHg, 36 °C) by rotary evaporation to afford

N,N-dibenzyl-3-methylbut-2-enamide (**3**) as a slightly yellow solid (4.8 g, 89%) (Note 31) (Figure 11).

Figure 11. Dried product 3 (photo provided by checkers)

Notes

1. Prior to performing each reaction, a thorough hazard analysis and risk assessment should be carried out with regard to each chemical substance and experimental operation on the scale planned and in the context of the laboratory where the procedures will be carried out. Guidelines for carrying out risk assessments and for analyzing the hazards associated with chemicals can be found in references such as Chapter 4 of "Prudent Practices in the Laboratory" (The National Academies Press, Washington, D.C., 2011; the full text can be accessed free of charge at https://www.nap.edu/catalog/12654/prudent-practices-in-the-laboratory-handling-and-management-of-chemical. See also "Identifying and Evaluating Hazards in Research Laboratories" (American Chemical Society, 2015) which is available via the associated website "Hazard Assessment in Research Laboratories" at https://www.acs.org/content/acs/en/about/governance/committees/chemicalsafety/hazard-assessment.html. In the case of this procedure, the risk assessment should include (but not necessarily be limited to) an

evaluation of the potential hazards associated with isovaleryl chloride, *n*-butyllithium, triethylamine, dibenzylamine, zinc bromide, diethyl allyl phosphate, nickel (II) bromide dimethoxyethane adduct, *N*-cyclohexyl-2,6-diisopropylaniline, 2,6-diisopropylaniline, cyclohexanone, sodium borohydride, acetic acid, ethyl acetate, hexane, diethyl ether, tetrahydrofuran, 1,2-dichloroethane, and methanol, as well as the proper procedures for working with and quenching pyrophoric materials.

2. Sodium borohydride (>98%) was obtained from Sigma-Aldrich and was used as received (checkers). Sodium borohydride (98%) was obtained from AK Scientific and was used as received (submitters).

3. 1,2-Dichloroethane (anhydrous, 99.8%) was obtained from Sigma-Aldrich and was used as received.

4. Acetic acid (>99.5%) was obtained from TCI and was used as received (checkers). Acetic acid (>99.5%) was obtained from J.T. Baker and was used as received (submitters).

5. 2,6-Diisopropylaniline (90%) was obtained from TCI and was used as received.

6. Cyclohexanone (>99.0%) was obtained from TCI and was used as received (checkers). Cyclohexanone (99.8%) was obtained from Acros and was used as received (submitters).

7. In the TLC analysis of the crude product, the R_f value of product **1** in 25% Et_2O in hexane was 0.74. The spot of the product on the silica gel plate [TLC silica gel 60 F_{254}, purchased from Merck KGaA (checkers)] was visualized under UV light (254 nm) and then stained with phosphomolybdic acid [FUJIFILM Wako Pure Chemical Corporation (checkers)] (Figure 12).

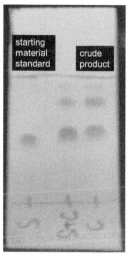

Figure 12. TLC analysis of the crude product in step A
(photo provided by checkers)

8. Sodium hydroxide (>93%) was purchased from FUJIFILM Wako Pure Chemical Corporation and was used as received (checkers).
9. Anhydrous Na_2SO_4 (>99%) was purchased from Nacalai Tesque, Inc. and was used as received (checkers).
10. Silica gel (Silica gel 60N, spherical and neutral, 0.050–0.060 mm) was purchased from Kanto Chemical Co., Inc. and used as received (checkers).
11. The column with a 12 cm diameter × 50 cm height was packed with silica gel (1.08 kg) using 5% EtOAc in hexane (1.5 L). Then, the crude material was loaded onto the column. At that point, fraction collection (500-mL fractions) was begun, and elution was continued with 6.0 L of 5% EtOAc in hexane. The desired product was obtained in fractions 6-9 (Figure 13).

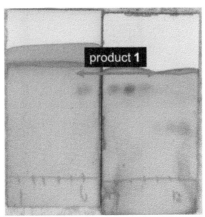

Figure 13. TLC analysis of column fractions in step A
(photo provided by checkers)

12. Product **1** exhibited the following properties: mp = 73 °C; R_f 0.42 (5% EtOAc in hexane); IR (cm^{-1}) 2960, 2928, 2852, 1447, 1383, 1318, 1085, 795, 756; ^{1}H NMR (500 MHz, CDCl$_3$) δ: 1.11–1.22 (m, 5H), 1.25 (d, J = 6.9 Hz, 12H), 1.65 (ad, J = 12 Hz, 1H), 1.78 (ad, J = 12 Hz, 2H), 2.01 (ad, J = 12 Hz, 2H), 2.76–2.81 (m, 1H), 2.96 (br s, 1H), 3.29 (sep, J = 6.9 Hz, 2H), 7.03–7.05 (m, 1H), 7.06–7.11 (m, 2H); ^{13}C NMR (125 MHz, CDCl$_3$) δ: 24.4, 26.1, 26.2, 28.0, 34.8, 59.6, 123.0, 123.5, 142.0; HRMS (ESI) m/z [M+H]$^+$ calc'd for C$_{18}$H$_{30}$N: 260.2373; found 260.2380. Purity was determined to be 98.9% by qNMR using 25.9 mg of **1** and 16.8 mg of 1,3,5-trimethoxybenzene (FUJIFILM Wako Pure Chemical Corporation, 99.8%) (checkers). When the reaction was carried out on a half-scale, 10.9 g (55%) of product **1** was obtained.
13. Isovaleryl chloride (98%) was purchased from Sigma-Aldrich and was used as received.
14. Diethyl ether (>99.5%) was purchased from Kanto Chemical Co., Inc. and was purified by Glass Contour solvent dispensing system (Nikko Hansen & Co., Ltd.) (checkers). Diethyl ether was dried over 3Å molecular sieves under an atmosphere of N$_2$ for 48 h before use in the reaction (submitters).
15. Triethylamine (>99.0%) was purchased from FUJIFILM Wako Pure Chemical Corporation and was distilled over CaH$_2$ and used immediately (checkers).
16. Dibenzylamine (97%) was obtained from Sigma-Aldrich and was used as received.

17. In the TLC analysis of the crude product, the R_f value of product **2** in 15% EtOAc in hexane was 0.35. The spot of the product on the silica gel plate [TLC silica gel 60 F_{254}, purchased from Merck KGaA (checkers)] was visualized under UV light (254 nm) and then stained with phosphomolybdic acid [FUJIFILM Wako Pure Chemical Corp. (checkers)] (Figure 14).

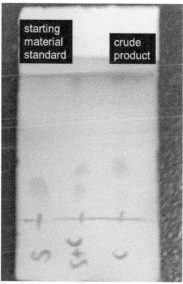

Figure 14. TLC analysis of the crude product in step B
(photo provided by checkers)

18. Ammonium chloride (>98.5%) was purchased from Nacalai Tesque, Inc. and used as received (checkers).
19. The column with a 6 cm diameter × 30 cm height was packed with silica gel (137 g) using 13% EtOAc in hexane (150 mL). Then, the crude material was loaded onto the column. At that point, fraction collection (100-mL fractions) was begun, and elution was continued with 2.1 L of 13% EtOAc in hexane. The desired product was obtained in fractions 8-18 (Figure 15).

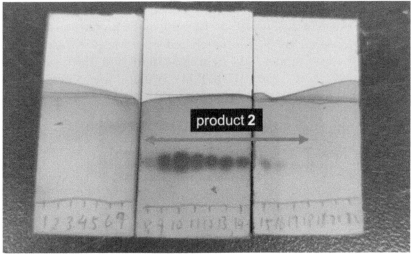

Figure 15. TLC analysis of column fractions in step B
(photo provided by checkers)

20. Product **2** exhibited the following properties: R_f 0.35 (15% EtOAc in hexane); IR (cm^{-1}) 2957, 2870, 1646, 1494, 1449, 1421, 1209, 732, 692; ^{1}H NMR (500 MHz, CDCl$_3$) δ: 0.98 (d, J = 7.4 Hz, 6H), 2.23–2.31 (m, 3H), 4.46 (s, 2H), 4.61 (s, 2H), 7.16 (d, J = 9.7 Hz, 2H), 7.23 (d, J = 9.7 Hz, 2H), 7.27–7.33 (m, 4H), 7.36–7.40 (m, 2H); ^{13}C NMR (125 MHz, CDCl$_3$) δ: 22.8, 25.9, 42.1, 48.0, 49.9, 126.4, 127.4, 127.6, 128.3, 128.6, 129.0, 136.7, 137.7, 173.1; HRMS (ESI) m/z [M+H]$^+$ calc'd for C$_{19}$H$_{23}$NO$^+$: 282.1852; found 282.1865. Purity was determined to be 98.6% by qNMR using 22.3 mg of **2** and 13.3 mg of 1,3,5-trimethoxybenzene (FUJIFILM Wako Pure Chemical Corporation, 99.8%) (checkers). When adjusted for purity the yield is 90.5% (5.52 g). When the reaction was carried out on a half-scale, 2.8 g (94%) of product **2** was obtained.
21. CyanH (36.2 g) was recrystallized from methanol (90 mL) and allowed to dry under reduced pressure overnight or for at least 12 hours.
22. Tetrahydrofuran (>99.0%) was purchased from Kanto Chemical Co., Inc. and purified by Glass Contour solvent dispensing system (Nikko Hansen & Co., Ltd.) (checkers).
23. *n*-Butyllithium solution was purchased from Kanto Chemical Co., Inc. (2.6 M in hexane) and was used as received (checkers). *n*-Butyllithium solution was purchased from Sigma-Aldrich (2.5 M in hexane) and was used as received (submitters).

24. *N,N*-Dibenzyl-3-methylbutanamide was dried under reduced pressure for 24 h prior to use. *N,N*-Dibenzyl-3-methylbutanamide was dissolved in THF and subsequently transferred to the addition funnel via syringe.
25. $ZnBr_2$ (98%) was purchased from Alfa Aesar as a white crystalline solid which contained large clumps of reagent. The crystals were lightly ground with a mortar and pestle to provide a finer reagent and subsequently dried using a heat gun (3 x) under reduced pressure (1–2 mmHg) immediately before use.
26. $NiBr_2$(dme) (97%) was purchased from Strem and was used as received.
27. Diethyl allyl phosphate was prepared according to the literature procedure[3e] and distilled under vacuum (60 °C at 225 mmHg) before use or used as received from Sigma-Aldrich (checkers). $NiBr_2$(dme) purchased from other sources resulted in lower yields (submitters).[3g]
28. Failure to premix $NiBr_2$(dme) and diethyl allyl phosphate or premixing for greater than 1 h resulted in decreased yields.
29. Upon TLC analysis, the R_f value of product **3** in 15% EtOAc/hexane was 0.25. The crude reaction mixture was compared to the starting material. The TLC silica gel plate [TLC silica gel 60 F_{254}, purchased from Merck KGaA (checkers)] was visualized under UV light (254 nm) and then stained with phosphomolybdic acid [FUJIFILM Wako Pure Chemical Corporation (checkers)] (Figure 16).

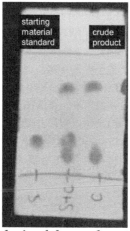

Figure 16. TLC analysis of the crude product in step C
(photo provided by checkers)

30. The column with a 8 cm diameter × 55 cm height was packed with silica gel (350 g) using 7% EtOAc in hexane (450 mL). Then, the crude material was loaded onto the column. At that point, fraction collection (300-mL fractions) was begun, and elution was continued with 7.2 L of 7% EtOAc in hexane and then 3.3 L of 15% EtOAc in hexane. The desired product was obtained in fractions 21-31 (Figure 17).

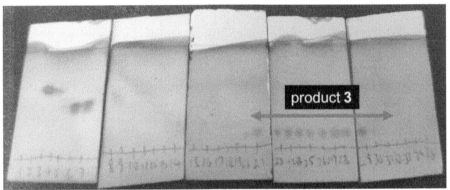

Figure 17. TLC analysis of column fractions in step B (photo provided by checkers)

31. Product **3** exhibited the following properties: mp = 55 °C; R_f 0.25 (15% EtOAc in hexane); IR (cm^{-1}) 3030, 2915, 1629, 1449, 1229, 1176; ^{1}H NMR (400 MHz, CDCl$_3$) δ: 1.82 (s, 3H), 2.03 (s, 3H), 4.45 (s, 2H), 4.59 (s, 2H), 5.93 (s, 1H), 7.16 (d, J = 7.8 Hz, 2H), 7.23–7.33 (m, 6H), 7.34–7.38 (m, 2H); ^{13}C NMR (100 MHz, CDCl$_3$) δ: 168.9, 148.4, 137.6, 136.9, 129.0, 128.7, 128.4, 127.7, 127.4, 126.9, 117.8, 50.4, 47.2, 26.6, 20.6; HRMS (ESI) *m/z* [M+H]$^+$ calc'd for C$_{19}$H$_{21}$NO$^+$: 280.1696; found 280.1687. Purity was determined to be 98.9% by qNMR using 20.9 mg of **3** and 12.6 mg of 1,3,5-trimethoxybenzene (FUJIFILM Wako Pure Chemical Corporation, 99.8%) (checkers). When adjusted for purity the yield is 4.74 g (88%). When the reaction was carried out on a half-scale, 2.4 g (88%) of product **3** was obtained.

Working with Hazardous Chemicals

The procedures in *Organic Syntheses* are intended for use only by persons with proper training in experimental organic chemistry. All hazardous materials should be handled using the standard procedures for work with chemicals described in references such as "Prudent Practices in the Laboratory" (The National Academies Press, Washington, D.C., 2011; the full text can be accessed free of charge at http://www.nap.edu/catalog.php?record_id=12654). All chemical waste should be disposed of in accordance with local regulations. For general guidelines for the management of chemical waste, see Chapter 8 of Prudent Practices.

In some articles in *Organic Syntheses*, chemical-specific hazards are highlighted in red "Caution Notes" within a procedure. It is important to recognize that the absence of a caution note does not imply that no significant hazards are associated with the chemicals involved in that procedure. Prior to performing a reaction, a thorough risk assessment should be carried out that includes a review of the potential hazards associated with each chemical and experimental operation on the scale that is planned for the procedure. Guidelines for carrying out a risk assessment and for analyzing the hazards associated with chemicals can be found in Chapter 4 of Prudent Practices.

The procedures described in *Organic Syntheses* are provided as published and are conducted at one's own risk. *Organic Syntheses, Inc.*, its Editors, and its Board of Directors do not warrant or guarantee the safety of individuals using these procedures and hereby disclaim any liability for any injuries or damages claimed to have resulted from or related in any way to the procedures herein.

Discussion

The α,β-dehydrogenation of carbonyl compounds introduces a synthetic handle that allows for a wide variety of transformations.[2] Recently, the Newhouse group developed a Ni-catalyzed dehydrogenation methodology that uses catalytic $NiBr_2(dme)$ with diethyl allyl phosphate as oxidant as an expansion on their previously reported palladium-catalyzed dehydrogenation methodology.[3] Development of this nickel-catalyzed dehydrogenation methodology addresses current limitations of allyl-palladium-catalyzed dehydrogenation chemistry. These general conditions have been applied to the dehydrogenation of ketones, esters, nitriles, and amides using LiCyan as base when a $ZnBr_2$ additive is employed, and can be applied to cyclic ketones, lactones, and lactams when Zn(TMP)$_2$ is used as base. Importantly, this nickel-catalyzed α,β-dehydrogenation methodology has provided access to substrates previously inaccessible through palladium-catalyzed α,β-dehydrogenation methodology.

The identity of the base used in these reactions had an important effect on the outcome of the reactions. While LiTMP could be used in the Ni-catalyzed dehydrogenation, LiCyan provided the best yields and tolerated a wider variety of substrates. Thus, the development of a new reagent was necessary to access a wide variety of products.

N-Cyclohexyl-2,6-diisopropylaniline, or CyanH, was an adequate precursor for the anilide base, LiCyan. After deprotonation with n-BuLi, LiCyan successfully deprotonated the α-proton of amides and provided high yields of dehydrogenated products using allyl-Ni catalysis. In addition to acting as a base, we believe that Cyan is important in the catalytic reaction itself and may have a role in changing the coordination sphere of the metal.

A selected substrate scope for these reactions is shown in Table 1 and Table 2. In this *Org. Synth.* article, we describe the nickel-catalyzed α,β-dehydrogenation of an amide using a hindered lithium anilide base. In addition, we detail the preparation of CyanH, the aniline precursor to the anilide base used for the dehydrogenation of these substrates.

Table 1. Representative scope of α,β-dehydrogenated carbonyls using allyl-Ni catalysis and LiCyan

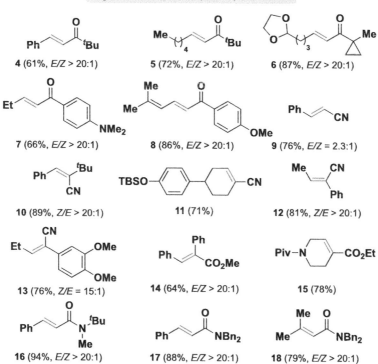

Table 2. Representative scope of α,β-dehydrogenated carbonyls using allyl-Ni catalysis and Zn(TMP)$_2$

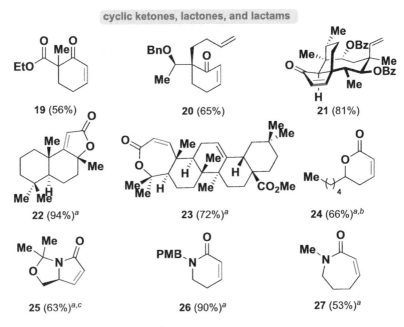

[a] 1,4-dioxane was used as solvent. [b] Reaction was conducted at 23 °C. [c] 1.0 equiv of base was used.

References

1. Department of Chemistry, Yale University, 225 Prospect Street, New Haven, CT 06520-8107, United States. Email: timothy.newhouse@yale.edu. ORCID: 0000-0001-8741-7236. Yale University and the National Science Foundation (CAREER, 1653793; GRFP to A.T.; Bristol Myers Squibb Graduate Fellowship to D.H.; ACS Petroleum Research Fund) are acknowledged for financial support.

2. For reviews on α,β-dehydrogenation, see: (a) Muzart, J. *Eur. J. Org. Chem.* **2010**, 3779–3790. (b) Stahl, S. S.; Diao, T. *Comp. Org. Synth.* **2014**, *7*, 178–212. (c) Turlik, A.; Chen, Y.; Newhouse, T. R. *Synlett* **2016**, *27*, 331–336. (d) Iosub, A. V.; Stahl, S. S. *ACS Catal.* **2016**, *6*, 8201–8213. (e) Hirao, T. *J. Org. Chem.* **2019**, *84*, 1687–1692. (f) Chen, H.; Liu, L.; Huang, T.; Chen, J.; Chen, T. *Adv. Synth. Catal.* **2020**, *362*, 3332–3346. (g) Wang, C.; Dong, G. *ACS Catal.* **2020**, *10*, 6058–6070.

3. (a) Chen, Y.; Romaire, J.; Newhouse, T. R. *J. Am. Chem. Soc.* **2015**, *137*, 5875–5878. (b) Chen, Y.; Turlik, A.; Newhouse, T. R. *J. Am. Chem. Soc.* **2016**, *138*, 1166–1169. (c) Zhao, Y.; Chen, Y.; Newhouse, T. R. *Angew. Chem. Int. Ed.* **2017**, *56*, 13122–13125. (d) Chen, Y.; Huang, D.; Zhao, Y.; Newhouse, T. R. *Angew. Chem. Int. Ed.* **2017**, *56*, 8258–8262. (e) Huang, D.; Zhao, Y.; Newhouse, T. R. *Org. Lett.* **2018**, *20*, 684–687. (f) Szewczyk, S. M.; Zhao, Y.; Sakai, H.; Dube, P.; Newhouse, T. R. *Tetrahedron* **2018**, *74*, 3293–3300. (g) Huang, D.; Szewczyk, S. M.; Zhang, P.; Newhouse, T. R. *J. Am. Chem. Soc.* **2019**, *141*, 5669–5674.

Appendix
Chemical Abstracts Nomenclature (Registry Number)

Isovaleryl chloride; (108-12-3)
Dibenzylamine; (103-49-1)
n-Butyllithium; (109-72-8)
Zinc bromide; (7699-45-8)
Nickel (II) bromide dimethoxyethane adduct; (28923-39-9)
2,6-Diisopropylaniline; (24544-04-5)
Cyclohexanone; (108-94-1)
Sodium borohydride; (16940-66-2)
Acetic acid; (64-19-7)
1,2-Dichloroethane; (107-06-2)

Alexandra Bodnar was born in Allentown, Pennsylvania and received her B.S. in Biochemistry from the University of Notre Dame in 2019. At Notre Dame, she conducted research under the direction of Professor Brandon Ashfeld, developing transition metal-catalyzed cycloadditions. Ali is currently a second year Ph.D. student in the Newhouse Group working on developing novel reaction methodologies. She is also interested in the synthesis of complex natural products and the use of computational strategies to access these molecules.

Aneta Turlik was born in Białystok, Poland, and grew up in Brooklyn, NY. In 2013, she received her Bachelor's degree at Barnard College, where she conducted research under the supervision of Professor Christian M. Rojas. She completed her Ph.D. at Yale University in the laboratory of Professor Timothy R. Newhouse (2019). During her time in the Newhouse Group, her research focused on the total synthesis of a diterpene natural product, as well as the development of palladium-catalyzed dehydrogenation reactions.

David Huang was born in New York City and received his B.A. in Chemistry from Princeton University in 2015, working under the supervision of Professor Erik Sorensen in the areas of C-H functionalization and base metal catalysis. He then completed his Ph.D. in 2020 at Yale University with Professor Timothy Newhouse, focusing on the development of palladium and nickel-catalyzed methods for the functionalization of enolates and heteroarenes using unconventional oxidants.

Will Butcher was born outside of Philadelphia, Pennsylvania and received his Bachelor's degree at Bucknell University in 2013. At Bucknell he conducted research under the supervision of Professor Eric Tillman in polymer chemistry. He completed his Master's degree in 2020 at Yale University in the Newhouse Group, where he worked on the total synthesis of complex natural products.

Joanna Lew was born in Durham, North Carolina. She attended North Carolina School of Science and Mathematics before matriculating to Yale as a member of the Class of 2017. During her time at Yale, she performed research in the Newhouse Laboratory.

Tim Newhouse was born in New Hampshire and grew up in northern New England. He received his B.A. in Chemistry from Colby College (2005) in Waterville, ME, where he was mentored by Prof. Dasan M. Thamattoor. After moving to La Jolla, CA, he completed his Ph.D. at The Scripps Research Institute with Prof. Phil S. Baran (2010). During his time at Scripps, he also worked in the laboratories of Prof. Donna G. Blackmond. He then returned to the east coast for postdoctoral studies with Prof. E.J. Corey at Harvard University. He started at Yale in 2013, and is an Associate Professor at Yale University in the Department of Chemistry.

Aoi Takeuchi was born and raised in Kochi, Japan in 1997. He received his Bachelor's degree in Pharmaceutical Sciences from the University of Tokyo in 2020. He continued his graduate studies at the same university under the supervision of Prof. Masayuki Inoue. His studies currently focus on development of synthetic and analytical methodology for structurally complex peptidic natural products.

Hiroaki Itoh was born in Mie, Japan, in 1985. He received his B.Sc. degree in Pharmaceutical Sciences from the University of Tokyo in 2008, and he received his Ph.D. from the same university under the supervision of Prof. Masayuki Inoue. After working for FUJIFILM Corporation for two years, he was appointed as an assistant professor in the Graduate School of Pharmaceutical Sciences at the University of Tokyo. His research interests include the synthesis and chemical biology of biologically active natural products and their analogues with a particular focus on peptidic natural products and related molecules.

Synthesis of *tert*-Alkyl Phosphines: Preparation of Di-(1-adamantyl)phosphonium Trifluoromethanesulfonate and Tri-(1-adamantyl)phosphine

Thomas Barber and Liam T. Ball[1]*

School of Chemistry, University of Nottingham, Nottingham NG7 2RD, U.K.

Checked by Praveen Kumar Gajula, Leila Terrab, Srinivasarao Tenneti, Gopal Sirasani, and Chris Senanayake

A.

1-Ad–OH (**1**) → 1-Ad–OAc (**2**)
Ac$_2$O, DMAP, pyridine
60 °C, 20 h

B.

1-Ad–OAc (**2**) → [1-Ad$_2$PH$_2$]$^+$ OTf$^-$ (**3**)
PH$_3$ (generated *ex situ*)
Me$_3$SiOTf, CH$_2$Cl$_2$
−78 °C to 50 °C, 4 h

C.

3 → 1-Ad$_3$P (**4**)
1) 1-adamantyl acetate **2**, DBU, CH$_2$Cl$_2$, 30 min
2) Me$_3$SiOTf, 24 h
3) NEt$_3$, 30 min

Procedure (Note 1)

A. *1-Adamantyl acetate* (**2**). A 100 mL single-necked, round-bottomed flask equipped with a Teflon-coated stir bar (25 mm × 12 mm, oval) is charged sequentially with 1-adamantanol (15.21 g, 100 mmol, 1 equiv) (Note 2), 4-dimethylaminopyridine (1.22 g, 10.0 mmol, 0.1 equiv) (Note 3), pyridine (20 mL) (Note 4) and acetic anhydride (14.2 mL, 15.3 g, 150 mmol, 1.5 equiv) (Notes 5 and 6). The flask is fitted with an air-cooled condenser, the top of which is open to air with a 22-gauge needle (Figure 1). The flask is then lowered into an oil bath pre-heated to 60–70 °C (as measured with a thermocouple) and is stirred (500 rpm) at that temperature for 20 h (Note 7). The initial colorless suspension becomes a light-yellow solution after *ca* 10 min and an orange/red solution after 20 h (Figure 1A-C).

Figure 1. Step A reaction mixture after heating at 60 °C for A) 1 min, B) 10 min, C) 20 h (photo provided by checkers)

The flask is removed from the oil bath and allowed to cool in the air until the internal temperature reaches 23–27 °C, as measured with a thermometer. Stirring (500 rpm) is maintained as diethyl ether (30 mL) (Note 8) is added to the reaction flask. The resulting orange/red solution is poured into a 500 mL Erlenmeyer flask containing a Teflon-coated stir bar (40 mm × 10 mm, rod) and a saturated aqueous solution of sodium hydrogen carbonate (200 mL) (Note 9). Diethyl ether (2 × 30 mL) is used to transfer all material from the reaction flask to the Erlenmeyer flask. The biphasic mixture is stirred (300 rpm) for 5 min to control the rate of effervescence; the stirring rate is then increased to 750 rpm (5 min) and finally 1400 rpm until effervescence is no longer observed (ca 5 min). The biphasic mixture is poured into a 500 mL separatory funnel, the organic layer is separated and the aqueous layer is extracted with diethyl ether (2 × 100 mL). The combined organic layers are returned to the separatory funnel and washed with a saturated aqueous solution of copper(II) sulfate (3 × 50 mL) (Notes 10 and 11). The organic layer is transferred to a 500 mL Erlenmeyer flask and anhydrous sodium sulfate (10 g) (Note 12) is added. After swirling for 20 s, the suspension is vacuum-filtered through a fritted glass funnel (70 mm diameter sinter, medium porosity) into a 500 mL round-bottomed flask. Diethyl ether (2 × 20 mL) is used to wash the Erlenmeyer flask and filter cake. The yellow/green filtrate is concentrated by rotary evaporation (40 °C, 525–75 mmHg) to give a green liquid which solidifies on standing (Figure 2a) (Note 13). This material is transferred to a 50 mL kugelrohr distillation flask (B14 joint) using a glass Pasteur pipette (Note 14). The distillation flask is attached to a kugelrohr apparatus via two 50 mL receiver bulbs (B14 joints at both ends); the distillation flask and first receiver bulb are positioned within the kugelrohr oven, and the second receiver bulb is cooled in a dry ice/acetone bath. Rotation is started (20 rpm) and the apparatus is gradually evacuated to between $2–5×10^{-2}$ mmHg via a vacuum manifold. The crude material is distilled by heating the oven to 80 °C (30 min) and then to 85 °C (15 min) while maintaining the pressure between $2–5 × 10^{-2}$ mmHg. The distillate is collected in the final, cooled bulb as an off-white liquid that quickly solidifies (Figure 2b). Once distillation is complete, the apparatus is returned to atmospheric pressure, and the purified material is transferred to a pre-weighed storage vial by gentle warming with a heat gun, giving 1-adamantyl acetate **2** (17.35–17.55 g, 89.3–90.3 mmol, 89.4–90.4%, >99% purity) as a low melting solid (Figure 2b,c) (Notes 14, 15 and 16).

Figure 2. Step A product A) crude, B) solid in kugelrohr receiving bulb, C) after melting and transfer to vial for storage (photos provided by submitters)

B. *Di-(1-adamantyl)phosphonium trifluoromethanesulfonate* (**3**). Both chambers of a 400 mL CO-ware reactor (Notes 17 and 18) are equipped with Teflon-coated stir bars (19 mm × 10 mm, oval) (Figure 3). Chamber A is charged with zinc phosphide (78% pure, 10.0 mmol, 3.32 g, 0.5 equiv) (Note 19), and chamber B is charged with 1-adamantyl acetate **2** (11.73 g, 60.0 mmol, 3 equiv) (Notes 14 and 20). Both necks of the reactor are fitted with CO-ware PTFE-faced silicone septa (Note 21), PTFE stabilizing disks (Note 22), and screw-on lids; a 22-gauge needle (Note 23) connected *via* Luer lock adapter to a Schlenk line is inserted through the septum on chamber B (Note 24).

Figure 3. CO-ware reactor for step B with chambers labelled, after charging with zinc phosphide (chamber A) and 1-adamantyl acetate (chamber B) (photo provided by submitters)

The reactor is evacuated (1×10^{-2} mmHg) (Note 25) and backfilled with anhydrous dinitrogen (3 cycles). Anhydrous, degassed dichloromethane (40 mL) (Note 26) is added to chamber B by syringe (22-gauge needle), and degassed deionised water (20 mL) (Note 27) is added to chamber A by syringe (22-gauge needle). The nitrogen inlet needle is removed from chamber B, rendering the CO-ware reactor a sealed system. Both chambers of the reactor are then cooled to –78 °C (dry ice/acetone bath) without stirring, until the water in chamber A is frozen solid (Note 28). Stirring is commenced (750 rpm), then trimethylsilyl trifluoromethanesulfonate (3.62 mL, 4.44 g, 20.0 mmol, 1 equiv) (Note 29) is added to chamber B by syringe (22-gauge needle) in a single portion. The reactor is removed from the cooling bath and 10 M aqueous HCl (20 mL, 200 mmol, 10 equiv) (Notes 30 and 31) is added immediately to chamber A by syringe (22-gauge needle) in a single portion (Note 28). Stirring is maintained (750 rpm) as both chambers of the reactor are warmed in a water bath set to 50 °C (as measured by thermocouple); effervescence is observed in chamber A as the ice melts and phosphine gas (PH_3, 20.0 mmol, 1 equiv) is generated (Note 32). The mixture is stirred for 4 h at 50 °C. The reactor is then removed from the heating bath and cooled to room temperature with stirring.

The gas scrubber apparatus (Figure 4) consists of three 50 mL, two-necked, round-bottomed flasks (primary neck: B24; secondary neck: B14). Flask 1 is empty, its primary neck is fitted with a rubber septum, and its secondary neck is fitted with an inlet adapter connected by PVC tubing to a Luer lock adapter fitted with a needle (22-gauge, 12 cm). The inlet adapter is secured in the secondary neck of flask 1 with a Keck clip. Flasks 2 and 3 are assembled as flask 1, but contain commercial concentrated sodium hypochlorite solution (30 mL; 10–14% w/w); their primary necks are fitted with rubber septa and the secondary neck of flask 3 is left open to air. The needle leading from flask 1 is inserted through the septum of flask 2 so that it reaches to the bottom of the sodium hypochlorite solution; the needle leading from flask 2 is inserted through the septum of flask 3 so that it reaches to the bottom of the sodium hypochlorite solution. A cannula is constructed from PVC tubing fitted with Luer lock adapters at each end; both Luer lock adapters are fitted with needles (22-gauge, 12 cm) (Figure 4b). Needle 1 of the cannula is inserted through the septum of flask A, and the scrubber system is purged with dinitrogen via needle 2 of the cannula for 10 min (Notes 34 and 35).

Needle 2 of the cannula is then inserted into chamber B of the CO-ware reactor, so that the gas released from the CO-ware reactor passes through the scrubber system. After the initial gas release has subsided (*ca* 10 s), a nitrogen inlet needle (Note 36) is inserted into chamber A of the CO-ware reactor and the entire system is flushed with nitrogen (Note 37) for 10 min to ensure any residual PH_3 is completely purged through the bleach baths. The scrubber system is then disconnected from the CO-ware reactor.

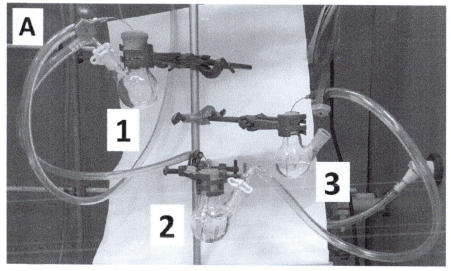

Figure 4. Gas scrubbing system for step B. A) bleach bubbler setup with flasks labelled, B) PVC tubing cannula (photos provided by submitters)

The solution in chamber B of the CO-ware reactor is transferred by syringe (22-gauge needle) to a 250 mL round-bottomed flask, and dichloromethane (2 × 20 mL portions) is used to ensure complete transfer. The colorless solution is concentrated by rotary evaporation (40 °C, 525–75 mmHg) and the colorless phosphonium salt starts to crystallize from the remaining liquid (Note 38). A Teflon-coated stir bar (16 mm, cross) is added to the round-bottomed flask. Diethyl ether (40 mL) (Note 8) is then added in a single portion, and the mixture is stirred (500 rpm) for 15 min until the suspended solid appears completely free flowing. The crystalline solid is collected by vacuum filtration on a fritted glass funnel (30 mm diameter sinter, medium porosity). Diethyl ether (3 × 20 mL portions) is used to wash the flask and filter cake, then air is drawn through the filter cake for 20 min to give di-(1-adamantyl)phosphonium trifluoromethanesulfonate **3** (8.92 g, 19.7 mmol, 98.5%, >98% purity) as a colorless crystalline solid (Figure 5) (Notes 39, 40, and 41).

Figure 5. Step B product A) on fritted funnel, B) in vial for storage (photos supplied by submitters)

C. *Tri-(1-adamantyl)phosphine* (**4**). A 100 mL single-necked round-bottomed flask equipped with a Teflon-coated stir bar (16 mm, cross) is charged with di-(1-adamantyl)phosphonium trifluoromethanesulfonate **3** (6.79 g, 15.0 mmol, 1 equiv) and 1-adamantyl acetate **2** (3.21 g, 16.5 mmol, 1.1 equiv) (Figure 6a) (Note 14). The neck is fitted with a rubber septum and a needle (22-gauge) connected *via* Luer lock adapter to a dual-bank Schlenk manifold is inserted into the flask. The vessel is evacuated (2 × 10^{-1} mmHg) (Note 25) and backfilled with anhydrous dinitrogen (3 cycles). Anhydrous, degassed dichloromethane (40 mL) (Note 26) is added to the mixture by syringe (22-gauge needle) and stirring (500 rpm) is commenced at room temperature (Figure 6b) (Note 42). 1,8-Diazabicyclo[5.4.0]undec-7-ene (2.24 mL, 2.28 g, 15.0 mmol, 1 equiv) (Note 43) is added by syringe (22-gauge needle), affording a colorless solution (Figure 6c), which is stirred (500 rpm) at room temperature for 30 min. Trimethylsilyl trifluoromethanesulfonate (3.26 mL, 4.00 g, 18.0 mmol, 1.2 equiv) (Note 29) is added by syringe (22-gauge needle) and the resulting colorless solution (Figure 6d) is stirred (500 rpm) for 24 h at room temperature.

Figure 6. Step C reaction mixture on additions of A) 1 and 2, B) degassed, anhydrous dichloromethane, C) 1,8-diazabicyclo[5.4.0]undec-7-ene, D) trimethylsilyl trifluoromethanesulfonate
(photos supplied by submitters)

Triethylamine (10.5 mL, 7.59 g, 75.0 mmol, 5 equiv) (Notes 44 and 45) is added at room temperature, and a colorless suspension forms immediately. Stirring is maintained for 30 min. The nitrogen inlet needle is then removed, the flask is opened to air and the suspended solid is immediately (Note 46) collected by vacuum filtration on a fritted glass funnel (30 mm diameter sinter, medium porosity). Ethanol (3 × 20 mL) (Notes 47 and 48) is used to wash the flask and filter cake. Air is drawn through the filter cake for 10 min, then the product is dried *in vacuo* (1 × 10^{-1} mmHg, 16 h) to give tri-(1-adamantyl)phosphine **4** (5.1 g, 11.69 mmol, 78%, >98% purity) as a colorless amorphous solid (**Figure 7**) (Notes 49, 50, 51, and 52).

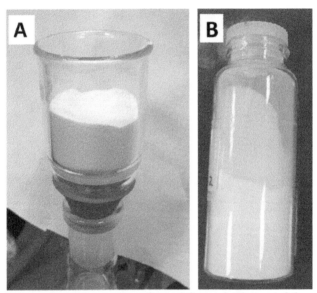

Figure 7. Step C product A) on fritted funnel, B) in vial for storage (photos supplied by submitters)

Notes

1. Prior to performing each reaction, a thorough hazard analysis and risk assessment should be carried out with regard to each chemical substance and experimental operation on the scale planned and in the context of the laboratory where the procedures will be carried out. Guidelines for carrying out risk assessments and for analyzing the

hazards associated with chemicals can be found in references such as Chapter 4 of "Prudent Practices in the Laboratory" (The National Academies Press, Washington, D.C., 2011; the full text can be accessed free of charge at https://www.nap.edu/catalog/12654/prudent-practices-in-the-laboratory-handling-and-management-of-chemical. See also "Identifying and Evaluating Hazards in Research Laboratories" (American Chemical Society, 2015) which is available via the associated website "Hazard Assessment in Research Laboratories" at https://www.acs.org/content/acs/en/about/governance/committees/chemicalsafety/hazard-assessment.html. In the case of this procedure, the risk assessment should include (but not necessarily be limited to) an evaluation of the potential hazards associated with 1-adamantanol, 4-dimethylaminopyridine, pyridine, acetic anhydride, diethyl ether, sodium hydrogen carbonate, copper sulfate pentahydrate, concentrated sulfuric acid, sodium sulfate, 1,3,5-trimethoxybenzene, chloroform-d, zinc phosphide, phosphine gas, dichloromethane, deionized water, trimethylsilyl trifluoromethanesulfonate, concentrated hydrochloric acid, sodium hypochlorite solution, 1,8-diazabicyclo[5.4.0]undec-7-ene, triethylamine, ethanol, benzene-d_6, calcium hydride, as well as the proper procedures for performing reactions in sealed vessels under pressure and vacuum distillations. Step B involves the generation of phosphine gas, which is pyrophoric and highly toxic; this procedure must be performed with extreme caution and only by trained scientists. Step B also involves the use of zinc phosphide, which is highly toxic; it must be handled with care and precautions must be taken to avoid it coming into contact with acid (which will lead to the generation of phosphine gas).

2. 1-Adamantanol (Ambeed; 97%) was used as received.
3. 4-Dimethylaminopyridine (Oakwood; 99%) was used as received.
4. Pyridine (Sigma Aldrich; 99.8 %) was used as received.
5. Acetic anhydride (Oakwood; 99%) was used as received.
6. The use of fewer than 1.5 equiv of acetic anhydride results in incomplete conversion of 1-adamantanol within 20 h, and therefore reduced yields of 1-adamantyl acetate **2**.
7. Reaction progress is monitored by TLC analysis on silica gel with cyclohexane:ethyl acetate (9 : 1) as eluent. TLC plates are visualized by staining with aqueous basic KMnO$_4$ solution (1.5 g KMnO$_4$, 10 g K$_2$CO$_3$ and 1.25 mL 10% wt/v NaOH in 200 mL water) followed by gentle

warming with a heat gun. R_f (1-adamantanol) = 0.18, R_f (1-adamantyl acetate **2**) = 0.67. (Figure 8).

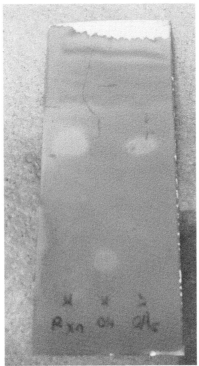

Figure 8. TLC of step A crude reaction mixture after 20 h. Spots from left to right: Rxn = reaction mixture; OH = 1-adamantanol **1** (starting material), OAc = 1-adamantyl acetate **2** (authentic material from a previous batch) (photo supplied by submitters)

8. Diethyl ether (Sigma-Aldrich; >99.8%) was used as received.
9. Sodium hydrogen carbonate (Oakwood Chemicals; >99%) was used as received.
10. Saturated copper sulfate solution is prepared from copper(II) sulfate pentahydrate (*ca* 160 g) (Sigma-Aldrich; >98%) and deionized water (500 mL) containing 2-3 drops of conc. sulfuric acid (Oakwood Chemicals; >98%).
11. The submitters observed that small amounts of a flocculent solid were formed during the second $CuSO_4$ wash. This material is removed during subsequent drying and filtration over sodium sulfate.

12. Anhydrous sodium sulfate (Oakwood Chemicals; >99%) was used as received.
13. Pure 1-adamantyl acetate **2** has a melting point of 31–32 °C. Whether the crude material solidifies depends on its purity and the ambient temperature.
14. The submitters found that due to its low melting point, 1-adamantyl acetate **2** is most conveniently weighed and handled as a liquid following gentle warming with a heat gun.
15. Characterization data for 1-adamantyl acetate **2**: ^{1}H NMR (400 MHz, CDCl$_3$): 1.60–1.71 (m, 6H), 1.96 (s, 3H), 2.08–2.12 (m, 6H), 2.12–2.19 (m, 3H); ^{13}C NMR (101 MHz, CDCl$_3$) δ: 22.9, 30.9, 36.4, 41.5, 80.4, 170.5. HRMS calcd. for C$_{12}$H$_{18}$O$_2$+Na$^+$: 217.1199 [M+Na]$^+$; found (ESI$^+$): 217.1192. mp 31–32 °C. IR (ATR, neat): 2910, 2853, 1731, 1456, 1367, 1354, 1241, 1059, 1016, 864 cm^{-1}
16. The purity of 1-adamantyl acetate **2** was determined to be >99% in both runs (99.9 in the example given) by quantitative ^{1}H NMR spectroscopy in chloroform-*d* with 1,3,5-trimethoxybenzene (Sigma Aldrich; 99.8%) as the internal standard (a mixture of 25.74 mg of 1-adamantyl acetate **2** and 12.45 mg of 1,3,5-trimethoxybenzene was used to determine purity of >99%).
17. The 400 mL CO-ware reactors can be purchased from Sigma Aldrich (catalog no.: STW6-1EA) or SyTracks (catalog no.: STW6).
18. Glassware used for pressurized reactions must be checked for signs of damage (including scratches, chips, and star-cracks) before use.
19. Zinc phosphide (Sigma-Aldrich; technical grade, *ca* 80% purity by weight) can be used as received, although isolated yields may be affected by using material of unknown titre. Alternatively, the purity of the zinc phosphide can be measured by volumetric gas titration (*ACS Catal.* **2020**, *10*, 5454–5461) before use. The submitters found that commercial, *ca* 80% purity by weight zinc phosphide (Sigma-Aldrich, technical grade) titrated to 78% purity.
20. The use of fewer than 3 equiv of 1-adamantylacetate **2** (*i.e.*, 1.5 equiv per P-C bond formed) leads to incomplete conversion of PH$_3$. This results in diminished yields and requires that the unreacted PH$_3$ gas must be quenched, presenting an increased safety risk.
21. PTFE-faced silicone septa for CO-ware reactors can be purchased from Sigma Aldrich (catalog no.: 743968-10EA or STW3-100EA) or SyTracks (catalog no.: STW3).

22. PTFE stabilizing disks for CO-ware reactors can be purchased from Sigma Aldrich (catalog no.: 743852-2EA or STW2-2EA) or SyTracks (catalog no.: STW2).
23. 22-Gauge needles are used to minimize damage to the CO-ware septa so that their integrity is retained once phosphine gas is generated within the sealed reactor.
24. It is important that the needle used for evacuating and back-filling the reactor is inserted into chamber B; if the needle is inserted into chamber A, the flow of nitrogen during backfilling can blow zinc phosphide throughout the reactor and into chamber B.
25. If the 1-adamantyl acetate **2** is a liquid at the beginning of this process, it may bubble during the first evacuation cycle, and then solidify during subsequent cycles.
26. Dichloromethane (Sigma Aldrich; anhydrous, >99.8%) was used as received.
27. Deionized water was degassed by sparging vigorously with dinitrogen for at least 15 min before use.
28. The water in chamber A *must* be frozen before addition of hydrochloric acid. If the water is not frozen, immediate formation of phosphine gas causes a rapid pressure increase within the apparatus; this can force the plunger from the syringe, leading to uncontrolled release of phosphine gas.
29. Trimethylsilyl trifluoromethanesulfonate (Oakwood Chemicals; 99%) was used as received.
30. Aqueous hydrochloric acid (10 M) was prepared by slow addition of concentrated HCl (24.6 mL, 300 mmol; Oakwood Chemicals, 36.5–38% w/w) to deionized water (5.4 mL). The diluted acid was degassed by sparging vigorously with dinitrogen for at least 15 min before use. Note that the use of a stainless-steel needle for degassing may lead to decolorization of the acid solution, but that this does not affect the outcome of the reaction.
31. The excess hydrochloric acid ensures complete protonolysis of the 10 mmol zinc phosphide (Zn_3P_2), which requires 60 mmol (6 equiv) of the acid to generate 20 mmol (2 equiv) of phosphine gas.
32. Gas evolution continues for *ca* 15–20 min, and is accompanied by a change in the color of the suspension in chamber A from gray to brown (Figure 9).

Figure 9. CO-ware reactor for step B, after A) 2 min and B) 20 min in the 50 °C water bath, showing suspension color change in chamber A (left hand side) (photos supplied by submitters)

33. Aqueous sodium hypochlorite solution (Sigma-Aldrich; 10–14% w/w) was used as received.
34. To achieve nitrogen flow through the scrubber apparatus, needle 2 of the cannula is inserted through the septum of a Schlenk flask that is connected to a nitrogen manifold (*ca* 1 atm overpressure). The Schlenk flask is prepared in advance by fitting a septum in the neck, then evacuating (2 × 10^{-1} mmHg) and backfilling with nitrogen (3 cycles).
35. This purging stage should be used to check that the gas flow passes efficiently through the sodium hypochlorite solution contained in flasks 2 and 3 of the scrubber system.
36. The inlet needle and tubing is purged for 10 min before use with nitrogen from a manifold at *ca* 1 atm overpressure.

37. Nitrogen gas is supplied by a manifold at *ca.* 1 atm overpressure.
38. In the submitters' experience, di-(1-adamantyl)phosphonium salt 3 has never failed to crystallize during rotary evaporation. However, by analogy to other di-(*tert*-alkyl)phosphonium salts (*ACS Catal.* **2020**, *10*, 5454–5461), if crystallization does not occur during rotary evaporation it should initiate upon addition of diethyl ether and stirring vigorously.
39. Characterization data for di-(1-adamantyl)phosphonium trifluoromethanesulfonate 3: ^{1}H NMR (400 MHz, CDCl$_3$) δ: 1.77–1.86 (br. m, 12H), 2.07-2.15 (br. m, 6H), 2.15–2.23 (br. m, 12H), 5.98 (d, *J* = 469.6 Hz, 2H); ^{13}C{^{1}H} NMR (101 MHz, CDCl$_3$) δ: 27.5 (d, *J* = 10.4 Hz), 35.4 (d, *J* = 34.1 Hz), 35.5 (d, *J* = 2.0 Hz), 39.1 (d, *J* = 2.0 Hz), 120.8 (q, *J* = 321.5 Hz); ^{19}F NMR pdf (376 MHz, CDCl$_3$) δ: −78.32 (s); ^{31}P NMR (162 MHz, CDCl$_3$) δ: 13.47 (t, *J* = 465.6 Hz). HRMS calcd. for C$_{20}$H$_{32}$P$^+$: 303.2236 [M-OTf]$^+$; found (ESI$^+$): 303.2248. m.p.: >280 °C (decomposition). IR (ATR, neat): 2906, 2854, 2424, 2400, 1450, 1345, 1305, 1276, 1245, 1221, 1154, 1022, 973, 944, 898, 755 ,634, 571, 515, 489, 447, 417 cm^{-1}. [*Note: one of the peaks of the doublet at 35.4 ppm in the ^{13}C{^{1}H} NMR spectrum overlaps with the doublet at 35.5 ppm, but can be resolved with additional spectral processing*].
40. A reaction performed on half-scale provided 4.46 g (99%) of the product.
41. The purity of di-(1-adamantyl)phosphonium trifluoromethanesulfonate 3 prepared in this way was determined to be >98% from both runs (98.9% in the example given) by quantitative ^{1}H NMR spectroscopy in chloroform-*d* with 1,3,5-trimethoxybenzene (Sigma Aldrich; 99.8%) as the internal standard (a mixture of 5.65 mg of di(1-adamanty)phosphonium trifluoromethanesulfonate 3 and 4.15 mg of 1,3,5-trimethoxybenzene was used to determine purity).
42. The reaction mixture is a colorless suspension containing a small amount of undissolved solid material.
43. 1,8-Diazabicyclo[5.4.0]undec-7-ene (Sigma Aldrich; 99%) was dried by stirring over calcium hydride (Sigma Aldrich) for 4 h, then distilled from calcium hydride in flame-dried apparatus at 1.5–2 × 10^{-1} mmHg. The distilled reagent was stored under anhydrous dinitrogen in a flame-dried vessel sealed with a J-Young's tap closure.
44. Triethylamine (Sigma Aldrich; ≥99%) was dried by stirring over calcium hydride (Sigma Aldrich) for 4 h, then distilled from calcium hydride in flame-dried apparatus under anhydrous dinitrogen. The distilled

reagent was stored under anhydrous dinitrogen in a flame-dried vessel sealed with a J-Young's tap closure.
45. The use of excess triethylamine drives the equilibrium deprotonation of the initially-formed tri-(1-adamantyl)phosphonium triflate salt, and hence precipitation of tri-(1-adamantyl)phosphine **4**.
46. The dichloromethane supernatant begins to discolor to yellow (Figure 10) immediately on exposure to air, so it is important to begin the filtration and washing procedure quickly.

Figure 10. Discoloration of dichloromethane supernatant (photo is of collected dichloromethane/ethanol filtrate) (photo supplied by submitters)

47. Ethanol (Sigma Aldrich; >99.8%) was used as received.
48. Although the submitters found that washing with three portions of ethanol was sufficient to completely remove the yellow color from the product, additional washes may be used as required.
49. Tri-(1-adamantyl)phosphine **4** is air stable in the solid state, but oxidizes in solution and reacts with chloroform. Samples for NMR spectroscopic analysis must therefore be prepared and sealed under an inert atmosphere using an appropriate solvent (*e.g.*, benzene-d_6). The benzene-d_6 solvent (Sigma-Aldrich; 99.6 %D) was used as received. The NMR sample was prepared in a glove box under inert atmosphere.
50. Characterization data for tri-(1-adamantyl)phosphine **4**: ^{1}H NMR (400 MHz, C_6D_6) δ: 1.63–1.71 (m, 9H), 1.73–1.80 (m, 9H), 1.93 (brs, 9H), 2.37 (brs, 18H); ^{13}C NMR (101 MHz, C_6D_6) δ: 29.8 (d, *J* = 7.1 Hz), 37.4, 41.6 (d, *J* = 36.6 Hz), 43.2 (br); ^{31}P NMR (162 MHz, C_6D_6) δ: 59.22. HRMS calcd. for $C_{30}H_{46}P^+$: 437.3332 [M+H]$^+$; found (ESI$^+$): 437.3343. mp >280 °C (decomposition). IR (ATR, neat): 2900, 2874, 2844, 1449, 1341, 1297, 1254, 1180, 1101, 1037, 966, 937, 928, 877, 817, 645, 482, 415 cm^{-1}.
51. Reactions performed on half-scale provided 2.5 g (76%) of the product.
52. The purity of tri-(1-adamantyl)phosphine **4** was determined to be 98% from both runs (98.1 in the example shown) by quantitative ^{1}H NMR pdf spectroscopy in benzene-d_6 (Note 48) with 1,3,5-trimethoxybenzene (Sigma Aldrich; 99.8%) as the internal standard (a mixture of 4.5 mg of tri-(1-adamantyl)phosphine **4** and 4.0 mg of 1,3,5-trimethoxybenzene was used to determine purity).

Working with Hazardous Chemicals

The procedures in *Organic Syntheses* are intended for use only by persons with proper training in experimental organic chemistry. All hazardous materials should be handled using the standard procedures for work with chemicals described in references such as "Prudent Practices in the Laboratory" (The National Academies Press, Washington, D.C., 2011; the full text can be accessed free of charge at http://www.nap.edu/catalog.php?record_id=12654). All chemical waste should be disposed of in accordance with local regulations. For general guidelines for the management of chemical waste, see Chapter 8 of Prudent Practices.

In some articles in *Organic Syntheses*, chemical-specific hazards are highlighted in red "Caution Notes" within a procedure. It is important to recognize that the absence of a caution note does not imply that no significant hazards are associated with the chemicals involved in that procedure. Prior to performing a reaction, a thorough risk assessment should be carried out that includes a review of the potential hazards associated with each chemical and experimental operation on the scale that is planned for the procedure. Guidelines for carrying out a risk assessment and for analyzing the hazards associated with chemicals can be found in Chapter 4 of Prudent Practices.

The procedures described in *Organic Syntheses* are provided as published and are conducted at one's own risk. *Organic Syntheses, Inc.*, its Editors, and its Board of Directors do not warrant or guarantee the safety of individuals using these procedures and hereby disclaim any liability for any injuries or damages claimed to have resulted from or related in any way to the procedures herein.

Discussion

Phosphine ligands play an indispensable role in homogeneous catalysis, providing chemists with the ability to finely tune catalyst activity, stability and selectivity. Among the many extant phosphine classes, those bearing tertiary alkyl substituents have emerged as especially privileged in fields as diverse as polymerization,[2] strong-bond activation[3] and cross-coupling.[4]

Despite their privileged status, the synthesis of *tert*-alkyl phosphines remains challenging. The vast majority of syntheses depend on step-wise addition of *tert*-alkyllithium or Grignard reagents to PCl_3 (Scheme 1A); primary and secondary phosphines are then obtained by reduction of the corresponding *P*-chloro phosphines with $LiAlH_4$, whereas tertiary phosphines are accessed by completing the third alkylation with excess nucleophile and a (catalytic) metal additive.[5] While the synthesis of di(adamantanoid) phosphines notably avoids the use of organometallic reagents, it still employs highly reactive air and moisture sensitive reagents (Scheme 1B).[6]

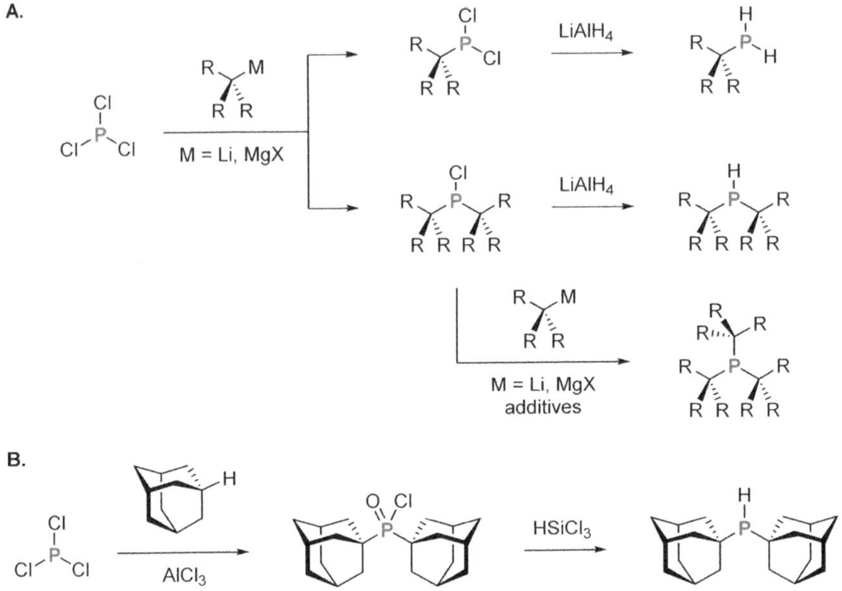

A.

B.

Scheme 1. *tert*-Alkyl phosphines are conventionally synthesized over multiple steps from PCl₃, using highly reactive reagents

The need to handle highly reactive reagents agents, use toxic PCl$_3$, and manipulate, purify and isolate air-sensitive intermediates over multiple steps renders the *de novo* synthesis of *tert*-alkyl substituted phosphines non-trivial. Combined with the lack of commercially available organometallic reagents or adamantanes, these practical difficulties have limited the diversity of the *tert*-alkyl substituents that can be installed at phosphorus.

We recently developed a general and practical method for the synthesis of di-*tert*-alkyl phosphine building blocks *via* the S$_N$1 alkylation of phosphine gas (Scheme 2A).[7] This umpolung strategy uses esters as benign, bench-stable alkylating agents which are readily prepared from commercially available tertiary alcohols. Although phosphine gas is pyrophoric and toxic, it is used as the limiting reagent and is prepared on demand within a sealed system by protonolysis of zinc phosphide, a cheap, air and water-stable solid. The di-*tert*-alkyl phosphine product is formed as the corresponding air-stable phosphonium salt, which is isolated from the reaction mixture in high purity by filtration under air.

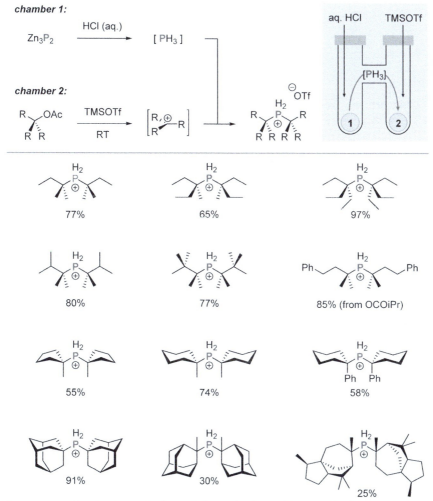

Scheme 2. Synthesis of air-stable, odorless di-*tert*-alkylphosphonium salts *via* S_N1 alkylation of *ex situ* generated phosphine gas. Yields are of isolated, purified materials

Our S_N1 alkylation methodology can be applied successfully to a range of diverse *tert*-alkyl esters, thereby providing facile access to novel di-*tert*-alkyl phosphonium building blocks (Scheme 2B). These salts can be used conveniently in ligand synthesis: treatment with base releases the corresponding phosphine *in situ*, prior to P-functionalization *via* established methods (Scheme 3).

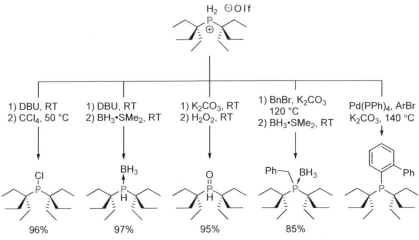

Scheme 3. Use of di-*tert*-alkylphosphonium salts in common *P*-functionalization reactions. Yields are of isolated, purified materials

One tri-*tert*-alkyl phosphine that has garnered much recent attention is tri(1-adamantyl)phosphine **4**.[2c,2d,8] This bulky, electron rich phosphine had eluded synthesis by conventional methods until 2016, when Carrow demonstrated that it could be accessed efficiently by S_N1 alkylation of commercial di(1-adamantyl)phosphine.[9] We recognized that we could adapt Carrow's pioneering procedure to start from our stable di(1-adamantyl)phosphonium salt **3**; in this way we would provide a practical route to substantial amounts of a best-in-class ligand that avoids using an expensive, air-sensitive secondary phosphine.

To showcase the scalability and reproducibility of our S_N1 alkylation methodology, and the versatility of its products in synthesis, we therefore chose to make di(1-adamantyl)phosphonium triflate **3** and subsequently tri(1-adamantyl)phosphine **4**.

References

1. School of Chemistry, University of Nottingham, Nottingham NG7 2RD, U.K. e-mail: liam.ball@nottingham.ac.uk (ORCID: 0000-0003-3849-9006). This work was supported by the Engineering and Physical Sciences Research Council (EPSRC) Centre for Doctoral Training in

Sustainable Chemistry (grant number EP/S022236/1) through a Ph.D. studentship to T.B.
2. For reviews, see: (a) Leone, A. K.; Mueller, E. A.; McNeil, A. J. *J. Am. Chem. Soc.* **2018**, *140*, 15126–15139. (b) Baker, M. A.; Tsai, C.-H.; Noonan, K. J. T. *Chem. Eur. J.* **2018**, *24*, 13078–13088. For selected examples, see: (c) Dong, J.; Guo, H.; Hu, Q.-S. *ACS Macro Lett.* **2017**, *6*, 1301–1304. (d) Kocen, A. L.; Brookhart, M.; Daugulis, O. *Nature Comm.* **2019**, *10*, 1–6. (e) Collier, G. S.; Reynolds, J. R. *ACS Macro Lett.* **2019**, *8*, 931–936.
3. For reviews, see: (a) Reed-Berendt, B. G.; Polidano, K.; Morrill, L. C. *Org. Biomol. Chem.* **2019**, *17*, 1595–1607. (b) Irrgang, T.; Kempe, R. *Chem. Rev.* **2019**, *119*, 2524–2549. (c) Tanabe, Y.; Nishibayashi, Y. *Coord. Chem. Rev.* **2019**, *381*, 135–150. For selected examples, see: (d) Ben-Ari, E.; Gandelman, M.; Rozenberg, H.; Shimon, L. J. W.; Milstein, D. *J. Am. Chem. Soc.* **2003**, *125*, 4714–4715. (e) Zhang, J.; Gandelman, M.; Shimon, L. J. W.; Rozenberg, H.; Milstein, D. *Organometallics* **2004**, *23*, 4026–4033. (f) Prechtl, M. H. G.; Hölscher, M.; Ben-David, Y.; Theyssen, N.; Loschen, R.; Milstein, D.; Leitner, W. *Angew. Chem. Int. Ed.* **2007**, *46*, 2269–2272. (g) Choi, J.; Choliy, Y.; Zhang, X.; Emge, T. J.; Krogh-Jespersen, K.; Goldman, A. S. *J. Am. Chem. Soc.* **2009**, *131*, 15627–15629. (h) Frech, C. M.; Shimon, L. J. W.; Milstein, D. *Organometallics* **2009**, *28*, 1900–1908. (i) Feller, M.; Ben-Ari, E.; Diskin-Posner, Y.; Carmieli, R.; Weiner, L.; Milstein, D. *J. Am. Chem. Soc.* **2015**, *137*, 4634–4637. (j) Heimann, J. E.; Bernskoetter, W. H.; Guthrie, J. A.; Hazari, N.; Mayer, J. M. *Organometallics* **2018**, *37*, 3649–3653.
4. For reviews, see: (a) Fleckenstein, C. A.; Plenio, H. *Chem. Soc. Rev.* **2010**, *39*, 694–711. (b) Magano, J.; Dunetz, J. R. *Chem. Rev.* **2011**, *111*, 2177–2250. (c) Ruiz-Castillo, P.; Buchwald, S. L. *Chem. Rev.* **2016**, *116*, 12564–12649. (d) Lavoie, C. M.; Stradiotto, M. *ACS Catal.* **2018**, *8*, 7228–7250. (e) Fu, G. C. *Acc. Chem. Res.* **2008**, *41*, 1555–1564.
5. (a) Wauters, I.; Debrouwer, W.; Stevens, C. V. *Beilstein J. Org. Chem.* **2014**, *10*, 1064–1096. (b) Kendall, A. J.; Tyler, D. R. *Dalton Trans.* **2015**, *44*, 12473–12483.
6. (a) Goerlich, J. R.; Schmutzler, R. *Phosphorus Sulfur Silicon Relat. Elem.* **1993**, *81*, 141–148. (b) Schwertfeger, H.; Machuy, M. M.; Würtele, C.; Dahl, J. E. P.; Carlson, R. M. K.; Schreiner, P. R. *Adv. Synth. Catal.* **2010**, *352*, 609–615.
7. Barber, T.; Argent, S. P.; Ball, L. T. *ACS Catal.* **2020**, *10*, 5454–5461.

8. (a) Chen, L.; Sanchez, D. R.; Zhang, B.; Carrow, B. P. *J. Am. Chem. Soc.* **2017**, *139*, 12418–12421. (b) Chen, L.; Francis, H.; Carrow, B. P. *ACS Catal.* **2018**, *8*, 2989–2994. (c) Dong, J.; Guo, H.; Peng, W.; H, Q.-S. *Tetrahedron* **2019**, *60*, 760–763.
9. Chen, L.; Ren, P.; Carrow, B. P. *J. Am. Chem. Soc.* **2016**, *138*, 6392–6395.

Appendix
Chemical Abstracts Nomenclature (Registry Number)

1-Adamantanol: tricyclo[3.3.1.1³,⁷]decan-1-ol; (768-95-6)
Acetic anhydride: acetic acid, anhydride; (108-24-7)
Triethylamine: ethanamine, *N,N*-diethyl-; (121-44-8)
4-Dimethylaminopyridine: 4-pyridinamine, *N,N*-dimethyl- (1122-58-3)
Pyridine: pyridine; (110-86-1)
Cyclohexane: cyclohexane; (110-82-7)
$KMnO_4$: permanganic acid, ($HMnO_4$), potassium salt; (7722-64-7)
K_2CO_3: carbonic acid, dipotassium salt; (584-08-7)
Sodium hydrogen carbonate: carbonic acid monosodium salt; (144-55-8)
Copper(II) sulfate pentahydrate, $CuSO_4$: sulfuric acid copper(2+) salt (1:1), pentahydrate; (7758-99-8)
Sulfuric acid: sulfuric acid; (7664-93-9)
Magnesium sulfate: sulfuric acid magnesium salt (1:1); (7487-88-9)
Zinc phosphide: zinc phosphide, (Zn_3P_2); (1314-84-7)
Trimethylsilyl trifluoromethanesulfonate: methanesulfonic acid, trifluoro-, trimethylsilyl ester; (27607-77-8)
Hydrochloric acid, HCl: hydrochloric acid; (7647-01-0)
Sodium hypochlorite: hypochlorous acid, sodium salt; (7681-52-9)
1,8-Diazabicyclo[5.4.0]undec-7-ene: pyrimido[1,2-a]azepine, 2,3,4,6,7,8,9,10-octahydro-; (6674-22-2)
Calcium hydride: calcium hydride, (CaH_2); (7789-78-8)
1,3,5-Trimethoxybenzene: (621-23-8)
Chloroform-*d*, $CDCl_3$: methane-d, trichloro-; (865-49-6)
Benzene-d_6, C_6D_6: benzene-d_6; (1076-43-3)
Tri-(1-adamantyl)phosphine: (897665-73-5)

Tom Barber obtained his MChem from Durham University, UK, following research with Dr Chris Coxon and Prof. Andy Whiting. Since 2016, he has been working towards his Ph.D. in organophosphorus chemistry and catalysis with Dr Liam Ball at the University of Nottingham.

Liam Ball obtained his MSc from the University of Bristol, UK. Following doctoral studies with Dr. Chris Russell and Prof. Guy Lloyd-Jones FRS FRSE at the University of Bristol (2009–2013), he moved to the University of Edinburgh, UK, as a postdoctoral researcher with Prof. Guy Lloyd-Jones FRS FRSE (2014–2015). In 2015, Liam was appointed Assistant Professor of Organic Chemistry at the University of Nottingham.

Praveen Kumar Gajula received his Bachelor's and Master's degrees from Hyderabad. He obtained his Ph.D. in synthetic organic chemistry in 2012 from CSIR-IICT, Hyderabad under the guidance of Dr. Tushar Kanti Chakraborty. His research was focused on amide-modified RNA mimics, total syntheses of natural products, anti-cancer compounds and their analogs thereof. In 2019, he joined Prof. Eriks Rozners research group at Binghamton University, optimization of RNA interference (RNAi) and CRISPR efficiency and specificity using chemically modified RNAs. In 2015, Praveen joined Sai Life Sciences, Hyderabad, India. He is currently working at TCG GreenChem, Inc. as a Senior Research Scientist in the department of process research and development.

Leila Terrab obtained her Ph.D. in synthetic organic chemistry in 2021 from the University of Pittsburgh under the guidance of Dr. Peter Wipf. Her doctoral studies focused on the synthesis of small molecule inhibitors of the Artemis endonuclease, as well as the synthesis of HSP70 agonists. In 2021, Leila joined TCG GreenChem, Inc. as a postdoctoral scientist in Process Chemistry.

Srinivasarao Tenneti received his Ph.D. in Chemistry under Dr. J. S. Yadav at Indian Institute of Chemical Technology, India in 2013. He worked with Professor Philip S. Low at Purdue University as a Postdoctoral Fellow, and in 2016, he joined Professor T. V. RajanBabu at The Ohio State University as a Postdoctoral Researcher, where he worked on synthesis of natural products using nickel-catalyzed asymmetric hydrovinylation methodology. He then moved to Univ. of Florida in 2019 as a Postdoctoral Associate under the supervision of Professor Robert W. Huigens III where his research was focused on synthesis of indole alkaloids. Currently, he is working as a senior scientist at TCG GreenChem, Inc.

Dr. Gopal Sirasani received his Bachelor's and Master's degrees in Hyderabad, India. He obtained his Ph.D. in synthetic organic chemistry in 2011 from Temple University, Philadelphia under the guidance of Prof. Rodrigo B. Andrade. His doctoral research was focused on developing novel methodologies, total syntheses of natural products and their analogs thereof. He got his post-doctoral training in the laboratory of Prof. Emily Balskus at Harvard University, where he developed biocompatible organic reactions utilizing microbially-generated reagents to realize transition metal catalysis in the presence of microbes. In 2013, Gopal began his industrial career at Melinta Therapeutics, New Haven, CT. He is currently working at TCG GreenChem, Inc. as a Director in the department of process research and development.

Preparation of 1-Benzyl-7-methylene-1,5,6,7-tetrahydro-4H-benzo[d]imidazol-4-one

Michael Morgen, Jasmin Lohbeck, and Aubry K. Miller[1]*

Cancer Drug Development Group, German Cancer Research Center (DKFZ), 69126 Heidelberg, Germany

Checked by Nageswara Rao Kalikinidi, Venumadhav Janganati, Gopal Sirasani, and Chris Senanayake

Procedure (Note 1)

A. *1-(1-Benzyl-5-iodo-1H-imidazol-4-yl)pent-4-en-1-one (2).* A 250 mL round-bottomed flask equipped with a 3.2 cm long oval stir bar is charged with 1-(5-iodo-1H-imidazol-4-yl)pent-4-en-1-one (1) (14.00 g, 50.71 mmol, 1.0 equiv) (Note 2) followed by DMF (100 mL) (Note 3). The flask is capped with a rubber septum and the resulting, orange-colored solution is magnetically stirred (500 rpm) and cooled to 0 °C using an ice bath. Potassium carbonate (10.51 g, 76.06 mmol, 1.5 equiv) (Note 4) is added in one portion followed by the addition of benzyl bromide (6.3 mL, 53.2 mmol, 1.05 equiv) (Note 5) dropwise using a syringe over a time period of 2 min. The resulting suspension is stirred at 0 °C for 2 h (Figure 1A). Analysis by TLC (Note 6) is performed to ensure that starting material is consumed. After **1** is entirely consumed the reaction mixture is transferred to a 1 L separatory funnel. The reaction vessel is sequentially rinsed with H_2O (50 mL) (Note 7) and EtOAc (50 mL) (Note 8), both of which are transferred to the separatory funnel. Additional H_2O (250 mL) and EtOAc (250 mL) are added to the separatory funnel. After mixing, the phases are separated, and the aqueous phase is extracted with EtOAc (2 × 200 mL) (Figure 1B). The combined organic phases are sequentially washed with water (2 × 300 mL) and a saturated aqueous NaCl solution (300 mL). The organic layer is dried with $MgSO_4$ (40 g) (Note 9), filtered through a fluted filter paper into a 1 L round-bottomed flask using a glass funnel (Note 10), and concentrated on a rotary evaporator (40 °C, 180 to 15 mmHg) (Note 11). The product is purified by crystallization by dissolving the solid in a minimal amount of EtOAc (100 mL) (Notes 12 and 13) on a gently boiling water bath. To the boiling hot solution is added, in portions, *n*-hexane (80 mL) (Note 14) to the point where the white precipitate that forms upon addition slowly disappears. The flask is then covered with a plastic cap, sealed with Parafilm, and the solution is allowed to cool to 24 °C on the bench and then at 4 °C in a refrigerator for 18 h (Figure 1C). The obtained solid is vacuum filtered through a round filter paper (ø = 9 cm) using a Büchner funnel and washed with a minimum amount of cold (4 °C) EtOAc/*n*-hexane 1:1 (approx. 20 mL) and then cold (4 °C) *n*-hexane (2 × 20 mL) (Note 15). The mother liquor is concentrated on a rotary evaporator (40 °C, 280 to 15 mmHg) and the remaining residue is crystallized as described above using a smaller solvent volume to obtain a second crop of material. The combined solids are dried under high vacuum (24 °C, 2×10^{-2} mm Hg) at 24 °C for 18 h to give 13.0 g (35.5 mmol, 70%) of **2** as an off-white solid (Notes 16, 17, and 18) (Figure 1D).

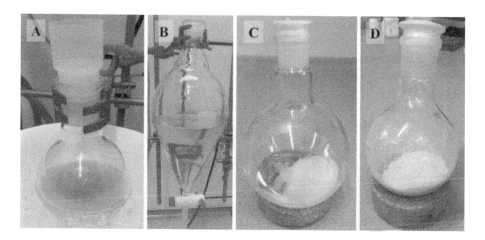

Figure 1. (A) Reaction mixture after addition of benzyl bromide; (B) Phases (EtOAc, top; aqueous, bottom) after shaking and allowing to separate; (C) Crystallization of 2 from a mixture of EtOAc and *n*-hexane; (D) Final product 2 dried under high vacuum (photos provided by submitters)

B. *1-Benzyl-7-methylene-1,5,6,7-tetrahydro-4H-benzo[d]imidazol-4-one (3)*. A 500 mL beaker (ø = 7 cm, h = 9 cm) (Note 19) equipped with a triangular prism-shaped stir bar (50 × 12 mm) is charged with tetra-*n*-butylammonium bromide (52.82 g, 163.84 mmol, 5.0 equiv) (Note 20) and tetra-*n*-butylammonium acetate (24.70 g, 81.92 mmol, 2.5 equiv) (Note 21). The flask is submerged in an oil bath preheated to 130 °C (Figure 2A). The salts are magnetically stirred (270 rpm) until a pale-yellow, homogenous melt is obtained (approximately 30 min) (Figure 2B) (Note 22). Then Pd(OAc)$_2$ (220.7 mg, 0.98 mmol, 0.03 equiv) (Note 23) is added in one portion to the melt (Note 24). The mixture is stirred for 1 min (Note 25), whereupon 1-(1-benzyl-5-iodo-1*H*-imidazol-4-yl)pent-4-en-1-one (**2**) (12.00 g, 32.77 mmol, 1.0 equiv) is added slowly but continuously in one portion to the melt (Notes 24 and 26).

The resulting dark brown melt (Figure 2C) is magnetically stirred (270 rpm) for 5–10 min, at which time TLC monitoring shows complete consumption of the starting material (Note 27). The reaction is stopped by pouring the hot melt into a 1 L Erlenmeyer flask that contains a mixture of EtOAc and water (300 mL each) and is equipped with a 5 cm long stir bar that is magnetically stirred (500 rpm) at 24 °C (Note 28) (Figure 2D). Complete transfer is assured by repeatedly rinsing the reaction vessel with EtOAc and H₂O from a squirt bottle. The phases are vigorously stirred for 5 min and then filtered through a firmly packed pad (2 cm) of Celite (Note 29) in a sintered glass Büchner funnel (ø = 10 cm, h = 15 cm; Porosity 4). (Figure 2E).

Figure 2: (A) The mixture of (*n*-Bu)₄NOAc and (*n*-Bu)₄NBr immediately after being immersed in the oil bath; (B) Appearance of the two salts after they have completely melted; (C) Appearance of the melt after adding Pd(OAc)₂ and 2; (D) Hot melt being poured into a stirring mixture of EtOAc and H₂O at 24 °C; (E) Filtration of the two phases through a pad of Celite using a sintered glass funnel (photos provided by submitters)

The Celite pad is washed with additional EtOAc (400 mL), and the phases are transferred into a 1 L separatory funnel and separated. The organic phase is washed with H₂O (6 × 200 mL) (Figure 3A) until TLC monitoring showed no remaining tetrabutylammonium salts in the organic layer (Notes 30, 31 and 32). The organic phase is then washed with 1 M aqueous NaOH (10 × 50 mL) (Figure 3B and 3C) (Note 33 and 34) followed by a final wash with saturated aqueous NaCl solution (200 mL). The organic phase is transferred into a 1 L round-bottomed flask equipped with a 3.2 cm long oval stir bar and capped with a rubber septum. To this red solution is added activated carbon (1.0 g) (Note 35). The mixture is stirred (500 rpm) at 24 °C for 18 h and afterwards filtered through a firmly packed (2 cm) pad of Celite (Figure 3D) (Note 36) in a sintered glass Büchner funnel (ø = 10 cm, h = 15 cm; Porosity 4). This Celite pad is washed with EtOAc (400 mL) and the obtained pale, yellow-colored solution is concentrated to roughly half its volume on a rotary evaporator (40 °C, 120 mmHg). To this solution is added *n*-hexane (400 mL), whereupon the product precipitates (Figure 3E) (Note 37). The resultant suspension is concentrated on a rotary evaporator (40 °C, 225 to 15 mmHg) to dryness (Note 38). The yellow solid is transferred into a sintered glass Büchner funnel (ø = 6 cm, h = 6 cm; Porosity 4) (Figure 3F) (Note 39), thoroughly washed with *n*-hexane (500 mL) at 24 °C (Note 40), and dried under high vacuum (24 °C, 2 × 10^{-2} mmHg) for 18 h. Imidazole **3** (4.3 g, 18.1 mmol, 55%) is obtained as a tan-colored solid (Figure 4) (Notes 18, 41 and 42). An aromatic by-product is also produced in the reaction (Note 43).

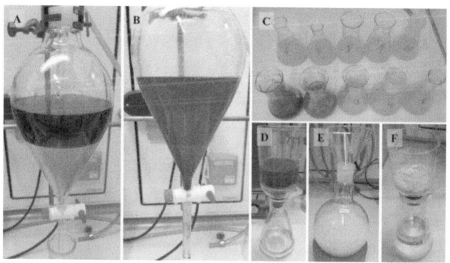

Figure 3. (A) Washing of the organic phase with water; (B) Subsequent washing of the organic phase with 1 M NaOH; (C) Appearance of the aqueous phases (1 – 10) from basic washing procedure; (D) Filtration of the organic phase through a pad of Celite after charcoal treatment; (E) Suspension formed after half concentration of the organic phase and addition of *n*-hexane; (F) Washing of **3** with *n*-hexane using a sintered glass frit (photos provided by submitters)

Figure 4. Appearance of pure product 3 (photo provided by submitters)

Notes

1. Prior to performing each reaction, a thorough hazard analysis and risk assessment should be carried out with regard to each chemical substance and experimental operation on the scale planned and in the context of the laboratory where the procedures will be carried out. Guidelines for carrying out risk assessments and for analyzing the hazards associated with chemicals can be found in references such as Chapter 4 of "Prudent Practices in the Laboratory" (The National Academies Press, Washington, D.C., 2011; the full text can be accessed free of charge at https://www.nap.edu/catalog/12654/prudent-practices-in-the-laboratory-handling-and-management-of-chemical. See also "Identifying and Evaluating Hazards in Research Laboratories" (American Chemical Society, 2015) which is available via the associated website "Hazard Assessment in Research Laboratories" at https://www.acs.org/content/acs/en/about/governance/committees/chemicalsafety/hazard-assessment.html. In the case of this procedure, the risk assessment should include (but not necessarily be limited to) an evaluation of the potential hazards associated with dimethylformamide, potassium carbonate, benzyl bromide, ethyl acetate, magnesium sulfate, sodium chloride, hydrochloric acid, *n*-hexane, palladium(II) acetate, tetra-*n*-butylammonium bromide, tetra-*n*-butylammonium acetate, sodium hydroxide, benzene, methanol, and silica gel.
2. This compound was synthesized as previously described in *Org. Synth.* **2021**, *98*, 171–193.

3. Dimethylformamide was obtained from Sigma Aldrich (≥99.8%) and was used as received.
4. Potassium carbonate (anhydrous, ≥99%) was obtained from Sigma Aldrich and was used as received.
5. Benzyl bromide (98.0%, reagent grade) was obtained from Sigma-Aldrich and was used as received.
6. For reaction monitoring thin-layer chromatography was performed using silica gel F_{254}-plates from Merck, Darmstadt, Germany and a 1:1 mixture of EtOAc/n-hexane as eluent. A small sample (~50 mL) was taken from the reaction mixture with a glass pipette equipped with a silicone bulb and quenched into H_2O (~1 mL) and EtOAc (~0.2 mL) in an Eppendorf tube. The two layers were agitated by pulling multiple times into the pipette and then were allowed to separate in the tube. The organic layer was used for TLC spotting. The starting material **1** shows an R_f value of 0.2, whereas the desired product **2** shows an R_f value of 0.7 (both visible under UV light). If analysis does not indicate complete consumption of **1**, another portion of benzyl bromide (0.35 mL, 5.07 mmol, 0.1 equiv) is added, and the reaction is stirred at the same temperature for an additional 1 h – the checkers found that addition of this additional portion of benzyl bromide was necessary.

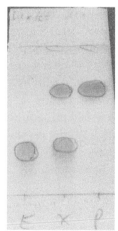

Figure 5. TLC monitoring of completed reaction using normal phase silica plates and 1:1 EtOAc/n-hexane as eluent (left lane: starting material; right lane: reaction mixture; middle lane: co-spot) (photo provided by submitters)

7. The submitters used deionized water from their in-house supply.
8. Ethyl acetate (≥99.8%) was obtained from Fisher Chemical and was used as received.
9. Magnesium sulfate (97%, pure, anhydrous) was obtained from Acros Organics and was used as received.
10. The remaining magnesium sulfate was washed with EtOAc (25 mL).
11. The crude product obtained by benzylation of imidazole **1** mainly consists of **2** and its regioisomer **4** which are formed in a ratio of 6:1–8:1 (Figure 6).

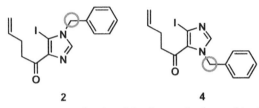

Figure 6. Regioisomers obtained by benzylation of imidazole 2

The ratio can be determined by ^{1}H NMR analysis of the crude product mixture after extraction, but before crystallization (diagnostic signal marked with red circle: desired product **2**, 5.30 ppm; regioisomer **4**, 5.50 ppm in DMSO-d_6). The two regioisomers have very similar R_f values, thus the desired (major) product is best purified by crystallization.

12. A wooden stick was placed in the flask to promote boiling and prevent bumping.
13. In some cases, the volume of hot EtOAc (100 mL) could not dissolve all of the solids. In these cases, the hot suspension was filtered through a fluted filter paper into a 500 mL round-bottomed flask using a glass funnel before continuing with the crystallization procedure. Analysis of the filtered white solid by ^{1}H NMR showed it to be the imidazolium salt **5** (Figure 7 and see discussion).

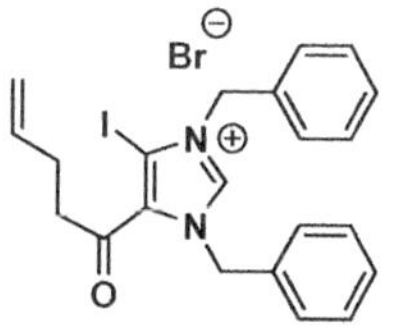

5

Figure 7. Poorly soluble imidazolium salt, which results from dibenzylation

14. *n*-Hexane (≥95%, HPLC grade) was obtained from Fisher Chemical and was used as received.
15. The solvent mixtures were cooled to 4 °C in a refrigerator prior to their use for washing the final compound.
16. The solid was sometimes obtained as a powder, sometimes fine needles, and sometimes thick plates (Figure 8). This variation appears to be dependent on the rate of crystallization. NMR analysis revealed that all forms are equally pure **2**. The product **2** exhibits the following analytical data: R_f = 0.70 (EtOAc/*n*-hexane 1:1). mp (uncorrected): 110–111 °C. ^{1}H NMR (600 MHz, DMSO-d_6) δ: 2.27 – 2.38 (m, 2H), 3.02 (dd, *J* = 7.8, 7.0 Hz, 2H), 4.94 (d, *J* = 10.4 Hz, 1H), 5.03 (d, *J* = 17.1Hz, 1H), 5.30 (s, 2H), 5.85 (ddt, *J* = 17.2, 10.2, 6.4 Hz, 1H), 7.14–7.15 (m, 2H), 7.30 – 7.31 (m, 1H), 7.35 – 7.37 (m, 2H), 8.23 (s, 1H). ^{13}C NMR (150 MHz, DMSO-d_6) δ: 27.7, 37.3, 50.4, 80.6, 114.9, 127.1, 127.8, 128.7, 136.2, 137.9, 141.3, 141.4, 194.5. LC-MS (M+H) calcd for $C_{15}H_{16}N_2OI$ (*m/z*): 367. HRMS–ESI (*m/z*) (provided by submitters): $[2M + Na]^+$ calcd for $C_{30}H_{30}I_2N_4NaO_2$: 755.0350, Found: 755.0357.

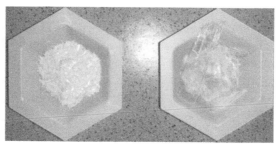

Figure 8. Appearance of 2 (left: needles; right: plates) from different crystallizations (photos provided by submitters)

DOI: 10.15227/orgsyn.098.0315

17. Purity was determined to be 98.75% by qNMR spectroscopy in DMSO-d_6 using 36 mg of compound (**2**) and 11.5 mg of mesitylene (96%) as an internal standard.
18. The product showed no decomposition after storage at 24 °C on the benchtop for one month.
19. The submitters suggest the use of a narrow beaker so that the melt has reasonable depth and so that stirring of the melt results in a vortex that facilitates the uptake and mixing of the added materials.
20. Tetra-*n*-butylammonium bromide (≥99.0%, reagent plus) was obtained from Sigma-Aldrich and was used as received.
21. Tetra-*n*-butylammonium acetate (≥90%, technical) was obtained from TCI Chemicals and was used as received.
22. The remaining salt residues sticking to the wall of the beaker were brought into the melt by carefully heating with a heat gun.
23. Palladium(II) acetate (Pd(OAc)$_2$ (recrystallized, 97%)) was obtained from Sigma-Aldrich and was used as received.
24. Care should be taken to carefully pour the material into the middle of the vortex to ensure that it is readily taken up by the melt and does not remain floating upon the surface.
25. The melt became pale brown with a few remaining undissolved particles of Pd(OAc)$_2$.
26. In the event that **2** was obtained as thick plates, the solid was finely ground using a mortar and pestle before addition.
27. Completion of the reaction occurred rapidly after dissolution of the starting material **2** into the melt. The submitters recommend to monitor the reaction by performing TLC-analyses every two minutes after the addition of **2**. A sample is taken by putting the tip of a glass pipette into the hot reaction mixture and the black residue remaining on the glass pipette is diluted with EtOAc and then used for TLC monitoring. Thin-layer chromatography was performed using silica gel 60 F_{254}-plates from Merck, Darmstadt, Germany and a 9:1 mixture of CH$_2$Cl$_2$/MeOH as eluent. The starting material **2** shows an R_f value of 0.91, whereas the desired product **3** shows an R_f value of 0.46 (both visible under UV light) (Figure 9).

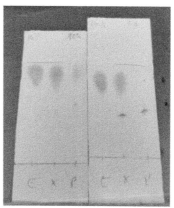

Figure 9. TLC monitoring of the Heck reaction immediately after adding 2 (left plate) and 5 min after the addition (right plate). Left lanes: starting material 2; Right lanes: reaction mixture diluted with EtOAc; middle lanes: co-spot. CH$_2$Cl$_2$/MeOH (9:1) used as eluent (photos provided by submitters)

28. The submitters wrapped paper towels around the beaker to facilitate pouring of the hot melt. This also assures that no oil sticking to the outside of the beaker is unintentionally transferred into the reaction mixture.
29. Celite Standard Super Cel was obtained from Carl Roth and was used as received.
30. The tetrabutylammonium salts don't exclusively partition into the aqueous layer when using EtOAc as an organic solvent, which necessitates multiple washes. The submitters suggest monitoring the organic layer for the presence of remaining salts by thin-layer chromatography (TLC) (CH$_2$Cl$_2$/MeOH 9:1). The salts appear as a streaky spot (visible with I$_2$ stain, but not UV) below the product spot (visible under UV and with I$_2$ stain). Washing should be repeated (usually 5–7 times) until no further tetrabutylammonium salts can be detected (Figure 10).

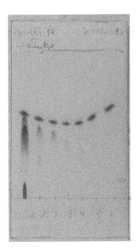

Figure 10. TLC monitoring of the washing procedure with H₂O. From left to right lane: organic layer after the first phase separation, then after the next six aqueous washes; CH$_2$Cl$_2$/MeOH 9:1 used as eluent; TLC plate is stained with iodine (photos provided by submitters)

31. Accurate phase separation is occasionally difficult due to cloudiness and bubble formation between the layers. The submitters prefer to keep a minimal amount of water in the funnel during each separation so as not to lose product during the washing procedure.
32. After the washing is complete, a small volume of the organic phase is removed and concentrated on a rotary evaporator (40 °C, 6 mmHg). The residue is analyzed by ^{1}H NMR spectroscopy in order to determine the ratio of desired product **3** and its aromatized isomer **6** (Figure 11 and see Discussion). This ratio varies between 5:1–8:1 (diagnostic signal marked with red circle: desired product **3**, 5.30 ppm; aromatized byproduct **6**, 5.53 ppm in CDCl$_3$).

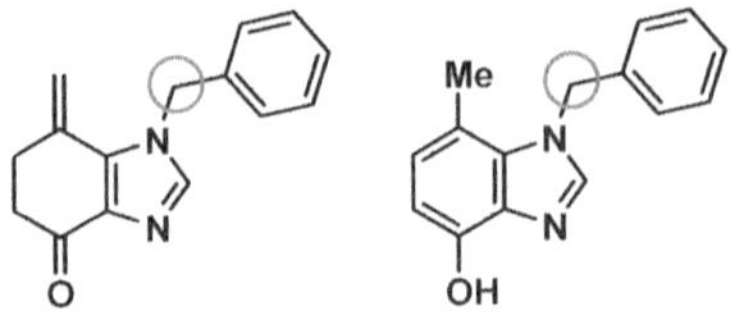

Figure 11. Products arising from Heck reaction of iodoimidazole **2**

33. Sodium hydroxide was obtained from VWR Chemicals Prolabo and was used as received.
34. These washes remove hydroxybenzimidazole **6** from the organic layer by deprotonating the hydroxyl group and helping to partially partition **6** into the aqueous layer. Ketone **3** also partially partitions into the aqueous layer during this procedure, but material loss is minimal. That both products partition into the aqueous phase is seen visually as depicted in Figure 3C: Upon standing for a few minutes at 24 °C, a fine off-white solid (**3** and **6**) precipitates from the aqueous phase.

The washing procedure can be monitored by thin-layer chromatography using silica gel 60 F_{254}-plates from Merck, Darmstadt, Germany and benzene/MeOH (9:1) as eluent (Figure 12). Byproduct **6** appears as a slightly faster eluting spot than **3** under these conditions. Small quantities of byproduct **6** visualize very well under UV and with I_2 staining, making it difficult to determine when to stop washing. The submitters have found that 10 washes consistently provide sufficiently pure samples of **3** (also see discussion), and recommend using this number of washes, but still monitoring via TLC for a visual analysis. After the 10 washes, a small volume from the organic phase can be concentrated and analyzed by ^{1}H NMR. If a **3**/**6** ratio ≥50:1 is not obtained after 10 washes, further washing with 1M NaOH must be performed.

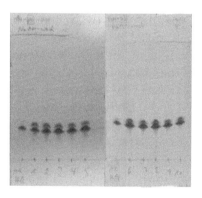

Figure 12. TLC monitoring of the basic washing procedure with 1M NaOH. From left to right lane: pure product **3** obtained in a different experiment, organic phase from 1st to 5th wash (left TLC plate) and organic phase from 6th to 10th wash (right TLC plate), benzene/MeOH 9:1 used as eluent; TLC plate is stained with iodine (photos provided by submitters)

35. Activated carbon was obtained from Sigma-Aldrich and was used as received.
36. Treatment of the organic phase with charcoal was done to decolorize the product by removing trace impurities such as palladium.
37. The submitters adopted this precipitation procedure due to difficulties in removing all traces of EtOAc from **3** when it was concentrated to a solid from an EtOAc solution, even after the resulting solid was dried at 100 °C under high vacuum for 24 h. This precipitation method reliably provided solvent-free **3** after drying the filtered solid under high vacuum.
38. If the product re-dissolves during concentration, the flask should be removed from the rotary evaporator and additional *n*-hexane should be added, to ensure that the product remains a solid throughout this process.
39. The majority of the solid can be scratched out of the flask with a spatula and transferred to the funnel by rinsing with *n*-hexane. Any solid remaining in the round-bottomed flask is negligible, but if a quantitative transfer is desired, it can be redissolved in EtOAc, transferred into a smaller (100 mL) round-bottomed flask, concentrated on a rotary evaporator (40 °C, 150 to 35 mmHg) to a solid, and then transferred to the Büchner funnel.
40. To ensure complete removal of trace EtOAc, the solid was crushed in the funnel with a spatula, then covered with *n*-hexane, the resulting suspension was well mixed with the spatula, and then vacuum was applied to pull the solvent through the funnel. This procedure was repeated (3-4 times) until 500 mL of *n*-hexane were consumed.
41. The product (**3**) exhibits the following analytical data: $R_f = 0.46$ (CH$_2$Cl$_2$/MeOH 9:1) and 0.19 (benzene/MeOH 9:1). mp (uncorrected): 147.0–149.0 °C. ^{1}H NMR (600 MHz, CDCl$_3$) δ: 2.70 (t, $J = 6.6$ Hz, 2H), 2.81 (t, $J = 6.0$ Hz, 2H), 5.13 (s, 1H), 5.26 (s, 1H), 5.33 (s, 2H), 7.08 (d, $J = 7.2$ Hz, 2H), 7.35 – 7.40 (m, 3H), 7.57 (s, 1H). ^{13}C NMR (151 MHz, CDCl$_3$) δ: 34.1, 38.6, 49.7, 112.2, 126.3, 128.4, 129.3, 132.9, 134.9, 138.0, 139.4, 141.9, 191.5. Purity was determined to be 89.6% by qNMR spectroscopy in CDCl$_3$ using 23.9 mg of compound (**3**) and 10.7 mg of dimethyl sulfone (99.65%) as an internal standard. LC-MS (M+H) calcd for C$_{15}$H$_{15}$N$_2$O (*m/z*): 239. HRMS–ESI (*m/z*) (provided by submitters): [2M + Na]$^+$ calcd for C$_{30}$H$_{28}$N$_4$NaO$_2{}^+$: 499.2104, found: 499.2109.
42. While **3** obtained via this extractive purification is 89.6% pure (based on qNMR), a sample that is entirely free of **6** and other minor impurities

can be prepared by crystallization: a sample of **3** (1.0 g, 4.19 mmol) is dissolved in hot EtOAc (~40 mL) on a water bath (100 °C), filtered while hot, and *n*-hexane (~15 mL) is then added while the solution is still on the water bath until the white precipitate which forms only slowly disappears. The mixture is cooled to 24 °C and then to −20 °C for 18 h. After filtration through a round paper filter (ø = 5.5 cm) using a Büchner funnel, washing with cold (4 °C) EtOAc/*n*-hexane 1:1 (13 mL) and then cold (4 °C) *n*-hexane (2 × 15 mL), and air drying for 1 h on the filter using house vacuum, an off-white solid (0.774 g, 3.23 mmol, 77.4%) was obtained (Figure 13). ^{1}H NMR (600 MHz, CDCl$_3$) δ: 2.70 (t, *J* = 6.6 Hz, 2H), 2.81 (t, *J* = 6.0 Hz, 2H), 5.13 (s, 1H), 5.26 (s, 1H), 5.34 (s, 2H), 7.08 (d, *J* = 7.2 Hz 2H), 7.33 – 7.40 (m, 3H), 7.57 (s, 1H). ^{13}C NMR (151 MHz, CDCl$_3$) δ: 34.1, 38.6, 49.7, 112.2, 126.3, 128.4, 129.3, 132.9, 134.9, 138.0, 139.4, 141.9, 191.5. Purity was determined to be 96.8% by qNMR spectroscopy in CDCl$_3$ using 24 mg of compound (**3**) and 9.7 mg of dimethyl sulfone (99.65%) as an internal standard.

From submitters   **From checkers**

Figure 13. Pure 3 obtained by crystallization from hot EtOAc/*n*-hexane

43. The submitters obtained an analytical sample of aromatized byproduct **6** as follows: The aqueous phases of the basic washing procedure (see Note 34) are combined and acidified by dropwise addition of concentrated HCl to pH~5–6. Then the aqueous phase is extracted with EtOAc (3 × 200 mL), whereupon the combined organic phases are washed with brine (300 mL) and concentrated on a rotary evaporator (40 °C, 280 to 15 mmHg). The obtained brown solid is suspended in hot EtOAc (50 mL). The suspension is filtered using a fluted filter paper in a 250 mL round-bottomed flask, and to the hot

solution of EtOAc is added *n*-hexane (50 mL). The flask is covered with a plastic cap (wrapped with Parafilm) and allowed to cool to 24 °C on the bench and then at –20°C in a freezer. The resulting yellow solid is filtered through a round filter paper (ø = 4.5 cm) using a Büchner funnel and washed with a minimum amount of cold (4 °C) EtOAc/*n*-hexane 1:1 (20 mL) and then cold (4 °C) *n*-hexane (2 × 20 mL) and air dried on the filter for 1 h using house vacuum (Figure 14).

Figure 14. Appearance of aromatized byproduct 6 (photo provided by submitters)

The aromatized byproduct 6 exhibits the following analytical data (provided by the submitters): R_f = 0.46 (CH$_2$Cl$_2$/MeOH 9:1) and 0.30 (benzene/MeOH 9:1). mp (uncorrected) 207–209 °C (decomposition). ^{1}H NMR (400 MHz, DMSO-d_6) δ 2.26 (s, 3H), 5.64 (s, 2H), 6.44 (d, *J* = 7.8 Hz, 1H), 6.67 (dd, *J* = 7.9, 1.0 Hz, 1H), 6.92 – 6.98 (m, 2H), 7.20 – 7.37 (m, 3H), 8.15 (s, 1H), 9.57 (br. s, 1H). ^{13}C NMR (101 MHz, DMSO-d_6) δ 17.4, 48.8, 106.5, 111.7, 125.2, 125.6, 127.5, 128.9, 133.9, 134.1, 139.2, 144.0, 147.6. HRMS–ESI (*m/z*): [2M + Na]$^+$ calcd for C$_{30}$H$_{28}$N$_4$NaO$_2$$^+$: 499.2104, Found: 499.2110.

Working with Hazardous Chemicals

The procedures in *Organic Syntheses* are intended for use only by persons with proper training in experimental organic chemistry. All hazardous materials should be handled using the standard procedures for work with chemicals described in references such as "Prudent Practices in the Laboratory" (The National Academies Press, Washington, D.C., 2011; the full text can be accessed free of charge at http://www.nap.edu/catalog.php?record_id=12654). All chemical waste should be disposed of in accordance with local regulations. For general guidelines for the management of chemical waste, see Chapter 8 of Prudent Practices.

In some articles in *Organic Syntheses*, chemical-specific hazards are highlighted in red "Caution Notes" within a procedure. It is important to recognize that the absence of a caution note does not imply that no significant hazards are associated with the chemicals involved in that procedure. Prior to performing a reaction, a thorough risk assessment should be carried out that includes a review of the potential hazards associated with each chemical and experimental operation on the scale that is planned for the procedure. Guidelines for carrying out a risk assessment and for analyzing the hazards associated with chemicals can be found in Chapter 4 of Prudent Practices.

The procedures described in *Organic Syntheses* are provided as published and are conducted at one's own risk. *Organic Syntheses, Inc.*, its Editors, and its Board of Directors do not warrant or guarantee the safety of individuals using these procedures and hereby disclaim any liability for any injuries or damages claimed to have resulted from or related in any way to the procedures herein.

Discussion

We required an efficient synthesis of ketone **3** and related substances for a project devoted to the synthesis of covalent methionine aminopeptidase 2 (MetAP2) inhibitors. Our first-generation library of substances were spiroepoxyimidazoles **7**, which were all synthesized from **1** via alkylation, Heck cyclization, and epoxidation with DMDO (Figure 15A). The compounds were designed to mimic the natural product ovalicin, a potent and selective covalent MetAP2 inhibitor (Figure 15B). Our second-generation inhibitors were based on a triazole scaffold, and we found them to be more potent and to have better physicochemical properties than the imidzoles (Figure 15C).[2]

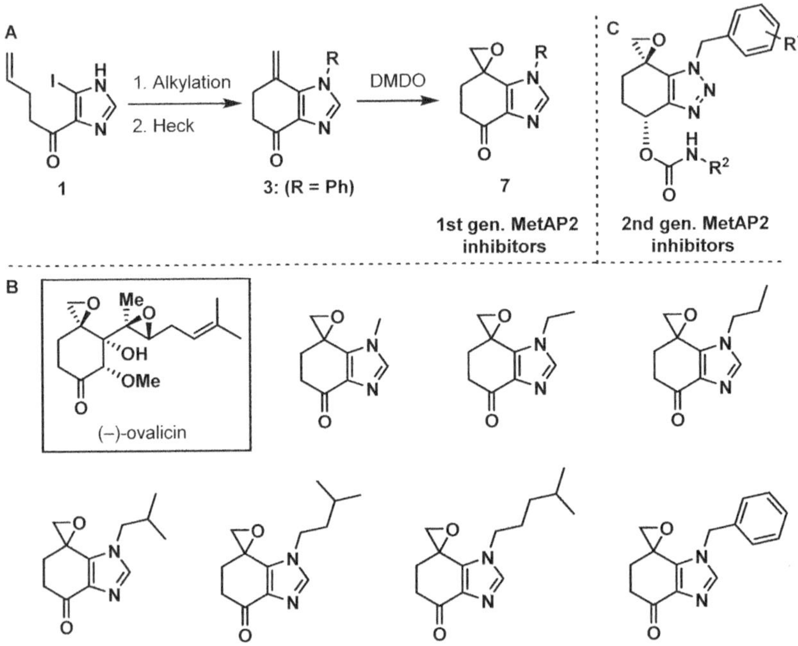

Figure 15. (A) Synthesis route to 1st generation MetAP2 inhibitors; (B) Small library of putative MetAP2 inhibitors inspired by ovalicin; (C) General structure of 2nd generation MetAP2 inhibitors

Outside of the context of our MetAP2 inhibitor project, compound 3 is an interesting building block due to the fact that it can potentially be derivatized in a variety of ways to generate a diverse collection of substances (Scheme 1). Besides oxidation to epoxide 7 as mentioned above, ketone 3 can also be transformed to aziridines 8,[3] undergo reductive amination to give amines 9,[4] or reduced to alcohol 10.[5] Alternatively, the exocyclic double bond can be reduced to a methyl group to afford ketone 11,[6] or undergo Huisgen cycloaddition reactions with 1,3-dipoles to afford spirocycles of type 12.[7] Many more manipulations can also be envisaged, including those that involve removal of the benzyl group and replacement with other substituents.[8]

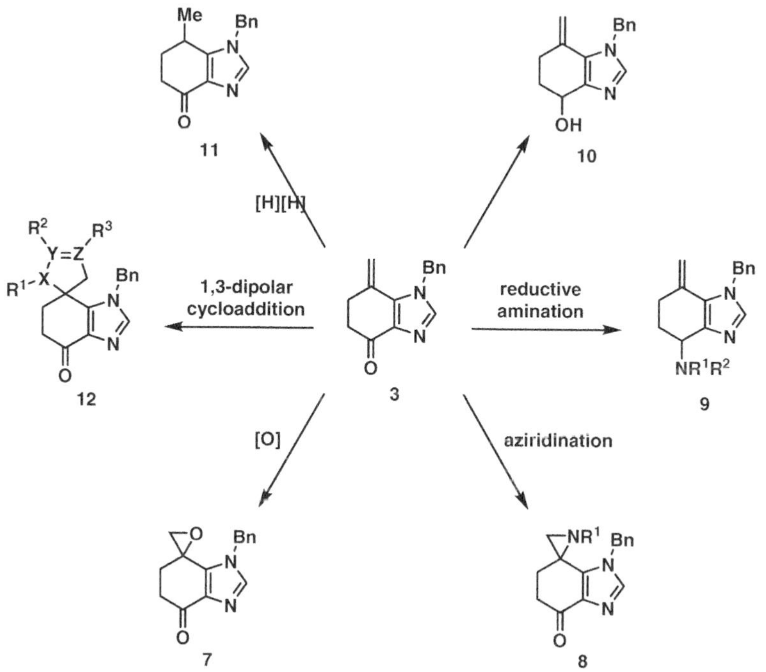

Scheme 1. Some possible transformations with annulated imidazole 3

We now discuss some noteworthy details from our explorations into the synthesis of **3**. In the first step, iodoimidazole **1** is alkylated with benzyl bromide. Because the two imidazole nitrogens in **1** are regiotopic, two products (**2** and **4**) can be expected and, indeed, are formed in the monobenzylation reaction (Scheme 2).

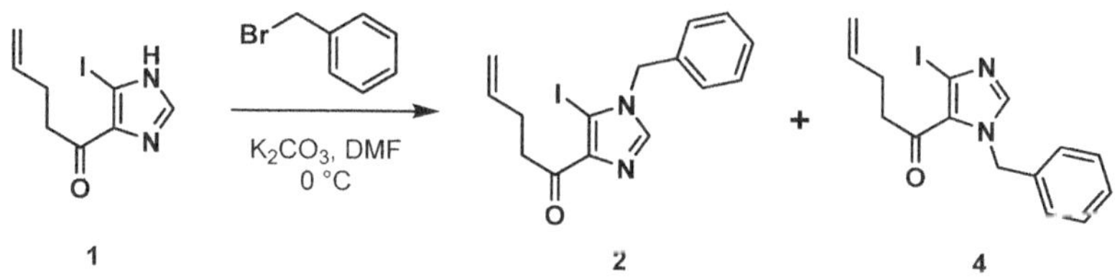

Scheme 2. Reaction of **1** with benzyl bromide produces two regioisomeric products

We investigated the influence of temperature, solvent, base, and electrophile on the reaction and found benzyl bromide with K_2CO_3 in DMF to give a clean, rapid conversion to the two products. Conducting the reaction at 0 °C gave the most consistent product ratios, with **2** and **4** formed in a ratio between 6:1 and 8:1. Because **2** is purified by crystallization, the resulting mother liquor is highly enriched in **4**. We speculated that if the product ratio obtained in the initial reaction reflected a thermodynamic ratio, we could potentially equilibrate the mother liquor mixture, via the intermediacy of the corresponding dibenzylated imidazolium salt (**5**), by heating with a catalytic amount of benzyl bromide (Scheme 3).[9] Attempts to perform such an equilibration revealed that **5** forms readily and is quite stable; therefore, no equilibration occurred. Moreover, running the initial benzylation reaction with larger equivalents of benzyl bromide (*e.g.* 1.2 equiv), even at lower temperatures, results in significant imidazolium salt formation. Compound **5** is highly crystalline and can cause problems during the crystallization of **2** (see Notes 11 and 13). Therefore, the submitters advise using only a slight excess of benzyl bromide (1.01–1.05 equivalents) for the alkylation of imidazole **1**.

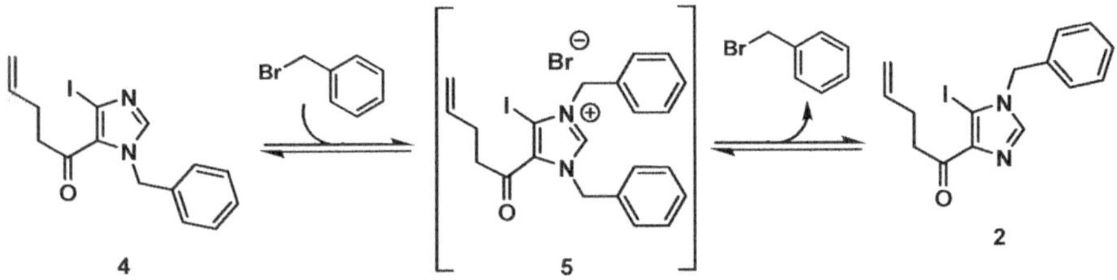

Scheme 3. Envisioned equilibration of **2** and **4** via imidazolium salt **5**

The subsequent intramolecular Heck cyclization with **2** proved challenging. Our initial attempts with "typical" Heck conditions, showed that only ligandless Jeffery conditions (Pd(OAc)$_2$, (*n*-Bu)$_4$NBr, Na$_2$CO$_3$, MeCN/H$_2$O, 70 °C) resulted in any reaction. However, long reaction times (several days) and multiple catalyst loadings were required to obtain useful quantities of product, and we never observed complete consumption of the starting material. We were, therefore, extremely pleased to find that the method of Nacci and co-workers,[11] which employs palladium colloids/nanoparticles in an ionic liquid of (*n*-Bu)$_4$NBr and (*n*-Bu)$_4$NOAc at 130 °C worked extremely well, with full conversion of the starting material in less than 10 minutes.

In addition to the desired product **3**, we observed its aromatized tautomer hydroxybenzimidazole **6** as a byproduct, typically with a **3**/**6** ratio between 5:1 and 8:1 (Scheme 4). We were pleased that **3** was stable enough to be isolated, but were concerned that isomerization of **3** to **6** was taking place during the reaction. If this was the case, small changes in the time spent in the reaction melt at 130 °C could produce variable product ratios. To investigate our concerns, we treated pure **3** under both strongly basic and acidic conditions, and found essentially no isomerization. We also found that performing the Heck reaction for longer reaction times, or under an inert anhydrous atmosphere, or with an added 10 equivalents of water had no effect on the **3**/**6** product ratio. We also re-subjected pure **3** to the reaction conditions for 1 h, but no **6** was detected by NMR of the reaction mixture upon work-up.

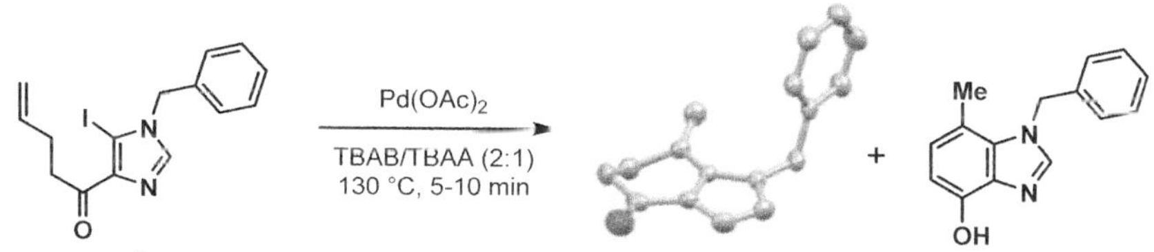

2 3 6

3/6 ratio
5:1 to 8:1

Scheme 4. Heck reaction of iodoimidazole 2 affords the desired product (3) (X-ray structure)[12] and its aromatized congener (6). Ellipsoids for 3 are shown at a 50% probability level; carbon: gray, nitrogen: blue, oxygen: red

Since the **3**/**6** ratio is unaffected by the presence of air or water, and **3** does not isomerize under the reaction conditions, we postulate that **6** is formed via a side reaction involving the intermediate palladium hydride species formed during the Heck reaction. The catalytic cycle starts with the oxidative addition of Pd(0) (Pd(OAc)$_2$ can be reduced to Pd(0) under these conditions to form colloids/nanoparticles, which are productive Heck catalysts)[13] into the I--Ar bond of **2** to form **13** (Scheme 5). After coordination, Pd(II) inserts into the double bond, giving intermediate **14**. *b*-Hydride elimination forms the expected product **3** and H-Pd-I. Reductive elimination of the latter regenerates the catalyst Pd(0), and produces HI, which is sequestered by acetate.

DOI: 10.15227/orgsyn.098.0315

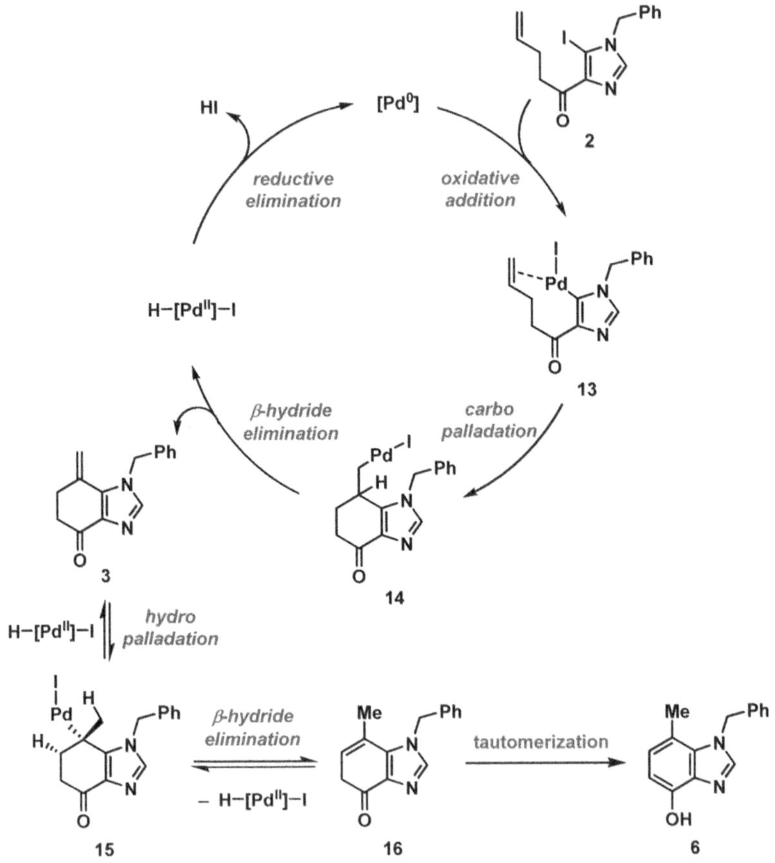

Scheme 5. Catalytic cycle of Heck reaction with iodoimidazole **2** affording product **3** (including postulated reaction path to hydroxybenzimidazole **6**)

If H-Pd-I re-inserts into **3** before undergoing reductive elimination, intermediate **15** could be formed.[14] A subsequent "productive" *syn-β*-hydride elimination from **15** would yield **16**, and after keto-enol tautomerization, hydroxybenzimidazole **6**. Compound **6**, due to its extended aromaticity, is assuredly more thermodynamically stable than **3** and will not revert to **3**. This mechanistic explanation for the conversion of **3** into **6** can only take place in the presence of a palladium hydride species, and is consistent with our observations, *i.e.* after consumption of the starting material (or if no starting material is present to begin with), palladium hydride species should not be formed, and isomerization of **3** to **6** cannot take place.

Isomers **3** and **6** are almost co-polar by TLC under all solvent systems that we tested, making chromatography ineffective as a means of purification. Crystallization was also problematic, due to the relatively low solubility and high crystallinity of **6**. We, therefore, employed a basic extraction procedure, which enabled enrichment of **3** in the organic phase due to the relatively low pK_a of hydroxybenzimidazole **6**. In one full-scale run of this reaction, we quantified the success of the extraction procedure by taking aliquots of the organic phase after the first, third, fifth, seventh and tenth washing step (see Note 34). These were concentrated on a rotary evaporator and measured by NMR to evaluate the 3/6 ratio at each step of this procedure. As can be seen in Figure 16, 10 washes were sufficient to bring the content of **6** to ~2%. Because each basic wash also leads to a loss of desired product **3** (monitored by TLC of the aqueous phase), the washing procedure should be stopped after ten cycles.

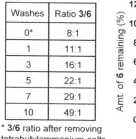
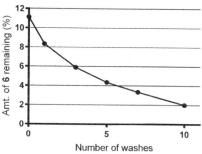

Figure 16. Product ratio 3/6 in the organic phase throughout the basic wash as determined by NMR in table and graph formats

References

1. Cancer Drug Development Group, German Cancer Research Center (DKFZ), Im Neuenheimer Feld 280, D-69120 Heidelberg, Germany. E-mail: aubry.miller@dkfz.de. ORCID for MM: 0000-0001-9539-6404; ORCID for AKM: 0000-0002-1761-4143; The Helmholtz Drug Initiative is gratefully acknowledged for financial support. We thank Frank Rominger (Organic Chemistry Institute, Heidelberg University) for X-ray data of compound **3**.

2. (a) Miller, A.; Jöst, C.; Klein, C.; "Preparation of spiroepoxide tetrahydrobenzotriazoles and -imidazoles useful as MetAP-ll-inhibitors", Eur. Pat. Appl. (2014), EP 2711367, A1 20140326. (b) Morgen, M.; Jöst, C.; Malz, M.; Janowski, R.; Niessing, D.; Klein, C. D.; Gunkel, N.; Miller, A. K. *ACS Chem. Biol.* **2016**, *11*, 1001–1011.
3. Jat, J. L.; Paudyal, M. P.; Gao, H.; Xu, Q.-L.; Yousufuddin, M.; Devarajan, D.; Ess, D. H.; Kürti, L.; Falck, J. R. *Science* **2014**, *343*, 61–65.
4. Miesel, J.; Benko, Z.; Durst, G.; Fitzpatrick, G. M.; Johnson, D. D.; Kaster, S. V.; Kemmitt, G. M.; Lo, W. C.; McKennon, M. J.; Orth, A. B.; Ricks, M. J.; Rogers, R. B.; Werk, T. L.; "Fungicidal Compositions and Methods, and Compounds and Methods for the Preparation thereof", WO 99/27783 (PCT/US98/25624).
5. (a) Kirmse, W.; Rode, K. *Chem. Ber.* **1987**, *120*, 839–846. (b) Fairhurst, R. A.; Marsilje, T. H.; Stutz, S.; Boos, A.; Niklaus, M.; Chen, B.; Jiang, S.; Lu, W.; Furet, P.; McCarthy, C.; Stauffer, F.; Guagnano, V.; Vaupel, A.; Michellys, P.-Y.; Schnell, C.; Jeay, S. *Bioorg. Med. Chem. Lett.* **2016**, *26*, 2057–2064.
6. (a) Ishiyama, J.; Maeda, S.; Takahashi, K.; Senda, Y.; Imaizumi, S. *Bull. Chem. Soc. Jpn.* **1987**, *60*, 1721–1726. (b) González-Avión, X. C.; Mouriño, A. *Org. Lett.* **2003**, *5*, 2291–2293.
7. (a) Howe, R. K.; Shelton, B. R. *J. Org. Chem.* **1990**, *55*, 4603–4607. (b) Zaki, M.; Oukhrib, A.; Akssira, M.; Berteina-Raboin, S. *RSC Adv.* **2017**, *7*, 6523–6529.
8. (a) Felix, A. M.; Heimer, E. P.; Lambros, T. J.; Tzougraki, C.; Meienhofer, J. *J. Org. Chem.* **1978**, *43*, 4194–4196. (b) Haddach, A. A.; Kelleman, A.; Deaton-Rewolinski, M. V. *Tetrahedron Lett.* **2002**, *43*, 399–402. (c) Graham, T. H. *Tetrahedron Lett.* **2015**, *56*, 2688–2690.
9. He, Y.; Chen, Y.; Du, H.; Schmid, L. A.; Lovely, C. J. *Tetrahedron Lett.* **2004**, *45*, 5529–5532.
10. Jeffery, T. *Tetrahedron* **1996**, *52*, 10113–10130.
11. (a) Calò, V.; Nacci, A.; Monopoli, A.; Cotugno, P. *Angew. Chem. Int. Ed.* **2009**, *48*, 6101–6103. (b) Calò, V.; Nacci, A.; Monopoli, A.; Cotugno, P. *Angew. Chem.* **2009**, *121*, 6217–6219.
12. The crystallographic data of **3** can be obtained free of charge from the joint Cambridge Crystallographic Data Centre and Fachinformationszentrum Karlsruhe Access Structures service www.ccdc.cam.ac.uk/structures (cif-file reference code: CCDC 1978198).

13. (a) Calò, V.; Nacci, A.; Monopoli, A. *Eur. J. Org. Chem.* **2006**, 3791–3802. (b) Kantam, M. L.; Reddy, P. V.; Srinivas, P.; Bhargava, S. *Tetrahedron Lett.* **2011**, *52*, 4490–4493.
14. For similar postulated mechanisms including a "Pd-H" re-addition in course of a Heck reaction see: (a) Battistuzzi, G.; Cacchi, S.; Farbizi, G. *Synlett* **2002**, *3*, 439–442. (b) Hou, Y.; Ma, J.; Yang, H.; Anderson, E. A.; Whiting, A.; Wu, N. *Chem. Commun.* **2019**, *55*, 3733–3736.

Appendix
Chemical Abstracts Nomenclature (Registry Number)

1-(5-Iodo-1*H*-imidazol-4-yl)pent-4-en-1-one: 4-Penten-1-one, 1-(5-iodo-1*H*-imidazol-4-yl)-; (1585257-59-5)

Potassium carbonate: Carbonic acid, potassium salt (1:2); (584-08-7)

Benzyl bromide: Benzene, (bromomethyl)-; (100-39-0)

1-(1-Benzyl-5-iodo-1*H*-imidazol-4-yl)pent-4-en-1-one: 4-penten-1-one, 1-[5-iodo-1-(phenylmethyl)-1*H*-imidazol-4-yl]-; (1585257-17-5)

Palladium(II) acetate: Acetic acid, palladium(2+) salt (2:1); (3375-31-3)

Tetra-*n*-butylammonium bromide: 1-Butanaminium, *N*,*N*,*N*-tributyl-, bromide (1:1); (1643-19-2)

Tetra-*n*-butylammonium acetate: 1-Butanaminium, *N*,*N*,*N*-tributyl-, acetate (1:1); (10534-59-5)

1-Benzyl-7-methylene-1,5,6,7-tetrahydro-4*H*-benzo[d]imidazole-4-one: 4*H*-Benzimidazol-4-one, 1,5,6,7-tetrahydro-7-methylene-1-(phenylmethyl)-; (1585257-23-3)

Michael Morgen received his Ph.D. in inorganic chemistry from the Ruprecht-Karls-University, Heidelberg in 2013 while working with Prof. Dr. Peter Comba on the design of novel bispidine-based multimodal imaging agents. Afterwards he joined the Cancer Drug Development Group of Dr. Aubry Miller and Dr. Nikolas Gunkel at the German Cancer Research Center (DKFZ). His current research interests lie in the design and synthesis of organic and metal-organic inhibitors for cancer-related drug targets.

Jasmin Lohbeck was trained as chemistry laboratory assistant at BASF in Ludwigshafen, Germany, from 2006–2009. She subsequently joined the chemistry lab of Dr. Aubry Miller at the German Cancer Research Center and specialized in preparative organic synthesis.

Aubry Miller leads the Cancer Drug Development Group at the German Cancer Research Center (DKFZ) in Heidelberg, Germany. His laboratory focuses on the discovery and biological characterization of chemical probes for use as tools and drug leads in cancer research.

Nageswara Rao Kalikinidi earned his Ph.D. in 2018 in Synthetic Organic Chemistry from the CSIR-IICT, India, under the guidance of Dr. Subhash Ghosh. His research focused on developing a synthetic route for the total synthesis of biologically active and architecturally complex marine natural products. In 2019, he joined in Prof. E. J. Corey research group at Harvard University, where his research was focused on the enantioselective epoxidation reactions with chiral azatetracycle catalysts. After completing his postdoctoral studies in 2021, Nageswara joined TCG GreenChem, Inc. as a post-doctoral Scientist. His research interests include the design and development of green and robust large-scale production of active pharmaceutical ingredients (APIs) and synthetic methodology of novel organic transformations using chiral phosphine ligands.

Venumadhav Janganati earned his Ph.D. in synthetic organic chemistry from National Institute of Technology, Warangal, India in 2009 under the guidance of Prof. B. Rajitha. His research focused on synthesis of bioactive heterocyclic molecules and developing novel methodologies by using new reagents. In his postdoctoral training at the University of Arkansas for Medical Sciences, USA, he designed and synthesized natural product (parthenolide) derivatives for anti-cancer therapy, and he was also involved in design and synthesis of lobelane derivatives, which focused on the development of new treatments for methamphetamine addiction. Currently working as a senior scientist in Process Chemistry at TCG GreenChem, Inc., his research interests include the process research and development and asymmetric catalysis.

Dr. Gopal Sirasani received his Bachelor's and Master's degrees in Hyderabad, India. He obtained his Ph.D. in synthetic organic chemistry in 2011 from Temple University, Philadelphia under the guidance of Prof. Rodrigo B. Andrade. His doctoral research focused on developing novel methodologies, total syntheses of natural products and their analogs thereof. He got his post-doctoral training in the laboratory of Prof. Emily Balskus at Harvard University, where he developed biocompatible organic reactions utilizing microbially generated reagents to realize transition metal catalysis in the presence of microbes. In 2013, Gopal began his industrial career at Melinta Therapeutics, New Haven, CT. He is currently working at TCG GreenChem, Inc. as a Director in the department of process research and development.

Catalytic Diazoalkane-Carbonyl Homologation: Synthesis of 2,2-Diphenylcycloheptanone and Other Quaternary or Tertiary Arylalkanones and Spirocycles by Ring Expansion

Jason S. Kingsbury,[1]* Victor L. Rendina, Jacob S. Burman, and Brittany A. Smolarski

Department of Chemistry, California Lutheran University, 60 W Olsen Rd., Thousand Oaks, CA, 91360

Checked by Junichi Taguchi, Haruka Fujino, and Masayuki Inoue

Procedure (Note 1)

2,2-Diphenylcycloheptanone (2). A three-necked 300-mL round-bottomed flask (central neck 29/32 joint, side necks 15/25 joint), equipped with a Teflon-coated magnetic stir bar (4.0 × 0.7 cm, rod-shaped) (Note 2) is charged with 1.0 g of scandium(III) triflate (Sc(OTf)$_3$, 2.0 mmol, 5.6 mol%) (Note 3) directly in air. One side neck and the central neck of the reaction vessel are then fitted with glass stoppers. The other side neck is fitted with a short-path distillation head (15/25 upper outer joint, 15/25 lower inner joint, 15/25 side inner joint). While the distillation head is topped with a 15/25 three-way cock connected to an argon inlet and a vacuum line, its branch is fitted with a single-necked 25-mL round-bottomed flask (15/25 joint) charged with 2.00 g of phosphorous pentoxide (P$_2$O$_5$) (Note 4) to absorb the water of hydrated Sc(OTf)$_3$. All joints are sealed with grease and Teflon tape (Figure 1A). After the entire apparatus is evacuated to

0.30 mmHg through the vacuum line, the flask is immersed in a silicone oil bath pre-heated to 150 °C. Keeping the reduced pressure below 0.30 mmHg, the catalyst powder is stirred at 150 °C for 12 h (Figures 1B and 1C). In the course of the drying, the formerly free-flowing white phosphorous pentoxide powder (Figure 1D) turns to a sticky white-pink solid containing phosphoric acid residues (Figure 1E).

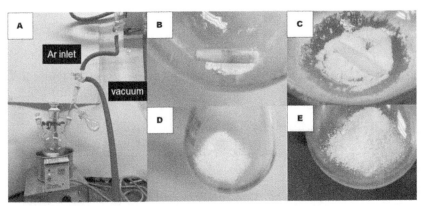

Figure 1. (A) Set-up for drying commercial Sc(OTf)$_3$ under a vacuum (150 °C, 0.30 mmHg); (B) Sc(OTf)$_3$ before drying; (C) Sc(OTf)$_3$ dried under a vacuum (150 °C, 0.30 mmHg) for 12 hours; (D) Phosphorous pentoxide before drying Sc(OTf)$_3$; (E) Phosphorous pentoxide after drying Sc(OTf)$_3$ (photos provided by checkers)

The assembly is next removed from the silicone oil bath, allowed to cool to 24 °C over the course of 20 min, and backfilled with argon. The distillation head is removed from the side-neck of the 300-mL flask, which is quickly fitted with a 15/25 three-way stopcock connected to an argon inlet and a vacuum line. Under argon flow, the two glass stoppers are removed from the flask. The uncapped side neck is fitted with a 15/25 rubber septum, and the central neck is fitted with a connecting adapter (upper outer joint 19/38, lower inner joint 29/32) and a 50-mL pressure-equalizing addition funnel, bearing an outer extra chamber for cooling (Note 5), which is topped with a 15/25 rubber septum. All junctures of the glassware are sealed with grease and Teflon tape. The flask is evacuated through the vacuum line (1.5 mmHg, 24 °C) for 1 min, after which the flask is backfilled with argon. This substitution of the gas is repeated three times. Then, the vacuum line is removed from the three-way cock, which is then connected to an argon outlet. Argon atmosphere is maintained throughout the reaction under a continuous argon flow. The assembly is positioned over a magnetic stir plate in a fume hood. (Figure 2A).

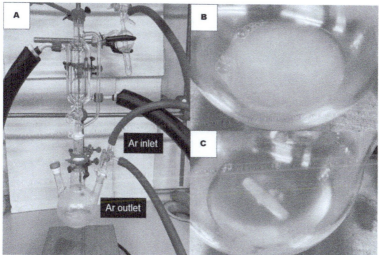

Figure 2. (A) Reaction set-up; (B) Reaction mixture after addition of toluene; (C) Reaction mixture after addition of cyclohexanone (photos provided by checkers)

To the flask is added toluene (32 mL to generate a 1.1 M solution) (Note 6) by a syringe through the rubber septum on the side neck. After suspending the catalyst in toluene (Figure 2B), 4.20 mL of cyclohexanone (3.98 g, 40.6 mmol, 1.13 equiv) (Note 7) is added by a syringe through the same rubber septum. The reaction mixture immediately turns from a white suspension to a pale yellow, homogeneous solution (Figure 2C). Then, the rubber septum on the top of the pressure-equalizing addition funnel is removed. In a separate single-necked 100-mL round-bottomed flask, 7.00 g of bright red diphenyldiazomethane (**1**, 36.0 mmol, 1.00 equiv) (Note 8) is dissolved in heptane (40 mL to generate a 0.90 M solution) (Note 9) at 0 °C, and the solution is transferred to the pressure-equalizing addition funnel using a 10-mL glass pipette. The rubber septum is reattached on the top of the funnel and sealed with Teflon tape. The solution inside the addition funnel, which is continuously cooled to 0 °C (Note 5) and shielded from light by aluminum foil, is added dropwise to the reaction mixture over 3 h at 24 °C (Notes 10 and 11) (Figure 3A). After the complete addition of the diazoalkane, the solution is allowed to stir for 3 h. The reaction mixture is light yellow in color but contains a brown, oily precipitate comprised of deactivated Sc(III) salts (Note 12) (Figure 3B).

Figure 3. (A) Diazoalkane addition to a solution of Sc(OTf)$_3$ and substrate using a pressure-equalizing addition funnel; (B) The reaction mixture at the end of diazoalkane addition (photos provided by the checkers)

After the addition funnel and the three-way stopcock are removed, the heterogeneous mixture is diluted with diethyl ether (30 mL) (Note 13). The stir bar is removed, and the resulting reaction mixture is transferred to a 300-mL separation funnel. Additional diethyl ether (15 mL) is used to wash the flask. The resultant mixture is then washed with pure water (100 mL) (Note 14) (Figure 4A). The pale brown organic layer is separated, and the aqueous layer is extracted with diethyl ether (30 mL). The organic layers are combined into a 300-mL Erlenmeyer flask and dried over sodium sulfate (80 g) (Note 15) for 10 min (Figure 4B). The drying agent is filtered through a plastic funnel with a cotton plug into a 300-mL round-bottomed flask and rinsed with diethyl ether (2 × 5 mL). The filtrate is concentrated by rotary evaporation (38 °C, 35 mmHg) and dried on a high vacuum (room temperature, 1.5 mmHg) for 10 min. These workup operations furnish 10.8 g of pale brown translucent oil that contains the crude homologation product (Figure 4C).

Figure 4. (A) The organic layer and the aqueous layer; (B) The organic layer dried over Na_2SO_4; (C) The crude pale brown translucent oil (photos provide by the checkers)

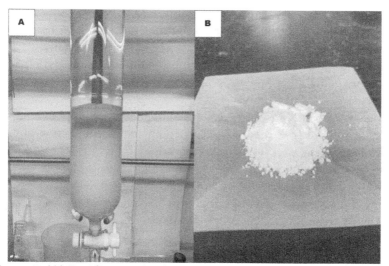

Figure 5. (A) Purification by flash column chromatography; (B) Title compound 2 (photos provided by the checkers)

The crude oil is purified by flash column chromatography on silica gel using hexane and diethyl ether as eluents (Notes 16 and 17) (Figure 5A) to provide 5.34 g (56%) of desired product **2** as a pale-yellow solid (Note 18) (Figure 5B).

Notes

1. Prior to performing each reaction, a thorough hazard analysis and risk assessment should be carried out with regard to each chemical substance and experimental operation on the scale planned and in the context of the laboratory where the procedures will be carried out. Guidelines for carrying out risk assessments and for analyzing the hazards associated with chemicals can be found in references such as Chapter 4 of "Prudent Practices in the Laboratory" (The National Academies Press, Washington, D.C., 2011; the full text can be accessed free of charge at https://www.nap.edu/catalog/12654/prudent-practices-in-the-laboratory-handling-and-management-of-chemical. See also "Identifying and Evaluating Hazards in Research Laboratories" (American Chemical Society, 2015) which is available via the associated website "Hazard Assessment in Research Laboratories" at https://www.acs.org/content/acs/en/about/governance/committees/chemicalsafety/hazard-assessment.html. In the case of this procedure, the risk assessment should include (but not necessarily be limited to) an evaluation of the health hazards associated with scandium(III) triflate, phosphorous pentoxide, toluene, diphenyldiazomethane, cyclohexanone, heptane, diethyl ether, sodium sulfate, hexane, 1,3,5-trimethoxybenzene, and silica gel.

2. All glassware and a Teflon-coated magnetic stir bar were oven-dried at 120 °C for 2 h and then cooled to room temperature in a desiccator cabinet prior to reaction set-up. The room temperature throughout this manuscript refers to temperature between 24 °C to 26 °C.

3. Sc(OTf)$_3$ (99%) was purchased from Sigma-Aldrich and dried rigorously using the apparatus pictured in Figure 1A. Higher grade lots of catalyst (99.995% trace metals basis) are available from multiple vendors, but the submitters found that all commercial supplies contained variable amounts of water due to the hygroscopic nature of Sc(OTf)$_3$. Attempts to use Sc(OTf)$_3$ in hydrated form (as received) can be met by substantial decomposition of the diazoalkanes, presumably due to the formation of triflic acid *in situ*. Once dry, the Sc(OTf)$_3$ powder weighed only 0.88 g by mass difference (1.8 mmol, catalyst loading of 5.0 mol % relative to the limiting reagent).

4. P$_2$O$_5$ (ACS Reagent-grade, >98%) was purchased from Sigma-Aldrich and used as received (submitters). P$_2$O$_5$ (Extra Pure Reagent, >97.0%) was purchased from Nacalai Tesque, Inc. and used as received (checkers).
5. The outer chamber of the pressure-equalizing funnel is connected to a chiller, purchased from Tokyo Rikakikai Co., Ltd, via inlet and outlet Teflon tubes insulated with polyurethane foam. Methanol is used as a circulating refrigerant, and the cooling temperature is kept at 0 °C.
6. PRA-grade toluene (99.8%) was purchased from Sigma-Aldrich and stored over vacuum-dried 3Å molecular sieves prior to use (submitters). Toluene (purity GC 99.9%) was purchased from FUJIFILM Wako Pure Chemical Corporation and purified by Glass Contour solvent dispensing system (Nikko Hansen & Co., Ltd.) (checkers).
7. Cyclohexanone was purchased from Sigma-Aldrich and stored over vacuum-dried 3Å molecular sieves prior to use (submitters). Cyclohexanone (99.8%) was purchased from Sigma-Aldrich and used as received (checkers).
8. Diphenyldiazomethane (**1**) was prepared by applying Javed and Brewer's method described in *Org. Synth.* **2008**, *85*, 189–195.[2b] Purity of **1** was determined to be 97.1% by qNMR using 16.1 mg of **1** and 15.2 mg of 1,3,5-trimethoxybenzene (FUJIFILM Wako Pure Chemical Corporation, >99.0%, used as received) as an internal standard, with relaxation time (D$_1$) set to 30 seconds (checkers).
9. HPLC-grade heptane was purchased from Pharmco-Aaper and used as received (submitters). Heptane (Anhydrous, 99%) was purchased from Sigma-Aldrich and used as received (checkers).
10. The submitters reported that not only a pressure-equalizing addition funnel but also a Razel R-99 syringe pump with a setting of '1.5' on the adjustable dial was available for the purpose of slow addition of **1** (Figure 6). The solution of **1** in heptane was pulled into a disposable gas tight syringe fitted with a 12-inch, 18 gauge needle and added to the reaction mixture at a rate of 10 mL/min at 23 °C.

Figure 6. Syringe pump diazoalkane addition to a solution of Sc(OTf)$_3$ (photo provided by submitters)

11. The most important feature of the slow addition, which helps to offset bimolecular decomposition of diazoalkane **1**, is that it be performed at a rate that is convenient but avoids any buildup of red color within the reaction mixture. The rate of productive diazo consumption (homologation vs. dimerization) may change as the reaction progresses.
12. TLC analysis of the reaction mixture is shown below. R_f values of benzophenone (side product) and diphenylcycloheptanone (desired product, **2**) in hexane/diethyl ether (10/1, v/v) are 0.40 and 0.34, respectively. These spots can be viewed by fluorescence on silica gel 60 F_{254} plates (TLC Silical gel 60 F_{254}, purchased from Merck KgaA) with UV light (254 nm) (Figure 7). The desired product **2** can be also stained with phosphomolybdic acid (PMA, FUJIFILM Wako Pure Chemical Corporation).

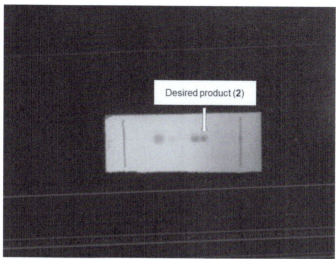

Figure 7. TLC analysis of the reaction mixture
(photo provided by checkers)

13. Uninhibited (ethanol-free) ACS reagent-grade diethyl ether was purchased from Pharmco-Aaper and used as received (submitters). Diethyl ether (>99.0%) was purchased from FUJIFILM Wako Pure Chemical Corporation and used as received (checkers).
14. Pure water (≤100%) was purchased from FUJIFILM Wako Pure Chemical Corporation and used as received (checkers).
15. Anhydrous Na_2SO_4 (>99.0%) was purchased from Nacalai Tesque, Inc. and used as received (checkers).
16. Flash column chromatography is performed on silica gel (Kanto Chemical Co., Inc., Silica gel 60 N, spherical and neutral, 0.040–0.050 mm) using compressed air. The mobile phase was purchased from Kanto Chemical Co., Inc. (hexane) and FUJIFILM Wako Pure Chemical Corporation. (diethyl ether) and used as received. The column with a 7.5 cm diameter × 15 cm height is wet packed with 300 g of silica gel in hexane (500 mL). Sea sand is added to the bottom and to the top of the column. The crude oil is loaded onto the column using 15 mL of toluene (>99.5%, FUJIFILM Wako Pure Chemical Corporation, used as received) (Figure 5A). At this point, fraction collection (250 mL × 20) is begun, and elution is continued with 5000 mL of hexane/diethyl ether (30/1, v/v). The desired product is obtained in fractions No. 14 through No. 17 as a pure form (Figure 8A). The fractions No. 9 through No. 13 are co-elution of the desired product **2**

containing benzophenone. These impure fractions are combined and concentrated by rotary evaporation (38 °C, 15 mmHg). The obtained yellow oil is purified again using column chromatography. The column with a 5.0 cm diameter × 15 cm height is wet packed with 150 g of silica gel in hexane (300 mL). Sea sand is added to the bottom and to the top of the column. The crude oil is loaded onto the column using 10 mL of toluene. At this point, fraction collection (50 mL × 60) is begun, and elution is continued with 1750 mL of hexane/diethyl ether (30/1, v/v) and then 1250 mL of hexane/diethyl ether (15/1, v/v). The desired product is obtained in fractions 33–52 as a pure form (Figure 8B). These fractions and the fractions 14–17 of the first purification are combined, concentrated by rotary evaporation (38 °C, 15 mmHg), and dried on a high vacuum (room temperature, 1.5 mmHg) for 4 h.

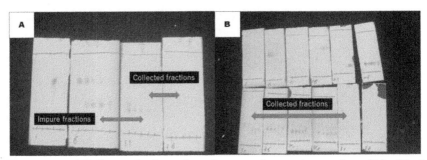

Figure 8. TLC analysis of column chromatography (A) First purification; (B) Second purification (checkers). The spots on the silica gel plate are visualized under UV light (254 nm) (photos provided by checkers)

17. The desired product (2,2-Diphenylcycloheptanone, **2**) shows the following characteristics: mp = 92–95 °C; IR (cm^{-1}) 3057, 2932, 2858, 1704, 1598, 1495, 1444, 1319, 1156, 1034, 751, 700; ^{1}H NMR (400 MHz, CDCl$_3$) δ: 1.72–1.73 (m, 4H), 1.78–1.80 (m, 2H), 2.59–2.66 (m, 4H), 7.15–7.30 (aromatic, 10H); ^{13}C NMR (100 MHz, CDCl$_3$) δ: 24.9, 26.5, 30.6, 37.3, 42.4, 65.0, 126.5 (2C), 128.0 (4C), 128.7 (4C), 144.3 (2C), 212.5; HRMS Calcd for C$_{19}$H$_{20}$O Na, [M+Na]$^+$: 287.1406, Found 287.1406; Anal. Calcd for C$_{19}$H$_{20}$O: C, 86.32; H, 7.63. Found: C, 86.14; H, 7.75; Purity was determined to be >98% by qNMR using 14.6 mg of **2** and 11.1 mg of 1,3,5-trimethoxybenzene (FUJIFILM Wako Pure Chemical Corporation, >99.0%) (checkers). Based on a purity of 98.5%, the amount of product formed is 5.25 g (55%).

18. A second run performed on the same scale provided **2** (5.30 g, 56% yield) in 99.0% purity as determined by qNMR.

Working with Hazardous Chemicals

The procedures in *Organic Syntheses* are intended for use only by persons with proper training in experimental organic chemistry. Diazo compounds are inherently toxic and irritating, especially on contact with the skin or eyes. Extreme caution and secondary containment must be applied when handling diazoalkane solutions, as there are constant risks associated with a spill or splash. Appropriate personal protective equipment (PPE), including safety glasses, a lab coat, and latex or nitrile gloves must be worn by practitioners. It is also prudent to manipulate diazo compounds behind a safety shield or with the sash of the fume hood lowered. All hazardous materials should be handled using the standard procedures for work with chemicals described in references such as "Prudent Practices in the Laboratory" (National Academies Press, Washington, D.C., 2011; the full text can be accessed free of charge at http://www.nap.edu/catalog.php?record_id=12654). All chemical waste should be disposed of according to local regulations. For general guidelines on the management of chemical waste, see Chapter 8 of Prudent Practices.

In some articles in *Organic Syntheses*, chemical-specific hazards are highlighted in red "Caution Notes" within a procedure. It is important to recognize that the absence of a caution note does not imply that no significant hazards are associated with the chemicals involved in that procedure. Prior to performing a reaction, a thorough risk assessment should be carried out that includes a review of the potential hazards associated with each chemical and experimental operation on the scale that is planned for the procedure. Guidelines for carrying out a risk assessment and for analyzing the hazards associated with chemicals can be found in Chapter 4 of Prudent Practices.

The procedures described in *Organic Syntheses* are provided as published and are conducted at one's own risk. *Organic Syntheses, Inc.*, its Editors, and its Board of Directors do not warrant or guarantee the safety of individuals using these procedures and hereby disclaim any liability for any injuries or damages claimed to have resulted from or related in any way to the procedures herein.

Discussion

The net or formal insertion of diazoalkyl carbon into non-polarized C-H and C-C bonds is a function of their proximity to carbonyl groups that engage in 1,2-addition/rearrangement. As such, the homologation of both aldehyde[3] and ketone[4] acceptors occurs with either a chain extension or ring expansion that is strategically powerful and often complexity-generating – especially in the case of substituted diazomethanes. For well over a century, the synthetic community has applied procedures for diazoalkane–carbonyl homologation that rely on traditional means for carbonyl activation, namely the presence of superstoichiometric amounts of B- and Al-based promoters[5] or general acid assistance based on water or alcohol co-solvents.[6] However, the past decade has seen a resurgence in the field of diazoalkane chemistry. The practicality of homologation reactions has improved with the development of Lewis acid catalysts that address previously unmet challenges, including low reactivity in the case of sterically hindered internal diazoalkanes, poor functional group tolerance, and greater regio- and stereochemical control.

In 2009, we reported a finding that commercial Sc(III) salts are uniquely potent catalysts for both ketone[7] and aldehyde[8] insertion reactions, focusing on application of non-carbonyl-stabilized diazo compounds as nucleophiles. A persistent goal in these investigations was to prepare both terminal and internal aryldiazomethanes without isolation using established methods for hydrazone oxidation.[9,10] As crystalline solids or high molecular weight oils, the stability of mono- and diaryldiazomethanes[11] is much higher than that of diazomethane – a poisonous, explosive gas that can only be handled under strict protocol as a dilute solution.[12] Reports show that diaryldiazomethanes have been applied industrially on-scale as donor-acceptor carbon sources for over twenty years without incident.[13] Procedure A was chosen to showcase the more challenging aspect of a diaryl quaternary carbon installation, yet it is applicable to many cycloalkanone ring sizes and a diversity of diazoalkane substituents (Table 1). α-Tertiary and –quaternary cycloalkanones *a–j* are all known substances that are fully characterized.[7] However, to our knowledge *ene*-bicyclopentanones **3a** and **3b** are α,α-diphenyl ketones prepared by catalytic carbonyl homologation for the first time with Procedure A.

Table 1. Scope and Generality of Sc-Catalyzed Ring Expansion Reactions

cyclobutanone as acceptor:

(1.0 equiv R₂CN₂, 1 mol % Sc(OTf)₃, 0.8 M in CH₂Cl₂)

a >98% yield

b 98% yield

c 95% yield

g 78% yield, 10 mol % Sc(TMHD)₃

f 95% yield

e 88% yield

d 92% yield

cyclohexanone, –heptanone, and –dodecanone acceptors:

(1.0 equiv R₂CN₂, 1–7 mol % Sc(OTf)₃, ~1.0 M in toluene)

h >98% yield

i 89% yield

j 84% yield, 7 mol % Sc(OTf)₃

[3.2.0]-bicycloheptenone acceptor:

(with 1.0 equiv Ph₂CN₂, 9 mol % Sc(OTf)₃, toluene)

3a major, 39% yield

3b minor, 10% yield (by ¹H NMR analysis)

Additional discussion points on Sc-catalyzed ring enlargements include the following: (1) Different substitution patterns and electronic modifications (see *d-f*) in the aryldiazoalkane reaction partner are readily tolerated. Even *ortho*-substituted nucleophiles give homologation products (*c*, *f*) in excellent yield; (2) The method also applies to aliphatic cyclic diazoalkanes and the production of spirocyclic structures. Synthesis of *g* is an example revealing a tolerance for acid-labile ketal protecting groups and whose optimal 78% yield was achieved by switching to commercial Sc(TMHD)₃,[7a] a more sterically hindered and less electron-poor catalyst; (3) Accommodation of further steric congestion is demonstrated by α-quaternary carbon installation with internal mono- and diarylated diazomethanes (→ *b*, *h*, *j*, and **2**). The greater crowding inherent to disubstituted diazo compounds does not limit their reactivity to cyclobutanone, as unstrained electrophiles are also suitable; (4) Absent from Table 1 is an example of cyclopentanone homologation. Since the desired 2-arylcyclohexanones are more reactive than starting material, the reactions afford a mixture of products derived from over-homologation. Nonetheless, six-membered structures can be accessed from substituted lower homologs (*i.e.*, *a-f*) by catalytic ring expansion with trimethylsilyldiazomethane. In such cases, 1,3-Brook rearrangement of α-trimethylsilyl cyclohexanone products gives enol silanes incapable of further reaction;[7b] (5) Diphenyldiazomethyl insertion with commercial *cis*-bicyclo[3.2.0]hept-2-en-6-one to give **3a** and **3b** in a 4:1 ratio illustrates a useful regioselectivity enabled by Sc(III)-catalysis. Predominant formation of **3a** is consistent with a steric model[14] in which the preferred 1,2-adduct (Sc-complexed diazonium betaine intermediate) situates the less substituted cyclobutane bond *anti*-coplanar to the dinitrogen leaving group.

All arylcyclopentanone syntheses in Table 1 are complete in < 1 h at –78 °C with low catalyst loading (0.5-1 mol %) and are free of byproducts derived from dimerization of the diazo compound. Reactions involving the merger of mono-aryl and mixed alky/aryl-diazomethanes with higher cycloalkanones also take place rapidly at –78 to –48 °C but require higher catalyst loadings (7 mol %, *j*, Table 1) for maximal efficiency. The optimal solvents for carbonyl homologation are toluene and dichloromethane; acetonitrile can also serve to completely dissolve anhydrous $Sc(OTf)_3$ powder. Coordinating ethers such as tetrahydrofuran and diethyl ether are less effective media, presumably due to competitive ligation to the trication in solution. Other strongly Lewis basic functions, such as free alcohols and amines, are not well tolerated, and O-H insertion products can be found for substrates containing hydroxyl groups. Some heteroaromatic functionality (*e.g.* furans, thiophenes, but not pyridines) has proven compatible with the Sc-catalyzed homologation of aldehydes.[8b]

In contrast to the features just mentioned, catalytic carbon insertion with diaryldiazomethanes is slower and best achieved at room temperature. The multigram synthesis of **2** (Procedure A) was optimized to permit subsequent testing of diazoalkane **4**[15] as a carbon nucleophile for spirocyclic annulation. As illustrated below in Scheme 1, *spiro*-fused dibenzo-bicycloundecanone ***k*** is a known intermediate in a patented synthesis of a rigid *ene*-ketone Michael acceptor showing μM affinity for mitochondrial permeability transition pore (MPTP).[16] As a polyprotein complex that regulates mitochondrial membrane permeability, MPTP has drawn attention as a mediator of energy metabolism, cellular Ca^{2+} homeostasis, and programmed cell death (apoptosis). In the published route to **k**, stepwise buildup of the quaternary α carbon and cyclopentanone ring was effected by a series of acidic and basic reaction conditions starting from the indicated dibenzonitrile (Scheme 1).[16] Therefore, we set out to test a direct, convergent synthesis of **k** as a proving ground for Sc-catalyzed ring expansion despite the crowded and conformationally constrained nature of cyclic diaryldiazoalkane **4**.

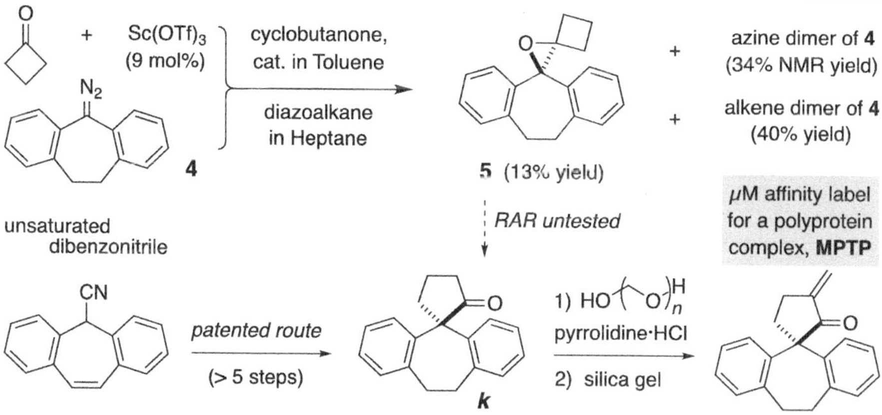

Scheme 1. Unusual Outcome in Targeting a Complex Spirocycle from 4

Unfortunately, slow addition of 5.6 g of the diazo compound (in heptane) to a premixed solution of 9 mol % catalyst and cyclobutanone (1 M in toluene) provided only the tetrasubstituted spirocyclic epoxide **5** in a low 13% yield following aqueous workup and chromatography. This outcome was reproducible over two separate runs involving the multigram scale, but in neither case was a cyclopentanone fraction detected by FTIR and ¹H NMR analysis. Instead, the major products were the corresponding alkene (40%) and azine (34%) dimers. Herein lies a stark reminder that upper limits to the reaction's capability can persist even with the benefit of 4C ring strain in the electrophile. We believe that the energetically allowed mode of diazoalkyl addition to cyclobutanone orients the Sc(III)-alkoxide *anti* to the diazonium cation, with alignment favoring closure to the epoxy byproduct **5**. Worse still, any bond rotation that would set up for the desired Wagner-Meerwein shift is associated with severe aryl/alkyl eclipsing interactions. Together with the fact that carbonyl 1,2-addition is reversible for highly resonance-stabilized diazo compounds, it is rather remarkable that strained spiroepoxide **5** could even be isolated. In our hands, the waxy material proved unpredictably labile depending on variables such as type of silica gel, time and temperature in solution, and exposure to trace impurities in solvents. It is possible that acidic conditions could be tested to effect rearrangement of **5** to the desired ketone *k*, but such efforts are arguably futile given the poor purified yield of **5**.

Several nuances behind the known synthesis of **4** and its exploitation as a donor-acceptor carbon source speak further to its ultimate test on efficiency. For instance, its hydrazone precursor is indefinitely stable in the absence of moisture but is not accessible under standard conditions (H_2NNH_2, EtOH, reflux) from dibenzosuberone because of an unfavorable equilibrium. In an elegant solution by Miao and coworkers, the suberone is instead transformed first to the thioketone (with Lawesson's reagent) before condensation with hydrazine.[15] In this manner, hydrazone formation is driven to completion by release of gaseous H_2S. In the same report, diazoalkane **4** was applied in a two-step Barton-Kellogg synthesis of tetrabenzoheptafulvalene dimer; severe crowding in the fjord region around the central alkene led to discrete *syn* and *anti* stereoisomers at 23 °C.[15] Miao and coauthors advance such overcrowded bis-tricyclic ethylenes – the major product in our synthesis of **5** – as promising substructures for photo- and thermally-induceable molecular switches.[15] As a final example of anomalous behavior attributed to the dibenzocycloheptane framework, the azine dimer of **4** experiences complete hydrolytic breakdown on silica, leading to recovery of mostly dibenzosuberone during isolation of **5**. Perhaps ongoing and continued tuning of lanthanide-based catalysts[17] for diazoalkane chemistry will be able to address the most elusive challenges of reactivity and regioselectivity in these transformations. In the meantime, we hope that our findings will only inspire and not deter other chemists from probing the upper limits of diazoalkane–carbonyl homologation, albeit with tempered expectations in structurally unusual or highly complex settings.

References

1. Swenson Science Center, Department of Chemistry, California Lutheran University, 60 West Olsen Road, Thousand Oaks, CA, 91360. Email: jkingsbu@callutheran.edu. ORCID (JSK): 0000-0001-7722-0069; ORCID

(BAS): 0000-0003-1306-716X; We thank the ACS Petroleum Research Fund (award no. 5001009) and Cal Lutheran University for financial support.
2. (a) Javed, M. I.; Brewer, M. *Org. Lett.* **2007**, *9*, 1789–1792. (b) Javed, M. I.; Brewer, M. *Org. Synth.* **2008**, *85*, 189–195.
3. Guttenberger, N.; Breinbauer, R. *Tetrahedron* **2017**, *73*, 6815–6829.
4. Candeias, N. R.; Paterna, R.; Gois, P. M. P. *Chem. Rev.* **2016**, *116*, 2937–2981.
5. (a) Maruoka, K.; Concepcion, A. B.; Yamamoto, H. *Synthesis* **1994**, 1283–1290. (b) Maruoka, K.; Concepcion, A. B.; Yamamoto, H. *Synlett* **1994**, 521–523.
6. (a) Gutsche, C. D. *Org. React.* **1954**, 364–429. (b) Gutsche, C. D.; Johnson, H. E. *Org. Synth.* **1955**, *35*, 91–94. (c) Krow, G. R. *Tetrahedron* **1987**, *43*, 3–38.
7. (a) Moebius, D. C.; Kingsbury, J. S. *J. Am. Chem. Soc.* **2009**, *131*, 878–879. (b) Dabrowski, J. A.; Moebius, D. C.; Wommack, A. J.; Kornahrens, A. F.; Kingsbury, J. S. *Org. Lett.* **2010**, *12*, 3598–3601. For catalytic enantioselective α-arylation of cycloalkanones, see: (c) Rendina, V. L.; Moebius, D. C.; Kingsbury, J. S. *Org. Lett.* **2011**, *13*, 2004–2007. (d) Rendina, V. L.; Kaplan, H. Z.; Kingsbury, J. S. *Synthesis* **2012**, *44*, 686–693.
8. (a) Wommack, A. J.; Moebius, D. C.; Travis, A. L.; Kingsbury, J. S. *Org. Lett.* **2009**, *11*, 3202–3205. (b) Wommack, A. J.; Kingsbury, J. S. *J. Org. Chem.* **2013**, *78*, 10573–10587.
9. For a review, see: (a) Moebius, D. C.; Rendina, V. L.; Kingsbury, J. S. *Top. Curr. Chem.* **2014**, *346*, 111–162. For rapid determination of diazoalkane titers in solution by ^{19}F NMR, see: (b) Rendina, V. L.; Kingsbury, J. S. *J. Org. Chem.* **2012**, *77*, 1181–1185.
10. For a lead reference on the preparation of aryldiazomethanes *in situ* for ester bond formation, see: Squitieri, R. A.; Shearn-Nance, G. P.; Hein, J. E.; Shaw, J. T. *J. Org. Chem.* **2016**, *81*, 5278–5284.
11. Davis, P. J.; Harris, L.; Karim, A.; Thompson, A. L.; Gilpin, M.; Moloney, M. G.; Pound, M. J.; Thompson, C. *Tetrahedron Lett.* **2011**, *52*, 1553–1556.
12. (a) Hudlicky, M. *J. Org. Chem.* **1980**, *45*, 5377–5378. (b) Moore, J. A.; Reed, D. E. *Org. Synth.* **1961**, *41*, 16–19.
13. Best, D.; Jenkinson, S. F.; Rule, S. D.; Higham, R.; Mercer, T. B.; Newell, R. J.; Weymouth-Wilson, A. C.; Fleet, G. W. J.; Petursson, S. *Tetrahedron Lett.* **2008**, *49*, 2196–2199.

14. (a) Seto, H.; Fujioka, S.; Koshino, H.; Hayasaka, H.; Shimizu, T.; Yoshida, S.; Watanabe, T. *Tetrahedron Lett.* **1999**, *40*, 2359–2362. (b) Kaplan, H. Z.; Rendina, V. L.; Kingsbury, J. S. *Molecules* **2017**, *22*, 1041.
15. Luo, J.; Song, K.; Gu, F. L.; Miao, Q. *Chem. Sci.* **2011**, *2*, 2029–2034.
16. Cesura, A.; Pinard, E. "MPTP Affinity Labels," European Patent Application no. 1 468 995 A1, **2004**, 10 pp.
17. Ladziata, U. *Arkivoc* **2014**, (i), 307–336.

Appendix
Chemical Abstracts Nomenclature (Registry Number)

Scandium(III) trifluoromethansulfonate: Trifluoromethansulfonic acid, scandium(3+) salt; (144026-79-9)
Phosphorous pentaoxide: Phosphorous(V) oxide; (1314-56-3)
Cyclohexanone: Oxocyclohexane; (108-94-1)
Diphenyldiazomethane: Diazodiphenylmethane; (883-40-9)
a,a-Diphenylcycloheptanone: 2,2-Diphenylcycloheptan-1-one (**2**)

Jason Kingsbury received his B.A. *summa cum laude* in Chemistry from Hamilton College and carried out undergraduate research with Professor Robin B. Kinnel, a veteran of marine natural products. From 1997-2003, he was an NSF predoctoral fellow in the labs of Amir H. Hoveyda at Boston College, where he synthesized new Ru-based catalysts for olefin metathesis. Following an NIH postdoctoral fellowship with E. J. Corey at Harvard University, he began his independent career in 2006 at Boston College before moving to California Lutheran University in 2013. His research interests include the design of new catalytic synthetic methods and strategies in targeting polycyclic natural products.

Victor Rendina received a B.Sc. in Chemistry with distinction from The Ohio State University, where he performed undergraduate research under the direction of Professor T. V. RajanBabu. In 2008, he joined the Kingsbury lab at Boston College, where he designed and tested tris(oxazoline) complexes in Sc(III)-catalyzed enantioselective diazoalkane–carbonyl homologation reactions. Upon receiving his Ph.D. in 2013, he went on to conduct process research in the pharmaceutical industry. In 2015, he worked to scale up energetic materials for a multi-national defense contractor. He is currently a lead software engineer for a smart home automation company in the greater Boston area.

Jacob Burman was raised in Chatsworth, CA and entered California Lutheran University in 2011 to pursue a B.Sc. degree in Chemistry. In his junior and senior years, Jacob played a vital role in the Kingsbury group's transition from the east coast, setting up new lab space and furthering scale-up studies on catalytic diazoalkyl carbon insertion. In 2015, he joined Professor Simon Blakey's team at Emory University to specialize in organometallic chemistry. His Ph.D. studies led to complementary methods of preparing regioisomeric allylic amines via Group IX MCp*-π-allyl intermediates. In 2019, he joined a discovery synthesis group at Raybow Pharmaceuticals in Brevard, North Carolina.

Brittany Smolarski was born in Thousand Oaks, CA and entered California Lutheran University as a Biology major. Her focus switched to Chemistry after taking organic chemistry in her sophomore year with Professor Kingsbury. She performed over two years of research as a Swenson and John Stauffer fellow, designing a new entry to dialkyl-aminonaphthylpyridium (DANPY) fluorophores and helping to optimize the article's title reaction. She graduated with distinction in 2016 and then joined Professor Dale Boger's lab at The Scripps Research Institute (La Jolla, CA). She is now a law student at the University of San Diego.

Junichi Taguchi was born in Saitama, Japan. He graduated from the University of Tokyo in 2021 with B.S. in Pharmaceutical Science. He is continuing his graduate studies at the University of Tokyo under the supervision of Prof. Masayuki Inoue. His research interests are in the area of the total synthesis of complex natural products.

Haruka Fujino received his Ph.D. (2019) from The University of Tokyo under the supervision of Prof. Masayuki Inoue. During the Ph.D. course, he spent 2 months as Visiting Student at the University of Chicago under the direction of Prof. Scott A. Snyder (2016). After carrying out postdoctoral research with Prof. Seth B. Herzon at Yale University (2019–2020), he was appointed as Assistant Professor in the Graduate School of Pharmaceutical Sciences at the University of Tokyo in 2020. His research interests include the total synthesis of bioactive and architecturally complex natural products.

C2 Amination of Pyridine with Primary Amines Mediated by Sodium Hydride in the Presence of Lithium Iodide

Jia Hao Pang, Derek Yiren Ong, and Shunsuke Chiba*[1]

Division of Chemistry and Biological Chemistry, School of Physical and Mathematical Sciences, Nanyang Technological University, Singapore 637371, Singapore

Checked by Simon Cooper and Sarah Reisman

Procedure (Note 1)

N-Butylpyridin-2-amine (**3**). An oven-dried, 250 mL three-necked round-bottomed flask fitted with a 25 × 15 mm Teflon-coated oval magnetic stir bar, a thermometer, rubber septum and a Liebig reflux condenser, is connected to a Schlenk line (Note 2). Sodium hydride (NaH) (3.00 g, 75.0 mmol, 3.0 equiv) (Note 3) is charged to the reaction vessel, after which it is evacuated and backfilled with argon three times. Anhydrous THF (Note 4) (35 mL) is added via a syringe. The reaction flask is suspended in a 22 °C water bath, and lithium iodide (LiI) (6.69 g, 50.0 mmol, 2.0 equiv) (Note 5) is introduced through the top of the reflux condenser, after which the Ar line is reintroduced with an out needle in place for 5 min to exchange the atmosphere. The addition of LiI to the reaction mixture resulted in an exotherm which causes the internal temperature to rise to 35 °C. The reaction mixture is allowed to cool to 25 °C under Ar atmosphere while stirring (Note 6), at which time *n*-butylamine (**2**) (5.00 mL, 50.6 mmol, 2.0 equiv) (Note 7) is added to the grey suspension in one portion using a syringe. Pyridine (**1**) (1.985 g, 25.1 mmol, 1.0 equiv) (Note 8) is weighed into a dry 4 mL vial and added via syringe in one portion. Tetrahydrofuran (THF) (15 mL) is used to rinse the vial and ensure complete transfer of the pyridine. (Figure 1). The grey suspension is heated at reflux temperature in a silicone oil bath for 24 h (Note 9), resulting in gradual formation of a greyish-yellow suspension (Figures 2 and 3) (Note 10).

Figure 1. Reaction setup (photo provided by submitters)

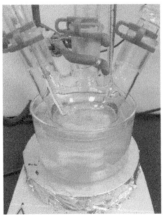

Figure 2. Color change observed from a grey suspension to a greyish-yellow suspension (photo provided by submitters)

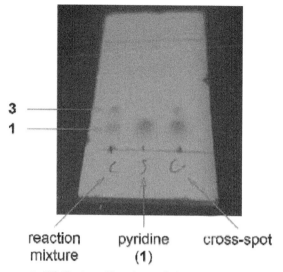

Figure 3. TLC visualization of the reaction mixture
(photo provided by submitters)

The reaction mixture is cooled to 10 °C (internal temperature) with an ice water bath. Cold water (15 mL) is added dropwise over 5 min before further addition of cold water (85 mL) in one portion. The layers are partitioned in a 500-mL separatory funnel. The aqueous layer is extracted with ethyl acetate (150 mL × 3), which is combined, dried over anhydrous sodium sulphate (55 g), and filtered with a Büchner funnel fitted with a fritted disk (medium porosity). The filtrate is concentrated on a rotary evaporator under reduced pressure (40 °C, 20 mmHg) to afford an orange solid (Figure 4).

Figure 4. Orange solids obtained after workup
(photo provided by submitters)

The crude compound is stored in a freezer at –20 °C until purification by column chromatography (Note 11) to afford *N*-butylpyridin-2-amine (**3**) (Notes 12, 13, and 14) as pale-yellow solids (3.42 g, 22.8 mmol, 91%) (Figure 5) (Note 15).

Figure 5. Pale-yellow solids of *N*-butylpyridin-2-amine (**3**) (photo provided by submitters)

Notes

1. Prior to performing each reaction, a thorough hazard analysis and risk assessment should be carried out with regards to each chemical substance and experimental operation on the scale planned and in the context of the laboratory where the procedures will be carried out. Guidelines for carrying out risk assessments and for analyzing the hazards associated with chemicals can be found in references such as Chapter 4 of "Prudent Practices in the Laboratory" (The National Academies Press, Washington, D.C., 2011; the full text can be accessed free of charge at https://www.nap.edu/catalog/12654/prudentpractices-in-the-laboratory-handling-and-management-of-chemical. See also "Identifying and Evaluating Hazards in Research Laboratories" (American Chemical Society, 2015) which is available via the associated website "Hazard Assessment in Research Laboratories" at https://www.acs.org/content/acs/en/about/governance/committees/chemicalsafety/hazard-assessment.html. In the case of this procedure, the risk assessment should include (but not necessarily be limited to) an evaluation of the potential hazards associated with sodium hydride, lithium iodide, pyridine, *n*-butylamine, hydrogen gas, sodium sulfate, THF, hexane and ethyl acetate.

2. The glass joints were greased thoroughly to minimize solvent evaporation over the course of the reaction.
3. Sodium hydride (60% dispersion in mineral oil) was purchased from Sigma Aldrich and used without any further purification. Submitter stored the reagent in an argon-filled glovebox.
4. Tetrahydrofuran (HiPerSolv CHROMANORM® for HPLC) was purchased from VWR Chemicals and was further purified using Pure Solv MD-5 solvent purification system by Innovative Technology before use.
5. Lithium iodide (anhydrous, beads, -10 mesh, 99.99% trace metal basis) was purchased from Sigma-Aldrich. The reagent was added to the reaction mixture immediately following opening of the ampule containing it.
6. The reaction was kept stirring at 500 rpm after the addition of tetrahydrofuran.
7. *n*-Butylamine (**2**) (99.5%) was purchased from Sigma Aldrich and used without any further purification.
8. Pyridine (**1**) (anhydrous, 99.8%) was purchased from Sigma Aldrich and used without any further purification.
9. The thermocouple was set to 85 °C to ensure a strong reflux. The oil bath temperature fluctuated between ±5 °C. The internal temperature of the reaction mixture was consistently at 66 °C.
10. The reaction was monitored by TLC (EMD Millipore™ TLC Silica Gel 60 F254, glass plates) on silica gel using hexanes/EtOAc (80:20) as eluent and visualization with UV light (Spectroline Longlife Filter Ultraviolet Light, ENF-240C, shortwave UV 254 nm). Product **3** had R_f = 0.42 and pyridine (**1**) had R_f = 0.22.
11. Column chromatography was performed using a Synthware chromatography column with a 500 mL reservoir: ID × Length (40.0 × 203 mm). The column was packed with 90 g of silica gel, high-purity grade, pore size 60 Å, 230–400 mesh particle size purchased from Sigma-Aldrich and conditioned with 500 mL hexanes. The crude mixture was mixed with silica (10 g) and ethyl acetate (10 mL), evaporated to dryness (40 °C, 20 mmHg), and dry-loaded onto the column. The column was eluted with hexanes/EtOAc (0.3 L, 95:5), followed by hexanes/EtOAc (1 L, 90:10) and hexanes/EtOAc (2 L, 80:20). Fractions sizes of 50 mL were collected. The fractions containing product show a UV-active/KMnO$_4$-active spot at R_f = 0.4 in hexanes/EtOAc (80:20). The compound was concentrated on a rotary

evaporator under reduced pressure (40 °C, 20 mmHg). The compound was placed under vacuum (0.75 mmHg) until it solidifies after which it was rinsed with hexanes (10 mL) and concentrated again on the rotary evaporator under reduced pressure (40 °C, 20 mmHg) to remove any residual solvents. The compound was further dried overnight under vacuum (room temperature, 0.75 mmHg).

12. *N*-Butylpyridin-2-amine (**3**) is bench stable. It displays the following characterization data: mp 38–39 °C; R_f = 0.42 (hexanes/EtOAc 80:20); ^{1}H NMR (400 MHz, CDCl$_3$) d: 0.95 (t, *J* = 7.3 Hz, 3H), 1.33 – 1.51 (m, 3H), 1.51 – 1.68 (m, 2H), 3.18 – 3.31 (m, 2H), 4.48 (s, 1H), 6.36 (dt, *J* = 8.4, 0.9 Hz, 1H), 6.54 (ddt, *J* = 6.9, 5.1, 0.9 Hz, 1H), 7.35 – 7.46 (m, 1H), 8.07 (ddd, *J* = 5.0, 1.9, 0.9 Hz, 1H); ^{13}C NMR (101 MHz, CDCl$_3$) d: 13.9, 20.2, 31.7, 42.0, 106.3, 112.6, 137.4, 148.2, 159.0; FTIR (neat) 3252, 3094, 3013, 2955, 2924, 2959, 1607, 1574, 1531, 1451, 1435, 1393, 1331, 1290, 1271, 1240, 1153, 1142, 1084, 980, 906, 840, 768, 748, 735, 637, 619, 600, 517 cm^{-1}.
13. Purity was determined to be 98.0% on the first run using qNMR (relaxation delay of 60 s) with 1,3,5-trimethoxybenzene as an internal standard.
14. Based upon the purity of 98%, the amount of material produced was 3.35 g (89%).
15. A second run by the checkers on a scale of 25.1 mmol provided 3.35 g (89%) of the same product with similar purity.

Working with Hazardous Chemicals

The procedures in *Organic Syntheses* are intended for use only by persons with proper training in experimental organic chemistry. All hazardous materials should be handled using the standard procedures for work with chemicals described in references such as "Prudent Practices in the Laboratory" (The National Academies Press, Washington, D.C., 2011; the full text can be accessed free of charge at http://www.nap.edu/catalog.php?record_id=12654). All chemical waste should be disposed of in accordance with local regulations. For general guidelines for the management of chemical waste, see Chapter 8 of Prudent Practices.

In some articles in *Organic Syntheses*, chemical-specific hazards are highlighted in red "Caution Notes" within a procedure. It is important to recognize that the absence of a caution note does not imply that no significant hazards are associated with the chemicals involved in that procedure. Prior to performing a reaction, a thorough risk assessment should be carried out that includes a review of the potential hazards associated with each chemical and experimental operation on the scale that is planned for the procedure. Guidelines for carrying out a risk assessment and for analyzing the hazards associated with chemicals can be found in Chapter 4 of Prudent Practices.

The procedures described in *Organic Syntheses* are provided as published and are conducted at one's own risk. *Organic Syntheses, Inc.*, its Editors, and its Board of Directors do not warrant or guarantee the safety of individuals using these procedures and hereby disclaim any liability for any injuries or damages claimed to have resulted from or related in any way to the procedures herein.

Discussion

It is known that the reaction of pyridine (**1**) with sodium amide (NaNH$_2$) gives 2-aminopyridine, which is known as the Chichibabin reaction (amination).[2,3] However, variants of the Chichibabin amination that engage primary alkyl amines as a nucleophile are scarce.[4]

Our group recently reported a concise protocol on the Chichibabin amination of pyridine (**1**) with primary alkyl amines using sodium hydride (NaH) in the presence of lithium iodide (LiI) in THF.[5,6] The optimization of the reaction conditions (Table 1) revealed that the reaction of pyridine (**1**) and *n*-butylamine (**2**) (2 equiv) could be facilitated in the presence of NaH (3 equiv) and LiI (2 equiv) in THF under reflux conditions (at 66 °C). The reaction could be completed within 18 h in 0.5 mmol scale of **1** (entry 1) and 24 h in 25 mmol scale of **1** (entry 2, see the detailed protocol above). Although higher reaction temperature (85 °C) in sealed conditions (in 0.5 mmol scale of **1**) allowed for completion of the process within 7 h (entry 3), the sealed reaction setting is not advisable, especially in the larger scale, since the process produces 2 molar equivalents of hydrogen (H$_2$) gas for each aminated product that is formed.

Table 1. Optimization of the reaction conditions

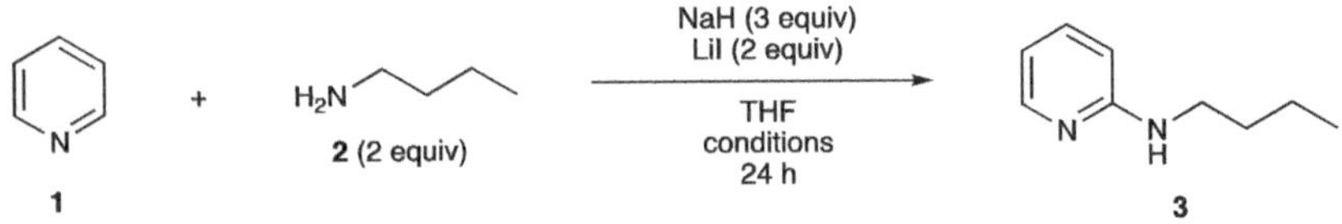

entry	amounts of **1** [mmol]	conditions	time [h]	Isolated yield of **3** [%]
1	0.5 mmol	65 °C, sealed	18	95
2	25 mmol	reflux	24	89
3	0.5 mmol	85 °C, sealed	7	93

The protocol could engage various sets of primary alkyl amines (Scheme 1). It should be noted that the present protocol is thus far proven unsuccessful for use of anilines and their derivatives.

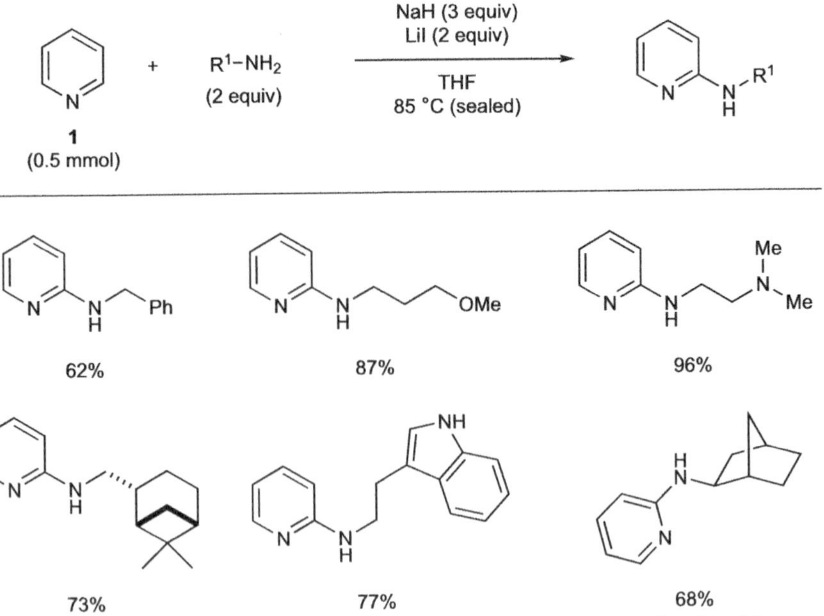

Scheme 1. Examples of C2-aminated pyridines synthesized by the present protocol

The present protocol allows for two-fold amination of 2,2'-bipyridine in good yield (Scheme 2).

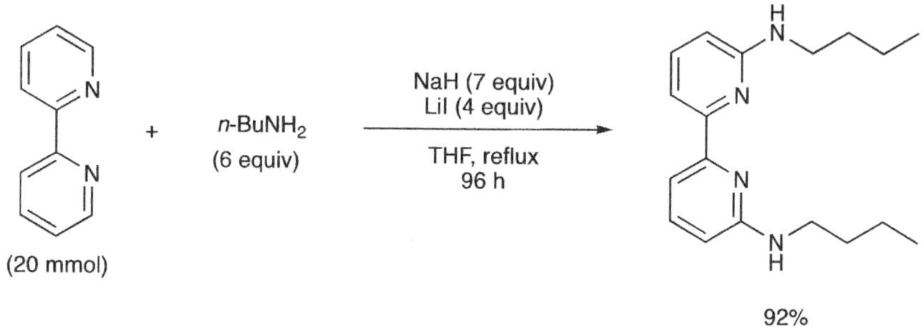

Scheme 2. Two-fold amination of 2,2'-bipyridine

References

1. Division of Chemistry and Biological Chemistry, School of Physical and Mathematical Sciences, Nanyang Technological University, Singapore 637371, Singapore. shunsuke@ntu.edu.sg; ORCID: 0000-0003-2039-023X; this work was supported by Nanyang Technological University, the Singapore Ministry of Education (Academic Research Fund Tier 2: MOE2019-T2-1-089).
2. Chichibabin, A. E.; Zeide, O. A. *J. Russ. Phys. Chem. Soc.* **1914**, *46*, 1216–1236.
3. For reviews on the Chichibabin reaction, see: a) van der Plas, H. C. *Adv. Heterocycl. Chem.* **2004**, *86*, 1–40. (b) McGill, C. K.; Rappa, A. *Adv. Heterocycl. Chem.* **1988**, *44*, 1–79.
4. (a) Breuker, J.; van der Plas, H. C. *Recl. Trav. Chim. Pays-Bas.* **1983**, *102*, 367–372. (b) Vajda, T.; Kovács, K. *Recl. Trav. Chim. Pays-Bas* **1961**, *80*, 47–56. (c) Bergstrom, F. W.; Sturz, H. G.; Tracy, H. W. *J. Org. Chem.* **1946**, *11*, 239–246.
5. Pang, J. H.; Kaga, A.; Roediger, S.; Lin, M. H.; Chiba, S. *Asian J. Org. Chem.* **2019**, *8*, 1058–1060.
6. For use of NaI I-iodide composites as the unprecedented Brønsted bases, see: (a) Pang, J.-H.; Ong, D. Y.; Watanabe, K.; Takita, R.; Chiba, S. *Synthesis* **2020**, *52*, 393–398. (b) Pang, J. H.; Kaga, A.; Chiba, S. *Chem. Commun.* **2018**, *54*, 10324–10327. (c) Kaga, A.; Hayashi, H.; Hakamata, H.; Oi, M.; Uchiyama, M.; Takita, R.; Chiba, S.; *Angew. Chem. Int. Ed.* **2017**, *56*, 11807–11811. (d) Huang, Y.; Chan, G. H.; Chiba, S. *Angew. Chem. Int. Ed.* **2017**, *56*, 6544–6547.

Appendix
Chemical Abstracts Nomenclature (Registry Number)

Sodium hydride; (7646-69-7)
Lithium iodide; (10377-51-2)
Tetrahydrofuran; (109-99-9)
Pyridine; (110-86-1)
n-Butylamine; (109-73-9)

Jia Hao Pang completed his undergraduate studies at Nanyang Technological University (NTU) Singapore in 2016 before beginning his PhD work in the laboratory of Shunsuke Chiba at NTU. He is currently focusing on chemistry of main group metal hydrides for methodology development.

Derek Yiren Ong completed his undergraduate studies at Nanyang Technological University (NTU) Singapore in 2013. After working as an NMR technician at NTU for several years, he started his Ph.D. work in the laboratory of Shunsuke Chiba and earned his Ph.D. in 2020. He is currently working at JEOL Asia Pte. Ltd. as an application specialist in NMR.

Shunsuke Chiba earned his Ph.D. in March 2006 from the University of Tokyo under the supervision of Prof. Koichi Narasaka. In 2007, he embarked on his independent career as the faculty of Nanyang Technological University (NTU) Singapore, where he is currently Professor of Chemistry. His research group focuses on methodology development in the area of synthetic chemistry and catalysis.

Simon Cooper completed his undergraduate studies at the University of San Francisco in 2015. He then joined Prof. Todd Hyster's laboratory developing biocatalytic methods at Princeton University and completed his PhD in 2020. He is currently working on natural product synthesis as a postdoctoral fellow with Prof. Sarah Reisman.

Synthesis and Acylation of 1,3-Thiazinane-2-thione

Stuart C. D. Kennington, Oriol Galeote, Miguel Mellado-Hidalgo, Pedro Romea,[1*] and Fèlix Urpí[1*]

Department of Inorganic and Organic Chemistry, Section of Organic Chemistry, and Institute of Biomedicine (IBUB), University of Barcelona, Carrer Martí i Franqués 1-11, 08028 Barcelona, Catalonia, Spain

Checked by Zhaobin Han and Kuiling Ding

Procedure (Note 1)

A. *3-Ammoniopropylsulfate (1)*. An oven-dried single-necked 100 mL round-bottomed flask (14/23 joint), equipped with a 2.5-cm Teflon-coated magnetic stirbar, is charged with 3-amino-1-propanol (11.5 mL, 150 mmol, 1 equiv) (Note 2) and anhydrous dichloromethane (35 mL) (Note 3). A 50 mL pressure-equalizing addition funnel (14/23 joint) equipped with a $CaCl_2$ tube is attached to the round-bottomed flask and is then charged with chlorosulfonic acid (10.5 mL, 159 mmol, 1.06 equiv) (Note 4) using a 20 mL glass luer-lock syringe. The flask is immersed in an ice/water bath and the solution is stirred for 5 min. The chlorosulfonic acid is added dropwise over 30 min, allowing the fumes to escape. A white precipitate is formed during the addition. Once the addition is complete, the reaction is stirred at 0 °C for 20 min and left to warm slowly to room temperature over 30 min (Note 5). Once at room temperature, the reaction mixture is stirred for 1 h. The resulting mixture is filtered through a 70 mm diameter Number 3 Glass filter funnel with a Büchner setup. A bent spatula and methanol (25 mL) (Note 6) are used to remove remaining product from the flask walls. The mixture in the filter funnel is triturated with methanol (40 mL, then 2 × 20 mL) (Note 6), using a spatula to break up the lumps each time. The resulting white solid is broken up into a coarse powder and transferred to a 100 mL round-bottomed flask (29/32 joint), where it is placed on a rotary evaporator (40 °C, 12 mmHg pressure) for 1 h. The resulting white solid is ground to a fine white powder using a 10 cm diameter glass mortar and pestle, retransferred to a 100 mL round-bottomed flask and dried on a high vacuum line (25 °C, 0.1 mmHg pressure) for 2 h giving the title compound **1** (20.72 g, 134 mmol, 89% yield) (Note 7) as a fine white powder.

Figure 1. Addition set-up (left), trituration (middle) and final product (right) (photos provided by submitters)

B. *1,3-Thiazinane-2-thione (2)*. An oven-dried single-necked 250 mL round-bottomed flask (29/32 joint), equipped with a 4-cm Teflon-coated magnetic stirbar, is charged with 3-ammoniopropylsulfate (**1**) (18.70 g, 121 mmol, 1 equiv) and absolute ethanol (15 mL) (Note 8) at room temperature. The resulting solution is stirred at 25 °C for 3 min and neat carbon disulfide (9.6 mL, 160 mmol, 1.3 equiv) (Note 9) is added in one portion using a 10 mL syringe.

Separately, KOH beads (14.84 g, 265 mmol, 2.2 equiv) (Note 10) are weighed in a 250 mL conical flask equipped with a 4-cm Teflon-coated magnetic stir bar, and the KOH is dissolved in 1:1 ethanol/water (100 mL). The mixture is stirred at 25 °C, and the resulting solution is transferred with a funnel to a 250 mL pressure-equalizing addition funnel (29/32 joint) attached to the round-bottomed flask. The neck of the addition funnel is sealed with a rubber septum, the system is purged with a nitrogen flow for 5 min, and a nitrogen atmosphere is maintained via a needle inserted into the septum. The KOH solution is added dropwise to the carbon disulfide solution in the round-bottomed flask over 30 min at room temperature to give a yellow solution. The addition funnel is replaced by a reflux condenser sealed with a rubber septum, and a nitrogen is provided by a needle inserted through the septum. The reaction mixture is heated to reflux in an aluminum heating block (70 °C) for 1 h under a N_2 atmosphere and allowed to cool to room temperature slowly to give a fluffy white precipitate in the solution, which is further cooled to 0 °C with an ice/water bath for 15 min.

The mixture is filtered using a 70 mm diameter Number 3 glass filter funnel with a Büchner setup. The flask is rinsed with cold deionized water (3 × 15 mL) with the washings being added to the filter funnel. The solid in the filter funnel is dried under vacuum (20 mmHg) for 15 min, after which the receiving flask is changed for a fresh one. The solid is washed with dichloromethane (3 × 35 mL) (Note 11), each time breaking up the solid with a spatula and mixing thoroughly before applying the vacuum. The combined organic extracts are dried over $MgSO_4$ (40 g) (Note 12) and concentrated under reduced pressure (12 mmHg) to give pure crystalline powder of 1,3-thiazinane-2-thione **2** (6.78 g, 42% yield) (Note 13). The remaining solid (Note 14) from the filter funnel is transferred to a 250 mL round-bottomed flask (29/32 joint) equipped with a 4-cm Teflon-coated magnetic stir bar. This round-bottomed flask is charged with dichloromethane (150 mL) (Note 11), a reflux condenser sealed with a rubber septum is attached to the flask, the system is purged with a nitrogen flow for a couple of minutes, and a nitrogen atmosphere is maintained through the septum. The resulting mixture is stirred and heated to reflux for 1 h, and the warm solution is filtered through a Number 3 glass filter funnel with a Büchner setup. The solid is washed with dichloromethane (2 × 50 mL) (Note 11), each time breaking up the solid with a spatula and mixing thoroughly before applying the vacuum (20 mmHg) as before. The combined filtrates are dried over $MgSO_4$ (35 g) (Note 12) and concentrated (12 mmHg) to give pure thiazinanethione **2** (3.34 g, 21% yield). The combined thiazinanethione **2** weighs 10.12 g (63% overall yield) (Note 15).

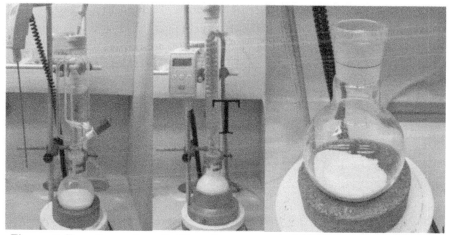

Figure 2. Reaction set-up (left), reflux set-up (middle) and final product (right) (photos provided by submitters)

C. *N-Propanoyl-1,3-thiazinane-2-thione* (**3**). An oven-dried single-necked 250 mL round-bottomed flask (29/32 joint), equipped with a 4-cm Teflon-coated magnetic stirbar, is charged with 1,3-thiazinane-2-thione (10.64 g, 80 mmol, 1 equiv). The flask is sealed with a rubber septum and the system is flushed with a flow of nitrogen. The flask is charged with anhydrous dichloromethane (80 mL) (Note 3) via syringe and immersed in an ice/water bath. The resulting solution is stirred for 2 min and freshly distilled triethylamine (14.5 mL, 104 mmol, 1.3 equiv) (Note 16) is added dropwise over 2 min. The solution is stirred for 2 min and propionyl chloride (8.4 mL, 96 mmol, 1.2 equiv) (Note 17) is carefully added over 20 min, which produces a yellow solution. The ice/water bath is removed, and the reaction mixture is allowed to warm to room temperature for 2 h. The resulting dark yellow/orange mixture is cooled with an ice/water bath, quenched with the addition of a saturated solution of NH_4Cl (25 mL) (Note 18) in one portion, and allowed to stir for 5 min.

The mixture is transferred to a 500 mL separating funnel. The flask is rinsed with dichloromethane (4 × 40 mL) (Note 11) and water (3 × 40 mL), which are added to the separating funnel. The mixture is shaken vigorously and the layers separated. The aqueous layer is extracted with another portion of dichloromethane (40 mL) (Note 11). The combined organic extracts are washed with 2 M NaOH (120 mL) (Note 19), dried over $MgSO_4$ (25 g), and filtered, with the flask and solid being rinsed with dichloromethane (3 × 40 mL) (Note 11), which is added to the filtrate. The solution is concentrated under reduced pressure (12 mmHg) and the resulting residue is submitted to column chromatography on silica gel (60 Å) using hexanes (Note 21) and ethyl acetate (Note 22) as eluent. Chromatographic purification (Note 23) provided pure acylated product 3 (12.74 g, 84% yield) (Note 24) (Figure 3).

Figure 3. Appearance of product 3 (photo provided by submitters)

Notes

1. Prior to performing each reaction, a thorough hazard analysis and risk assessment should be carried out with regard to each chemical substance and experimental operation on the scale planned and in the context of the laboratory where the procedures will be carried out. Guidelines for carrying out risk assessments and for analyzing the hazards associated with chemicals can be found in references such as Chapter 4 of "Prudent Practices in the Laboratory" (The National Academies Press, Washington, D.C., 2011; the full text can be accessed free of charge at https://www.nap.edu/catalog/12654/prudent-practices-in-the-laboratory-handling-and-management-of-chemical. See also "Identifying and Evaluating Hazards in Research Laboratories" (American Chemical Society, 2015) which is available via the associated website "Hazard Assessment in Research Laboratories" at https://www.acs.org/content/acs/en/about/governance/committees/chemicalsafety/hazard-assessment.html. In the case of this procedure, the risk assessment should include (but not necessarily be limited to) an evaluation of the potential hazards associated with chlorosulfonic acid, carbon disulfide, triethylamine, propionyl chloride, sodium hydroxide, potassium hydroxide, Celite®, silica gel, dichloromethane, methanol, hexane, and ethyl acetate.
2. 3-Amino-1-propanol (99%) was purchased from Acros Organics and used as received. The checkers purchased 3-Amino-1-propanol (>99%) from Tokyo Chemical Industry Co., Ltd. and was used as received.
3. Dichloromethane (99%) was purchased from Acros Organics and was freshly distilled from CaH_2. The checkers purchased dichloromethane (99%) from Sinopharm Chemical Reagent Co., Ltd. and the solvent was freshly distilled from CaH_2.
4. Chlorosulfonic acid (99%) was purchased from Acros Organics and used as received. The checkers purchased chlorosulfonic acid (>97%) from Tokyo Chemical Industry Co., Ltd. and used the reagent as received.
5. If the reaction becomes too vigorous it can be kept cooled.
6. Methanol (99%) was purchased from Acros Organics and used as received. The checkers purchased methanol (99%) from Sinopharm Chemical Reagent Co., Ltd. and used the solvent as received.
7. 3-Ammoniopropylsulfate (**1**) has the following physical and spectroscopic properties: mp 205–207 °C; IR (film): 3126, 3064, 2976,

1627, 1530, 1195, 1169, 1065, 923, 759, 574 cm^{-1}; ^{1}H NMR (400 MHz, DMSO-d$_6$) δ: 1.74–1.86 (m, 2H), 2.76–2.90 (m, 2H), 3.81 (t, J = 6.0 Hz, 2H), 7.64 (br s, 3H); ^{13}C NMR (100.6 MHz, DMSO-d$_6$) δ: 27.3, 36.7, 63.2; HRMS (+ESI) m/z calcd for C$_3$H$_{10}$NO$_4$S [M + H]$^+$ 156.0325, found 156.0322. The purity of **1** was determined to be 97% by ^{1}H qNMR using 10.8 mg of ethylene carbonate (>99% purity) as an internal standard and 18.7 mg of compound **1**. Based on this purity, the actual amount of **1** formed in the reaction was 20.1 g. A second reaction on identical scale provided 20.84 g (89%, uncorrected for purity) of the product **1**.

8. Absolute ethanol (99%) was purchased from Acros Organics and used as received. The checkers purchased absolute ethanol (99%) from Sinopharm Chemical Reagent Co., Ltd. and used the solvent as received.
9. Carbon disulfide was purchased from Sigma Aldrich (ACS reagent, 99.9%) and used as received. The checkers purchased carbon disulfide (>98%) from Tokyo Chemical Industry Co., Ltd. and used it as received.
10. KOH (98%) beads was purchased from Panreac and used as received. The checkers purchased KOH (99%) from Sinopharm Chemical Reagent Co., Ltd. and used it as received.
11. Dichloromethane (99%) was purchased from Acros Organics and used as received. The checkers purchased dichloromethane (99%) from Sinopharm Chemical Reagent Co., Ltd. and used the solvent as received.
12. Anhydrous MgSO$_4$ was purchased from Panreac and used as received. The checkers purchased anhydrous MgSO$_4$ (99%) from Sinopharm Chemical Reagent Co., Ltd. and used the drying agent as received.
13. 1,3-Thiazinane-2-thione (**2**) has the following physical and spectroscopic properties: mp 138–140 °C [lit.[2] mp 132–133 °C]; IR (film, cm^{-1}): 3140, 3049, 3000, 2918, 2848, 1543, 1426, 1353, 1331, 1273, 1182, 1084, 1011, 897, 748, 625; ^{1}H NMR (400 MHz, CDCl$_3$) δ: 2.13–2.23 (m, 2H), 3.00 (t, J = 6.0 Hz, 2H), 3.48 (t, J = 5.6 Hz, 2H), 8.75 (br s, 1H); ^{13}C NMR (100.6 MHz, CDCl$_3$) δ: 20.5, 30.1, 44.3, 194.5; HRMS (+ESI) m/z calcd for C$_4$H$_8$NS$_2$ [M + H]$^+$ 134.0093, found 134.0091. The purity of **2** was determined to be 99% by ^{1}H qNMR using 16.4 mg of 1,3,5-dimethoxybenzene (>99% purity) as an internal standard and 14.3 mg of compound **2**.
14. The remaining solid is a thick paste containing impurities formed in the reaction.

15. The quantities of both crops can vary, but the overall yield and the characterization data for the two crops are consistent. A second reaction on identical scale provided 10.06 g (62%) of the product **2**.
16. Triethylamine (99%) was purchased from Fluorochem and was freshly distilled over CaH_2. The checkers purchased triethylamine (99%) from Sinopharm Chemical Reagent Co., Ltd. and it was freshly distilled over CaH_2.
17. Propionyl chloride (99%) was purchased from Acros Organics and used as received. The checkers purchased propionyl chloride (>98%) from Tokyo Chemical Industry Co., Ltd. and it was used as received.
18. Ammonium chloride was purchased from Panreac and used as received. The checkers purchased ammonium chloride (99%) from Sinopharm Chemical Reagent Co., Ltd. and it was used as received.
19. NaOH (98%) beads were purchased from Panreac and they were used as received. The checkers purchased NaOH (99%) from Sinopharm Chemical Reagent Co., Ltd. and used it as received.
20. Silica gel was purchased from Sigma Aldrich. The checkers purchased silica gel (SiliaFlash P60, particle size: 40-63 μm, pore size 60 Å) from SILICYCLE.
21. Hexanes (99%) was purchased from VWR International and used as received. The checkers purchased hexanes (99%) from Tansoole.
22. Ethyl acetate was purchased from Panreac and used as received. The checkers purchased ethyl acetate (99%) from Tansoole.
23. A 6 cm diameter column with a length of 25 cm contained silica (*ca* 400 g) (Note 20). The silica is first compacted with 90:10 hexanes/ethyl acetate (1 L) (Notes 21 and 22) and the surface levelled. The crude residue is dissolved in ethyl acetate (4 mL) (Note 22), diluted in hexanes (8 mL) (Note 21), and added onto the compacted column. After adsorption, the flask that contained the crude residue is washed with 90:10 hexanes/ethyl acetate (3 × 8 mL, or until the flask is no longer yellow), each time waiting until the liquid is adsorbed. The walls of the column are then washed with 90:10 hexanes/ethyl acetate (3 × 8 mL) (Notes 21 and 22). Once all of the yellow product is adsorbed on the silica, a thick layer of sand is added to protect the silica. The column is eluted with 90:10 hexanes/ethyl acetate (*ca* 4.5 L) until all of the yellow color (Figure 4) has left the column and the eluent runs clear. The product is collected in *circa* 150 30 mL–test tubes. Contents of the tubes are assessed by TLC (80:20 hexanes/ethyl acetate; R_f = 0.39) (Figure 5), and those containing pure product are sequentially added to a 1 L

round-bottomed flask and concentrated (20 mmHg, 40 °C). Once the product is concentrated, the resulting thick yellow oil is diluted with dichloromethane (50 mL) (Note 11), and the solution transferred to a 100 mL round-bottomed flask. The solution is concentrated under reduced pressure (20 mmHg) and kept under high vacuum (0.1 mmHg) at room temperature for 4 h to afford the product (**3**).

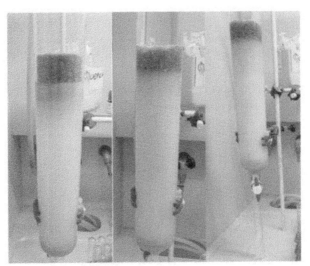

Figure 4. Product 3 (yellow) eluting down the silica column
(photos provided by submitters)

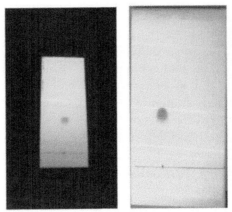

Figure 5. Image of a TLC plate (UV) of a pure sample
(photos provided by checkers)

24. *N*-Propanoyl-1,3-thiazinane-2-thione (**3**) has the following physical and spectroscopic properties: yellow–orange oil; R_f 0.39 (80:20 hexanes/EtOAc); IR (film): 2975, 2934, 2877, 1700, 1469, 1374, 1344, 1302, 1198, 1166, 1020, 966, 919 cm^{-1}; ^{1}H NMR (400 MHz, CDCl$_3$) δ: 1.22 (t, *J* = 7.2 Hz, 3H), 2.19–2.30 (m, 2H), 3.04 (t, *J* = 6.8 Hz, 2H), 3.09 (q, *J* = 7.2 Hz, 2H), 3.91–3.95 (m, 2H); ^{13}C NMR (100.6 MHz, CDCl$_3$) δ: 10.1, 22.9, 32.1, 32.7, 46.6, 179.0, 202.9; HRMS (+ESI) *m/z* calcd for C$_7$H$_{12}$NOS$_2$ [M + H]$^+$ 190.0355, found 190.0354. The purity of **3** was determined to be >98.5% by ^{1}H qNMR using 16.8 mg of 1,3,5-dimethoxybenzene (>99% purity) as an internal standard and 22.6 mg of compound **3**. Based on this purity, the actual amount of **3** formed in the reaction was 12.57 g. A second reaction on identical scale provided 12.68 g (84%, uncorrected for purity) of the product **3**.

Working with Hazardous Chemicals

The procedures in *Organic Syntheses* are intended for use only by persons with proper training in experimental organic chemistry. All hazardous materials should be handled using the standard procedures for work with chemicals described in references such as "Prudent Practices in the Laboratory" (The National Academies Press, Washington, D.C., 2011; the full text can be accessed free of charge at http://www.nap.edu/catalog.php?record_id=12654). All chemical waste should be disposed of in accordance with local regulations. For general guidelines for the management of chemical waste, see Chapter 8 of Prudent Practices.

In some articles in *Organic Syntheses*, chemical-specific hazards are highlighted in red "Caution Notes" within a procedure. It is important to recognize that the absence of a caution note does not imply that no significant hazards are associated with the chemicals involved in that procedure. Prior to performing a reaction, a thorough risk assessment should be carried out that includes a review of the potential hazards associated with each chemical and experimental operation on the scale that is planned for the procedure. Guidelines for carrying out a risk assessment and for analyzing the hazards associated with chemicals can be found in Chapter 4 of Prudent Practices.

The procedures described in *Organic Syntheses* are provided as published and are conducted at one's own risk. *Organic Syntheses, Inc.*, its Editors, and its Board of Directors do not warrant or guarantee the safety of individuals using these procedures and hereby disclaim any liability for any injuries or damages claimed to have resulted from or related in any way to the procedures herein.

Discussion

Chiral heterocycles are widespread platforms from which a variety of stereoselective transformations can be carried out.[3] Indeed, chiral oxazolidinones (Scheme 1) introduced by Evans in the 80s still hold a prominent position among the most efficient chiral auxiliaries[4-6] and are currently employed for the synthesis of natural products.[7] Furthermore, structurally similar oxazolidinethiones and thiazolidinethiones (Scheme 1) have also played a crucial role in stereoselective synthesis.[8-11] Irrespective of the high stereocontrol provided by such auxiliaries, their prevalence is also due to their straightforward synthesis from proteinogenic α-amino acids (Scheme 1).[5,6,8,12]

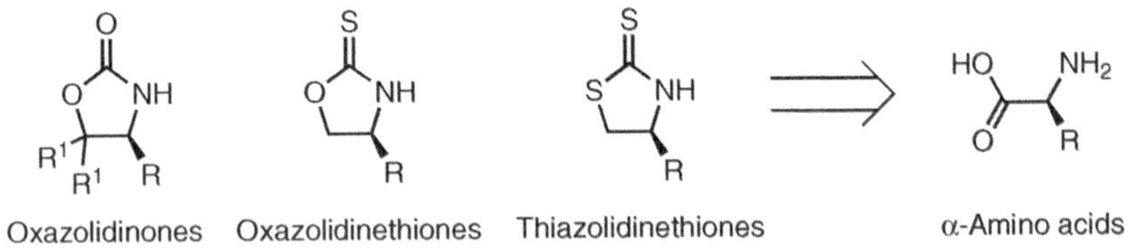

Scheme 1. Heterocycles used as chiral auxiliaries

Importantly, the quest for new catalytic and asymmetric transformations adhered to the atom economy principle[13] have also aroused the interest in the achiral counterparts of such heterocycles. Indeed, Evans reported the crucial role played by achiral *N*-propanoyl-1,3-thiazolidine-2-thione in enantioselective aldol reactions catalyzed by a chiral nickel(II) complex (Scheme 2).[14]

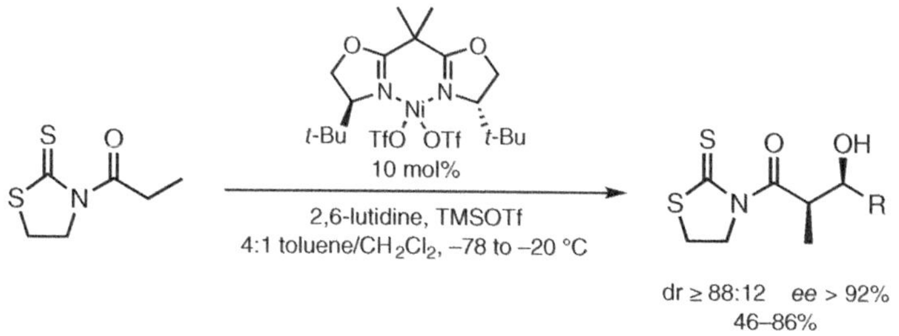

Scheme 2. Achiral *N*-propanoyl-1,3-thiazolidine-2-thione in enantioselective aldol reaction

Five membered achiral scaffolds may thus mimic the performance of their chiral counterparts shown in Scheme 1. Surprisingly, there is a lack of similar reactions based in the corresponding six membered heterocycles.[2] In this context, we have recently reported that *N*-propanoyl-1,3-thiazinane-2-thione undergoes highly efficient alkylation reactions with benzhydryl

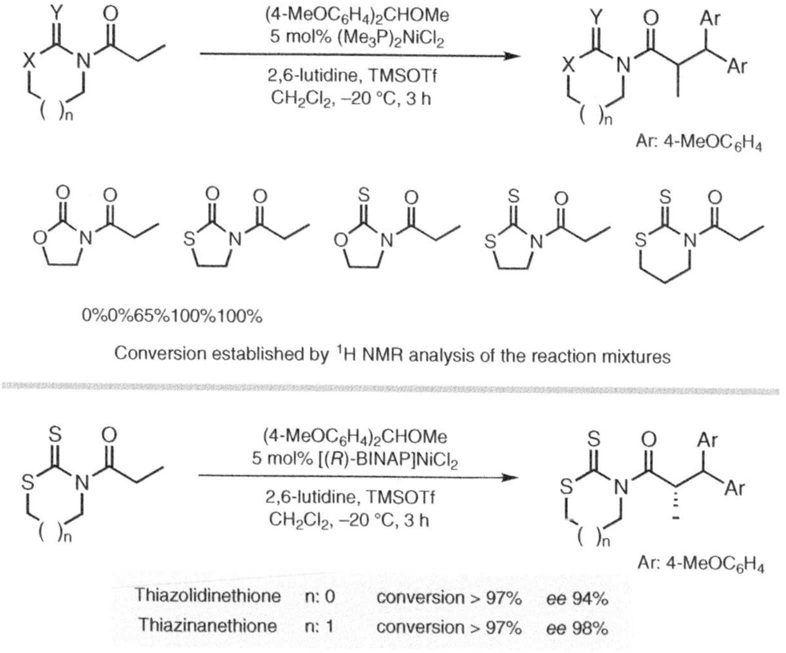

| Thiazolidinethione | n: 0 | conversion > 97% | ee 94% |
| Thiazinanethione | n: 1 | conversion > 97% | ee 98% |

Scheme 3. Alkylation of *N*-propanoyl heterocycles with (4-MeOC$_6$H$_4$)$_2$CHOMe

methyl ethers, slightly more enantioselective than the parallel thiazolidinethione scaffold (Scheme 3).[15] Therefore, a six-membered thiazinane heterocycle may be a valuable platform for asymmetric synthesis.

Acylation of 1,3-thiazinane-2-thione with acyl chlorides at 10–15 mmol provides the corresponding *N*-acyl thiazianethiones with yields up to 93% (Scheme 4. Part A). Alternatively, such acylations can be carried out by coupling of 1,3-thiazinane-2-thione and carboxylic acids with EDC at 5–10 mmol scale (Scheme 4. Part B).

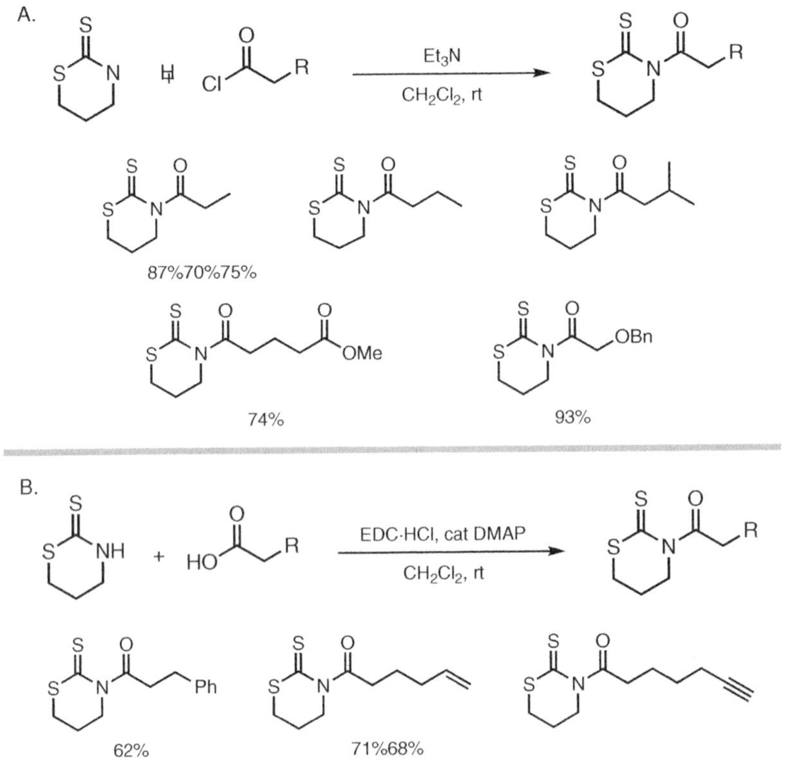

Scheme 4. Acylation of 1,3-thiazinane-2-thione

References

1. Secció de Química Orgànica, Departament de Química Inorgànica i Orgànica & Institut de Biomedecina de la Universitat de Barcelona (IBUB), Universitat de Barcelona, 08028 Barcelona, Catalonia, Spain. Email: pedro.romea@ub.edu; mailto:felix.urpi@ub.edu. (ORCID (PR): 0000-0002-0259-9155; ORCID (FU) 0000-0003-4289-6506); Financial support from the Spanish Ministerio de Ciencia, Innovación y Universidades (MCIU)/Agencia Estatal de Investigación (AEI)/Fondo Europeo de Desarrollo Regional (FEDER, UE) (Grant No. PGC2018-094311-B-I00), and the Generalitat de Catalunya (2017SGR 271) as well as a doctorate studentship to S. C. D. K. (FI, Generalitat de Catalunya) are acknowledged.
2. Obata, N. *Bull. Chem. Soc. Jpn.* **1977**, *50*, 2187–2188.
3. *Asymmetric Synthesis – The Essentials*. Editors: Christmann, M.; Bräse, S. Wiley-VCH: Weinheim, **2008**.
4. (a) Evans D. A.; Bartroli, J.; Shih, T. L. *J. Am. Chem. Soc.* **1981**, *103*, 2127–2129. (b) Evans, D. A.; Ennis, M. D.; Mathre, D. J. *J. Am. Chem. Soc.* **1982**, *104*, 1737–1739.
5. (a) Gage, J. R.; Evans, D. A. *Org. Synth.* **1990**, *68*, 77–81. (b) Gage, J. R.; Evans, D. A. *Org. Synth.* **1990**, *68*, 83–90.
6. See also: (a) Davies, S. G.; Sanganee, H. J. *Tetrahedron: Asymmetry* **1995**, *6*, 671–674. (b) Hintermann, T.; Seebach, D. *Helv. Chim. Acta* **1998**, *81*, 2093-2126. (c) Brenner, M.; La Vecchia, L.; Leutert, T.; Seebach, D. *Org. Synth.* **2003**, *80*, 57–65.
7. For recent contributions, see: (a) Yu, G.; Jung, B.; Lee, H.-S.; Kang, S. H. *Angew. Chem. Int. Ed.* **2016**, *55*, 2573–2576. (b) Lu, Z.; Zhang, X.; Guo, Z.; Chen, Y.; Mu, T.; Li, A. *J. Am. Chem. Soc.* **2018**, *140*, 9211–9218.
8. (a) Gálvez, E.; Romea, P.; Urpí, F. *Org. Synth.* **2009**, *86*, 70–80. (b) Gálvez, E.; Romea, P.; Urpí, F. *Org. Synth.* **2009**, *86*, 81–91. (c) Crimmins, M. T.; Christie, H. S.; Hughes, C. O. *Org. Synth.* **2011**, *88*, 364–376.
9. (a) Crimmins, M. T.; King, B. W.; Tabet, E. A. *J. Am. Chem. Soc.* **1997**, *119*, 7883–7884. (b) Crimmins, M. T.; King, B. W.; Tabet, E. A.; Chaudhary, K. *J. Org. Chem.* **2001**, *66*, 894–902. (c) Crimmins, M. T.; She, J. *Synlett* **2004**, 1371–1374.
10. For the reaction of titanium(IV) enolates of chiral *N*-acyl thiazolidinethiones, see: (a) Cosp, A.; Romea, P.; Talavera, P.; Urpí, F.; Vilarrasa, J.; Font-Bardia, M.; Solans, X. *Org. Lett.* **2001**, *3*, 615–617.

(b) Larrosa, I.; Romea, P.; Urpí, F.; Balsells, D.; Vilarrasa, J.; Font-Bardia, M.; Solans, X. *Org. Lett.* **2002**, *4*, 4651–4654. (c) Checa, B.; Gálvez, E.; Parello, R.; Sau, M.; Romea, P.; Urpí, F.; Font-Bardia, M.; Solans, X. *Org. Lett.* **2009**, *11*, 2193–2196. (d) Kennington, S. C. D.; Romo, J. M.; Romea, P.; Urpí, F. *Org. Lett.* **2016**, *18*, 3018–3021.
11. For the reaction of chiral *N*-acyl thiazolidinethiones catalyzed by nickel(II) complexes, see: (a) Romo, J. M.; Gálvez, E.; Nubiola, I.; Romea, P.; Urpí, F.; Kindred, M. *Adv. Synth. Catal.* **2013**, *355*, 2781–2786. (b) Fernández-Valparís, J.; Romo, J. M.; Romea, P.; Urpí, F.; Kowalski, H.; Font-Bardia, M. *Org. Lett.* **2015**, *17*, 3540–3543. (c) Fernández-Valparís, J.; Romea, P.; Urpí, F.; Font-Bardia, M. *Org. Lett.* **2017**, *19*, 6400–6403.
12. Delaunay, D.; Toupet, L.; Le Corre, M. *J. Org. Chem.* **1995**, *60*, 6604–6607.
13. Trost, B. M. *Science* **1991**, *254*, 1471–1477.
14. Evans, D. A.; Downey, C. W.; Hubbs, J. L. *J. Am. Chem. Soc.* **2003**, *125*, 8706–8707.
15. Kennington, S. C. D.; Taylor, A. J.; Romea, P.; Urpí, F.; Aullón, G.; Font-Bardia, M.; Ferré, L.; Rodrigalvarez, J. *Org. Lett.* **2019**, *21*, 305–309.

Appendix
Chemical Abstracts Nomenclature (Registry Number)

3-Amino-1-propanol: 1-Propanol, 3-amino-; (156-87-6)
Chlorosulfonic acid: Chlorosulfuric acid; (7790-94-5)
3-Ammoniopropylsulfate: 1-Propanol, 3-amino-, 1-(hydrogen sulfate); (1071-29-0)
Carbon disulfide: Carbon disulfide; (75-15-0)
1,3-Thiazinane-2-thione: 2*H*-1,3-Thiazine-2-thione, tetrahydro-; (5554-48-3)
Triethylamine: Ethanamine, *N*,*N*-diethyl-; (121-44-8)
Propionyl chloride: Propanoyl chloride; (79-03-8)
N-Propanoyl-1.3-thiazinane-2-thione: 1-Propanone, 1-(dihydro-2-thioxo-2*H*-1,3-thiazin-3(4*H*)-yl)-; (2138126-72-2)

Stuart C. D. Kennington, born in 1992 in Cambridgeshire, England, received his MChem degree from the University of Warwick in 2015. He is currently carrying out his Ph.D. thesis under the supervision of Prof. Fèlix Urpí and Pedro Romea at the University of Barcelona with a FI scholarship from the Generalitat de Catalunya. His research focuses on new catalyzed asymmetric synthesis methodologies and their application to the total synthesis of natural products.

Oriol Galeote, born in Barcelona in 1997, received his Degree in Chemistry from the University of Barcelona in 2019. He collaborated as an undergraduate internship in the direct and asymmetric construction of carbon–carbon bonds from N-acyl-1,3-thiazinane-2-thiones catalyzed by nickel(II) complexes under the supervision of Prof. Fèlix Urpí and Pedro Romea. He is currently enrolled in the Master in Organic Chemistry of the University of Barcelona.

Miguel Mellado-Hidalgo, born in Barcelona in 1996, received his Degree in Chemistry from the University of Barcelona in 2018. Then, he enrolled in the Master in Organic Chemistry in the same university, joining the group of Prof. Fèlix Urpí and Pedro Romea to study new direct and enantioselective aldol reactions from N-acyl-1,3-thiazinane-2-thiones catalyzed by nickel(II) chiral complexes. Currently, he is carrying out his PhD Thesis under their supervision, focusing his research on new catalytic and asymmetric methods and their application to the synthesis of natural products.

Pedro Romea completed his B.Sc. in Chemistry at the University of Barcelona and followed Ph.D. studies from 1987 to 1991 under the supervision of Professor Jaume Vilarrasa at the same University of Barcelona. Then, he joined the group of Professor Ian Paterson at the University of Cambridge (UK), where he participated in the total synthesis of oleandolide. Back to the University of Barcelona, he became Associate Professor in 1993. His research interests have focused on the development of new synthetic methodologies and their application to the stereoselective synthesis of naturally occurring molecular structures.

Fèlix Urpí received his B.Sc. in Chemistry at the University of Barcelona and completed Ph.D. studies under the guidance of Professor Jaume Vilarrasa at the University of Barcelona in 1988. He then worked as a NATO postdoctoral research associate in titanium enolate chemistry with Professor David A. Evans, at Harvard University in Cambridge, MA. He moved back to the University of Barcelona and he became Associate Professor in 1991, where he holds a chair of Full Professor in Organic Chemistry since 2017. His research interests have focused on the development of new synthetic methodologies and their application to the stereoselective synthesis of naturally occurring molecular structures.

Dr. Zhaobin Han received his B.S. degree in chemistry from Nanjing University in 2003. He received his Ph.D. degree from Shanghai Institute of Organic Chemistry under the supervision of Prof. Kuiling Ding and Prof. Xumu Zhang in 2009, working on development of novel chiral ligands for asymmetric catalysis. At present he is an Associate Professor in the same institute, and his current research interests focus on the development of efficient catalytic methods based on homogeneous catalysis.

Preparation of (Bis)Cationic Nitrogen-Ligated I(III) Reagents: Synthesis of [(pyridine)₂IPh](OTf)₂ and [(4-CF₃-pyridine)₂IPh](OTf)₂

Bilal Hoblos and Sarah E. Wengryniuk[1]*

Temple University, Department of Chemistry, 1901 North 13th Street, Philadelphia, PA, 19122, United States

Checked by Jeffrey T. Kuethe and Kevin R. Campos

Procedure (Note 1)

A. *[(Pyridine)₂IPh](OTf)₂* (**2**). A 100 mL three-necked (24/40) round-bottomed flask containing an egg-shaped Teflon-coated stir bar (12.5 mm × 25 mm) (Note 2) is evacuated (Note 3) and flame-dried (Note 4). After cooling to room temperature, the flask is backfilled with argon and charged with diacetoxyiodobenzene (PhI(OAc)₂) ; (2.00 g, 6.21 mmol, 1 equiv)

(Note 5), and the three necks are equipped with rubber septa. A positive pressure of argon is provided through a needle inserted into one of the septa. Diethyl ether (62 mL) (Note 6) is then added via syringe to generate a white, cloudy solution (Figure 1a). Stirring is initiated (Note 7), and, at room temperature, trimethylsilyl trifluoromethanesulfonate (2.25 mL, 12.4 mmol, 2 equiv) (Note 8) is then added via syringe over a period of 10 sec.

Figure 1. Appearance of reaction while stirring (a) before addition of TMSOTf, (b) immediately after full addition of TMSOTf (photos provided by checkers when using a single-necked flask)

The resulting mauve-colored, slightly cloudy mixture is stirred for 10 minutes, over which time the solution becomes homogenous and more yellow in appearance (Figure 1b). Pyridine (1.0 mL, 12.4 mmol, 2 equiv) (Note 9) is added via syringe over a period of 20 sec. Immediately upon addition of pyridine, the desired product precipitates as an off-white solid (Figure 2a). Stirring is continued for 15 min, which is the time required for any precipitate or residue stuck to the side of the flask to be liberated (Note 10). The septum on the leftmost neck of the reaction flask is replaced with a 24/40 glass vacuum adapter connected to a Schlenk manifold flowing a positive pressure of argon.

A second 100-mL three-necked (24/40), round-bottomed flask equipped with two glass stoppers and a glass vacuum adapter is evacuated (Note 3) and flame-dried (Note 4). A positive pressure of argon is provided via the adapter, and the flask is allowed to cool to room temperature. The stopper on the leftmost neck of the empty three-necked flask is removed and replaced by a dry double male joint filter stick (Note 11). The septum on the central neck of the reaction flask is then replaced with the other, shallow end of the filter stick (Figure 2b). The ground glass joints are wrapped with parafilm and/or secured with Keck clamps. The entire apparatus is then flipped, and the suspension is allowed to fall into the filter stick. The receiving flask is submerged in a −78 °C dry ice/acetone bath to limit filtrate evaporation. Argon flow to the reaction flask is increased and a vacuum is applied to the receiving flask via the corresponding vacuum adapters connected to the Schlenk manifold. After the solvent has fully passed through the material into the receiving flask, the vacuum line is closed, and the three-necked reaction flask is removed from the top of the filter stick to allow for the addition of rinsing solvent (diethyl ether (Note 6)) and agitation of the product cake. Between rinses, a dry one-necked flask is placed on the top of the filter stick (Figure 2b). The flask is then replaced with a female 24/40 glass vacuum adapter flowing a positive pressure of argon, and the vacuum line on the receiving flask is then opened, to allow solvent to pass through the filter. This process is performed twice with diethyl ether (2 × 25 mL), and the product is allowed to dry on the filter for an additional 2 min.

Figure 2. (a) Reaction mixture immediately after addition of pyridine (illustrated using a one-necked flask), and (b) reaction product during filtration (photos provided by checkers)

The product is quickly transferred to a pre-weighed, flame-dried, single-necked (24/40) 100 mL round-bottomed flask for drying. The flask is evacuated and left under constant vacuum for 1 h (Note 3), then backfilled with argon to reveal the product as a free-flowing, off-white powder (3.81 g, 91%) (Notes 12 and 13). The product is transferred to a flame-dried product vial (20 mL) for long term storage in a desiccator (Note 14).

B. *[(4-(Trifluoromethyl)pyridine)$_2$IPh](OTf)$_2$ (3)*. A 100-mL three-necked (24/40) round-bottomed flask containing an egg-shaped Teflon-coated stir bar (12.5mm × 25mm) (Note 2) is evacuated (Note 3) and flame-dried (Note 4). After cooling to room temperature, the flask is backfilled with argon and charged with diacetoxyiodobenzene (PhI(OAc)$_2$; 2.00 g, 6.21 mmol, 1 equiv) (Note 5), and the three necks are equipped with rubber septa, one of which is pierced with a needle that is delivering a positive pressure of argon. Diethyl ether (62 mL) (Note 6) is then added via syringe to generate a white, cloudy solution. Stirring is initiated (Note 7), then, at room temperature, trimethylsilyl trifluoromethanesulfonate (2.25 mL, 12.4 mmol, 2 equiv) (Note 8) is added via syringe over a period of 10 sec.

The resulting mauve colored, slightly cloudy mixture is stirred for 10 min, over which time the solution becomes homogeneous and yellow in appearance. 4-(Trifluoromethyl)pyridine (1.44 mL, 12.4 mmol, 2 equiv) (Note 9) is added by syringe over a period of 20 sec. Immediately upon addition of the heterocycle, the off-white product precipitates (Figure 3). Stirring is continued for 15 min, which is the time required for any precipitate or residue stuck to the side of the flask to be liberated (Note 10). The septum on the leftmost neck of the reaction flask is replaced with a 24/40 glass vacuum adapter connected to a Schlenk manifold flowing a positive pressure of argon.

To effect isolation of the product (Note 15) a second 100-mL three-necked (24/40), round-bottomed flask is equipped with two glass stoppers and a vacuum adapter. The flask is evacuated (Note 3) and flame-dried (Note 4). A positive pressure of argon is provided via the adapter, and the flask is allowed to cool to room temperature. The stopper on the leftmost neck of the empty three-necked flask is removed and replaced by the shallow end of a dry double male end adapter filter stick (Note 11). The septum on the central neck of the reaction flask is then replaced with the other end of the filter stick (Figure 2b). The ground glass joints are wrapped with parafilm and/or secured with Keck clamps. The entire apparatus is then flipped, and the suspension is allowed to fall into the filter stick, and the receiving flask is submerged in a –78 °C dry ice/acetone bath to limit filtrate evaporation. Argon flow to the reaction flask is increased and a vacuum is applied to the receiving flask via the corresponding vacuum adapters connected to the Schlenk line. After the solvent has fully passed through the material into the receiving flask, the vacuum line is closed, and the three-necked reaction flask is removed from the top of the filter stick to allow for the addition of rinsing solvent (diethyl ether (Note 6)) and agitation of the product cake. Between rinses, a dry one-necked flask is placed on the top of the filter stick (Figure 2b). The flask is then replaced with a female 24/40 glass vacuum adapter flowing a positive pressure of argon from the Schlenk manifold, and the vacuum line on the receiving flask is then opened, to allow solvent to pass through the filter. This process is performed twice with diethyl ether (2 × 25 mL), and the product is allowed to dry on the filter for an additional 2 min.

The product is quickly transferred to a pre-weighed, flame-dried, single-necked (24/40) 100 mL round-bottomed flask for drying. The flask is evacuated and left under constant vacuum for one h (Note 3), then backfilled with argon to reveal the product as a free-flowing, off-white powder (4.40 g, 89%) (Notes 16 and 17). The product is transferred to a flame-dried product vial (20 mL) for long term storage in a desiccator (Note 15).

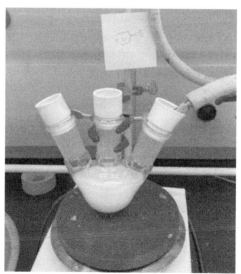

Figure 3. Reaction mixture immediately after addition of 4-CF$_3$-pyridine

Notes

1. Prior to performing each reaction, a thorough hazard analysis and risk assessment should be carried out with regard to each chemical substance and experimental operation on the scale planned and in the context of the laboratory where the procedures will be carried out. Guidelines for carrying out risk assessments and for analyzing the hazards associated with chemicals can be found in references such as Chapter 4 of "Prudent Practices in the Laboratory" (The National Academies Press, Washington, D.C., 2011; the full text can be accessed free of charge at https://www.nap.edu/catalog/12654/prudent-practices-in-the-laboratory-handling-and-management-of-chemical. See also "Identifying and Evaluating Hazards in Research Laboratories" (American Chemical Society, 2015) which is available via the associated website "Hazard Assessment in Research Laboratories" at https://www.acs.org/content/acs/en/about/governance/committees/chemicalsafety/hazard-assessment.html. In the case of this procedure, the risk assessment should include (but not necessarily be limited to) an evaluation of the potential hazards associated with (diacetoxyiodobenzene (PhI(OAc)$_2$), diethyl ether, trimethylsilyl trifluoromethanesulfonate, pyridine, 4-(trifluoromethyl)pyridine, dichloromethane, calcium hydride, calcium sulfate, acetone, dry ice, and methanol.
2. The shape of the stir bar is important as a thick slurry forms after the addition of the heterocycle, and the slurry can stick to the inner walls of the flask. The shape of the stir bar allows for greater surface contact with the flask and more uniform stirring of the slurry.
3. An Edwards RV12 direct-drive pump was used to create the vacuum (2-6 mmHg) during the evacuation of the glassware, as well as for the removal of residual solvent from products.
4. The checkers dried the reaction flask in an oven heated at 120 °C for 24 h prior to use and cooled the flask under an atmosphere of nitrogen. The submitters flame-dried the glassware. The flask (under vacuum via a needle through a rubber septum) is subjected to a torch flame from a bernzomatic-brand butane torch for 20 sec, ensuring all surfaces of the glass receive nearly equal heat. The flask is allowed to cool under vacuum (Note 3).
5. Diacetoxyiodobenzene, PhI(OAc)$_2$, was purchased by the submitters from Oakwood Chemical (98% purity) and used without further

purification. The checkers purchased PhI(OAc)$_2$ from Chem-Impex Int'L, Inc. (99.4%, HPLC).

6. The checkers purchased anhydrous diethyl ether from Sigma Aldrich in Sure/Seal™ bottles which were used as received. The submitter's solvents, including diethyl ether, were purchased from Fisher Scientific (HPLC grade passed through an activated alumina column). Diethyl ether is the preferred solvent for the examples provided herein. However, if the heterocycle used is found to be poorly soluble in diethyl ether, dichloromethane should be substituted since low yields and byproduct formation are observed when the heterocycle is not completely solubilized.

7. The stir plate used was a Corning PC351 and stir rate was around 600 rpm throughout the experiment

8. Trimethylsilyl trifluoromethanesulfonate (TMSOTf) was purchased by the submitters from Oakwood Chemical (99% reagent grade). The reagent was distilled over calcium hydride and stored over activated 3Å molecular sieves (Sigma Aldrich). The checkers purchased TMSOTf from Sigma Aldrich in sealed ampules and were used as received.

9. Pyridine was purchased by the submitters from Fisher Scientific (>99% ACS grade), distilled over calcium hydride, and stored over activated 3Å molecular sieves. The checkers purchased anhydrous pyridine from Sigma Aldrich in Sure/Seal™ bottles which were used as received. Both submitters and checkers purchased 4-CF$_3$-pyridine (99%) from Oakwood. The material was purified and stored in the same fashion as described above for pyridine.

10. In some cases, the stir bar was unable to reach the solids that were stuck to the walls of the flask. In this case, the flask was removed from the clamp and manually agitated (swirled and lightly shaken) until the walls of the flask were sufficiently cleared of all residue.

11. Complex glassware was dried by placement in an oven (150 °C) and was cooled in a glass desiccator filled with calcium sulfate (Drierite) dessicant.

12. Characterization data for *Py*-HVI (**2**): ^{1}H NMR (500 MHz, CD$_3$OD) δ: 7.67 (t, 2H, *J* = 7.7 Hz), 7.80 (t, 1H, *J* = 7.4 Hz), 7.95 (t, 4H, *J* = 5.7 Hz), 8.31 (d, 2H, *J* = 8.0 Hz), 8.48 (t, 2H, *J* = 7.8 Hz), 8.79 (d, 4H, *J* = 5.7 Hz); ^{13}C NMR (125 MHz, CD$_3$OD) δ: 121.7 (q, *J* = 396.2 Hz), 122.4, 128.2, 132.8, 134.5, 136.4, 144.5; 146.0;. ^{19}F NMR (470 MHz, CD$_3$OD) δ: –80.08; IR (ATR): 3084, 2360, 1605, 1242, 1155, 1025, 753, 681, 630. Elemental

Analysis: calc'd; C: 32.74, H: 2.29, N: 4.24, Found; C: 31.95, H: 2.38, N: 4.06 (submitter's data) Due to instability and rapid decomposition, the checkers were unable to obtain purity data. Due to inherent instability, melting point was not determined.

13. Two additional reactions checked on the same scale provided 3.82 g (93%) and 3.81 g (93%), respectively.
14. The product should be stored in a cool, dry desiccator to avoid degradation into an undesired μ_2-oxo-species (Figures 4 and 5). Degradation from exposure to moisture is accelerated for *N*-HVIs possessing more sterically hindered or electron-deficient heterocyclic ligands. This degradation is pictured below for the *2-OMe-Py*-HVI, in which a small sample was placed on a watch glass and degradation was monitored both qualitatively, by color change (Figure 4), and by ^{1}H-NMR spectroscopy (Figure 5) at 5-min intervals. Partial degradation was observed after just 5 min, and complete disappearance of desired product was observed within 10 min of exposure to air inside a fume hood.

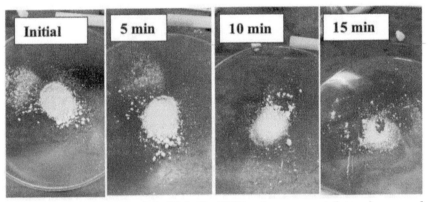

Figure 4. Degradation of *2-OMe-Py*-HVI when exposed to air on a dry watch glass at various time points (photos provided by submitter)

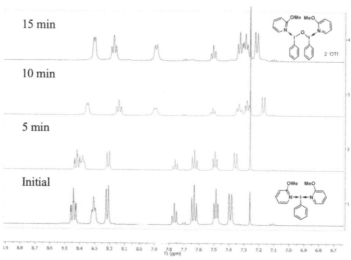

Figure 5. Degradation of 2-*MeO-Py*-HVI monitored by ¹HNMR at 5 min intervals (Figure provided by submitters)

15. The submitters used the following procedure in an air- and moisture-free glovebox for the isolation of the product. The flask containing the precipitated product is evacuated and purged with argon 5 times, and the rubber septum is securely fixed to the neck with electrical tape. The flask and its contents are then transferred to a glovebox (Note 18) for inert filtration under a nitrogen atmosphere (Figure 6a). Contents of the flask are poured onto a 4-5 µm glass fritted filter (Figure 6b) (Note 19), which is placed on a 250 mL filter flask, at which time suction is applied via vacuum (Figure 6c) (Note 20). The submitters used an air- and moisture-free glovebox (<0.5 ppm each) for the isolation of the product. The flask is evacuated and purged with argon 5 times, and the rubber septum is securely fixed to the neck with electrical tape. The flask and its contents are then transferred to a glovebox (Note 18) for inert filtration under a nitrogen atmosphere (Figure 4a). Contents of the flask are poured onto a 4-5 µm glass fritted filter (Figure 6b) (Note 19), which is placed on a 250 mL filter flask, at which time suction is applied via vacuum (Figure 6c) (Note 20). The reaction flask is rinsed with diethyl ether (25 mL) (Note 6) to remove any residual solids, and these solids are collected on the filter. The combined solids are rinsed with additional diethyl ether (35 mL). Following filtration, the damp product is transferred via spatula to a pre-weighed, single-necked (24/40) 100

mL round-bottomed flask for drying (Figure 6d). The flask is evacuated and left under constant vacuum for 1 h to reveal the product as a free-flowing, off-white powder (4.40 g, 89%) (Notes 14 and 15). The product can then be transferred to a flame-dried product vial (20 mL) for long term storage in moisture-free conditions such as a glovebox or desiccator (Note 14).

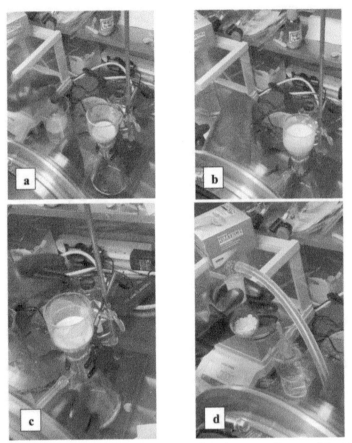

Figure 6. *N*-HVI transferred into glovebox in 250 mL round-bottomed flask prior to filtration; (b) solution of *N*-HVI suspended in solvent poured onto filter prior to application of suction; (c) *N*-HVI after removal of solvent via vacuum filtration and (d) *N*-HVI product transferred to 100 mL round-bottomed flask to further dry under vacuum (photos provided by submitters)

16. Characterization data for 4-CF$_3$-Py-HVI (**3**): ^{1}H NMR (500 MHz, CD$_3$OD) δ: 7.75 (t, 2H, J = 7.8 Hz), 7.88 (t, 1H, J = 7.5 Hz), 8.14 (d, 4H, J = 6.3 Hz), 8.41 (d, 2H, J = 7.4 Hz), 9.01 (d, 4H, J = 6.3 Hz); ^{13}C NMR (125 MHz, CD$_3$OD) δ: 121.3 (q, J = 316.2 Hz), 121.7, 123.0 (q, J = 271.0 Hz), 123.4 (q, J = 3.8 Hz), 132.9, 135.1, 137.1, 143.8 (q, J = 35.0 Hz), 147.3; ^{19}F NMR (470 MHz, CD$_3$OD) δ: –66.82, –80.09; IR (ATR): 3100, 3060, 3032, 1433, 1320, 1250, 1145, 1024, 840, 632. Elemental Analysis: calc'd; C: 30.17, H: 1.65, N: 3.52. Found; C: 29.65, H: 1.66, N: 3.21 (submitter's data) Due to instability and rapid decomposition, the checkers were unable to obtain purity data. Due to inherent instability, melting point was not determined.
17. A second reaction performed the same scale provided 4.43 g (90%) of the same product.
18. The glovebox used is a Vacuum Atmospheres NexGen system. A glovebox is required to obtain high yields of chemically pure N-HVIs containing more sterically hindered or electron deficient heterocyclic ligands. The presence of trace water is known to cause rapid degradation, and therefore any exposure to atmospheric moisture during filtration or transfer for storage leads to depreciation of yield and reagent quality.
19. Due to the fine nature of the powder product, a fritted filter (150 or 250 mL) with small pore size (4–5 μm) is required to capture all solids. Use of a 60 μm pore size led to loss of a significant amount of material through the filter.
20. The vacuum used for glovebox filtration is a Welch Systems belt drive vacuum pump (8–11 mmHg).

Working with Hazardous Chemicals

The procedures in *Organic Syntheses* are intended for use only by persons with proper training in experimental organic chemistry. All hazardous materials should be handled using the standard procedures for work with chemicals described in references such as "Prudent Practices in the Laboratory" (The National Academies Press, Washington, D.C., 2011; the full text can be accessed free of charge at http://www.nap.edu/catalog.php?record_id=12654). All chemical waste should be disposed of in accordance with local regulations. For general

guidelines for the management of chemical waste, see Chapter 8 of Prudent Practices.

In some articles in *Organic Syntheses*, chemical-specific hazards are highlighted in red "Caution Notes" within a procedure. It is important to recognize that the absence of a caution note does not imply that no significant hazards are associated with the chemicals involved in that procedure. Prior to performing a reaction, a thorough risk assessment should be carried out that includes a review of the potential hazards associated with each chemical and experimental operation on the scale that is planned for the procedure. Guidelines for carrying out a risk assessment and for analyzing the hazards associated with chemicals can be found in Chapter 4 of Prudent Practices.

The procedures described in *Organic Syntheses* are provided as published and are conducted at one's own risk. *Organic Syntheses, Inc.*, its Editors, and its Board of Directors do not warrant or guarantee the safety of individuals using these procedures and hereby disclaim any liability for any injuries or damages claimed to have resulted from or related in any way to the procedures herein.

Discussion

Hypervalent iodine compounds have emerged as a versatile, non-toxic, environmentally benign class of reagents class that can often serve as alternatives to transition metal-mediated transformations.[2] In particular, I(III) species serve as mild oxidants, group transfer agents, and electrophilic activators that enable a myriad of synthetic transformations. These reagents feature a central aryl iodine that shares a hypervalent 3c-4e$^-$ bond with two X-ligands, which are most commonly either carboxylates (i.e. $PhI(OAc)_2$, $PhI(OTFA)_2$) or halogens (i.e. $PhICl_2$).

Over the past several years, our laboratory has been exploring a class of (bis)cationic nitrogen-ligated I(III) reagents possessing two datively bound aromatic nitrogen heterocycles as X-ligands, which we have termed N-HVIs. The first examples of these reagents were reported in 1994 by Weiss,[3] but the subsequent 20 years saw almost no explorations into their synthetic utility.[4-8] N-HVI's undergo facile ligand exchange with nucleophiles, show enhanced oxidation potentials, and the nitrogen heterocycles provide excellent handles for tuning reactivity and selectivity. Several groups have reported on their utility in a variety of oxidative transformations, pi-bond activation, or in accessing high-valent metal complexes, representative examples of which are shown in Figure 7. Efforts in our laboratory have found that N-HVIs can activate alcohols for oxidative ring expansions to access medium-ring ethers,[9,10] or enable mild, equatorial-selective oxidation to the corresponding carbonyls,[11] transformations not accessible with other I(III) species. More recent studies by our group have unlocked their potential as "heterocyclic group transfer" reagents for the synthesis of diverse (heteroaryl)onium salts.[12]

In light of ever-expanding suite of transformations enabled by N-HVIs, the disclosure of a detailed protocol for their synthesis and isolation that includes several useful updates to the original procedure seemed timely. The procedure above highlights the preparation of two members of this class of compounds. Procedure A is generally applied to the synthesis of N-HVI reagents with heterocyclic ligands that are (1) non-sterically hindered, (2) electron neutral, or (3) electron-rich. Procedure B is generally applied to the synthesis of N-HVI reagents with heterocyclic ligands that are (1) sterically hindered and (2) electron deficient.

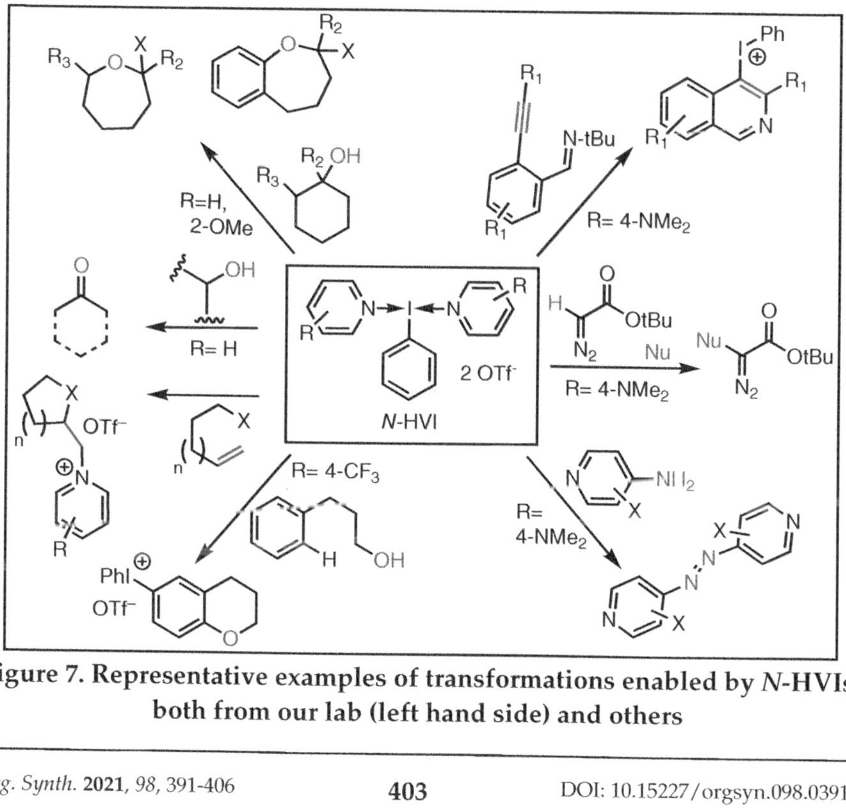

Figure 7. Representative examples of transformations enabled by N-HVIs both from our lab (left hand side) and others

Following Weiss' original report, N-HVIs can be readily accessed from commercial PhI(OAc)$_2$ via activation with a silyl triflate followed by addition of the heterocycle of choice.[3] This leads to precipitation of the N-HVI as the bistriflate salt which can then be collected via filtration. A key feature in working with N-HVIs is their moisture sensitivity, which varies depending on the steric and electronic nature of the heterocyclic ligand. There are conflicting reports regarding the stability of N-HVIs, however we have found that, with proper technique, a wide variety of N-HVIs can be reliably synthesized and handled either on the benchtop or using inert atmosphere conditions. In this report, we detail two strategies for the synthesis and isolation of N-HVIs with varying levels of bench stability including both a benchtop and a glovebox isolation. Furthermore, it was found that substitution of diethyl ether for dichloromethane improves the isolation and subsequent stability of the N-HVIs shown in this report. It is our aim that with this *Organic Syntheses* protocol we can demystify the synthesis and handling of N-HVIs, opening the door to more development and application of this valuable reagent class in organic synthesis.

References

1. Temple University, Department of Chemistry, 1901 N. 13th Street, Philadelphia, Pennsylvania, 19122, United States. Email: sarahw@temple.edu ORCID: 0000-0002-4797-0181. The authors are grateful to the National Institutes of Health NIGMS (R01-GM123098) and the National Science Foundation (CAREER 1752244) for financial support of this work and would like to acknowledge the Donors of the American Chemical Society Petroleum Research Fund for their support of this research (DNI-56603).
2. Yoshimura, A.; Zhdankin, V. V. *Chem. Rev.* **2016**, *116*, 3328–3435.
3. Weiss, R.; Seubert J. *Angew. Chem. Int. Ed.* **1994**, *33*, 891–893.
4. Corbo, R.; Dutton, J. L. *Coord. Chem. Rev.* **2018**, *375*, 69–79.
5. De Mico, A.; Margarita, R.; Piancatelli, G. *Gazz. Chim. Ital.* **1995**, *215*, 325.
6. Zhdankin, V. V.; Maydanovych, O.; Herschbach, J.; Bruno, J.; Matveeva, E. D.; Zefirov, N. S. *J. Org. Chem.* **2003**, *68*, 1018–1023.
7. Kniep, F.; Walter, S. M.; Herdtweck, E.; Huber, S. M. *Chem. - Eur. J.* **2012**, *18*, 1306–1310.

8. Yuan, Z.; Cheng, R.; Chen, P.; Liu, G.; Liang, S. H. *Angew. Chem., Int. Ed.* **2016**, *55*, 11882–11886.
9. Kelley, B. T.; Walters, J. C.; Wengryniuk, S. E. *Org. Lett.* **2016**, *18*(8), 1896–1899.
10. Walters, J. C.; Tierno, A. F.; Dubin, A. H.; Wengryniuk, S. E. *Eur. J. Org. Chem.* **2018**, *12*, 1460–1464.
11. Mikhael, M.; Adler, S. A.; Wengryniuk, S. E. *Org. Lett.* **2019**, *21*(15) 5889–5893.
12. Tierno, A. F., Walters, J. C., Vazquez-Lopez, A., Xiao X., Wengryniuk, S. E. *Chem. Sci.*, **2021**, *12*, 6385–6392.

Appendix
Chemical Abstracts Nomenclature (Registry Number)

Diacetoxyiodobenzene: Iodine, bis(acetato-κO)phenyl-; (3240-34-4)
Trimethylsilyl trifluoromethanesulfonate: Methanesulfonic acid, 1,1,1-trifluoro-, trimethylsilyl ester; (27607-77-8)
Pyridine: pyridine; (110-86-1)
2-Methoxypyridine: Pyridine, 2-methoxy-; (1628-89-3)
4-(Dimethylamino)pyridine: 4-Pyridinamine, *N*,*N*-dimethyl-; (1122-58-3)
4-(Trifluoromethyl)pyridine: Pyridine, 4-(trifluoromethyl)-; (3796-24-5)
Py-HVI , [(pyridine)$_2$IPh](OTf)$_2$: Iodine(2+), phenylbis(pyridine)-, 1,1,1-trifluoromethanesulfonate (1:2); (156002-39-0)
4-*CF$_3$-Py*-HVI, [(4-(trifluoromethyl)pyridine)$_2$IPh](OTf)$_2$: (No CAS Registry number found)

Bilal Hoblos obtained his BS in chemistry from The Pennsylvania State University, Behrend Campus in 2016. He then moved to Temple University in Philadelphia, PA to pursue a PhD under the guidance of Dr. Sarah Wengryniuk. His doctoral work focuses on developing divergent total syntheses of the heliannuol family of natural products utilizing via hypervalent iodine mediated oxidative rearrangements developed in Dr. Wengryniuk's laboratory.

Dr. Sarah Wengryniuk obtained her B.S. in chemistry and biology from Winthrop University in Rock Hill, SC in 2007. She received her Ph.D. in 2012 under the guidance of Prof. Don Coltart at Duke University where she was supported as an NSF Graduate Fellow. After completing postdoctoral training in the laboratory of Prof. Phil S. Baran at The Scripps Research Institute as an NIH Ruth L. Kirchstein fellow, she began her independent career at Temple University in 2015 where she is currently an assistant professor. Her laboratory works on the development of novel reverse-polarity transformations enabled by hypervalent iodine reagents.

Jeff Kuethe studied chemistry at Middle Tennessee State University where he received a Bachelor of Science in 1993. He then joined the group of Professor Albert Padwa at Emory University in Atlanta, Georgia where he received a Ph.D. in 1998. He continued as a postdoctoral fellow in the group of Professor Daniel Comins at North Carolina State University before joining the Department of Process Research at Merck & Co., Inc., Rahway, New Jersey in 2000. His research interests include Process research, synthetic methodology, heterocyclic chemistry, alkaloid and natural product synthesis, and tandem transformations.

Preparation of 6-(Triethylsilyl)cyclohex-1-en-1-yl Trifluoromethanesulfonate as a Precursor to 1, 2-Cyclohexadiene

Ryo Nakura, Kazuki Inoue, Mayu Itoh, Atsunori Mori, and Kentaro Okano[*1]

Department of Chemical Science and Engineering, Kobe University, 1-1 Rokkodai, Nada, Kobe 657-8501, Japan

Checked by Zhixun Wang and Kevin Campos

Procedure (Note 1)

A. *(Cyclohex-1-en-1-yloxy)triethylsilane (2)*. A 250-mL three-necked round-bottomed flask is equipped with a Teflon-coated magnetic stir bar (3.5 × 1.5 cm), an inlet adapter with a two-way stopcock, and two rubber septa, one of which with a thermocouple inserted. After charged with sodium iodide (9.0 g, 60 mmol, 1.2 equiv) (Notes 2 and 3), the flask is evacuated and backfilled with nitrogen and is charged with anhydrous acetonitrile (80 mL) (Note 4), cyclohexanone (4.9 g, 50 mmol) (Note 5), and triethylamine (8.4 mL, 60 mmol, 1.2 equiv) (Note 6). Triethylsilyl chloride (9.2 mL, 55 mmol, 1.1 equiv) (Note 7) is added to the flask via a syringe through the septum (Figure 1). After stirring at ambient temperature for 3 h (Note 8), the reaction is quenched with water (50 mL).

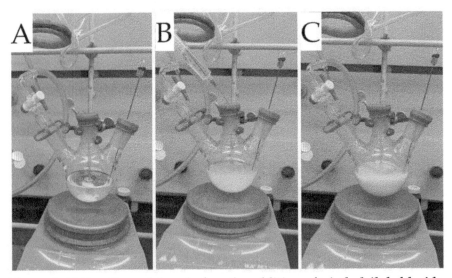

Figure 1. A) Reaction mixture before the addition of triethylsilyl chloride, B) Reaction mixture during the addition of triethylsilyl chloride, C) Reaction mixture after the addition of triethylsilyl chloride (photos provided by checkers)

The reaction mixture is transferred into a 500-mL separatory funnel. After partitioning between *n*-hexane (100 mL) and water twice, the combined organic extracts are washed with sat. aq. NaCl solution (1 × 100 mL), dried over Na_2SO_4 (25 g), and filtered. The filtrate is concentrated on a rotary evaporator under reduced pressure (15 °C, 10 mmHg), and the residue is dried *in vacuo* to afford 10.3–11.5 g of a crude product as a colorless oil. The crude product is azeotroped with anhydrous THF (2 × 25 mL) on a rotary evaporator under reduced pressure (40–50 °C, 100 to 20 mmHg) to remove volatiles and water. The resulting crude product (10.84 g) is obtained as a colorless oil in 93% purity (Note 9). The material obtained through this method is used in Step B (Notes 10 and 11) (Figure 2).

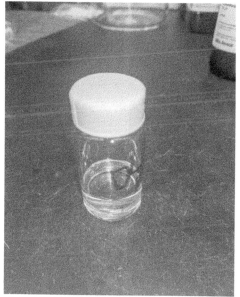

Figure 2. (Cyclohex-1-en-1-yloxy)triethylsilane (2) obtained after azeotropic removal with anhydrous THF (photo provided by checkers)

B. *6-(Triethylsilyl)cyclohex-1-en-1-yl trifluoromethanesulfonate (5).* An oven-dried 1-L three-necked flask equipped with two 125-mL pressure-equalizing dropping funnels with a rubber septum, a Teflon-coated magnetic stir bar (4.5 × 1.5 cm), and a rubber septum inserted with a nitrogen gas inlet and a thermocouple is charged with *t*-BuOK (7.01 g, 62.5 mmol, 2.5 equiv) (Note 12) and anhydrous *n*-hexane (100 mL) (Note 13). To the suspension is added LDA (1.0 M in THF/hexanes, 62.5 mL, 62.5 mmol, 2.5 equiv) (Note 14) dropwise during 5 min through the pressure-equalizing dropping funnel, and the resulting mixture is stirred at ambient temperature for 30 min (Figure 3).

Figure 3. A) LDA being transferred into the dropping funnel via cannula, B) Reaction mixture after addition of LDA, C) Reaction mixture after addition of (cyclohex-1-en-1-yloxy)triethylsilane and stirring at ambient temperature for 30 min (photos provided by checkers)

To the reaction mixture is added (cyclohex-1-en-1-yloxy)triethylsilane (**2**) (5.73 g of 93% purity, 25.0 mmol) via a syringe during 1 min through the septum (Figure 3). After stirring at ambient temperature for 1.0–2.5 h (Note 15), the reaction mixture is diluted with anhydrous THF (100 mL) (Note 16). After the flask is cooled to –78 °C with a dry ice-acetone bath, Comins' reagent (19.6 g, 50.0 mmol, 2.0 equiv) (Note 17) in anhydrous THF (75 mL) is added to the flask dropwise (Note 18) through the other pressure-equalizing dropping funnel (Figure 4).

Figure 4. A) Reaction mixture before addition of Comins' reagent in THF, B) Reaction mixture after addition of Comins' reagent in THF at −78 °C, C) Reaction mixture during the warming to ambient temperature (photos provided by checkers)

The cooling bath is removed after 10 min. After stirring at ambient temperature for 1 h (Note 19), the reaction is quenched with water (100 mL) (Figure 5). The reaction mixture is transferred into a 1-L separatory funnel. After partitioning between *n*-hexane (150 mL) and water three times, the combined organic extracts are washed with sat. aq. NaCl solution (1 × 150 mL), dried over Na_2SO_4 (25 g), and filtered. The filtrate is concentrated on a rotary evaporator under reduced pressure (15 °C, 10 mmHg), and the residue is dried *in vacuo* to afford 12.4–13.5 g of a crude product as a brown oil, which is purified by column chromatography on silica gel while eluting with *n*-hexane (Note 20) to provide 6-(triethylsilyl)cyclohex-1-en-1-yl trifluoromethanesulfonate (**5**) as a colorless oil (7.03 g, 82%) (Note 21) (Figure 6).

Figure 5. Reaction mixture after quench with water (photo provided by checkers)

Figure 6. 6-(Triethylsilyl)cyclohex-1-en-1-yl trifluoromethanesulfonate (5) (photo provided by checkers)

C. *9,10-Diphenyl-1,2,3,9,9a,10-hexahydro-9,10-epoxyanthracene* (**8**). A 500 mL one-necked round-bottomed flask equipped with a Teflon-coated magnetic stir bar (4.5 × 1.5 cm) is charged with 1,3-diphenylisobenzofuran (6.08 g, 22.5 mmol, 1.5 equiv) (Note 22), 6-(triethylsilyl)cyclohex-1-en-1-yl trifluoromethanesulfonate (**5**) (5.5 g of 97% purity, 15.5 mmol, 1 equiv), and THF (165 mL). A thermocouple is inserted into the reaction mixture. Tetrabutylammonium fluoride (1.0 M in THF, 22.5 mL, 22.5 mmol, 1.5 equiv) (Note 23) is added to the flask over a period of 15 min via a syringe (Note 24) (Figure 7), after which the thermocouple is removed.

Figure 7. Reaction mixture after addition of tetrabutylammonium fluoride (photo provided by submitters)

After stirring at ambient temperature (22–24 °C) for 1.5 h (Note 25), the reaction is quenched with water (150 mL). The reaction mixture is transferred into a 1-L separatory funnel. After partitioning between Et$_2$O (60 mL) and water twice, the combined organic extracts are washed with sat. aq. NaCl solution (1 × 100 mL), dried over Na$_2$SO$_4$ (20 g), and filtered. The filtrate is concentrated on a rotary evaporator under reduced pressure (15 °C, 10 mmHg), and the residue is dried *in vacuo* to afford 13.9–15.0 g of a crude product as a yellow solid, which is purified by column chromatography on silica gel (Note 26) to provide 9,10-diphenyl-1,2,3,9,9a,10-hexahydro-9,10-epoxyanthracene (**8**) as a pale yellowish green solid (3.24–3.30 g, 60–61%) (Note 27) (Figure 8).

Figure 8. Concentrated product after flash column chromatography (photo provided by checkers)

The resulting solid is recrystallized from boiling ethanol (200 mL) (Note 28) in a 500-mL one-necked round-bottomed flask equipped with a reflux condenser and a Teflon-coated magnetic stir bar (4.5 × 1.5 cm). Crystallization is allowed to occur at room temperature over 30 min, then at 0 °C. The crystals are collected by filtration and washed with cold (0 °C) ethanol (20 mL) to provide the title compound **8** (*endo/exo* >99/1) as a colorless solid (2.43 g, 45%) (Note 29) (Figure 9).

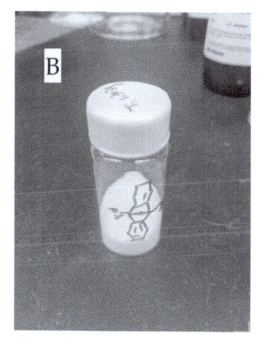

Figure 9. A) Filtration of product (8), and B) appearance of recrystallized product (photos provided by checkers)

Notes

1. Prior to performing each reaction, a thorough hazard analysis and risk assessment should be carried out with regard to each chemical substance and experimental operation on the scale planned and in the context of the laboratory where the procedures will be carried out. Guidelines for carrying out risk assessments and for analyzing the hazards associated with chemicals can be found in references such as Chapter 4 of "Prudent Practices in the Laboratory" (The National Academies Press, Washington, D.C., 2011; the full text can be accessed free of charge at https://www.nap.edu/catalog/12654/prudent-practices-in-the-laboratory-handling-and-management-of-chemical. See also "Identifying and Evaluating Hazards in Research Laboratories" (American Chemical Society, 2015) which is available via the associated website "Hazard Assessment in Research Laboratories" at

https://www.acs.org/content/acs/en/about/governance/committees/chemicalsafety/hazard-assessment.html. In the case of this procedure, the risk assessment should include (but not necessarily be limited to) an evaluation of the potential hazards associated with cyclohexanone, triethylsilyl chloride, sodium iodide, triethylamine, acetonitrile, lithium diisopropylamide, potassium tert-butoxide, hexane, tetrahydrofuran, Comins' reagent, tetrabutylammonium fluoride, 1,3-diphenylisobenzofuran, as well as the proper procedures for setting up experimental operations.

2. Sodium iodide (>99.5%) was purchased from Wako Pure Chemical Industries, Ltd. and used pre-dried at 90 °C (bath temperature) with an oil-bath for 18 h under ca. 1 mmHg (Figure 10).

Figure 10. Setup of pre-drying of sodium iodide
(photos provided by submitters)

3. Commercially available anhydrous sodium iodide (>99.5%), purchased from Sigma-Aldrich Co., LLC by the checkers, works as well as pre-dried material described in Note 2.
4. Acetonitrile (>99.8%, water content: < 0.001%) was purchased from FUJIFILM Wako Pure Chemical Corporation and used as received without further purification.
5. Cyclohexanone (>98.0%) was purchased from Nacalai Tesque Co., Inc. and used as received without further purification.
6. Triethylamine (>98.0%) was purchased from Nacalai Tesque Co., Inc. and used as received without further purification.
7. Triethylsilyl chloride (>97.0%) was purchased from Tokyo Kasei Kogyo Co., Inc. and used as received without further purification.

8. The reaction typically requires 3 h to consume all the cyclohexanone and is monitored by TLC analysis on Merck silica gel 60 F_{254} plates developing with *n*-hexane/methyl acetate (9:1). The R_f values of cyclohexanone (**1**) and (cyclohex-1-en-1-yloxy)triethylsilane (**2**) are 0.16 and 0.68, respectively (stained with an ethanol solution of *p*-anisaldehyde). After dipping the TLC plate to the *p*-anisaldehyde solution, the chromatogram is stained by heating.

9. (Cyclohex-1-en-1-yloxy)triethylsilane (**2**): colorless oil; R_f = 0.84 (*n*-hexane/ethyl acetate = 9:1); Merck silica gel 60 F_{254} plates (stained with an ethanol solution of *p*-anisaldehyde). After dipping the TLC plate to

Figure 11. TLC of cyclohexanone (left), reaction mixture (right), and their combination spot (middle). Mobile phase: *n*-hexane/ethyl acetate = 9:1 (photo provided by checkers)

the *p*-anisaldehyde solution, the chromatogram is stained by heating (Figure 11); IR (neat): 2933, 2876, 1667, 1187, 1004, 742, 727 cm^{-1}; ^{1}H NMR (500 MHz, CDCl$_3$) δ: 0.65 (q, *J* = 7.9 Hz, 6H), 0.98 (t, *J* = 7.9 Hz, 9H), 1.48 – 1.53 (m, 2H), 1.60 – 1.70 (m, 2H), 1.92 – 2.07 (m, 4H), 4.87 (ddd, *J* = 3.9, 2.6, 1.3 Hz, 1H); ^{13}C NMR (126 MHz, CDCl$_3$) δ: 5.1, 6.7, 22.4, 23.2, 23.9, 29.9, 103.9, 150.4; The purity of the sample was determined to be 93% by ^{1}H qNMR using 32.7 mg of 1,3,5-trimethoxybenzene (Note 30) as an internal standard and 58.9 mg of (cyclohex-1-en-1-yloxy)triethylsilane (**2**), therefore the amount of product in the reaction is 10.08 g. A second reaction on the identical scale provided 11.05 g of the unpurified product with 93% purity.

10. The azeotropic drying method described in Step A was developed as an alternative to column chromatography. The checkers observed that (cyclohex-1-en-1-yloxy)triethylsilane (**2**) undergoes decomposition on triethylamine-basified silica gel during flash column chromatography (described in Note 11), which led to lower yield and purity (79% yield, 85% purity).

11. The submitters reported the chromatographic purification of the crude material. The crude product was dissolved in 0.5% (v/v) triethylamine in cyclohexane (30 mL) and was charged onto a column (diameter = 10 cm, height = 5 cm) of 95-gram silica gel. The column was eluted with 0.5% triethylamine in cyclohexane, and 90-mL fractions were collected. Fractions 8–30 were combined and concentrated on a rotary evaporator under reduced pressure (40 °C, 10 mmHg). The submitters used silica gel purchased from FUJIFILM Wako Pure Chemical Corporation (pH = 6.5–7.5, particle size 0.063–0.212 mm), which was used as received.

12. *t*-BuOK (>95.0%) was purchased from Tokyo Kasei Kogyo Co., Inc. and used as received without further purification.

13. Hexane (>96.0%, water content: < 30 ppm) was purchased from Nacalai Tesque Co., Inc. and used as received without further purification

14. LDA (1.0 M in THF/hexanes) was purchased from Sigma-Aldrich Co., LLC and used as received without further purification. Slightly excess amount (1.5 equiv) of LDA resulted in incomplete consumption of the starting silyl enol ether.

15. The reaction typically requires 1.0–2.5 h to consume all the (cyclohex-1-en-1-yloxy)triethylsilane and is monitored by TLC analysis on Merck silica gel 60 F_{254} plates developing with *n*-hexane/methyl acetate (9:1). The R_f values of (cyclohex-1-en-1-yloxy)triethylsilane (**2**) and 2-(triethylsilyl)cyclohexan-1-one are 0.78 and 0.25, respectively (stained with an ethanol solution of *p*-anisaldehyde). After dipping the TLC plate to the *p*-anisaldehyde solution, the chromatogram is stained by heating (Figure 12).

Figure 12. TLC of (cyclohex-1-en-1-yloxy)triethylsilane (left), reaction mixture (right), and their combination spot (middle). Mobile phase: *n*-hexane/methyl acetate = 9:1 (photo provided by submitters)

16. THF (>99.5%, water content: < 10 ppm) was purchased from Kanto Chemical Co., Inc. and further dried by passing through a solvent purification system (Glass Contour) prior to use.
17. Comins' reagent (>99.0%) was purchased from Oakwood Products, Inc. and used as received without further purification.
18. An exotherm (from –72 to –64 °C) was observed during the 15 min addition time at the reported scale. A larger scale reaction than that described in this procedure may require a longer addition time to control the exotherm.
19. The reaction is monitored by TLC analysis on Merck silica gel 60 F_{254} plates developing with *n*-hexane/ethyl acetate (9:1). The R_f values of (cyclohex-1-en-1-yloxy)triethylsilane and 6-(triethylsilyl)cyclohex-1-en-1-yl trifluoromethanesulfonate (**5**) are 0.90 and 0.50, respectively (stained with an ethanol solution of *p*-anisaldehyde). After dipping the TLC plate to the *p*-anisaldehyde solution, the chromatogram is stained by heating (Figure 13). The submitters recommend the use of 2-(triethylsilyl)cyclohexan-1-one to monitor the reaction progress.

Figure 13. TLC of (cyclohex-1-en-1-yloxy)triethylsilane (2, left), cyclohexanone (1, right), reaction mixture containing 5 (middle), the combination of 2 and reaction mixture (2nd from left), the combination of 1 and reaction mixture (2nd from right). Mobile phase: *n*-hexane/ethyl acetate = 9:1 (photo provided by checkers)

20. The crude product is dissolved in *n*-hexane/CH_2Cl_2 = 1:1 (10 mL) and is charged onto a column (diameter = 6.5 cm, height = 9.0 cm) of 59-gram silica gel. The column is eluted with *n*-hexane, and 60-mL fractions are collected. Fractions 7–69 are combined and concentrated on a rotary evaporator under reduced pressure (40 °C, 5 mmHg). The submitters used silica gel purchased from FUJIFILM Wako Pure Chemical Corporation (pH = 6.5–7.5, particle size 0.063–0.212 mm), which was used as received. The checkers used silica gel purchased from Fisher Scientific (pH = 4–5, particle size 0.040–0.063 mm), which was used as received. The yield of **5** was not significantly reduced, when acidic silica gel (pH = 4–5) was used as a stationary phase by the checkers.

21. 6-(Triethylsilyl)cyclohex-1-en-1-yl trifluoromethanesulfonate (**5**): colorless oil; R_f = 0.36 (*n*-hexane); Merck silica gel 60 F_{254} plates (stained with an ethanol solution of *p*-anisaldehyde). After dipping the TLC plate to the *p*-anisaldehyde solution, the chromatogram is stained by heating; IR (film): 2954, 2878, 1413, 1201, 1141, 1016, 891, 730 cm^{-1}; ^{1}H NMR (500 MHz, CDCl$_3$) δ: 0.64 – 0.70 (m, 6H), 0.97 (t, *J* = 8.0 Hz, 9H), 1.43 – 1.54 (m, 1H), 1.62 – 1.73 (m, 2H), 1.90 – 2.01 (m, 1H), 2.02 – 2.15 (m, 2H), 2.15 – 2.27 (m, 1H), 5.65 (ddd, *J* = 5.2, 3.3, 1.9 Hz, 1H); ^{13}C NMR (126 MHz, CDCl$_3$) δ: 2.9, 7.4, 21.4, 24.2, 25.5, 26.0, 115.1, 118.7 (q, $^1J_{C-F}$ = 321 Hz), 153.2; ^{19}F NMR (471 MHz, CDCl$_3$) d: –73.50; The purity of the

sample (>97%) was determined by ^{1}H qNMR using 23.3 mg of 1,3,5-trimethoxybenzene (Note 30) as an internal standard and 42.7 mg of 6-(triethylsilyl)cyclohex-1-en-1-yl trifluoromethanesulfonate (**5**). Based on the purity of the sample, 6.82 g (79%) of **5** was prepared. A second reaction on the identical scale provided 6.88 g (80%, uncorrected for purity) of the product.

22. 1,3-Diphenylisobenzofuran (>97.0%) was purchased from Tokyo Kasei Kogyo Co., Inc. and used as received without further purification.
23. Tetrabutylammonium fluoride (1.0 M in THF) was purchased from Sigma-Aldrich Co., LLC and used as received without further purification.
24. A small exotherm (+5 °C) was observed during 15 min addition time at current scale.
25. The reaction typically requires 1.5 h to consume all the 6-(triethylsilyl)cyclohex-1-en-1-yl trifluoromethanesulfonate (**5**) and is monitored by TLC analysis on Merck silica gel 60 F_{254} plates developing

Figure 14. TLC of (A) 6-(triethylsilyl)cyclohex-1-en-1-yl trifluoromethanesulfonate and (B) the reaction mixture
(photo provided by submitters)

with *n*-hexane/CH_2Cl_2 (9:1). The R_f values of 6-(triethylsilyl)cyclohex-1-en-1-yl trifluoromethanesulfonate (**5**), *endo*-**8**, and *exo*-**8** are 0.60, 0.34, and 0.25, respectively (stained with an ethanol solution of *p*-anisaldehyde). After dipping the TLC plate to the *p*-anisaldehyde solution, the chromatogram is stained by heating (Figure 14).

26. The crude product is dissolved in CH$_2$Cl$_2$ (40 mL) and 30-gram silica gel is added. The resulting slurry is evaporated to dryness on a rotary evaporator (35 °C, 100 to 10 mmHg). The crude product mixture is charged onto a RediSep RF column (diameter = 5.24 cm, height = 18.9 cm, 220-gram silica gel). The column is eluted with linear gradient from 0 to 15% CH$_2$Cl$_2$ in *n*-hexane on a ISCO CombiFlash system, and 25-mL fractions are collected. Fractions 61–83 are combined (Figure 15) and concentrated on a rotary evaporator under reduced pressure (35 °C, 100 to 10 mmHg). This chromatography method provides product with *endo/exo* ratio of > 99/1.

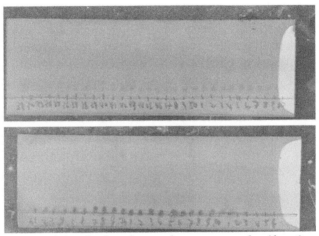

Figure 15. TLC of the flash column chromatography (fractions 31-86) (photo provided by checkers)

27. The ^{1}H NMR spectrum of the product shows no impurities; however, the submitters report that elemental analysis of this product did not satisfy purity criterion; therefore, recrystallization was determined to be necessary to provide colorless crystals with appropriate purity.
28. The recrystallization is performed with a Chemglass reaction block as the heat source.
29. *endo*-9,10-Diphenyl-1,2,3,9,9a,10-hexahydro-9,10-epoxyanthracene (*endo*-**8**): colorless crystals; R$_f$ = 0.34 (*endo*) (*n*-hexane/CH$_2$Cl$_2$ = 9:1); Merck silica gel 60 F$_{254}$ plates (stained with an ethanol solution of *p*-anisaldehyde). After dipping the TLC plate to the *p*-anisaldehyde solution, the chromatogram is stained by heating; mp 128–130 °C (ethanol); IR (film): 3026, 2928, 2844,

1445, 1305, 984, 748 cm^{-1}; ^{1}H NMR (500 MHz, CDCl$_3$) δ: 0.46 (dtd, J = 13.1, 11.7, 3.2 Hz, 1H), 1.53 – 1.67 (m, 1H), 1.71 – 1.90 (m, 2H), 2.09 – 2.32 (m, 2H), 2.95 – 3.16 (m, 1H), 5.68 (q, J = 3.2 Hz, 1H), 7.09 – 7.16 (m, 2H), 7.18 – 7.22 (m, 2H), 7.35 – 7.51 (m, 6H), 7.68 – 7.73 (m, 2H), 7.83 – 7.91 (m, 2H); ^{13}C NMR (126 MHz, CDCl$_3$) δ: 22.1, 25.0, 26.3, 48.3, 89.9, 90.3, 117.9, 120.3, 121.4, 125.8, 126.9, 127.1, 128.0, 128.3, 128.5, 128.6, 128.7, 135.0, 138.2, 144.4, 144.5, 148.4; The purity of the sample (>98.5%) was determined by ^{1}H qNMR using 25.8 mg of 1,3,5-trimethoxybenzene (Note 30) as an internal standard and 21.3 mg of *endo*-9,10-diphenyl-1,2,3,9,9a,10-hexahydro-9,10-epoxyanthracene (*endo*-8). Based on the purity (98.5%) of the sample, 2.39 g (44%) of *endo*-8 was prepared. A second reaction on the identical scale provided 2.51 g (48%, uncorrected for purity) of the product.

30. 1,3,5-Trimethoxybenzene (>99%) was purchased from FUJIFILM Wako Pure Chemical Corporation and used as received.

Working with Hazardous Chemicals

The procedures in *Organic Syntheses* are intended for use only by persons with proper training in experimental organic chemistry. All hazardous materials should be handled using the standard procedures for work with chemicals described in references such as "Prudent Practices in the Laboratory" (The National Academies Press, Washington, D.C., 2011; the full text can be accessed free of charge at http://www.nap.edu/catalog.php?record_id=12654). All chemical waste should be disposed of in accordance with local regulations. For general guidelines for the management of chemical waste, see Chapter 8 of Prudent Practices.

In some articles in *Organic Syntheses*, chemical-specific hazards are highlighted in red "Caution Notes" within a procedure. It is important to recognize that the absence of a caution note does not imply that no significant hazards are associated with the chemicals involved in that procedure. Prior to performing a reaction, a thorough risk assessment should be carried out that includes a review of the potential hazards associated with each chemical and experimental operation on the scale that is planned for the procedure. Guidelines for carrying out a risk assessment and for analyzing the hazards associated with chemicals can be found in Chapter 4 of Prudent Practices.

The procedures described in *Organic Syntheses* are provided as published and are conducted at one's own risk. *Organic Syntheses, Inc.*, its Editors, and its Board of Directors do not warrant or guarantee the safety of individuals using these procedures and hereby disclaim any liability for any injuries or damages claimed to have resulted from or related in any way to the procedures herein.

Discussion

Among various strained reactive intermediates, 1,2-cyclohexadiene (6) has attracted considerable attention due to the promising synthetic utility of the bent allene structure (Scheme 1). Since the report of cycloaddition of 1,3-diphenylisobenzofuran with 1,2-cyclohexadiene (6) by Wittig,[2] a number of papers regarding generation of 6 have been reported.[3] Similar to generation of benzyne from 2-silylphenyl triflate with fluoride ion,[4] 6-silyl cyclic enol triflate 9 has been the superior precursor for generating 1,2-cyclohexadiene (6) so far, because of the mild generation conditions (CsF, rt) and broad functional group compatibilities (Scheme 1).[5] The generated 1,2-cyclohexadiene (6) undergoes [2+2], [4+2], and [3+2] cycloaddition reactions to provide the corresponding cycloadducts 10–12.

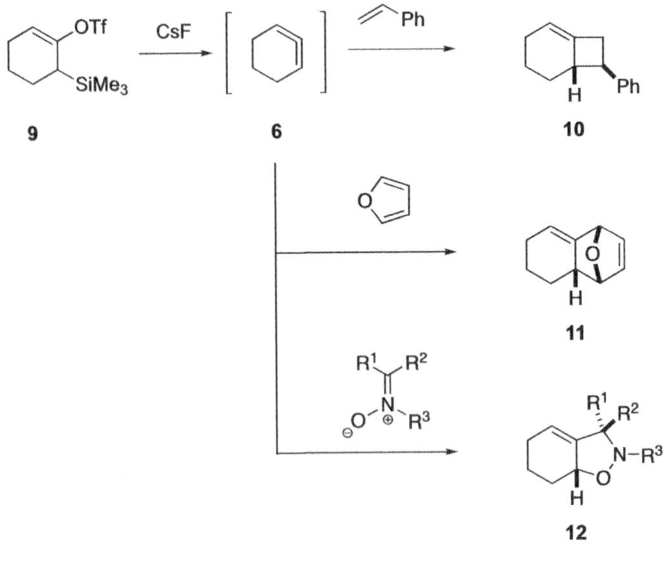

Scheme 1. Generation of 1,2-cyclohexadiene (6) and its reactions

In spite of the synthetic utility of this compound, the reported method[5a] requires multi-step synthesis (Scheme 2). The synthesis commenced with bromination of cyclohexenone (13) followed by protection of the carbonyl moiety as its acetal 14 to introduce a silyl group via the corresponding alkenyllithium. Acidic aqueous workup provided 2-silylcyclohexenone 15, which was then reduced with L-Selectride and treated the resultant lithium enolate with water to provide α-silylcyclohexanone 16. This cyclohexanone underwent kinetic deprotonation with LDA, and the corresponding lithium enolate 17 reacted with PhNTf$_2$ to give silylated cyclohexenyl triflate 9, which is the precursor for 1,2-cyclohexadiene (6).

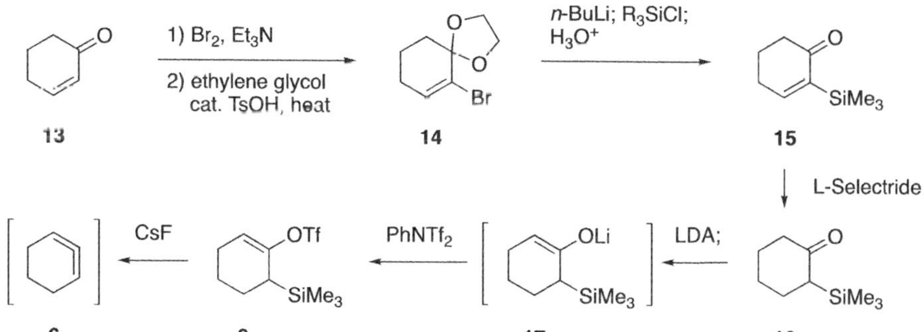

Scheme 2. First generation synthesis of silylated enol triflate 9

The modified synthesis starts from silylation of the inexpensive cyclohexanone (1) to provide silyl enol ether 2 (Scheme 3).[6] According to the independent reports from Kuwajima[7] and Corey,[8] a combination of LDA and t-BuOK allows for the smooth transfer of the silyl group to the allylic position via the allyllithium, providing the corresponding enolate 3. Subsequent triflation with Comins' reagent[9] provides the precursor 5 for 1,2-cyclohexadiene (1) in a single flask from silyl enol ether 2 on a 5-gram scale.

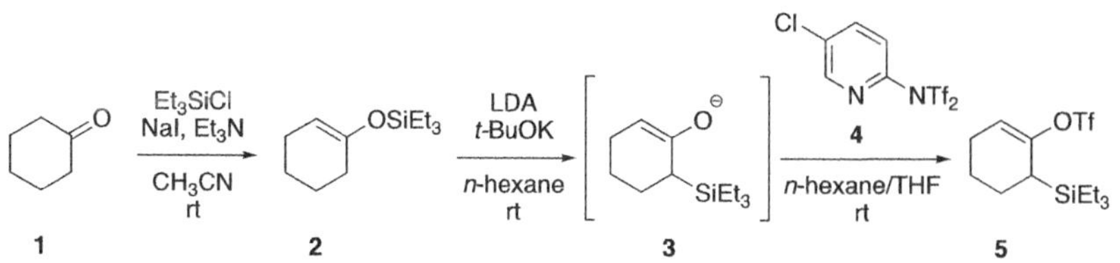

Scheme 3. Modified synthesis of silylated enol triflate 5

References

1. Department of Chemical Science and Engineering, Kobe University, 1-1 Rokkodai, Nada-ku, Kobe 657-8501, Japan. E-mail: okano@harbor.kobe-u.ac.jp. ORCID: 0000-0003-2029-8505. These studies were supported by JSPS KAKENHI Grant Numbers JP16K05774 in Scientific Research (C), JP19H02717 in Scientific Research (B), Tonen General Sekiyu Research Development Encouragement & Scholarship Foundation, the Harmonic Ito Foundation, and Foundation for Interaction in Science & Technology.
2. (a) Wittig, G.; Fritze, P. *Angew. Chem., Int. Ed. Engl.* **1966**, *5*, 846. (b) Wittig, G.; Fritze, P. *Justus Liebigs Ann.* **1968**, *711*, 82–87.
3. (a) Moore, W. R.; Moser, W. R. *J. Org. Chem.* **1970**, *35*, 908–912. (b) Bottini, A. T.; Corson, F. P.; Fitzgerald, R.; Frost, K. A., II *Tetrahedron* **1972**, *28*, 4883–4904. (c) Bottini, A. T.; Hilton, L. L.; Plott, J. *Tetrahedron* **1975**, *31*, 1997–2001. (d) Balci, M.; Jones, W. M. *J. Am. Chem. Soc.* **1980**, *102*, 7607–7608. (e) Wentrup, C.; Gross, G.; Maquestiau, A.; Flammang, R. *Angew. Chem., Int. Ed. Engl.* **1983**, *22*, 542–543. (f) Christl, M.; Schreck, M. *Angew. Chem., Int. Ed. Engl.* **1987**, *26*, 449–451. (g) Nendel, M.; Tolbert, L. M.; Herring, L. E.; Islam, M. N.; Houk, K. N. *J. Org. Chem.* **1999**, *64*, 976–983. (h) Christl, M.; Fischer, H.; Arnone, M.; Engels, B. *Chem. Eur. J.* **2009**, *15*, 11266–11272. (i) Hioki, Y.; Mori, A.; Okano, K. *Tetrahedron* **2020**, *76*, 131103.
4. Himeshima, Y.; Sonoda, T.; Kobayashi, H. *Chem. Lett.* **1983**, 1211–1214.
5. (a) Peña, D.; Iglesias, B.; Quintana, I.; Pérez, D.; Guitián, E.; Castedo, L. *Pure Appl. Chem.* **2006**, *78*, 451–455. (b) Quintana, I.; Peña, D.; Pérez, D.; Guitián, E. *Eur. J. Org. Chem.* **2009**, 5519–5524. (c) Barber, J. S.; Styduhar, E. D.; Pham, H. V.; McMahon, T. C.; Houk, K. N.; Garg, N. K. *J. Am. Chem. Soc.* **2016**, *138*, 2512–2515. (d) Chari, J. V.; Ippoliti, F. M.; Garg, N. K. *J. Org. Chem.* **2019**, *84*, 3652–3655. Generation of 1,2-cyclohexadiene from 6-trimethylsilyl-1-bromocyclohexene with CsF: (e) Shakespeare, W. C.; Johnson, R. P. *J. Am. Chem. Soc.* **1990**, *112*, 8578–8579. Generation of 1,2-cyclohexadiene from 6-trimethylsilyl-1-bromocyclohexene with TBAF: (f) Sütbeyaz, Y.; Ceylan, M.; Seçen, H. *J. Chem. Res., Synop.* **1993**, 293.
6. (a) Inoue, K.; Nakura, R.; Okano, K.; Mori, A. *Eur. J. Org. Chem.* **2018**, 3343–3347. (b) Nakura, R.; Inoue, K.; Okano, K.; Mori, A. *Synthesis* **2019**, *51*, 1561–1564.

7. Kuwajima, I.; Takeda, R. *Tetrahedron Lett.* **1981**, *22*, 2381–2384.
8. Corey, E. J.; Rücker, C. *Tetrahedron Lett.* **1984**, *25*, 4345–4348.
9. (a) Comins, D. L.; Dehghani, A. *Tetrahedron Lett.* **1992**, *33*, 6299–6302. (b) Comins, D. L.; Dehghani, A.; Foti, C. J.; Joseph, S. P. *Org. Synth.* **1997**, *74*, 77–83.

Appendix
Chemical Abstracts Nomenclature (Registry Number)

Sodium iodide: Sodium iodide; (7681-82-5)
Acetonitrile: Acetonitrile; (75-05-8)
Cyclohexanone: Cyclohexanone; (108-94-1)
Triethylamine: Ethanamine, *N*,*N*-diethyl-; (121-44-8)
Triethylsilyl chloride: Silane, chlorotriethyl-; (994-30-9)
t-BuOK: 2-Propanol, 2-methyl-, potassium salt (1:1); (865-47-4)
Hexane: Hexane; (110-54-3)
LDA: 2-Propanamine, *N*-(1-methylethyl)-, lithium salt (1:1); (4111-54-0)
THF: Furan, tetrahydro-; (109-99-9)
Comins' reagent: Methanesulfonamide, *N*-(5-chloro-2-pyridinyl)-1,1,1-trifluoro-*N*-[(trifluoromethyl)sulfonyl]-; (145100-51-2)
1,3-Diphenylisobenzofuran: Isobenzofuran, 1,3-diphenyl-; (5471-63-6)
Tetrabutylammonium fluoride: 1-Butanaminium, *N*,*N*,*N*-tributyl-, fluoride (1:1); (429-41-4)
1,3,5-trimethoxybenzene: Benzene, 1,3,5-trimethoxy-; (621-23-8)

Ryo Nakura was born in Fukuoka in 1994. He received his B.S. in 2017 and M.S. in 2019 from Kobe University under the supervision of Professors Atsunori Mori and Kentaro Okano. During his M.S. studies, he worked on development of facile preparation of cycloallene intermediates and its application to the synthesis of functionalized cyclohexenes. Currently, he works for Mitsubishi Chemical Corporations.

Kazuki Inoue was born in Osaka in 1994. He received his B.S. in 2017 and M.S. in 2019 from Kobe University under the direction of Professors Atsunori Mori and Kentaro Okano. During his M.S. studies, he worked on generation of strained cycloalkynes and its application to the synthesis of functionalized poly(cycloalkyne)s. Currently, he works for Sumitomo Bakelite Co., Ltd.

Mayu Itoh was born in Kochi in 1996. She received his B.S. in 2019 from Kobe University, where she carried out undergraduate research in the laboratories of Professor Atsunori Mori. In the same year, she then began her graduate studies at the Graduate School of Chemical Science and Engineering, Kobe University under the supervision of under the supervision of Professors Atsunori Mori and Kentaro Okano. Her research focuses on photo-redox reaction of organotin compounds.

Atsunori Mori was born in Aichi in 1959. He received B.S. (1982), M.S. (1984), and Ph.D. (1987) degrees from Nagoya University under the direction of Professor Hisashi Yamamoto. After postdoctoral study in U.C. Berkeley with Peter Vollhardt (1987–1988), he started his academic career as an Assistant Professor at the University of Tokyo (1988) and Japan Advanced Institute of Science and Technology (1993). He was appointed as an Associate Professor in 1995 at Chemical Resources Laboratory of Tokyo Institute of Technology. Since 2005, he has been Professor at the Graduate School of Engineering, Kobe University.

Kentaro Okano was born in Tokyo in 1979. He received his B.S. in 2003 from Kyoto University under the supervision of Professor Tamejiro Hiyama. He then moved to the laboratories of Professor Tohru Fukuyama at the University of Tokyo. In 2007, he joined the faculty at Tohoku University in Professor Hidetoshi Tokuyama's group. In 2014, he visited Professor Amir Hoveyda's laboratories at Boston College as a visiting researcher. In 2015, he moved to Kobe University, where he is currently Associate Professor of Department of Chemical Science and Engineering. His research interest is natural product synthesis based on the development of new synthetic methodologies.

Zhixun Wang joined the Process Research Department of Merck & Co., Inc. in 2019. His research focuses on the synthesis of therapeutic candidates for large-scale production. He received his B. S. degree from Fudan University, and then obtained his Ph.D. under the supervision of Professor Liming Zhang at University of California, Santa Barbara with a research focus on homogeneous gold catalysis. Zhixun then moved to New Haven, CT, where he did postdoctoral work on synthesis of pleuromutilin analogs and structure elucidation of colibactin with Professor Seth Herzon at Yale University.

Discussion Addendum for:
Intra- and Intermolecular Kulinkovich Cyclopropanation Reactions of Carboxylic Esters with Olefins: Bicyclo[3.1.0]hexan-1-ol and *trans*-2-benzyl-1-methylcyclopropan-1-ol

Jin Kun Cha[1]*

Department of Chemistry, Wayne State University, 5101 Cass Ave, Detroit, Michigan 48202

Original Article: Kim, S.-H.; Sung, M. J.; Cha, J. K. *Org. Synth.* **2003**, *80*, 111–119.

The low-valent titanium-mediated cyclopropanation (the Kulinkovich reaction; eq 1) of esters provides easy access to 1,2-*cis*-alkylcyclopropanols, complementing the venerable Simmons-Smith cyclopropanation of silylenol ethers or derivatives. The original *Organic Syntheses* article describes a

$$\text{R}^1\text{C(O)OR}^3 \xrightarrow[\text{Ti(O-}i\text{-Pr})_4]{\text{R}^2\text{CH}_2\text{CH}_2\text{MgX (>2 equiv)}} \text{cyclopropanol with R}^1, \text{R}^2, \text{OH} \quad \text{(eq 1)}$$

convenient variant by making use of cyclopentyl or cyclohexyl Grignard reagents as a sacrificial reagent in the presence of a titanium alkoxide (eq 2). Juxtaposition of a strained three-membered ring and a hydroxyl functionality makes cyclopropanols well suited for subsequent elaboration, especially in carbon-carbon bond forming reactions. The Kulinkovich reaction has been the subject of several reviews, including two recent articles.[2,3] In light of these reviews, this discussion addendum presents a brief overview of selected developments in the Kulinkovich cyclopropanation with particular emphasis on ring opening of functionalized cyclopropanols in C C bond forming reactions, including applications in natural product synthesis. Also briefly discussed are related uses of in situ generated Kulinkovich intermediates (titanacyclopropanes). These examples underscore the synthetic utility of cyclopropanols as a family of attractively functionalized, yet readily accessible homologous enol equivalents.

$$\text{R}^1\text{C(O)OR}^3 + \text{CH}_2=\text{CHR}^2 \xrightarrow[\text{Ti(O-}i\text{-Pr})_4 \text{ or ClTi(O-}i\text{-Pr})_3]{\text{xs } c\text{-C}_5\text{H}_9\text{MgCl or } c\text{-C}_6\text{H}_{11}\text{MgCl}} \text{cyclopropanol} \quad \text{(eq 2)}$$

[Kulinkovich Intermediate: i-Pr-O-Ti(O-i-Pr)(R^2) ↔ i-Pr-O-Ti(O-i-Pr)(R^2)]

Useful Developments in the Kulinkovich Cyclopropanation

Enantioselective approaches to the Kulinkovich reaction of esters were documented in the literature by employing TADDOL derivatives, but generally applicable asymmetric methods, especially for the olefin-exchange mediated variant, remain a missing link.[4,5] Satisfactory levels of diastereoselective cyclopropanation reactions are possible by employing

1-alkenes having a suitable adjacent stereocenter[6] or secondary homoallylic alcohols (Figure 1).[7] The directing effect of the hydroxyl group in the latter substrates leads to the formation of 1,2-*trans*-alkylcyclopropanols via seven-membered cyclic titanates to override the intrinsic *cis*-alkyl stereochemical preference of the original Kulinkovich cyclopropanation.

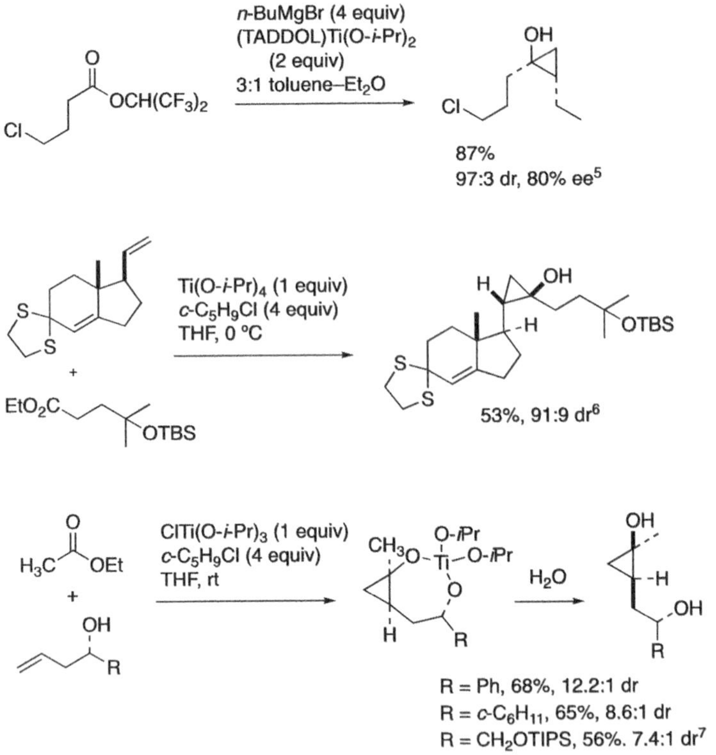

Figure 1. Enantioselective and diastereoselective approaches

Other carboxylic acid derivatives (such as tertiary amides, cyclic carbonates, and nitriles) also undergo the Kulinkovich reactions to afford the corresponding heteroatom-substituted cyclopropanes (Figure 2).[2,3] In the case of nitriles, the use of homoallylic alcohols is required for the olefin exchange-mediated variant to countermand stronger affinity of a nitrile toward low-valent titanium species.[8a-c,9] On the other hand, imides and *N*-acylpyrroles having a non-basic nitrogen instead produce the titanacyclopentanes, arising from preferential addition of the less substituted Ti–C bond of the presumed Kulinkovich intermediate to the respective carbonyl group.[8d]

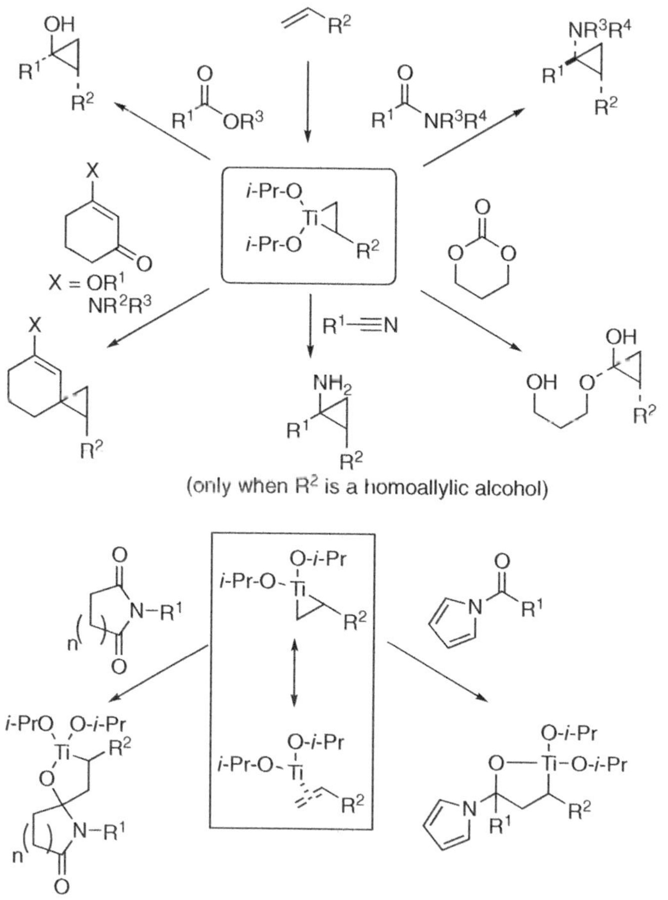

Figure 2. The Kulinkovich cyclopropanation of carboxylic acid derivatives and related coupling reactions

Ring Opening of Cyclopropanols

Cyclopropanols function as a homoenol surrogate in C–C bond forming reactions driven by release of ring strain. A notable exception notwithstanding,[10] ring opening reactions of cyclopropanols were limited primarily to non-carbon electrophiles (e.g., halogens) until recently. Electrophilic attack at cyclopropanols by tethered oxocarbenium ions provides a unique approach to common structural motifs such as appropriately functionalized carbocycles (Figure 3),[11] oxepanes[12] and tetrahydropyrans (Figure 4).[13]

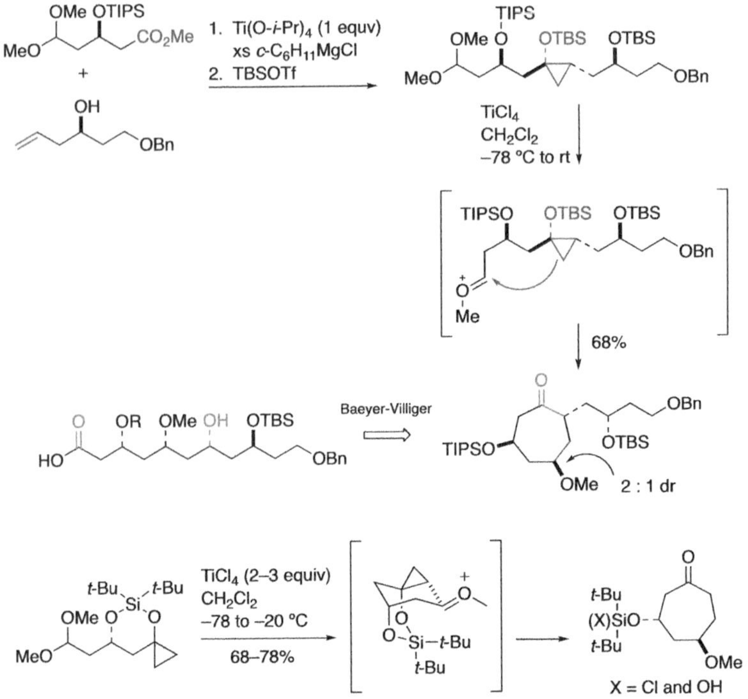

Figure 3. Seven-membered ring formation

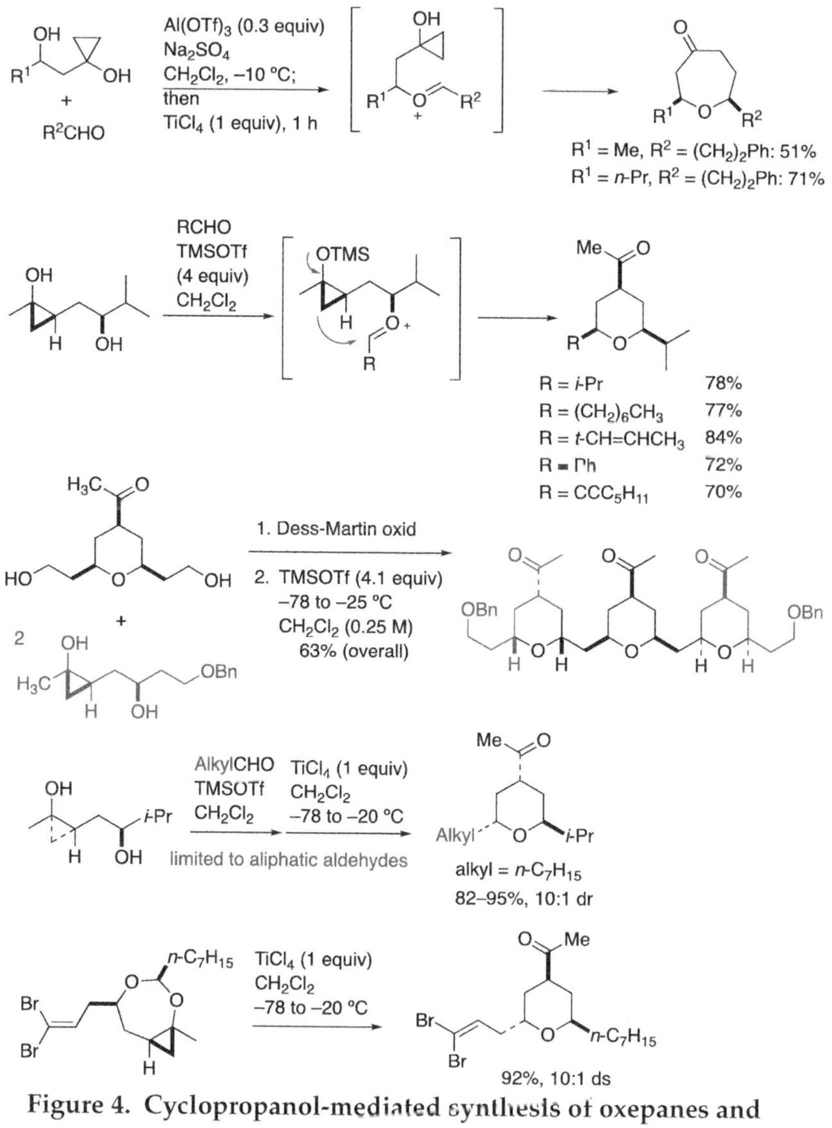

Figure 4. Cyclopropanol-mediated synthesis of oxepanes and tetrahydropyrans

The temporary tethering strategy with the homoallylic alcohol functionality is not only effective for the aforementioned cyclopropanation (Figures 1 and 3), but also broadly applicable to coupling of two unsaturated fragments (such as alkynes and imines) by the action of the

Kulinkovich intermediate, as elegantly developed by the Micalizio group (Figure 5).[14]

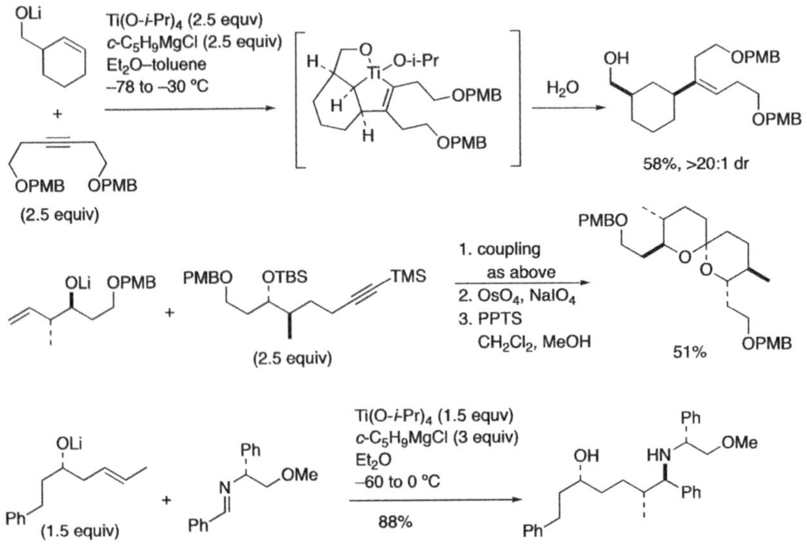

Figure 5. Coupling of homoallylic alcohols with alkynes and imine

Allylic alcohols are also versatile substrates that readily react with the ethyl Grignard reagent-derived Kulinkovich intermediate to afford *syn*-S_N2'-type ethylation products, accompanied by elimination of a "Ti=O" moiety (Figure 6).[15a] Importantly, the presumed intermediates can be trapped with suitable electrophiles such as aldehydes, halogens, and molecular oxygen.[15b] High levels of diastereoselectivity are obtained with cyclic allylic alcohols and (Z)-allylic alcohols. This reaction is reminiscent of the Dzhemilev–Hoveyda reaction, which employs the Negishi reagent in the place of the Kulinkovich reagent.[16] An advantage of this method is an efficient trapping with carbon-based electrophiles (e.g., aldehydes). Styrenes and vinylsilanes also undergo coupling with allylic alcohols in the presence of the cyclopentyl Grignard reagent under the typical Kulinkovich reaction conditions (Figure 6).[15a]

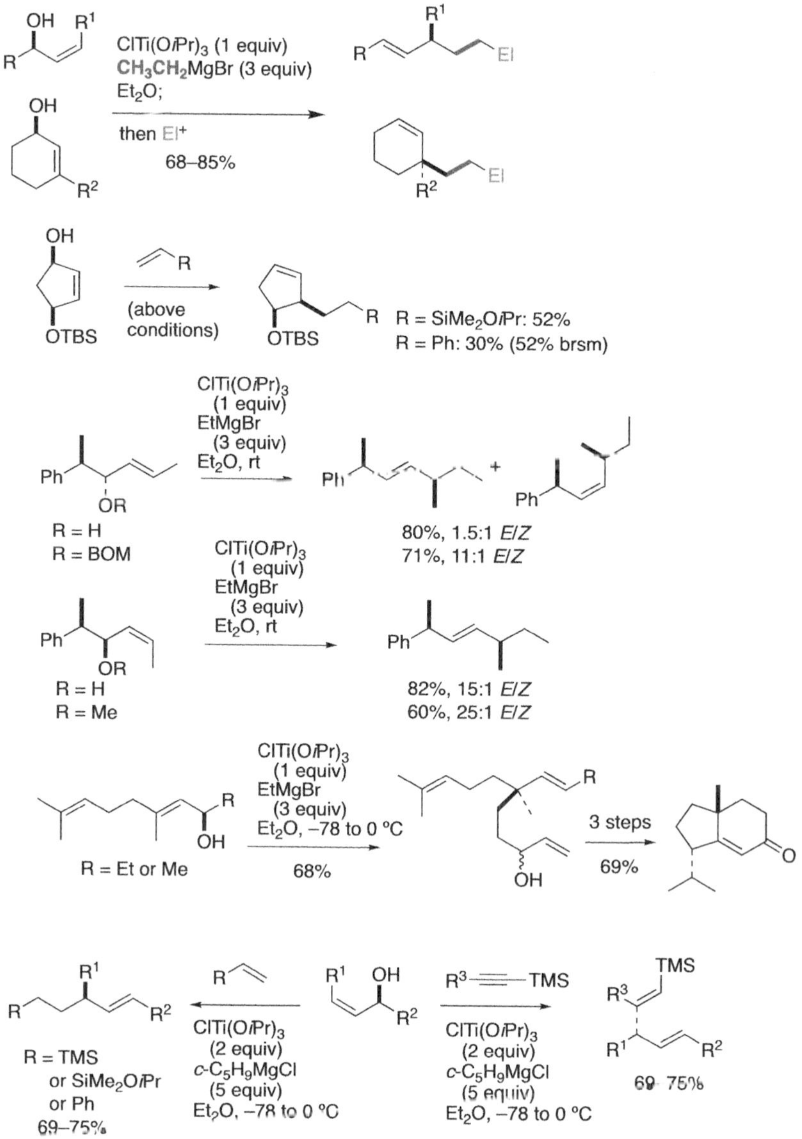

Figure 6. Stereoselective alkylation of allylic alcohols by ethylation and trapping

The corresponding coupling reaction of allylic alcohols with imines results in diastereoselective allylation of aromatic and aliphatic imines to afford homoallylic amines diastereoselectively (Figure 7).[17] Direct use of an attractively functionalized allylic alcohol as an allylating reagent without pre-derivatization obviates the use of preformed organometallic reagents or activated imine derivatives.

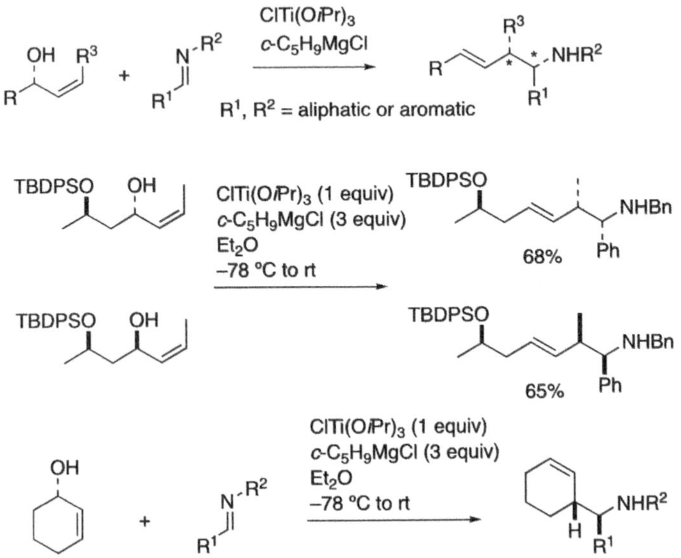

Figure 7. Diastereoselective synthesis of homoallylic amines

R^1	R^2	yield
Ph	Ph	90%
	p-MeOC$_6$H$_4$	75%
	Bn	78%
	CH$_2$-o-MeOC$_6$H$_4$	69%
	n-Bu	76%
2-furyl	Bn	55%
i-Pr	Bn	40%
n-C$_7$H$_{15}$	Bn	60%

By combining ring opening of cyclopropanols with transmetalation of a suitable transition metal, the otherwise unfavorable equilibrium can be shifted to generate in situ β-keto homoenolates, which open new avenues to transition metal-catalyzed transformations with sp^3, sp^2, and sp electrophiles (eq 3). Additionally, the formation of the presumed 5-membered keto chelate intermediate is advantageous in promoting facile reductive elimination (subsequent to transmetalation) in preference to the competing beta-hydride

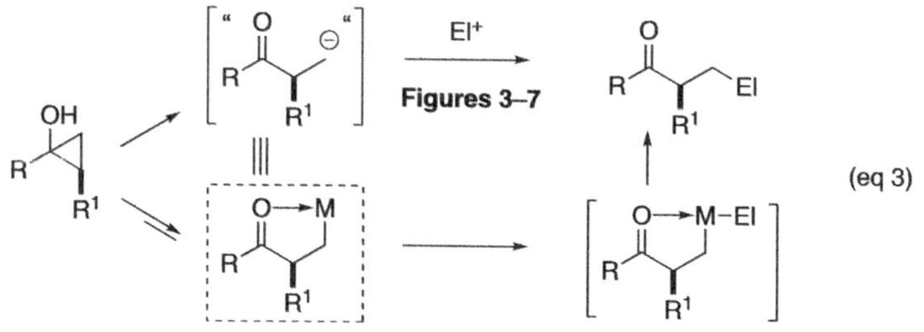

$$\text{(eq 3)}$$

elimination, as well as minimizing epimerization of the alpha-stereocenter. Transition metal-catalyzed cross-coupling reactions of cyclopropanols for the C–C bond formation have been an active area of research: they include: S_N2' alkylation, alkenylation, alkynylation, and acylation (Figure 8), in addition trifluoromethylation and related reactions.[3,18]

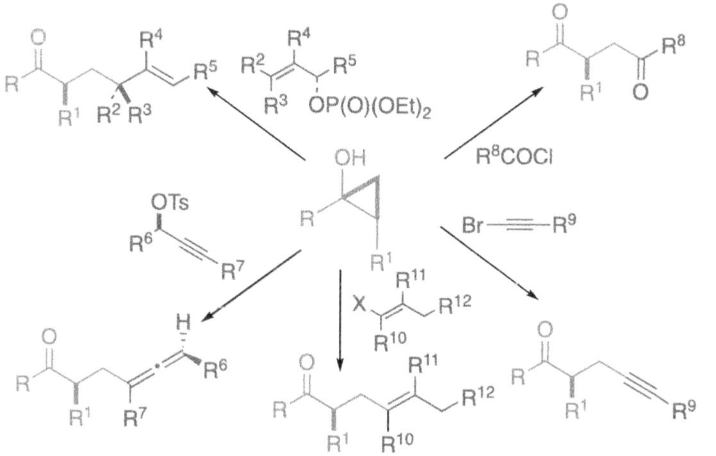

Figure 8. Transition metal-catalyzed cross-coupling reactions

As a representative example of this family of transition metal-catalyzed C–C bond forming reactions, alkenylation and acylation reactions of cyclopropanols are described briefly. The Orellana group reported Pd-catalyzed coupling of cyclopropanols with aryl halides, which proceeded cleanly without beta-hydride elimination (Figure 9).[19] This methodology is also amenable to intramolecular processes, providing an efficient route to the otherwise challenging medium-sized (seven- and eight-membered)

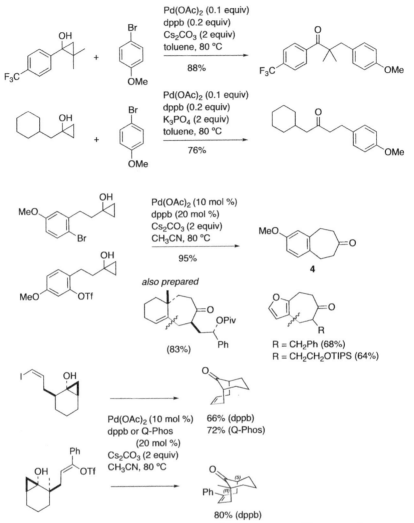

Figure 9. Inter- and intramolecular cross-coupling reactions of aryl or alkenyl halides and derivatives

carbocycles. An efficient construction of seven- and eight-membered carbocycles can be attributed to facile ligand exchange between the Pd species [from oxidative addition of Pd(0)] and the tethered cyclopropanol to set the stage for beta-carbon elimination of the resulting Pd cyclopropoxide, followed by reductive elimination.[18d]

The Dai group developed an elegant carbonylative ring opening reaction of cyclopropanols having a tethered hydroxyl group to synthesize spirocyclic and fused bicyclic lactones of varying ring size (Figure 10).[20] The cyclopropanol substrates are readily available by the Kulinkovich cyclopropanation of the corresponding lactones.

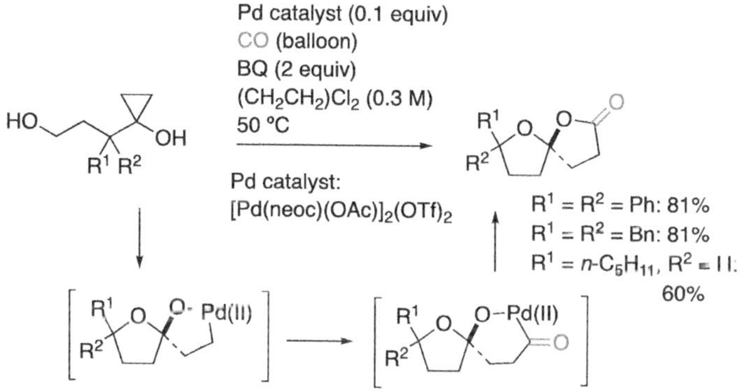

Figure 10. Pd-Catalyzed acylation of cyclopropanols: catalytic carbonylative spirolactonization of hydroxyl-tethered cyclopropanols

Conversion of cyclopropanols to the corresponding cyclopropylamines was achieved by the Rousseaux group by trapping of the presumed zinc homoenolates with amines (Figure 11).[21a] Zinc homoenolates can also be generated by treatment of alpha-chloroaldehydes with $CH_2(ZnI)_2$ by the method of Matsubara.[21b] These studies demonstrate facile interconversion between zinc cyclopropoxides and the corresponding homoenolates for in situ trapping of the latter intermediates.

This Discussion Addendum focuses on regioselective ring opening of cyclopropanols at the unsubstituted carbon. The alternative mode of ring opening relies on generation of beta-keto alkyl radical intermediates. Trapping of the latter has led to a spate of useful transformations. The chemistry of beta-keto radicals has a long rich history, but it is beyond the scope of this Addendum.[2,3] Also excluded are vinyl cyclopropanols and donor-acceptor cyclopropanols.

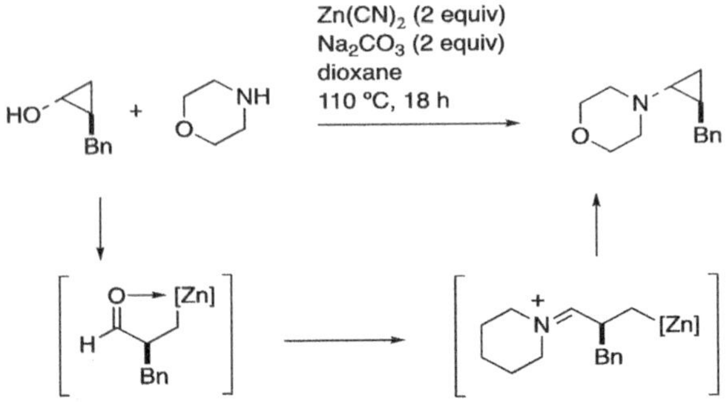

Figure 11. Conversion of cyclopropanols to cyclopropylamines

Application in Natural Product Synthesis

A short total synthesis of paeonilide, a monoterpenoid isolated from *Paeonia delavayi*, was reported by Dai and coworkers: the fused bicyclic lactone moiety was assembled in an appealing manner by Pd-catalyzed carbonylative lactonization of a highly functionalized cyclopropanol (Figure 12).[20b]

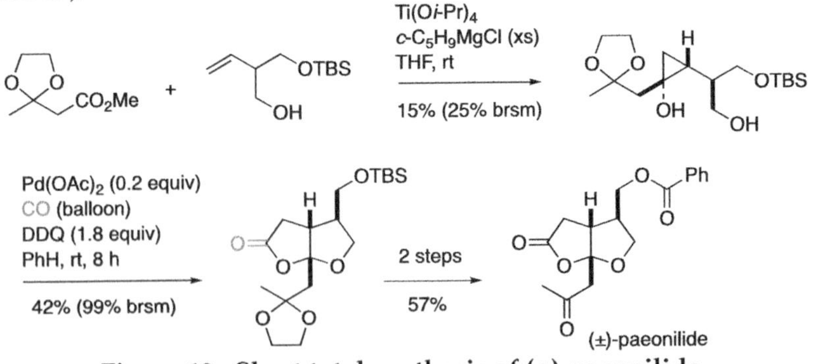

Figure 12. Short total synthesis of (±)-paeonilide

The sequential orchestration of the olefin exchange-mediated cyclopropanation and *anti*-S_N2' alkylation of the resulting cyclopropanol with a propargyl tosylate was central to a concise total synthesis of alkaloid 205B containing a deceptively simple array of stereocenters.[22] Starting from two easily accessible coupling partners, this alkaloid was synthesized in four straightforward steps (Figure 13). Also noteworthy was uncommon solvent effect (THF vs Et_2O)[22a] in LAH reduction of the six-membered imine intermediate to afford the *trans*-2,6-disubstituted piperidine stereoselectively.

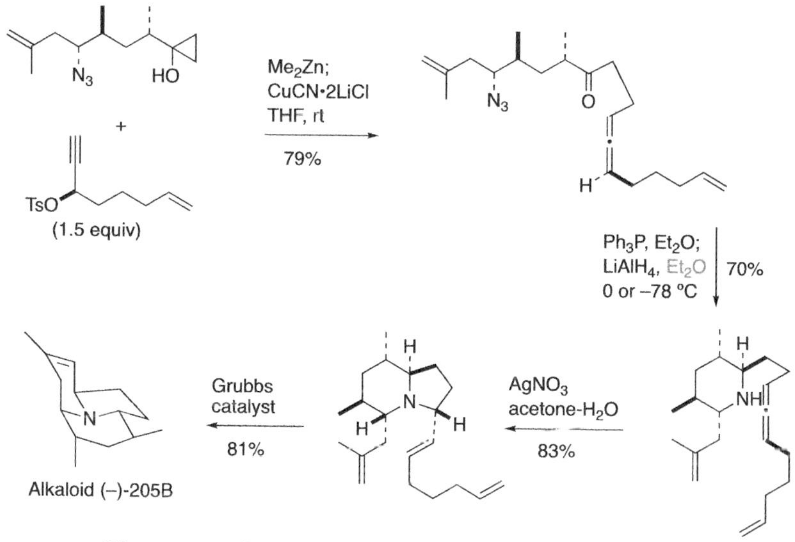

Figure 13. Concise synthesis of alkaloid (–)-205B

In conclusion, the Kulinkovich cyclopropanation reaction provides ready access to attractively functionalized cyclopropanols. The olefin exchange-mediated variant is a versatile method for coupling of two segments – esters and terminal alkenes. Cross-coupling reactions of cyclopropanols and appropriate partners allow a rapid increase in molecular complexity through an expedient bond connection under mild conditions and with operational simplicity.

References

1. Department of Chemistry, Wayne State University, 5101 Cass Ave, Detroit, Michigan 48202. ORCID (0000-0003-4038-3213). We thank the National Science Foundation (1665331) for financial support.
2. (a) Kulinkovich, O. G.; de Meijere, A. *Chem. Rev.* **2000**, *100*, 2789–2834. (b) Kulinkovich, O. G. *Chem. Rev.* **2003**, *103*, 2597–2632. (c) Kulinkovich, O. G. *Russ. Chem. Bull., Int. Ed.* **2004**, *53*, 1065–1086. (d) Wolan, A.; Six, Y. *Tetrahedron* **2010**, *66*, 15–61. (e) Sato, F.; Urabe, H.; Okamoto, S. *Chem. Rev.* **2000**, *100*, 2835–2886. (f) Sato, F.; Urabe, H. In *Titanium and zirconium in organic synthesis*; Marek, I. Ed.; Wiley-VCH: New York, 2002, pp 319. (g) For a closely related Discussion Addendum for facile

syntheses of aminocyclopropanes: de Meijere, A.; Kozhushkov, S. I. *Org. Synth.* **2018**, *95*, 289–309.
3. (a) Cha, J. K.; Kulinkovich, O. G. *Org. React.* **2012**, *77*, 1–159. (b) McDonald, T. R.; Mills, L. R.; West, M. S.; Rousseaux, S. A. *Chem. Rev.* **2021**, *121*, 3–79.
4. Corey, E. J.; Rao, S. A.; Noe, M. C. *J. Am. Chem. Soc.* **1994**, *116*, 9345–9346.
5. (a) Kulinkovich, O. G.; Kananovich, D. G.; Lopp, M.; Snieckus, V. *Adv. Synth. Catal.* **2014**, *356*, 3615–3626. (b) Konik, Y. A.; Kananovich, D. G.; Kulinkovich, O. G. *Tetrahedron* **2013**, *69*, 6673–6678. (c) Iskryk, M.; Barysevich, M.; Oseka, M.; Adamson, J.; Kananovich, D. *Synthesis* **2019**, *51*, 1935–1948.
6. Barysevich, M. V.; Kazlova, V. V.; Kukel, A. G.; Liubina, A. I.; Hurski, A. L.; Zhabinskii, V. N.; Khripach, V. A. *Chem. Commun.* **2018**, *54*, 2800–2803.
7. Quan, L. G.; Kim, S.-H.; Lee, J. C.; Cha, J. K. *Angew. Chem. Int. Ed.* **2002**, *41*, 2160–2162.
8. (a) Sung, M. J.; Lee, C.-W.; Cha, J. K. *Synlett* **1999**, 561–562. (b) Kim, S.-H.; Kim, S.-I.; Lai, S.; Cha, J. K. *J. Org. Chem.* **1999**, *64*, 6771–6775. (c) Santra, S.; Masalov, N.; Epstein, O. L.; Cha, J. K. *Org. Lett.* **2005**, *7*, 5901–5904. (d) Bobrov, D. N.; Kim, K.; Cha, J. K. *Tetrahedron Lett.* **2008**, *49*, 4089–4091. (d) Astashko, D.; Lee, H. G.; Bobrov, D. N.; Cha, J. K. *J. Org. Chem.* **2009**, *74*, 5528–5532.
9. Addition of a Lewis acid is necessary to induce the cyclopropane formation for application of the Kulinkovich cyclopropanation conditions to nitriles and imides: (a) Bertus, P.; Szymoniak, J. *Chem. Commun.* **2001**, 1792–1793. (b) Bertus, P.; Szymoniak, J. *Synlett* **2007**, 1346. (c) Bertus, P.; Szymoniak, J. *Org. Lett.* **2007**, *9*, 659–662.
10. Carey, J. T.; Knors, C.; Helquist, P. *J. Am. Chem. Soc.* **1986**, *108*, 8313–8314.
11. Epstein, O. L.; Lee, S.; Cha, J. K. *Angew. Chem. Int. Ed.* **2006**, *45*, 4988–4991.
12. O'Neil, K. E.; Kingree, S. V.; Minbole, K. P. C. *Org. Lett.* **2005**, *7*, 515–517.
13. (a) Lee, H. G.; Lysenko, I.; Cha, J. K. *Angew. Chem. Int. Ed.* **2007**, *46*, 3326–3328. (b) Parida, B. B.; Lysenko, I.; Cha, J. K. *Org. Lett.* **2012**, *14*, 6258–6261.
14. Richard, H. A.; Micalizio, G. C. *Chem. Sci.* **2011**, *2*, 573–589 and references therein.
15. (a) Lysenko, I. L.; Kim, K.; Lee, H. G.; Cha, J. K. *J. Am. Chem. Soc.* **2008**, *130*, 15997–16002. (b) Das, P. P.; Lysenko, I. L.; Cha, J. K. *Angew. Chem. Int. Ed.* **2011**, *50*, 9459–9461.

16. Hoveyda, A. H. In *Titanium and zirconium in organic synthesis*; Marek, I. Ed.; Wiley-VCH: New York, 2002, pp 181 and references therein.
17. Lysenko, I. L.; Lee, H. G.; Cha, J. K. *Org. Lett.* **2009**, *11*, 3132–3134.
18. (a) Das, P. P.; Belmore, K.; Cha, J. K. *Angew. Chem. Int. Ed.* **2012**, *51*, 95179520. (b) Parida, B. B.; Das, P. P.; Niocel, M.; Cha, J. K. *Org. Lett.* **2013**, *15*, 1780–1783. (c) Murali, R. V. N. S.; Nagavaram, N. R.; Cha, J. K. *Org. Lett.* **2015**, *17*, 3854–3856. (d) Ydhyam, S.; Cha, J. K. *Org. Lett.* **2015**, *17*, 5820–5823.
19. (a) Rosa, D.; Orellana, A. *Org. Lett.* **2011**, *13*, 110–113. (b) Rosa, D.; Orellana, A. *Chem. Commun.* **2013**, *49*, 5420–5422. (c) Rosa, D.; Orellana, A. *Chem. Commun.* **2012**, *48*, 1922–1924. (d) Cheng, K.; Walsh, P. J. *Org. Lett.* **2013**, *15*, 2298–2301.
20. (a) Davis, D. C.; Walker, K. L.; Hu, C.; Zare, R. N.; Waymouth, R. M.; Dai, M. *J. Am. Chem. Soc.* **2016**, *138*, 10693–10699. (b) Cai, X.; Liang, W.; Liu, M.; Li, X.; Dai, M. *J. Am. Chem. Soc.* **2020**, *142*, 13677–13682. (c) Ma, K.; Yin, X.; Dai, M. *Angew. Chem. Int. Ed.* **2018**, *57*, 15209–15212. (d) Cai, X.; Liang, W.; Dai, M. *Tetrahedron* **2019**, *75*, 193–208.
21. (a) Mills, L. R.; Barrera Arbelaez, L. M.; Rousseaux, S. A. L. *J. Am. Chem. Soc.* **2017**, *139*, 11357–11360. (b) West, M. S.; Mills, L. R.; McDonald, T. R.; Lee, J. B.; Ensan, D.; Rousseaux, S. A. L. *Org. Lett.* **2019**, *21*, 8409–8413.
22. (a) Nagavaram, N. R.; Cha, J. K. *J. Am. Chem. Soc.* **2015**, *137*, 2243–2246. (b) Nagavaram, N. R.; Parida, B. B.; Cha, J. K. *Org. Lett.* **2014**, *16*, 6208–6211.

Jin K. Cha was born and raised in Korea. After graduating from Seoul National University in Korea, he obtained his D. Phil. degree from the University of Oxford under the supervision of the late Professor Sir Jack E. Baldwin. The first two years of his doctoral work were performed at MIT. After postdoctoral research (1981-1983) with Professor Y. Kishi at Harvard University, he started his independent academic career and is now professor of chemistry at Wayne State University. His research group initiated the low-valent titanium-mediated cyclopropanation and related reactions at The University of Alabama, Tuscaloosa, AL.

Synthesis of Chiral Aziridine Ligands for Asymmetric Alkylation with Alkylzincs: Diphenyl((S)-1-((S)-1-phenylethyl)aziridin-2-yl)methanol

Siyuan Sun and Pavel Nagorny[*1]

Chemistry Department, University of Michigan, 930 N. University Ave., Ann Arbor, MI 48109, USA

Checked by Matthew Winston and Kevin Campos

Procedure (Note 1)

A. *Ethyl (S)-1-((S)-1-phenylethyl)aziridine-2-carboxylate (1)*. An oven-dried (Note 2) 250-mL single-necked, round-bottomed flask (24/40 joint) is equipped with a Teflon coated magnetic stir bar (16 × 32 mm, egg-shape). The flask is sealed with a rubber septum, connected to a Schlenk line with a needle adapter and subsequently cooled to room temperature (Note 3). (S)-(-)-1-Phenylethanamine (3.0 mL, 23.0 mmol, 1.0 equiv) (Note 4), triethylamine (4.65 g, 6.41 mL, 46.0 mmol, 2.0 equiv) (Note 5) and toluene

(46 mL) (Note 6) are added to the flask via syringes under nitrogen atmosphere (Figure 1A).

An oven-dried (Note 2) 100-mL single-necked, round-bottomed flask (14/20 joint) is sealed with a rubber septum, connected to a Schlenk line with a needle adapter, and subsequently cooled to room temperature (Note 3). Ethyl 2,3-dibromopropanoate (5.98 g, 3.34 mL, 23.0 mmol, 1.00 equiv) (Note 7) and toluene (46 mL) (Note 6) are added to the flask via syringes under nitrogen atmosphere (Note 8). The resulting clear solution is added to

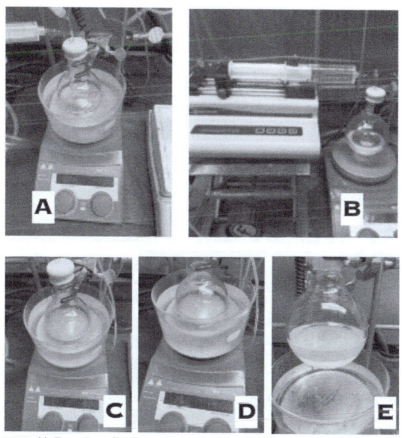

Figure 1. A) Reaction flask set-up after the addition of toluene and (S)-(-)-1-phenylethanamine; B) addition of ethyl 2,3-dibromopropanoate; C) reaction mixture at the beginning of heating; D) reaction mixture after 6 hours of heating; E) reaction mixture after settling (photos provided by submitters)

the 250-mL flask with a syringe pump (Note 9) with a 60-mL syringe over 60 min (Figure 1B). The suspension is stirred (Note 10) for 5 min at room temperature (Note 3) and then heated to 90 °C (bath temperature) in an oil bath (Note 11) (Figure 1C). After 6 h (Note 12) (Figure 1D), the reaction mixture is removed from the oil bath and cooled to room temperature (Note 3), at which time the solution naturally separates into two layers (Figure 1E).

Deionized water (50 mL) (Note 13) is added to the reaction mixture, the stir bar is removed, and the biphasic mixture is transferred to a 250-mL separatory funnel. An additional portion of ethyl acetate (50 mL) (Note 14) is used to rinse the reaction flask and then poured into the separatory funnel. The organic layer is collected, and the aqueous layer is extracted with ethyl acetate (2 × 50 mL) (Note 14). The combined organic extracts are washed with saturated sodium chloride solution (50 mL) (Note 15) and dried with sodium sulfate (25 g) (Note 16). The solution is filtered (Note 17) into a 500-mL single-necked round-bottomed flask (24/40 joint) with ethyl acetate washings (3 × 10 mL) (Note 14) and then concentrated with the aid of a rotary evaporator (Note 18) to afford a crude, yellow oily mixture. The crude material is purified by chromatography on silica (Note 19) to afford ethyl (S)-1-((S)-1-phenylethyl)aziridine-2-carboxylate **1** (2.20 g, 43%, 97% purity) (Notes 20 and 21) as a yellow oil (Figure 2A) and ethyl (R)-1-((S)-1-phenylethyl)aziridine-2-carboxylate **2** (2.13 g, 42%, 99% purity) (Notes 22 and 23) as a yellow oil (Figure 2B).

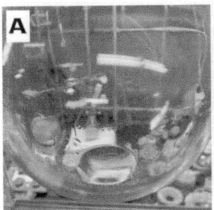

Figure 2. A) Product 1; B) Product 2 (photos provided by submitters)

B. *Diphenyl((S)-1-((S)-1-phenylethyl)aziridin-2-yl)methanol* (**3**). An oven-dried (Note 2) 100-mL single-necked, round-bottomed flask (14/20 joint) is equipped with a Teflon coated magnetic stir bar (9 × 12 mm, octagon). The flask is sealed with a rubber septum, connected to a Schlenk line with a needle adapter and subsequently cooled to room temperature (Note 3). Phenylmagnesium chloride solution (5.45 g, 26.9 mL, 39.8 mmol, 4.0 equiv) (Note 24) is added via a syringe under nitrogen atmosphere and then cooled to 0 °C (Note 25) while stirring (Note 10) (Figure 3A).

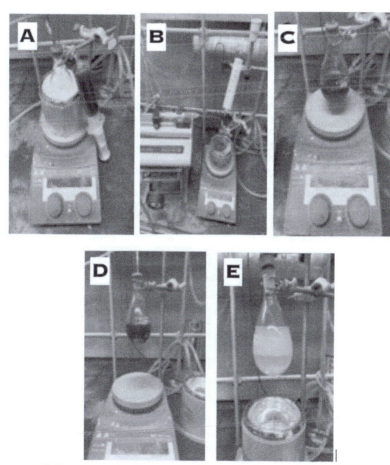

Figure 3. A) Reaction flask set-up after addition of PhMgCl solution; B) addition of starting material (1); C) reaction solution after the addition; D) reaction solution before work-up; E) reaction mixture after work-up (photos provided by submitters)

An oven-dried (Note 2) 50-mL single-necked, round-bottomed flask (14/20 joint) is sealed with a rubber septum, connected to a Schlenk line with a needle adapter, and subsequently cooled to room temperature (Note 3). Ethyl 1-((S)-1-phenylethyl)aziridine-2-carboxylate (**1**) (2.18 g, 9.94 mmol, 1.00 equiv.) and THF (26 mL) (Note 26) are added via syringes under nitrogen atmosphere (Note 8). The resulting yellow solution is added to the 100-mL flask with the syringe pump (Note 9) at 0 °C (Note 26) over 40 minutes (Figure 3B) (Note 27). The solution is slowly warmed to room temperature (Note 3) (Figure 3C) and stirred (Note 10) for 12 h (Note 28) (Figure 3D). The reaction flask is cooled to 0 °C (Note 24) and saturated ammonium chloride solution (20 mL) (Notes 29 and 30) (Figure 3E) and 1N HCl solution (30 mL) (Note 31) are added slowly. The biphasic mixture is transferred to a 250-mL separatory funnel (Note 13). The organic layer is collected, and the aqueous layer is extracted with ethyl acetate (4 × 40 mL) (Note 14). The combined organic phases are dried with sodium sulfate (30 g) (Note 16), filtered (Note 18) into a 500-mL single-necked round-bottomed flask (24/40 joint) with ethyl acetate washings (3 × 15 mL) (Note 14). The solution is concentrated with the aid of a rotary evaporator (Note 18) to afford crude, light yellow solids. The crude product is purified by chromatography on silica (Note 32) to afford diphenyl((S)-1-((S)-1-phenylethyl)aziridin-2-yl)methanol **3** (2.97 g, 91%, 99% ee, 99% purity) (Notes 33, 34, and 35) as a white solid (Figure 4).

Figure 4. Product 3 (photo provided by submitter)

Notes

1. Prior to performing each reaction, a thorough hazard analysis and risk assessment should be carried out with regard to each chemical substance and experimental operation on the scale planned and in the context of the laboratory where the procedures will be carried out. Guidelines for carrying out risk assessments and for analyzing the hazards associated with chemicals can be found in references such as Chapter 4 of "Prudent Practices in the Laboratory" (The National Academies Press, Washington, D.C., 2011; the full text can be accessed free of charge at https://www.nap.edu/catalog/12654/prudent-practices-in-the-laboratory-handling-and-management-of-chemical. See also "Identifying and Evaluating Hazards in Research Laboratories" (American Chemical Society, 2015) which is available via the associated website "Hazard Assessment in Research Laboratories" at https://www.acs.org/content/acs/en/about/governance/committees/chemicalsafety/hazard-assessment.html. In the case of this procedure, the risk assessment should include (but not necessarily be limited to) an evaluation of the potential hazards associated with (S)-(–)-1-phenylethylamine, ammonium chloride, chloroform, 1,3,5-trimethoxybenzene, ethyl 2,3-dibromopropanoate, ethyl acetate, hexanes, hydrochloric acid, magnesium sulfate anhydrous, methylene chloride, phenylmagnesium chloride, silica, sodium bicarbonate, sodium chloride, sodium sulfate anhydrous, THF, toluene, triethylamine, as well as the proper procedures for working with dry ice and under an inert atmosphere.
2. Unless otherwise reported, all glassware was dried in a 120 °C oven prior to use and then brought down to room temperature under an inert atmosphere.
3. The room temperature throughout this manuscript refers to temperatures between 22 °C and 23 °C. Room temperature in the checker's lab was 21 °C.
4. (S) (–)-1-Phenylethylamine (98%, 98% ee) was purchased from Sigma-Aldrich and used as received.
5. Triethylamine (99%) purchased from Sigma Aldrich under SureSeal is sufficiently dry (KF < 200 ppm) and does not require distillation over NaH. The submitters purchased triethylamine (99%) from Fisher Scientific and distilled the liquid under nitrogen from sodium hydride before the use.

6. The checkers purchased toluene from Sigma Aldrich (SureSeal) and used the material as received. The submitters purchased toluene (Certified ACS) from Fisher Scientific and purified it by pressure filtration under nitrogen through activated alumina prior to use.
7. Ethyl 2,3-dibromopropanoate (for synthesis, >98%) was purchased from Sigma-Aldrich and used as received.
8. The submitters report that the flask and its contents were sonicated for 30 sec. Branson® Ultrasonic Bath (115 Vac, 60 Hz) was used with 2.8 L (0.75-gal) tank filled with water at room temperature.
9. The Fisherbrand™ syringe pump was setup with a built-in syringe size table for Air-Tite™ All-Plastic Norm-Ject™ Syringes. A minor exotherm from 21 °C to 23 °C was observed throughout addition.
10. IKA RET basic hot plate stirrer (115V, 620W, 50-60 Hz) and Cole-Parmer IKA C-Mag hot plate stirrer (115V, 1000W, 50-60 Hz) were used. Unless indicated otherwise, 500 rpm was used for stirring.
11. The submitter's studies used Fisher Chemical™ silicone oil for the oil bath. Unless specified differently, the oil in the oil bath should cover the reaction mixture in the reaction flask while heating. Unless otherwise reported, the temperatures throughout this manuscript refer to temperatures of oil in oil baths which were detected by the stirring plates' temperature probes. The checkers confirmed that aluminum heating mantels were also suitable heat sources.
12. The reaction can be monitored by TLC (SiO$_2$, hexanes/EtOAc 4/1, starting material **a**: R$_f$ 0.17, starting material **b**: R$_f$ 0.62, product **1**: R$_f$ 0.42, product **2**: R$_f$ 0.30; UV-C 254 nm) to observe complete consumption of starting material **a** (S refers to starting materials **a** and **b**. C refers to co-spot of reaction mixture and starting materials. R refers to the reaction mixture.).

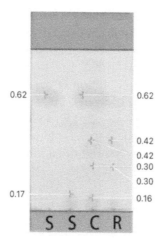

Figure 5. TLC monitoring of Step A (photo provided by submitters)

13. The quality of the deionized water was not determined.
14. Ethyl acetate (Certified ACS) was purchased from Fisher Scientific and used as received.
15. Sodium chloride (Crystalline/Certified ACS) was purchased from Fisher Scientific and added to a bottle of deionized water until solids crashed out.
16. Sodium sulfate anhydrous (Granular/Certified ACS) was purchased from Fisher Scientific and used as received.
17. To wash the filter cake effectively, vacuum was turned off between separate washing cycles, washing solvent was added and the resultant mixture was stirred thoroughly with a stainless-steel spatula before the washing solvent was removed by vacuum suction.
18. BUCHI™ Rotavapor™ Scholar with Dry Ice Cold Trap Condenser was connected to Heidolph™ Valve-Regulated Vacuum Pump. Unless specified differently, water bath remained at 30 °C and the vacuum was regulated to 20 mmHg.
19. The crude material was loaded onto a slurry-packed (hexane) column (ID 42 mm) containing SiO_2 (150 g, 40 – 63 μm, 60 Å silica gel purchased from SiliCycle Inc.), and the flask was then rinsed with hexanes (7 mL) which was loaded afterwards. After loading, solvents were eluted under positive nitrogen pressure and fractions were taken in 25-mL tubes. The solvent system was switched to 900 mL of 8/1 hexane/EtOAc (ACS grade purchased from Fisher Scientific which was

used as received) and product **1** (R_f 0.42, hexane/EtOAc 4/1, v/v) eluted first and was typically removed with this mixture. Fractions 21 through 35 were combined, concentrated on a rotary evaporator (30 °C, 780 to 20 mmHg), and dried in vacuo (1–2 mmHg) at ambient temperature for 12 h. After elution of product **1**, the solvent system was switched to 800 mL of 5/1 hexane/EtOAc, and elution of the product **2** (R_f 0.30, hexane/EtOAc 4/1, v/v) was completed this solvent mixture. Fractions 45 through 60 were combined, concentrated on a rotary evaporator (30 °C, 780 to 20 mmHg), and dried in vacuo (1-2 mmHg) at ambient temperature for 12 h.

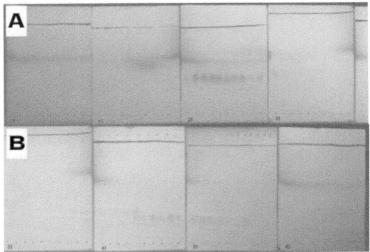

Figure 6. TLC analysis of the fractions. (Visualization with UV-C 254 nm) A) fractions 1 through 40; B) fractions 31 through 70 (photos provided by submitters)

20. The product (**1**) exhibited the following properties: $[\alpha]_D^{23}$ –80.43 (*c* 0.50, CHCl$_3$); R_f 0.42 (4/1, hexane/EtOAc, v/v); IR (film): 3062, 2978, 2929, 1741, 1725, 1601, 1494, 1448, 1410, 1384, 1281, 1233, 1182, 1086, 1028, 959, 763, 698 cm^{-1}; ^{1}H NMR (500 MHz, CDCl$_3$) δ: 1.32 (t, *J* = 7.1, 3H), 1.50 (d, *J* = 6.6 Hz, 3H), 1.61 (dd, *J* = 6.4, 0.9 Hz, 1H), 2.16 (dd, *J* = 3.2, 1.1 Hz, 1H), 2.22 (dd, *J* = 6.4, 3.2 Hz, 1H), 2.5 (q, *J* = 6.6 Hz, 1H), 4.16 – 4.32 (m, 2H), 7.24 – 7.29 (m, 1H), 7.31 – 7.38 (m, 2H), 7.41 (d, *J* = 7.3, 2H). ^{13}C NMR (126 MHz, CDCl$_3$) δ: 14.3, 23.3, 34.01, 38.3, 61.2, 70.1, 127.0, 127.4, 128.5, 143.6, 171.0; HRMS (ESI) *m/z* calcd for C$_{13}$H$_{17}$NO$_2$Na [M+Na]$^+$

242.1151, found 242.1148. Purity was determined by quantitative ^{1}H NMR spectroscopic analysis using 1,3,5-trimethoxybenzene as an internal standard to be 97% by weight. The corrected yield based on purity was 2.13 g (42%). (The enantiomeric excess (ee) of this product could not be determined by the available HPLC or SFC techniques, and the product ee is reported based on the ee of precursor measured after the Grignard reaction step.)

21. A second reaction on identical scale provided 1.92 g (42%) of the same compound.
22. The product **(2)** exhibited the following properties: $[\alpha]_D^{23}$ +41.70 (c 0.35, CHCl$_3$); R$_f$ 0.28 (4/1, hexanes/EtOAc, v/v); IR (film): 3061, 2976, 2928, 17439, 1493, 1447, 1412, 1282, 1235, 1184, 1089, 1028, 960, 759, 699 cm^{-1}; ^{1}H NMR (500 MHz, CDCl$_3$) δ: 1.23 (t, J = 7.1 Hz, 3H), 1.48 (d, J = 6.6 Hz, 3H), 1.79 (dd, J = 6.5, 1.0 Hz, 1H), 2.07 (dd, J = 6.5, 3.1 Hz, 1H), 2.35 (dd, J = 3.1, 1.0 Hz, 1H), 2.59 (q, J = 6.6 Hz, 1H), 4.17 (qd, J = 7.1, 3.0 Hz, 2H), 7.23 – 7.30 (m, 1H), 7.32 – 7.40 (m, 3H); ^{13}C NMR (126 MHz, CDCl$_3$) δ: 14.2, 23.6, 34.9, 37.2, 61.0, 69.8, 126.5, 127.2, 128.5, 143.8, 170.7. HRMS (ESI) m/z calcd for C$_{13}$H$_{17}$NO$_2$Na [M+Na]$^+$ 242.1157, found 242.1154. Purity was determined by quantitative ^{1}H NMR spectroscopic analysis using trimethoxybenzene as an internal standard to be 96% by weight.
23. A second reaction on identical scale provided 1.82 g (40%) of the same compound.
24. Phenylmagnesium chloride solution (2.0 M in THF) was purchased from Sigma-Aldrich and titrated to be 1.48M before use following the published procedures (Watson, S. C.; Eastham, J. F. *J. Organomet. Chem.* **1967**, *9*, 165–168).
25. The 0 °C temperature was reached and maintained by mixing water with ice.
26. Tetrahydrofuran (HPLC) was purchased from Fisher Scientific and purified by pressure filtration under nitrogen through activated alumina prior to use.
27. There is an exotherm in the first 5 min of addition, from 0.5 °C to 5.5 °C. After this exotherm, the solution cools back to 0–0.5 °C).
28. The reaction can be monitored by TLC (SiO$_2$, Hexane/EtOAc 4/1, starting material **1**: R$_f$ 0.42, product **3**: R$_f$ 0.78; UV-C 254 nm) to observe complete consumption of starting material **1** (S refers to starting material **1**. C refers to co-spot of reaction mixture and starting material. R refers to the reaction mixture.).

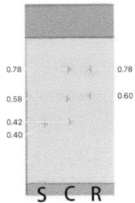

Figure 7. TLC monitoring of Step B

29. Ammonium chloride (Crystalline/Certified ACS) was purchased from Fisher Scientific and added to a bottle of deionized water until solids crashed out.
30. A very strong exotherm was observed. The temperature rises from 0.5 °C to 40.4 °C in 2–3 min if the ammonium chloride solution is added at rate of ~4 mL/min. It is recommended that an addition of <1 mL/min be performed using an addition funnel, with the expectation that a temperature rise to 25–30 °C will be observed. After ~5 mL of sat. aq. NH$_4$Cl are added, no further exotherm is observed.
31. Hydrochloric acid (ACS reagent, 37%) was purchased from Sigma-Aldrich and diluted to 1N with deionized water. No exotherm was observed with 1N HCl solution, when added at 10 mL/min.
32. The crude was loaded onto a slurry-packed (hexane) column (ID 42 mm) containing SiO$_2$ (150 g, 40 – 63 µm, 60 Å silica gels purchased from SiliCycle Inc.), and the flask was then rinsed with hexanes (10 mL) which was loaded afterwards. After loading, solvents were eluted under positive nitrogen pressure and fractions were taken in 25-mL tubes. The solvent system was switched to 420 mL of 20/1 hexane/EtOAc (ACS grade purchased from Fisher Scientific which was used as received) and followed by 450 mL of 15/1 hexane/EtOAc. Product **3** (R_f 0.78, hexane/EtOAc 4/1, v/v) eluted, fractions 36 through 52 were combined, concentrated on a rotary evaporator (30 °C, 780 to 20 mmHg), and dried in vacuo (1-2 mmHg) at ambient temperature for 12 h.

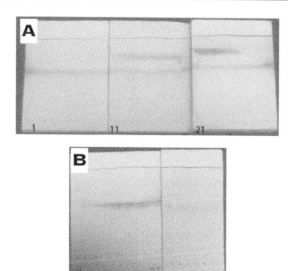

Figure 8. TLC analysis of the fractions. (Visualization with UV-C 254 nm) A) fractions 1 through 30; B) fractions 36 through 53

33. The product (3) exhibited the following properties: $[\alpha]_D^{23}$ -68.20 (c 0.10, CHCl$_3$); R$_f$ 0.66 (4/1, hexanes/EtOAc, v/v); mp 125.3–125.5 °C; IR (powder): 3351 (br), 3059, 3027, 2981, 2966, 2926, 1599, 1492, 1449, 1354, 1342, 1300, 1188, 1167, 1066, 1029, 1014, 981, 931, 768, 747, 691, 644, 638, 612 cm^{-1}; ^{1}H NMR (500 MHz, CDCl$_3$) δ: 0.91 (d, J = 6.2 Hz, 3H), 1.60 (d, J = 6.0 Hz, 1H), 1.99 (bs, 1H), 2.56 (bs, 1H), 2.75 (q, J = 6.2 Hz, 1H), 4.14 (s, 1H), 7.24 – 7.32 (m, 3H), 7.31 – 7.42 (m, 8H), 7.51 – 7.60 (m, 4H); ^{13}C NMR (176 MHz, CDCl$_3$) δ: 23.6, 30.8, 47.5, 68.2, 74.0, 126.1, 126.78, 126.8, 126.9, 127.3, 127.4, 128.1, 128.3, 128.5, 144.3, 145.0, 148.0; HRMS (ESI) m/z calcd for C$_{23}$H$_{23}$NONa [M+Na]$^+$ 352.1677, found 352.1673. Purity was determined by quantitative ^{1}H NMR spectroscopic analysis using 1,3,5-trimethoxybenzene as an internal standard to be 99% by weight.
34. The ee was determined to be 99% by HPLC analysis with a Waters Alliance e2695 Separations Module HPLC system equipped with a CHIRALPAK IA column (length 250 mm, I.D. 4.6 mm) (90:10 hexanes/isopropanol, 1.0 ml/min), tr = 5.7 min (S), 6.6 min (R).
35. A second reaction on a similar scale provided 2.46 g (93%) of the same product.

Working with Hazardous Chemicals

The procedures in *Organic Syntheses* are intended for use only by persons with proper training in experimental organic chemistry. All hazardous materials should be handled using the standard procedures for work with chemicals described in references such as "Prudent Practices in the Laboratory" (The National Academies Press, Washington, D.C., 2011; the full text can be accessed free of charge at http://www.nap.edu/catalog.php?record_id=12654). All chemical waste should be disposed of in accordance with local regulations. For general guidelines for the management of chemical waste, see Chapter 8 of Prudent Practices.

In some articles in *Organic Syntheses*, chemical-specific hazards are highlighted in red "Caution Notes" within a procedure. It is important to recognize that the absence of a caution note does not imply that no significant hazards are associated with the chemicals involved in that procedure. Prior to performing a reaction, a thorough risk assessment should be carried out that includes a review of the potential hazards associated with each chemical and experimental operation on the scale that is planned for the procedure. Guidelines for carrying out a risk assessment and for analyzing the hazards associated with chemicals can be found in Chapter 4 of Prudent Practices.

The procedures described in *Organic Syntheses* are provided as published and are conducted at one's own risk. *Organic Syntheses, Inc.*, its Editors, and its Board of Directors do not warrant or guarantee the safety of individuals using these procedures and hereby disclaim any liability for any injuries or damages claimed to have resulted from or related in any way to the procedures herein.

Discussion

Enantioselective alkylation is one of the most important transformations in the asymmetric catalysis towards biologically and pharmacologically valuable compounds.[2] Particularly, asymmetric addition of alkyl- and arylzinc species with chiral ligands to aromatic aldehydes has been extensively studied.[3] *N,N*-Dibutylnorephedrine (DBNE) is often considered as one of the most versatile ligands in this field.[4]

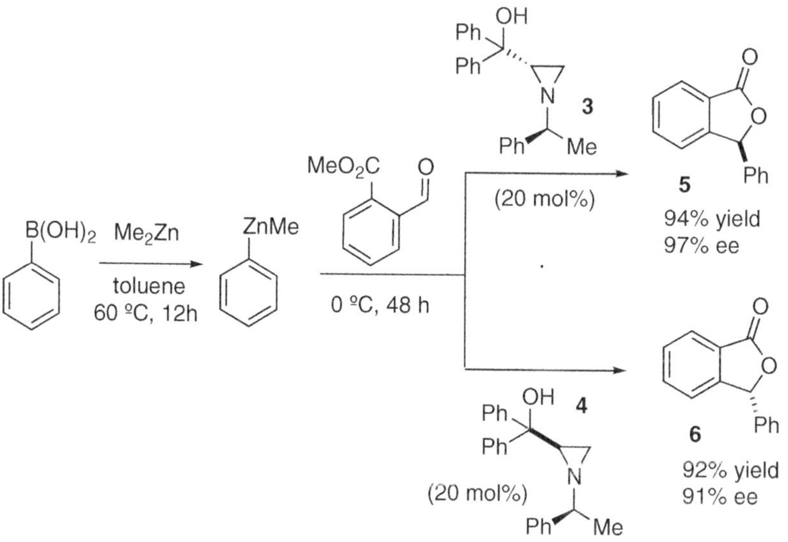

Scheme 1. Application to enantioselective synthesis of 3-aryl phthalides

Aziridines have gained significant amounts of attention in synthesis and drug discovery as they are featured in various natural products.[5] Recent years have seen a rise in the number of asymmetric catalysis with aziridine-based catalysts as well. Such examples include enantioselective arylation with ligand **3** as well as its diastereomer **4** derived from alkylation side-product **2**. For the reaction with aromatic aldehydes, ligand **3** often exhibits better selectivity profiles. One of the selected applications of this ligand involved the enantioselective formation of 3-aryl phthalides such as **5** and **6** (Scheme 1) starting with methyl 2-formylbenzoate and various arylboronic acids.[6] In our recent studies of novel C_2-symmetric SPIROL ligands[7], we discovered that the enantiopurity of benzylic alcohol is a key factor to the selectivity of diastereoselective spiroketalization. Commercially available (-)-DBNE would only yield up to 94% ee in this transformation while catalyst **3** could provide products in 99% ee when R is MOM (Scheme 2). The procedure described here is adapted from the previous report.[8] It is comprised of a two-step synthesis of chiral aziridine catalyst **3** from commercially available starting materials.

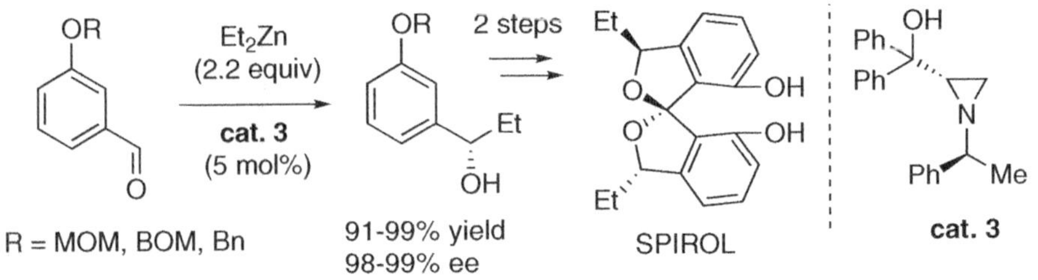

Scheme 2. Application to the synthesis of SPIROL C2-symmetric ligands

References

1. Department of Chemistry, University of Michigan, Ann Arbor, MI 48109, USA, E-mail: nagorny@umich.edu; ORCID: 0000-0002-7043-984X. Financial support from National Science Foundation (grant CHE-1955069 to PN) is gratefully acknowledged.
2. Kohler, M. C.;Wengryniuk, S. E.; Coltart, D. M. *Stereoselective Synthesis of Drugs and Natural Products*, Wiley, Hoboken, 2013; pp.1–31.
3. (a) Binder, C. M.; Singaram, B. *Org. Prep. Proced. Int.* **2011**, *43*, 139–208; (b) Lemire, A.; Côté, A.; Janes, M. K.; Charette, A. B. *Aldrichimica Acta*, **2009**, *42*, 71–83.
4. Soai, K.; Yokoyama, S.; Hayasaka, T. *J. Org. Chem.* **1991**, *56*, 4264–4268.
5. (a) Degennaro, L.; Trinchera, P.; Luisi, R. *Chem. Rev.* **2014**, *114*, 7881–7929; (b) Sweeney, J. B. *Chem. Soc. Rev.* **2002**, *31*, 247–258; (c) Singh, G. S. *Advances in Heterocyclic Chemistry*; Elsevier, **2019**; Vol. 129, pp 245–335.
6. Song, X.; Hua, Y.-Z.; Shi, J.-G.; Sun, P.-P.; Wang, M.-C.; Chang, J. *J. Org. Chem.* **2014**, *79*, 6087–6093.
7. Arguelles, A. J.; Sun, S.; Budaitis, B. G.; Nagorny, P. *Angew. Chem., Int. Ed.* **2018**, *57*, 5325–5329.
8. Wang, M.; Wang, Y.; Li, G.; Sun, P.; Tian, J.; Lu, H. *Tetrahedron Asymmetry* **2011**, *22*, 761–768.

Appendix
Chemical Abstracts Nomenclature (Registry Number)

(S)-(-)-1-Phenylethanamine: S-1-Phenylethylamine; (2627-86-3)
Triethylamine; (121-44-8)
Ethyl 2,3-dibromopropanoate: Ethyl 2,3-dibromopropionate; (3674-13-3)
Toluene: Benzene, methyl-; (108-88-3)
Ethyl acetate; (141-78-6)
Phenylmagnesium chloride; (100-59-4)

Siyuan Sun was born in Suzhou, China. He obtained a B.S. degree from the Purdue University, where he studied the monoterpene indole alkaloids synthesis under the direction of Prof. Mingji Dai. He is currently pursuing his Ph.D. in the laboratory of Prof. Pavel Nagorny at the University of Michigan, Ann Arbor where he studies the synthesis and catalysis of novel SPIROL-based ligands for asymmetric catalysis.

Dr. Pavel Nagorny received his B.S. degree in chemistry in 2001 from the Oregon State University. After earning his Ph.D. degree in chemistry from Harvard University in 2007, he spent three years as a postdoctoral fellow with at the Memorial Sloan-Kettering Cancer Center. In 2010, Pavel joined the faculty of the University of Michigan. From 2014-2017 he was appointed as a William R. Roush Assistant Professor in Chemistry, and in 2017 he was promoted to the rank of Associate Professor with tenure. Pavel's research group interests range from natural product synthesis to asymmetric catalysis, organocatalysis and carbohydrate chemistry.

Matthew S. Winston joined Small Molecule Process Research and Development at Merck & Co., Inc. in 2018. His research focuses on using mechanistic analysis to solve critical problems in process chemistry. He received his A.B. degree from Harvard University, and his Ph.D. under the supervision of Professor John E. Bercaw at the California Institute of Technology. He was then a NIH postdoctoral fellow with Professor F. Dean Toste at the University of California-Berkeley, and a Glenn T. Seaborg Fellow at Los Alamos National Laboratory.

Large-Scale Preparation of Oppolzer's Glycylsultam

Upendra Rathnayake,[†] H. Ümit Kaniskan,[‡] Jieyu Hu,[‡] Christopher G. Parker,[‡] and Philip Garner*[,†1]

[†]Department of Chemistry, Washington State University, Pullman, Washington 99164-4630, United States
[‡] Department of Chemistry, Case Western Reserve University, 10900 Euclid Avenue, Cleveland, Ohio 44106-7078. United States

Checked by Bogdan R. Brutiu, Martina Drescher, Daniel Kaiser and Nuno Maulide

Procedure (Note 1)

A. *Bromoacetylsultam* **2**. To a 3 L three-necked, round-bottomed flask (24/40 joints) equipped with a 6.5 cm egg-shaped Teflon-coated magnetic stir bar, 60 mL graduated pressure-equalizing dropping funnel (Note 2), fitted with an argon inlet (Note 3) and a rubber septum are added (2S)-bornane-2,10-sultam **(1)** (30.0 g, 0.139 mol, 1.00 equiv) (Note 4) and dry THF (800 mL) (Note 5). The reaction mixture is stirred until **1** is completely dissolved and the resulting solution is cooled in a –78 °C dry ice-acetone bath for 45 min. At this point, 1.6 M *n*-butyllithium in hexanes (97.0 mL, 0.155 mol, 1.12 equiv) (Notes 6 and 7) is transferred by cannula into the pressure-equalizing dropping funnel and then added dropwise over 30 min to the stirring (340 rpm) reaction mixture at –78 °C (Note 8) (Figure 1). The resulting solution is stirred at –78 °C for an additional 1 h.

Figure 1. Reaction set-up for the addition of *n*-butyllithium to the sultam solution (photo provided by submitter)

To a 500 mL single-necked round-bottomed flask fitted with a rubber septum, equipped with an argon inlet and charged with dry THF (150 mL) is added bromoacetyl bromide (13.6 mL, 31.5 g, 0.156 mol, 1.12 equiv) (Note 9) *via* a syringe in one portion. This solution is added to the sultam lithium salt mixture at –78 °C *via* cannula over 1.5 h, during which time the solution turned yellow (Figure 2).

Figure 2. Reaction set-up for the addition of bromoacetyl bromide to the sultam lithium salt mixture (photo provided by submitter)

The resulting mixture is further stirred (180 rpm) for 2 h at –78 °C. Analysis by TLC shows complete consumption of sultam **1** (Note 10). The dry ice-acetone bath is removed, and the reaction mixture is diluted with water (500 mL) (Note 11) over 10 min and brought to 23–25 °C over 30 min. Approximately, half of the reaction mixture is transferred to a 2 L separatory funnel, and the aqueous phase is separated and extracted with diethyl ether (3 × 500 mL) (Note 12). Then the process is repeated with the remaining half of the reaction mixture. The combined organic layers are dried over 300 g of anhydrous Na_2SO_4 (Note 13), decanted and concentrated by rotary evaporation (35 °C, starting with 50 mmHg and gradually lowering to 10 mmHg). The product is further dried under high vacuum at 0.1 mmHg to afford crude **2** (51.4 g, 86%, 78% purity by qNMR) as an amber colored solid (Notes 14, 15, 16, and 17) (Figure 3).

B. *Azidoacetylsultam* **3**. To a 1 L three-necked, round-bottomed flask (24/40 joint) equipped with a 6.5 cm egg-shaped Teflon-coated magnetic stir bar, fitted with an argon inlet and a rubber septum are added crude **2** (51.4 g, 78% purity, 0.119 mol, 1.00 equiv) (Note 18) and reagent grade DMF (350 mL) (Note 19). To this reaction mixture, NaN_3 (9.98 g, 0.154 mol, 1.29 equiv) (Notes 20 and 21) is added in one portion and stirred for 8 h at 23–25 °C under argon (Figure 4).

Figure 3. Crude bromoacetylsultam 2 (photo provided by submitter)

Figure 4. Reaction mixture after the addition NaN$_3$ (photo provided by submitter)

During the course of the reaction the color of the solution turns dark-brown. The progress of the reaction is monitored by ^{1}H NMR (Note 22).

Once the starting material **2** completely disappears, the reaction is diluted with water (150 mL) and transferred to a 2 L separatory funnel, and the aqueous phase is extracted with diethyl ether-hexanes (1:1, v/v) (4 × 500 mL). Approximately half of the combined organic layers is transferred to a 2 L separatory funnel and washed with water (2 × 350 mL). The washing process is repeated with the remaining half of the combined organic layers. The organic layers are recombined and dried over anhydrous Na_2SO_4 (300 g), filtered over cotton, and concentrated by rotary evaporation (35 °C, 10 mmHg). The product is further dried under high vacuum (0.1 mmHg) to afford crude **3** (35.2 g, 83%, 84% purity by qNMR) as a pale-yellow solid (Notes 23, 24, 25, and 26) (Figure 5).

Figure 5. Crude azidoacetylsultam **3** (photo provided by submitter)

C. *Glycylsultam* **4**. To a 2 L three-necked, round-bottomed flask (24/40 joint) equipped with a 6.5 cm egg shaped Teflon-coated magnetic stir bar, fitted with rubber septa and a three-way glass adaptor (24/40) connected to an argon inlet and a vacuum inlet (130 mm Hg, Note 27) are added crude **3** (35.2 g, 84% purity, 0.099 mol, 1.00 equiv) (Notes 28 and 29) and HPLC grade MeOH (800 mL) (Note 30). To this reaction mixture, conc. HCl (12.1 M, 20.1 mL, 0.243 mol, 2.43 equiv) (Note 31) is added at 23–25 °C over 5 min. The argon flow sweeping through the round-bottomed flask is increased and a slurry consisting of 10 wt.% Pd-C (2.50 g) (Note 32) in water (25 mL) (Note 33) prepared in a 50 mL beaker is added to the reaction mixture over 15 min. Any remaining 10 wt.% Pd-C in the beaker and in the round-bottomed flask wall is rinsed down and added to the reaction mixture with another portion of water (5-10 mL). The third neck is capped with a septum and the suspension is stirred under argon. While the stirring is maintained the flask is evacuated by switching to the vacuum (130 mmHg) just until the solvent starts to bubble, then back-filled with argon. This vacuum-refilling cycle is repeated three times.

Then the argon-vacuum adaptor is replaced with a three-way glass adaptor (24/40 joint) fitted with a hydrogen balloon and a vacuum inlet (130 mmHg). The stirring reaction flask is evacuated, while keeping the balloon closed off, until the solvent starts to bubble, and then back-filled with hydrogen by closing the vacuum and opening the balloon to the flask. This process is repeated four times. The reaction mixture is stirred under hydrogen for another 24 h (Figure 6), during which period the hydrogen balloon is refilled every 6 h and the flask is subjected to a cycle of vacuum evacuation and backfilling with hydrogen (Note 34).

Figure 6. Reaction set-up for the hydrogenolysis of azidoacetylsultam 3 (photo provided by submitter)

After 24 h, the adaptor with the hydrogen balloon is replaced with the argon-vacuum adaptor. Hydrogen is removed under vacuum and backfilled with argon, repeating the cycle three times. Then another portion of slurry consisting of 10 wt.% Pd-C (2.50 g) in water (25 mL) is added (Note 35) over 15 min following the procedure explained above. The reaction setup is assembled for the hydrogenolysis reaction as previously described, and the reaction was placed under a hydrogen atmosphere for another 12 h, at which point TLC analysis showed complete consumption of the starting material (Note 36). The hydrogen balloon is removed, and the crude reaction mixture is filtered under vacuum (130 mmHg), through a pad of Celite (150 g) (Note 37) packed into a Büchner funnel (outer diameter = 10.5 cm, height = 5 cm), always taking care to maintain the solvent level above the Celite bed (Note 38). Then the Celite is washed with methanol (800 mL), and the organic layer is concentrated by rotary evaporation (35 °C, 10 mmHg). The product is subjected to high vacuum (0.1 mmHg) until it reaches a constant weight (Note 39) to afford crude **5** (38.8 g) as a colorless solid (Figure 7a) (Notes 40 and 41).

The crude ammonium salt **5** (2.50 g, Note 42) is dissolved in H_2O (125 mL) and transferred to a 1 L separatory funnel, and the aqueous phase is extracted with CH_2Cl_2 (125 mL) to remove any neutral material. The aqueous phase is transferred to a 500 mL round-bottomed flask that contains a stir bar, chilled to 0 °C in an ice-water bath and neutralized by adding $NaHCO_3$ (sat. aq.) in 5–10 drop increments *via* a pasture pipette while gently stirring (80 rpm). The pH is monitored after each series of addition using pH paper (Note 43). Once the pH reaches 7.2–7.4, the solution is transferred to a 1 L separatory funnel, and the aqueous phase is extracted with CH_2Cl_2 (3 × 250 mL). The organic layers are combined and dried over anhydrous Na_2SO_4 (200 g), decanted, and concentrated by rotary evaporation (starting with 50 mmHg and gradually lowering to 10 mmHg), while maintaining the water bath temperature at 17 °C until the final volume reaches approximately 1–2 mL. The remaining solvent is removed by rotary evaporation at 50 mmHg, leading to rapid precipitation of the amine (Note 44). The product is subjected to high vacuum (0.1 mmHg) to afford glycylsultam **4** (1.49 g, 84%, 98% purity) as a colorless solid (Figure 7b) (Notes 45, 46 and 47).

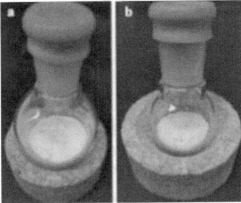

Figure 7. (a) Glycylsultam ammonium chloride 5, (b) Glycylsultam 4 (photo provided by submitter)

Notes

1. Prior to performing each reaction, a thorough hazard analysis and risk assessment should be carried out with regard to each chemical substance and experimental operation on the scale planned and in the context of the laboratory where the procedures will be carried out. Guidelines for carrying out risk assessments and for analyzing the hazards associated with chemicals can be found in references such as Chapter 4 of "Prudent Practices in the Laboratory" (The National Academies Press, Washington, D.C., 2011; the full text can be accessed free of charge at https://www.nap.edu/catalog/12654/prudent-practices-in-the-laboratory-handling-and-management-of-chemical. See also "Identifying and Evaluating Hazards in Research Laboratories" (American Chemical Society, 2015) which is available via the associated website "Hazard Assessment in Research Laboratories" at https://www.acs.org/content/acs/en/about/governance/committees/chemicalsafety/hazard-assessment.html. In the case of this procedure, the risk assessment should include (but not necessarily be limited to) an evaluation of the potential hazards associated with (2S)-bornane-2,10-sultam, tetrahydrofuran, 1.6 M n-Butyllithium, hexanes, bromoacetyl bromide, diethyl ether, sodium sulfate, N,N-dimethylformamide, sodium azide, methanol, conc. HCl, 10 wt.% palladium on carbon, sodium bicarbonate and dichloromethane as well as the proper

procedures for handling, addition, cannulation and quenching of highly reactive n-butyllithium and bromoacetyl bromide, handling and addition of potentially explosive sodium azide and hydrogenolysis reaction using palladium on carbon.

2. A 100 mL dropping funnel would be ideal for the experiment since it reduces the measuring error caused by filling *n*-butyllithium two times to make up the required volume.

3. The reaction is very sensitive to moisture and air. Hence, it was carried out with extreme care, under an inert (argon) environment. All the joints in the experimental setup were sealed using parafilm tape.

4. (2S)-Bornane-2,10-sultam (98%) was purchased from AK scientific and used as received. The checkers purchased this compound from Fluorochem.

5. Tetrahydrofuran (certified ACS grade) was purchased from Fisher Scientific and was distilled from Na/benzophenone under argon. The reaction is very sensitive to water, therefore care was taken to use dry solvents. The checkers purchased anhydrous THF from TCI.

6. 1.6 M *n*-Butyllithium in hexanes was purchased from Sigma-Aldrich and used as received.

7. Weight of the 1.6 M *n*-butyllithium used is given based on the *n*-butyllithium moles.

$$\text{wt} = \frac{1.60 \text{ mol} \times 97.00 \text{ mL} \times 64.06 \text{ g/mol}}{1000.00 \text{ mL}} = 9.94 \text{ g}$$

8. After complete addition of *n*-butyllithium in hexanes, the pressure-equalizing dropping funnel can be quickly replaced with a rubber septum at an increased argon flow.

9. Bromoacetyl bromide (≥98%) was purchased from Sigma-Aldrich and used as received.

10. The reaction was monitored by TLC using silica gel HL TLC plates, purchased from Sorbent Technologies, Inc. The plate was developed using (3:1, v/v) hexanes-ethyl acetate as the mobile phase and visualized using cerium ammonium molybdate stain (Figure 8). The starting material forms a deep-blue colored spot and the product forms a gray-blue colored spot on the TLC upon heating with the stain. The R_f value of the starting material **1** was 0.26 and the R_f of the product **2** was 0.41.

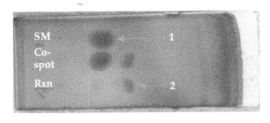
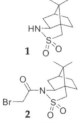

Figure 8. TLC analysis of the reaction mixture at 2 h, after addition of bromoacetyl bromide (photo provided by submitter)

11. Addition of water was done slowly and while the reaction mixture was chilled. Even though water can react vigorously with the small amount of excess bromoacetyl bromide and/or n-butyllithium remaining in the reaction mixture, such was not observed during the dilution.
12. Diethyl ether, anhydrous (BHT stabilized, certified ACS grade) was purchased from Fisher Scientific and used as received.
13. Sodium sulphate, anhydrous was purchased from MilliporeSigma and used as received.
14. The crude product is of sufficient purity for use in Step B, therefore purification is not required at this stage. The presence of a small amount of bromoacetic acid was occasionally observed in the product by ^{1}H NMR and/or a crude weight is occasionally determined to be greater than the theoretical value. If this is the case, partitioning the crude material between NaHCO$_3$ (sat. aq.) (100 mL) and diethyl ether (500 mL) can address those issues. In fact, this route to synthesize Oppolzer's glycylsultam does not require any chromatographic or crystallization purification apart from a simple organic wash after the hydrogenolysis at the final step. Hence crude **2** was taken directly to the next step.
15. Characterization data for the crude product **2** (data provided by checkers): ^{1}H NMR (400 MHz, CDCl$_3$) δ: 4.32 (d, J = 13.1 Hz, 1H), 4.19 (d, J = 13.1 Hz, 1H), 3.90 (dd, J = 7.6, 5.1 Hz, 1H), 3.49 (app q, J = 13.8 Hz, 2H), 2.19–2.02 (m, 2H), 2.00–1.82 (m, 3H), 1.53–1.31 (m, 2H), 1.14 (s, 3H), 0.97 (s, 3H); ^{13}C NMR (101 MHz, CDCl$_3$) δ: 164.6, 65.6, 52.8, 49.1, 48.0, 44.6, 38.0, 32.9, 27.6, 26.5, 20.8, 20.0; IR (neat): 2989, 2959, 2884, 2360, 2341, 2256, 1696, 1616, 1541, 1507, 1481, 1456, 1435, 1411, 1395, 1373, 1327, 1313, 1301, 1258, 1232, 1205, 1167, 1152, 1132, 1110, 1084, 1059, 1039, 987, 952, 939, 911, 888, 868, 843, 805, 774 cm^{-1}; HRMS (ESI) calcd for C$_{12}$H$_{18}$NO$_3$SBrNa [M + Na]: 358.0083. Found: 358.0092. Purity

of the crude product (**2**) was assessed to be 78% based on qNMR analysis using 1,3,5-trimethoxybenzene as an internal standard. The checkers performed a second run on half scale, yielding 23.7 g (89% yield, 88% purity). The submitters reported the following result on a full scale reaction: 47.7 g, 82% yield, 81% purity.

16. The major impurity present in crude **2** is the unreacted camphor sultam **1** (Figure 8), which is completely inert towards ancillary transformations. But if further purification is required, flash chromatography can be performed to obtain pure **2** as follows: Silica gel 60 Å was purchased from Sorbent Technologies, Inc. The flash column (dimensions: 13 mm inner diameter, 203 mm in height) is wet packed with silica gel (15 g; 150 mm, height of the silica bed in the column) using a solution of (4:1, v/v) hexanes-ethyl acetate. Crude bromoacetylsultam **2** (59 mg, dissolved in 1 mL of methylene chloride) is wet loaded onto the packed silica column. The column is eluted with 100 mL of (4:1, v/v) hexanes-ethyl acetate and then continued with 200 mL of (3:1, v/v) hexanes-ethyl acetate until the product completely elutes off the column (Figure 9). The fraction collection is begun (5-mL fractions) immediately after starting the solvent elution and each fraction is analyzed by TLC to find the product (Figure 9). The plate was developed using (3:1, v/v) hexanes-ethyl acetate as the mobile phase and visualized using cerium ammonium molybdate stain followed by heating. Column fractions 7–9 were combined and concentrated by rotary evaporation (35 °C, started with 50 mmHg and gradually lowered to 10 mmHg) and then at 0.1 mmHg to afford purified **2** (49.5 mg).

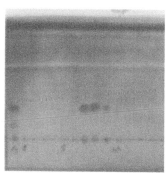

Figure 9. TLC analysis of the column chromatography fractions of **2** (photo provided by submitter)

17. Pure bromoacetylsultam **2** has the following characteristics (data provided by submitters): mp 105–107 °C; ^{1}H NMR (500 MHz, CDCl$_3$) δ: 4.34 (d, J = 12.8 Hz, 1H), 4.20 (d, J = 13.0 Hz, 1H), 3.91 (dd, J = 7.8, 5.0 Hz, 1H), 3.53 (d, J = 13.8 Hz, 1H), 3.46 (d, J = 13.9 Hz, 1H), 2.18–2.12 (m, 1H), 2.09 (dd, J = 14.0, 7.6 Hz, 1H), 1.96–1.87 (m, 3H), 1.47–1.40 (m, 1H), 1.39–1.32 (m, 1H), 1.16 (s, 3H), 0.98 (s, 3H); ^{13}C NMR (126 MHz, CDCl$_3$) δ: 164.7, 65.6, 52.9, 49.2, 48.0, 44.7, 38.1, 32.9, 27.7, 26.6, 20.9, 20.0; HRMS (ESI) m/z calcd for C$_{12}$H$_{19}$BrNO$_3$S [M+H]$^+$ 336.0269; found, 336.0282; IR (film): 3010, 2958, 2907, 2881, 1701, 1458, 1412, 1390, 1368, 1332, 1311, 1264, 1232, 1208, 1166, 1133, 1084, 1063, 1038, 988, 952, 939, 909, 889, 868, 804, 777, 745, 693, 632, 618, 563, 537, 529, 495, 456 cm^{-1}.
18. The number of moles of bromoacetylsultam **2** in the crude is calculated based on the purity percentage from the quantitative NMR data.

$$\text{Bromoacetylsultam mol} = \frac{51.4 \text{ g} \times 78}{336.2440 \text{g/mol} \times 100} = 0.119 \text{ mol}$$

19. *N,N*-Dimethylformamide (certified ACS grade) was purchased from VWR analytical and used as received. The checkers purchased DMF from Acros.
20. Sodium azide (>99.0%) was purchased from TCI chemicals and used as received. The checkers purchased this compound from Sigma Aldrich.
21. Solid sodium azide was added to the reaction using a plastic spatula. Metal spatulas can react with sodium azide and produce explosive metal azides. Hence, care was taken to minimize the contact with metals.
22. The reaction is difficult to monitor using TLC (Figure 10). The starting material **2** and the product **3** have the same R$_f$ value in the given solvent system. The TLC plate was developed using (3:1, v/v) hexanes-ethyl acetate mobile phase and visualized using ninhydrin stain. The R$_f$ value of both **2** and **3** was 0.40. Despite having the same Rf value, the compounds stain differently when visualized using the ninhydrin stain. Starting material (**2**) forms a faint pink colored spot and the product (**3**) forms an orange-pink colored spot on the TLC upon heating.

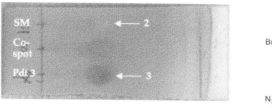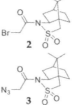

Figure 10. TLC analysis of the starting material 2 and the product 3 (photo provided by submitter)

Hence, the reaction was monitored using ^{1}H NMR. In most parts of the spectrum, the NMR chemical shifts of the starting material 2 and the product 3 are quite similar. Distinctive differences were mostly observed between the AB quartets in 4.50-3.30 ppm range, hence, these signals were used to monitor the progress of the reaction. The reaction usually goes to the completion within 5–6 h. To confirm the complete conversion, ^{1}H NMR analysis was performed on the reaction mixture after 8 h, including NMR analysis of the reaction mixture that was spiked with the starting material (Figure 11). To prepare the NMR sample, approximately 0.3 mL of the reaction mixture was removed and partitioned between 0.5 mL of water and 0.5 mL of (1:1, v/v) diethyl ether-hexanes in a 4 mL vial. The organic phase was removed using a pasture pipette and concentrated under vacuum (10 mmHg). The NMR sample was prepared in CDCl$_3$ and monitored using a Varian 400 MHz spectrometer.

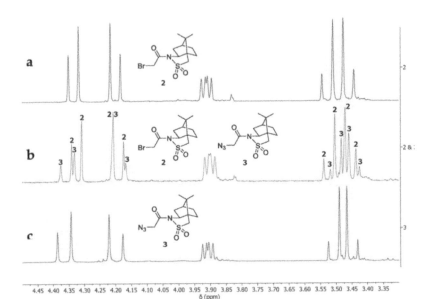

Figure 11. NMR analysis of (a) starting material 2, (b) reaction mixture after 8 h, spiked with the starting material, (c) product 3 (provided by submitter)

23. Purification was not required at this stage and crude **3** was directly taken to the next step.
24. Characterization data for the crude product **3** (data provided by checkers): ^{1}H NMR (400 MHz, CDCl$_3$) δ: 4.35 (d, J = 17.3 Hz, 1H), 4.18 (d, J = 17.3 Hz, 1H), 3.90 (dd, J = 7.7, 5.0 Hz, 1H), 3.50 (d, J = 14.1 Hz, 1H), 3.45 (d, J = 14.2 Hz, 1H), 2.32–2.14 (m, 1H), 2.10 (dd, J = 14.0, 7.9 Hz, 1H), 2.00–1.83 (m, 3H), 1.47–1.31 (m, 2H), 0.97 (s, 3H), 1.12 (s, 3H); ^{13}C NMR (101 MHz, CDCl$_3$) δ: 166.7, 65.4, 52.8, 51.4, 49.4, 48.0, 44.7, 38.2, 32.9, 26.5, 20.8, 19.9; IR (film): 2961, 2359, 2337, 2160, 2109, 1992, 1703, 1541, 1458, 1415, 1376, 1332, 1269, 1238, 1220, 1167, 1137, 1117, 1066, 1040, 985, 941, 872, 808, 781 cm^{-1}; HRMS (ESI) calcd for C$_{12}$H$_{18}$N$_4$O$_3$SNa [M + Na]: 321.0992. Found: 321.1001. The purity of the crude product **3** was assessed to be 84% based on qNMR analysis using 1,3,5-trimethoxybenzene as an internal standard. The checkers performed a second run on half scale, yielding 17.7 g (92% yield, 96% purity). The submitters reported the following result: 36.3 g, 88% yield, 83% purity.

Percentage yield calculation,

$$\text{Theoretical yield} = \frac{47.6 \text{ g} \times 81 \times 298.361 \text{g/mol}}{336.2440 \text{g/mol} \times 100} = 34.21 \text{ g}$$

$$\text{Percent yield} = \frac{36.3 \text{ g} \times 83 \times 100}{34.21 \text{ g} \times 100} = 88 \%$$

25. The major impurity that is present in crude **3** is again the unreacted camphorsultam **1** carried over from step A, which is completely inert towards ancillary transformations. But if desired, flash chromatography can be performed to obtain pure **2** as follows:
The flash column (dimensions: 13 mm inner diameter, 203 mm in height) is wet packed with silica gel (15 g) using a solution of (4:1, v/v) hexanes-ethyl acetate (150 mm, height of the silica bed in the column). Azidoacetylsultam **3** (52.0 mg, dissolved in 1 mL of methylene chloride) is wet loaded onto the packed silica column. The column is eluted with 100 mL of (4:1, v/v) hexanes-ethyl acetate and then continued with 200 mL of (3:1, v/v) hexanes-ethyl acetate until the product completely eluted off the column. The fraction collection is begun (5-mL fractions) immediately after starting the solvent elution and each fraction is analyzed by TLC to find the product (Figure 12). The plates were developed using (3:1, v/v) hexanes-ethyl acetate as the mobile phase and visualized using both ninhydrin and cerium ammonium molybdate stain followed by heating (camphorsultam **1** is insensitive to ninhydrin stain).
Column fractions 7–10 were combined and concentrated by rotary evaporation (35 °C, starting with 50 mmHg and gradually lowering to 10 mmHg) and then at 0.1 mmHg to afford purified **3** (45.0 mg).

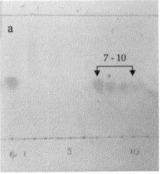
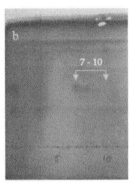

Figure 12. TLC analysis of the column chromatography fractions of 3, with (a) ninhydrin stain and (b) cerium ammonium molybdate stain (photos provided by submitter)

26. Pure azidoacetylsultam 3 has the following characteristics (data provided by submitters): mp 79–81 °C; ^{1}H NMR (500 MHz, CDCl$_3$) δ: 4.36 (d, J = 17.3 Hz, 1H), 4.20 (d, J = 17.3 Hz, 1H), 3.91 (dd, J = 7.9, 4.9 Hz, 1H), 3.51 (d, J = 13.9 Hz, 1H), 3.45 (d, J = 13.8 Hz, 1H), 2.21 (dq, J = 14.0, 3.8 Hz, 1H), 2.12 (dd, J = 14.0, 7.9 Hz, 1H), 1.96–1.87 (m, 3H), 1.47–1.41 (m, 1H), 1.40–1.34 (m, 1H), 1.14 (s, 3H), 0.98 (s, 3H); ^{13}C NMR (126 MHz, CDCl$_3$) δ: 166.7, 65.4, 52.8, 51.5, 49.4, 48.0, 44.7, 38.2, 32.9, 26.5, 20.9, 20.0 ; HRMS (ESI) m/z calcd for C$_{12}$H$_{19}$N$_4$O$_3$S [M+H]$^+$ 299.1178; found, 299.1192; IR (film): 2962, 2884, 2109, 1704, 1483, 1457, 1413, 1377, 1332, 1270, 1239, 1221, 1167, 1137, 1119, 1084, 1065, 1041, 983, 943, 923, 873, 855, 839, 807, 782, 758, 675, 634, 619, 553, 534, 495, 458 cm^{-1}.
27. House vacuum was used (130 mmHg) for this purpose.
28. Number of moles of azidoacetylsultam 3 in the crude is calculated based on the purity percentage obtained from the quantitative NMR data.

$$\text{Azidoacetylsultam mol} = \frac{35.2 \text{ g} \times 84}{298.3610 \text{g/mol} \times 100} = 0.099 \text{ mol}$$

29. The crude product was added as a solid to the reaction mixture. Larger pieces were broken gently with a glass rod before addition to the reaction.
30. HPLC grade MeOH (certified ACS grade) was purchased from Fisher Scientific and used as received.

31. Hydrochloric acid (36.5 to 38.0% (w/w), certified ACS grade) was purchased from Fisher Scientific and used as received.
32. Palladium on carbon (10 wt%) was purchased from Sigma-Aldrich and used as received. The checkers purchased this catalyst from TCI.
33. The hydrogenolysis reaction using palladium on carbon (10 wt%) creates a significant fire hazard if the activated catalyst is allowed to dry. Hence, palladium on carbon was added to the reaction mixture mixed with water as a slurry to minimize the pyrophoricity.
34. During hydrogenolysis of azides, gaseous nitrogen is produced, hence every time before changing the hydrogen balloon, a cycle of evacuating and back-filling was performed.
35. After 24 h, TLC analysis indicated that a substantial amount starting material was still present. Hence another portion of 10 wt% Pd-C (2.50 g) was added following the described procedure.
36. The checkers observed remaining azide until the total reaction time had reached 40 h. The reaction was monitored by TLC, developed using (3:1, v/v) hexanes-ethyl acetate as the mobile phase and visualized using ninhydrin stain (Figure 13). The starting material forms a pink spot and the product forms a yellow spot upon heating with the stain. The R_f value of the starting material **1** was 0.41. The product doesn't move with the solvent and remains at the origin.

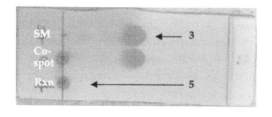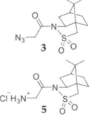

Figure 13. TLC analysis of the reaction mixture at 36 h (photo provided by submitter)

37. Celite 545 filtering aid (neutral or acidic – basic Celite leads to partial decomposition of the amine) was purchased form Fisher Scientific and used as received.
38. Care was taken to always maintain the solvent level above the Celite bed. Passing air through Pd-C, trapped on the Celite bed can creates a significant fire hazard, especially in the presence of methanol.

39. While concentrating, the product turned into a highly viscous liquid and then crystallized. After 24 h under 0.1 mmHg pressure, the large crystals were crushed using a spatula to facilitate the drying.
40. The submitters obtained 39.1 g of crude hydrochloride.
41. Glycylsultam in the free amine form is highly reactive. In solution it can react with itself and undergo nucleophile-induced deacylation to give back the parent sultam **1**, which is indicative of this decomposition pathway. Hence, trapping the amine as an acid salt is important, where in this case it is converted into the stable hydrochloride salt. The salt can be stored at –20 °C for an extended period without decomposition, and free amine can be regenerated easily by an aq NaHCO$_3$ neutralization.
42. Due to the excess amount of hydrochloric acid that was added to the reaction mixture during the hydrogenolysis, the actual weight of the ammonium salt **5** is higher than the theoretical value. Hence, the entire crude ammonium salt product obtained from the previous step was finely crushed and homogenized before weighing out 5.00 g for the aq. NaHCO$_3$ neutralization. For the yield determination, a calculation was performed to determine the amount of free glycylsultam **4** that can be obtained from the entire batch **5**, based on the value received for a 5.00 g sample (Note 46). The submitters used 5.00 g of crude material.
43. MColorpHast pH test strips (pH 6.5-10 and pH 5.2-7.2) were purchased from MilliporeSigma.
44. Due to the reactivity of the free amine, care was taken to perform the neutralization at 0 °C and to maintain the water bath temperature at 15–19 °C while concentrating the product under vacuum. When the neutralization and concentration were performed at higher temperatures, the product was observed to usually contain 3-5% of parent sultam **1** formed *via* deacylation.
45. Completion of the neutralization was confirmed by monitoring the pH of the reaction mixture. If needed, the product **4** can be visualized by TLC developed using (9:1, v/v) dichloromethane-methanol as the mobile phase and visualized using ninhydrin stain. The product forms a yellow spot with a R$_f$ value of 0.40 (Figure 14).

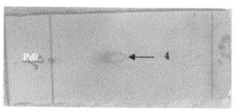

Figure 14. TLC analysis of the Oppolzer's glycylsultam 4 (photo provided by submitter)

46. Glycylsultam **4** has the following characteristics (data provided by checkers): mp 118 – 121 °C; $[\alpha]_{589}^{23}$ +115.6 (c 1.00, CHCl$_3$); ^{1}H NMR (400 MHz, CDCl$_3$) δ: 3.93–3.84 (m, 2H), 3.76 (d, J = 18.1 Hz, 1H), 3.49 (d, J = 13.8 Hz, 1H), 3.43 (d, J = 13.8 Hz, 1H), 2.16 (dd, J = 12.4, 4.9 Hz, 1H), 2.12 – 2.04 (m, 1H), 1.96 – 1.83 (m, 3H), 1.46 (d, J = 8.6 Hz, 2H), 1.42 (d, J = 8.9 Hz, 1H), 1.36 (t, J = 9.4 Hz, 1H), 1.14 (s, 3H), 0.97 (s, 3H); ^{13}C NMR (176 MHz, CDCl$_3$) δ: 173.1, 65.3, 52.9, 49.3, 48.0, 45.6, 44.8, 38.4, 33.0, 26.6, 20.9, 20.0; IR (film): 3412, 3353, 2970, 2920, 2887, 2360, 2176, 1688, 1541, 1457, 1420, 1377, 1317, 1269, 1235, 1218, 1164, 1130, 1113, 1083, 1061, 1041, 981, 940, 910, 874, 815, 766 cm^{-1}. Purity of (**4**) was assessed to be 99% based on qNMR analysis using 1,3,5-trimethoxybenzene as an internal standard. The checkers performed a second run on half scale, performing isolation on 1 g of crude material. This extraction yielded 0.58 g of product (99% purity) for a total yield of 83%.

47. The submitters reported the following result (based on 5.00 g of crude hydrochloride): 3.19 g, 89% yield, 98% purity.

Following are the calculations to determine the amount of free glycylsultam **4** present in the entire ammonium salt **5** batch (39.1 g), based on the neutralization data received for 5.00 g sample and the percent yield.

$$\text{Free glycylsultam in 39.1 g of ammonium salt} = \frac{3.19 \text{ g} \times 39.1 \text{ g}}{5.00 \text{ g}} = 24.9 \text{ g}$$

$$\text{Theoretical yield} = \frac{36.0 \text{ g} \times 83 \times 272.3630 \text{ g/mol}}{298.3610 \text{ g/mol} \times 100} = 27.3 \text{ g}$$

$$\text{Percent yield} = \frac{24.9 \text{ g} \times 98 \times 100}{27.3 \text{ g} \times 100} = 89\%$$

Working with Hazardous Chemicals

The procedures in *Organic Syntheses* are intended for use only by persons with proper training in experimental organic chemistry. All hazardous materials should be handled using the standard procedures for work with chemicals described in references such as "Prudent Practices in the Laboratory" (The National Academies Press, Washington, D.C., 2011; the full text can be accessed free of charge at http://www.nap.edu/catalog.php?record_id=12654). All chemical waste should be disposed of in accordance with local regulations. For general guidelines for the management of chemical waste, see Chapter 8 of Prudent Practices.

In some articles in *Organic Syntheses*, chemical-specific hazards are highlighted in red "Caution Notes" within a procedure. It is important to recognize that the absence of a caution note does not imply that no significant hazards are associated with the chemicals involved in that procedure. Prior to performing a reaction, a thorough risk assessment should be carried out that includes a review of the potential hazards associated with each chemical and experimental operation on the scale that is planned for the procedure. Guidelines for carrying out a risk assessment and for analyzing the hazards associated with chemicals can be found in Chapter 4 of Prudent Practices.

The procedures described in *Organic Syntheses* are provided as published and are conducted at one's own risk. *Organic Syntheses, Inc.*, its Editors, and its Board of Directors do not warrant or guarantee the safety of individuals using these procedures and hereby disclaim any liability for any injuries or damages claimed to have resulted from or related in any way to the procedures herein.

Discussion

A convenient and scalable preparation of chiral glycylsultam **4/ent-4** is described. The three-step procedure involves acylation of Oppolzer's camphor-derived sultam **1/ent-1** with bromacetyl bromide (reaction A), displacement of bromide by azide anion (reaction B), and catalytic reduction of the azide (reaction C). The entire procedure can be performed on a large scale and does not involve any chromatographic separation or crystallization.

Glycylsultam **4/ent-4** is the NC component in the [C+NC+CC] cycloaddition, a robust multicomponent reaction that enables the asymmetric synthesis of highly functionalized pyrrolidines (Figure 15). The camphorsultam moiety plays four important roles in this reaction: (1) it enhances the acidity of the glycyl α-protons, thus accelerating the formation of the intermediate azomethine ylide, (2) it determines which diastereotopic face of the azomethine ylide will be attacked by the dipolarophile, overriding resident substrate chirality, (3) it facilitates product characterization: since natural camphor is enantiomerically pure, one can readily assess the diastereomeric purity of the cycloadduct using NMR or HPLC, and (4) it activates the aminoacyl carbon towards nucleophilic attack.

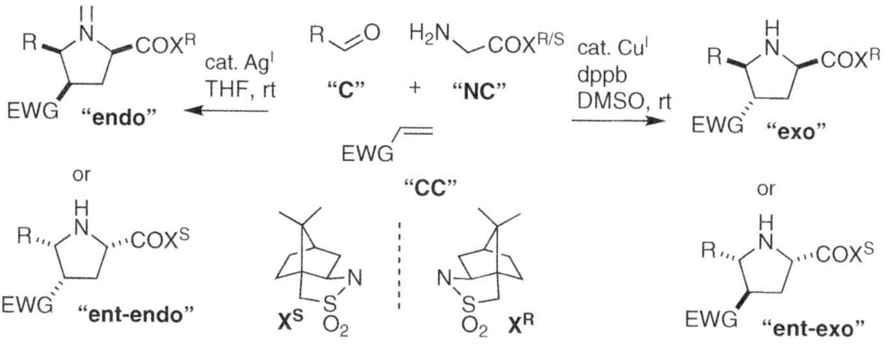

R = aliphatic, vinylic, aromatic, or heteroaromatic substituent
EWG = electron withdrawing substituent

Figure 15. Asymmetric multicomponent [C+NC+CC] synthesis of pyrrolidines. (X^R and X^S = antipodes of Oppolzer's camphorsultam)

Perhaps the feature that distinguishes the [C+NC+CC] cycloaddition from other azomethine ylide cycloadditions is the ability to employ structurally diverse enolizable aldehydes as the C component. This permits the use of structurally complex aldehydes. The aforementioned attributes have resulted in application of the asymmetric [C+NC+CC] cycloaddition to the synthesis of pyrrolidine-containing natural products and drugs (Table 1). Examples include cyanocycline A,[2] the neuraminidase inhibitor A-315675,[3] kaitocephalin,[4] quinocarcin,[5] and tetrazomine.[6] The asymmetric [C+NC+CC] reaction has also been used for the synthesis of a focused library of heteroaryl substituted pyrrolidines for fragment based drug discovery.[7]

Table 1. Application of the asymmetric [C+NC+CC] reaction to the synthesis of pyrrolidine containing natural products and drugs

During the course of these and related studies, our group required rather large quantities of both the R- and S -versions of Oppolzer's glycylsultam,[8] which serves as the "NC" component in the [C+NC+CC] coupling reaction. In this Organic Syntheses procedure, we report a simple and very practical three-step synthesis of Oppolzer's glycylsultam that is amenable to large-scale production.

Initially, we used a variation of Oppolzer's route to chiral α-amino acids which was published in the late 80s to prepare the glycylsultam.[9–10] This two-step process involved Me_3Al-mediated acylation of the parent sultam with $(MeS)_2C=NCH_2CO_2Me$[11] to give $(MeS)_2C=NCH_2COX^*$, followed by hydrolytic release of the primary amine. Chassaing and co-workers developed an alternative route to prepare the glycylsultam during their synthesis of isotopically labelled amino acids.[12] The ^{15}N-labeled hydrochloride glycylsultam salt was prepared by acylation of the sultam potassium salt with $BrOCCH_2Br$, followed by a S_N2 displacement with ^{15}N-phthalimide potassium salt and finally the N-deprotection. For ^{13}C-labeling, Chassaing's approach was similar to what Oppolzer reported[8] but used labelled $(Ph)_2C=NCH_2{}^{13}CO_2Et$ and $(Ph)_2C=N^{13}CH_2CO_2Et$ for the acylation of the parent sultam. Apart from this, an interesting alternative was reported, in which the parent sultam sodium salt was acylated with the mixed anhydride of N-Boc protected labeled glycine followed by removal of the Boc group to obtain glycylsultam hydrochloride salt. It was stated that this synthesis could be performed on a multi-hundred-gram scale, although no experimental details were given. Dogan and coworkers reported acylation of the parent sultam with azidoacetyl chloride to produce the corresponding azidoacetylsultam, which was subjected to a Staudinger reaction to give the desired glycylsultam.[13] Since, scale-up of this synthesis would involve large quantities of the potentially explosive azidoacetyl chloride,[14] a more atom-economical and safe process was desirable. Our group published an alternative route to afford the desired glycylsultam via the Delépine reaction.[15] The reaction cascade involves nucleophilic displacement of known $BrCH_2COX^*$ by the ammonia surrogate hexamethylenetetramine (HMTA), followed by acid decomposition to yield the glycylsultam ammonium salt. This process is relatively safe and environmentally benign but suffers from a non-ideal halide transfer reaction and two extensive heating steps which is undesirable on an increased scale.

In the route to Oppolzer's glycylsultam reported herein, the lithium salt of camphor-derived sultam **1** is acylated with bromoacetyl bromide at –78 °C to give the known *N*-bromoacetylsultam **2**. Except for a 20-fold increase in scale, the procedure was the same as that reported by Sweeney and coworkers.[16] Foregoing purification at this stage, the crude bromide **2** was treated with NaN$_3$ in DMF at room temperature to effect S$_N$2 displacement and produce the stable azidoacetylsultam **3**. After extractive workup, this material was subjected to catalytic hydrogenolysis in the presence of conc. HCl followed by NaHCO$_3$ neutralization to give the desired product **4** in good overall yield (64% on 30 g scale and 67% on 60 g scale). Each of these reactions gives solid crystalline products. The entire sequence can be performed on a large scale without the need for chromatographic purification at any stage. The R- and S- glycylsultams so obtained are suitable for use in our asymmetric multicomponent [C+NC+CC] coupling reactions.

References

1. Contact information: UR: Warren Center for Neuroscience Drug Discovery, Vanderbilt University, Nashville, Tennessee, 37232-0697, USA. E-mail: upendra.a.rathnayake@vanderbilt.edu. HÜK: Department of Pharmacological Sciences, Icahn School of Medicine at Mount Sinai, 1425 Madison Avenue, New York 10029, USA. E-mail: husnu.kaniskan@mssm.edu. JH: Johns Manville, 10100 West Ute Avenue Littleton, Colorado 80127, USA. E-mail: jieyu.hu@jm.com. CGP: Department of Chemistry, The Scripps Research Institute, 130 Scripps Way, 3B3 Jupiter, Florida 33458, USA. E-mail: cparker@scripps.edu. PG: Department of Chemistry, Washington State University, Pullman, Washington 99164-4630, USA. E-mail: ppg@wsu.edu. Phone: (+1) 509-335-7620. ORCID: 0000-0002-6503-9550. We thank the National Science Foundation for financial support. The authors dedicate this procedure to the memory of Professor Wolfgang Oppolzer.
2. Garner, P.; Kaniskan, H. Ü.; Keyari, C. M.; Weerasinghe, L. *J. Org. Chem.* **2011**, *76* (13), 5283–5294.
3. Garner, P.; Weerasinghe, L.; Youngs, W. J.; Wright, B.; Wilson, D.; Jacobs, D. *Org. Lett.* **2012**, *14* (5), 1326–1329.
4. Garner, P.; Weerasinghe, L.; Van Houten, I.; Hu, J. *Chem. Commun.* **2014**, *50* (38), 4908–4910.
5. Fang, S.-L.; Jiang, M.-X.; Zhang, S.; Wu, Y.-J.; Shi, B.-F. *Org. Lett.* **2019**, *21* (12), 4609–4613.

6. Qi, W.-Y.; Fang, S.-L.; Xu, X.-T.; Zhang, K.; Shi, B.F. *Organic Chemistry Frontiers* **2021**, *8*, 1802–1807.
7. Garner, P.; Cox, P. B.; Rathnayake, U.; Holloran, N.; Erdman, P. *ACS Med. Chem. Lett.* **2019**, *10* (5), 811–815.
8. Although Oppolzer did not actually report the parent glycylsultam (see references 9 and 10), we feel that it is appropriate to refer to it as "Oppolzer's glycylsultam" for descriptive reasons.
9. Oppolzer, W.; Moretti, R.; Thomi, S. *Tetrahedron Lett.* **1989**, *30* (44), 6009–6010.
10. Oppolzer, W.; Moretti, R.; Zhou, C. *Helv. Chim. Acta.* **1994**, *77* (8), 2363–2380.
11. Hoppe, D.; Beckmann, L. *Liebigs Annalen der Chemie* **1979**, 2066–2075.
12. Martin, A.; Chassaing, G.; Vanhove, A. *Isotopes in Environmental and Health Studies* **1996**, *32* (1), 15–19.
13. Dogan, Ö.; Öner, İ.; Ülkü, D.; Arici, C. *Tetrahedron: Asymmetry* **2002**, *13* (19), 2099-2104.
14. Nicolaides, E. D.; Westland, R. D.; Wittle, E. L. *J. Am. Chem. Soc.* **1954**, *76* (11), 2887–2891.
15. Isleyen, A.; Gonsky, C.; Ronald, R. C.; Garner, P. *Synthesis* **2009**, *08*, 1261–1264.
16. Sweeney, J. B.; Cantrill, A. A.; McLaren, A. B.; Thobhani, S. *Tetrahedron* **2006**, *62* (15), 3681–3693.

Appendix
Chemical Abstracts Nomenclature (Registry Number)

(2S)-Bornane-2,10-sultam: 10,10-dimethyl-3λ^6-thia-4-azatricyclo[5.2.1.0^{1,5}]decane 3,3-dioxide; (108448-77-7)
THF: Tetrahydrofuran; (109-99-9)
n-Butyllithium: Tetra-μ_3-butyl-tetralithium; (109-72-8)
Hexanes; (110-54-3)
Bromoacetyl bromide: 2-Bromoacetyl bromide; (598-21-0)
Diethyl ether: Ethoxyethane; (60-29-7)
Na_2SO_4: Sodium sulfate; (7757-82-6)
DMF: N,N-Dimethylformamide; (68-12-2)
NaN_3: Sodium azide; (26628-22-8)
MeOH: Methanol; (67-56-1)

HCl: Hydrochloric acid; (7647-01-0)
Pd-C: Palladium on carbon; (7440-05-3)
NaHCO$_3$: Sodium bicarbonate; (144-55-8)
CH$_2$Cl$_2$: Dichloromethane; (75-09-2)

Phil Garner received his Ph.D. degree from the University of Pittsburgh under the guidance of Paul Dowd. This was followed by postdoctoral work in Paul Grieco's laboratory at Indiana University. In 1983, he took up his first faculty position at Illinois Institute of Technology. He moved to Case Western Reserve University in 1985 where he established a research program that focused on development of new methodology. He headed west to Washington State University in 2007. His scientific accomplishments include a widely used serinal derivative that has come to bear his name. He currently holds the rank of Professor of Chemistry.

Upendra Rathnayake received his B.S. (Hons) in chemistry from university of Sri Jayewardenepura, Sri Lanka and M.S. (with a major component of research) in chemical and process engineering from the University of Moratuwa, Sri Lanka. After working as a scientist at Sri Lanka Institute of Nanotechnology (SLINTEC) for two and half years, he moved to Washington State University to pursue his PhD under the supervision of Prof. Philip Garner. His doctoral studies are focused on application of the asymmetric [C+NC+CC] coupling reaction to natural product synthesis.

H. Ümit Kaniskan earned his Ph.D. at Case Western Reserve University under the supervision of Dr. Philip Garner. He then pursued his postdoctoral studies in Dr. Movassaghi's group at Massachusetts Institute of Technology. Dr. Kaniskan then joined Dr. Jin's laboratory UNC at Chapel Hill and later at the Icahn School of Medicine at M. Sinai as a postdoctoral research associate. He is currently an Associate Professor in the Department of Pharmacological Sciences, and Assistant Director of Mount Sinai Center for Therapeutics Discovery at the Icahn School of Medicine at Mount Sinai. His research interests include development of inhibitors of protein methyltransferases and targeted protein degradation.

Jieyu Hu obtained her B.S. in Pharmacy from Zhejiang University and M. S. in Pharmacy from Fudan University in China, and Ph.D. in Organic Chemistry from Case Western Reserve University. After joining Corning Incorporated, she continued her organic synthesis career and then she focused on flow chemistry and polymer synthesis. Her career focus is in the field of materials science at Johns Manville.

Christopher Parker earned his B.S. in Chemistry from Case Western Reserve University (2007) working in the lab of Philip Garner and a Ph.D. in Chemistry from Yale University (2013) under the supervision of David A. Spiegel. He then carried out postdoctoral studies under the supervision of Benjamin F. Cravatt as a fellow of the American Cancer Society at The Scripps Research Institute. Chris is currently an Assistant Professor in the Department of Chemistry at Scripps Research. His group develops and applies chemical proteomic methods to discover useful chemical probes to investigate various protein targets relevant to human health.

Bogdan R. Brutiu received his M.Sc at the University of Vienna in 2019. He is currently a second-year graduate student in the group of Prof. Nuno Maulide. His research focuses on the chemistry of destabilized carbocations and C–C coupling reactions mediated by hydride transfer.

Martina Drescher is the lead technician of the Maulide group at the University of Vienna, where she has worked with several group leaders over the course of 38 years.

Daniel Kaiser received his Ph.D. at the University of Vienna in 2018, completing his studies under the supervision of Prof. Nuno Maulide. After a postdoctoral stay with Prof. Varinder K. Aggarwal at the University of Bristol, he returned to Vienna in 2020 to assume a position as senior scientist in the Maulide group. His current research focusses on the chemistry of destabilized carbocations and related high-energy intermediates.

Stereoselective [2+2] Cycloadditions: Synthesis of a Tri-*O*-Bn-D-Glucal-derived β-Lactam

Maria Varghese,[&] Hannah E. Caputo,[&] Ruiqing Xiao,[&] Anant Balijepalli,[#] Aladin Hamoud,[&] and Mark W. Grinstaff[1,&,#,*]

[&]Department of Chemistry and [#]Department of Biomedical Engineering, Boston University, Boston, MA, 02215

Checked by Kyohei Oga, Masanori Nagatomo, and Masayuki Inoue

Procedure (Note 1)

A. **(1S,3R,4S,5R,6R)-4,5-Bis(phenylmethoxy)-3-[(phenylmethoxy)methyl]-2-oxa-8-azabicyclo[4.2.0]octan-7-one (1)**. An oven-dried 500 mL three-necked round-bottomed flask is equipped with an oven-dried 3 cm Teflon-coated magnetic stir bar and oven-dried sodium carbonate (4.00 g, 37.7 mmol, 1.6 equiv) (Note 2). The right neck is fitted with an oven-dried 15/25 glass stopper (Note 3). The left neck is fitted with a 15/25 rubber septum, and the central neck is equipped with a 29/32 three-way cock connected to an argon inlet and a vacuum line. The flask is evacuated and backfilled with argon gas three times. Argon atmosphere is maintained throughout the reaction under argon flow (Figure 1A). Anhydrous toluene (40 mL) (Note 4) and chlorosulfonyl isocyanate (CSI) (2.72 mL, 4.42 g, 31.2 mmol, 1.3 equiv) (Notes 5 and 6) are added at room temperature (26 °C) to the 500 mL three-necked round-bottom flask using 60 mL and 6 mL disposable syringes, respectively. The reaction flask is cooled to –78 °C (bath temperature) in a dry ice and methanol bath within a 1 L Dewar cooling bath (Note 7).

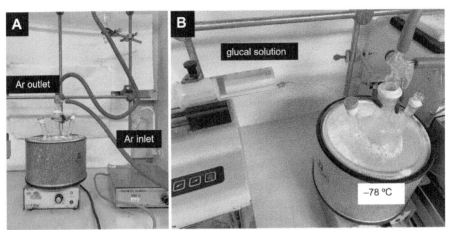

Figure 1. (A) Reaction set-up; (B) Addition of glucal solution at –78 °C (photos provided by checkers)

A solution of tri-*O*-benzyl-D-glucal (**2**, 10.0 g, 24.0 mmol) (Note 8) in toluene (40 mL) is added via a 60 mL disposable syringe and a syringe pump to the stirring CSI solution in the three-necked round-bottomed flask at –78 °C. The solution was added dropwise at a rate of 1.5 mL/min (Note 9) (Figure 1B) resulting in a yellow solution. After the addition is completed, the reaction mixture is warmed to –62 °C using a constant temperature bath with methanol (Note 10) (Figures 2A and 2B) and stirred for 3 h at this temperature.

The reaction mixture is cooled to –78 °C in dry ice and methanol bath (Figure 2C). Then the reaction mixture is diluted with anhydrous toluene (120 mL: 60 mL × 2) using a 60 mL disposable syringe and syringe pump over 30 min at a rate of 4 mL/min (Figure 2D). Red-Al® (>60 wt % solution in toluene, 10.9 mL, 7.08 g, 35.0 mmol, 1.46 equiv) (Notes 11 and 12) is added to the three-necked, round-bottomed flask using a 12 mL disposable syringe and syringe pump over 15 min at a rate of 0.6 mL/min (Figure 2E).

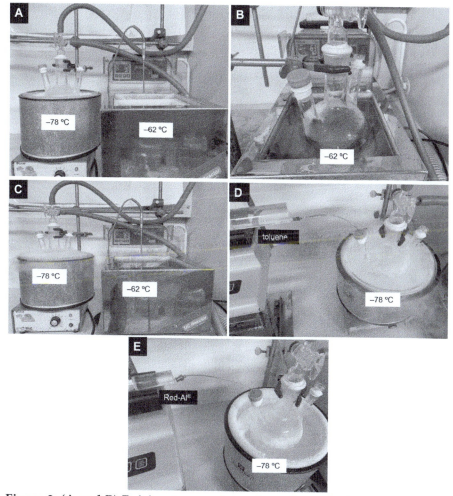

Figure 2. (A and B) Raising temperature of reaction solution to –62 °C; (C) Cooling temperature of reaction solution to –78 °C; (D and E) Addition of toluene and Red-Al® (photos provided by checkers)

After being stirred at –78 °C for 15 min, the reaction mixture is warmed to –62 °C and stirred for 10 min. The solution is warmed to –10 °C and stirred for 10 min in an ice and acetone bath (Note 13) (Figure 3A). The reaction mixture is effervescing and is slightly yellow at this point. The reaction is quenched by dropwise addition of deionized water (1.0 mL) over one min using a disposable syringe. The resultant mixture is stirred for an additional 15 min while warming up the mixture to 0 °C (Figure 3B) in ice water and then stirred another 15 min, at which time the solution is at 26 °C (Figure 3C) (Note 14).

Figure 3. Raising temperature of reaction solution to (A) –10 °C; (B) 0 °C; (C) 26 °C (photos provided by checkers)

The solution is then filtered via vacuum filtration with a Büchner funnel and filter paper (Note 15) to remove solids. The solid is washed with ethyl acetate (300 mL), and the filtrate is transferred to a 2 L round-bottomed flask. The resultant solid on the filter paper is transferred to a 500 mL separation funnel using deionized water (200 mL), and ethyl acetate (100 mL) is added to the separation funnel. The aqueous layer is extracted with ethyl acetate (100 mL × 2), and the organic layers are combined into the above 2 L round-bottomed flask. The solution is concentrated by rotary evaporation (37 °C, from 400 mmHg to 20 mmHg). The residue is transferred to a 500 mL separation funnel using ethyl acetate (200 mL). The organic layers are washed with 1 M HCl solution (200 mL), saturated sodium bicarbonate (200 mL), and brine (200 mL). The organic layers are dried over sodium sulfate (10 g) to remove residual water, and the drying agent is filtered via vacuum filtration with a Büchner funnel and filter paper (Note 15) into a 1 L round-bottomed flask. The filtrate is concentrated by rotary evaporation (37 °C, from 400 mmHg to 20 mmHg) to obtain a crude yellow oil (5.10 g).

The crude yellow oil is purified by flash column chromatography on silica gel (150 g) (Note 16) using ethyl acetate and n-hexane as eluents (Note 17). The desired product is obtained in fractions 16–22, combined, and concentrated by rotary evaporation (37 °C, from 400 mmHg to 10 mmHg) to yield a colorless oil. The oil is then placed under high vacuum (27 °C, 5 mmHg) for 20 h to afford 3.95 g (36%, 97.9% purity) of (1S,3R,4S,5R,6R)-4,5-Bis(phenylmethoxy)-3-[(phenylmethoxy) methyl]-2-oxa-8-azabicyclo[4.2.0]-octan-7-one (**1**) as a white solid (Figure 4) (Notes 18, 19, and 20).

Figure 4. Purified title compound **1** (photos provided by checkers)

Notes

1. Prior to performing each reaction, a thorough hazard analysis and risk assessment should be carried out with regard to each chemical substance and experimental operation on the scale planned and in the context of the laboratory where the procedures will be carried out. Guidelines for carrying out risk assessments and for analyzing the hazards associated with chemicals can be found in references such as Chapter 4 of "Prudent Practices in the Laboratory" (The National Academies Press, Washington, D.C., 2011; the full text can be accessed free of charge at https://www.nap.edu/catalog/12654/prudent-practices-in-the-laboratory-handling-and-management-of-chemical. See also "Identifying and Evaluating Hazards in Research Laboratories"

(American Chemical Society, 2015) which is available via the associated website "Hazard Assessment in Research Laboratories" at https://www.acs.org/content/acs/en/about/governance/committees/chemicalsafety/hazard-assessment.html. In the case of this procedure, the risk assessment should include (but not necessarily be limited to) an evaluation of the potential hazards associated with sodium carbonate, toluene, chlorosulfonyl isocyanate, dry ice, methanol, tri-*O*-benzyl-d-glucal, Red-Al® [sodium bis(2-methoxyethoxy)aluminum hydride solution (>60 wt% solution in toluene)], acetone, isopropanol, ethyl acetate, cyclohexane, hexanes, sodium bicarbonate, 1 M HCl solution, silica gel, $CDCl_3$, and duroquinone, as well as the proper procedures for preparation of the CSI reagent and quenching of the CSI and Red-Al®.

2. Sodium carbonate was purchased from Millipore Sigma (submitters) and Nacalai Tesque, Inc. (checkers).

3. A 500 mL three-necked, round-bottomed flask containing sodium carbonate (~9.0 g), a Teflon-coated stir bar, and two glass stoppers are dried overnight (12 h) in an oven at 120 °C. Upon removal from the oven, a stir bar is added to the round-bottomed flask, and the flask is fitted with two glass stoppers. Glassware is cooled to room temperature under a vacuum before proceeding.

4. Reagents and solvents used in this procedure, except chlorosulfonyl isocyanate (CSI), are commercially available and were used without further purification. This includes ethyl acetate, hexanes, and toluene purchased from Pharmco (submitters). Ethyl acetate (>99%) and *n*-hexane (>95%) were purchased from Kanto Chemical Co., Inc. and were used as received (checkers). Anhydrous toluene was purchased from FUJIFILM Wako Pure Chemical Corporation and purified by Glass Contour solvent dispensing system (Nikko Hansen & Co., Ltd.) (checkers).

5. The liquid reagent chlorosulfonyl isocyanate (CSI) is a highly toxic and corrosive chemical that reacts violently with water and is a lachrymator. Extreme care should be taken when using this reagent. To remove the contaminants (sulfur trioxide and hydrogen chloride) in commercially available CSI, the CSI is mixed with oven-dried sodium carbonate (120 °C, 12 h) and stored for at least one week in the refrigerator (0–4 °C) prior to use (checkers) (Figure 5). CSI was colorless before the reaction. The CSI is warmed up to room temperature before starting the reaction. The leftover CSI is diluted with cyclohexane and quenched by dropwise addition of isopropanol (submitters).

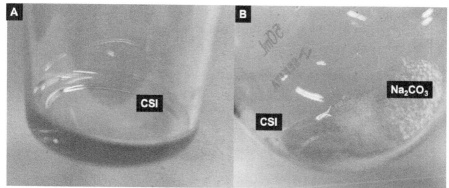

Figure 5. Color of CSI (A) Before purification; (B) After purification (photos provided by checkers)

6. Chlorosulfonyl isocyanate (98 %) was purchased from Sigma Aldrich (submitters and checkers).
7. In a 1 L Dewar cooling bath, approximately 500 g of dry ice and 250 mL of methanol were used to set the bath temperature at –78 °C. All temperatures reported in the procedure are bath temperatures.
8. Tri-O-benzyl-D-glucal (2) (98%) was purchased from Carbosynth (submitters and checkers).
9. Faster addition will decrease the yield and stereoselectivity because of the increase in the overall temperature of the reaction mixture leading to degradation of intermediate or a mixture of stereoisomers.
10. A constant temperature bath (UCR-150N-S, Techno Sigma Co., Ltd.) was used (checkers).
11. Red-Al®, sodium bis(2-methoxyethoxy)aluminum hydride solution (>60 wt% in toluene), was purchased from Sigma Aldrich (submitters and checkers).
12. Red-Al® is a highly toxic and flammable liquid that reacts violently with water. Extreme care should be taken when using this reagent. This is a highly dense reagent. Therefore a 21 G needle should be used for transferring the reagent. The leftover reagent in the syringe should be diluted with hexanes and quenched by dropwise addition of isopropanol (submitters).
13. In a 1 L Dewar cooling bath, approximately 200 g of ice and 200 mL of acetone were used to set the bath temperature at –10 °C.
14. TLC analyses of the reaction mixture are shown below (Figure 6) (checkers). The spot of the product can be visualized by fluorescence on

silica gel 60 F$_{254}$ plates (TLC Silica gel 60 F$_{254}$, purchased from Merck KGaA) with UV light (254 nm) and by anisaldehyde stain. R$_f$ = 0.33 (1:1 *n*-hexane : ethyl acetate).

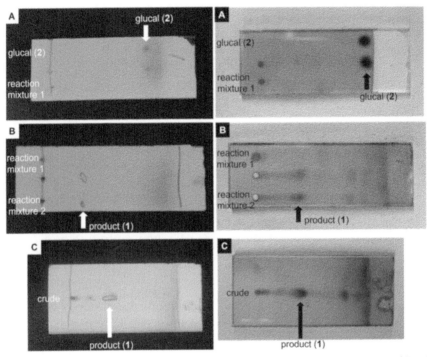

Figure 6. TLC analysis of reaction mixture (A) Before the addition of Red-Al®; (B) 40 minutes after the addition of RedAl®; (C) Crude mixture (photos provided by checkers)

15. The filter paper (No. 1, diameter: 9 cm) was purchased from Tokyo Roshi Kaisha, Ltd.
16. Silica gel (Silica gel 60N, spherical and neutral, 0.040-0.050 mm) was purchased from Kanto Chemical Co., Inc. and used as received (checkers).
17. The column chromatography is run as follows: A flash column (6 cm column diameter) is charged with sea sand to a height of 1 cm and then with silica gel (150 g) using a wet-pack method to give a column height of 13 cm. Sea sand with 1 cm minimum height is added to the to the top of the column. The crude oil is loaded onto the column using 15 mL of

n-hexane/ethyl acetate (2/1, v/v). At this point, fraction collection (200 mL × 25) is begun, and elution is continued with 2 L of *n*-hexane/ethyl acetate (2/1, v/v), then 4 L of *n*-hexane/ethyl acetate (1.5/1, v/v), then 4 L of *n*-hexane/ethyl acetate (1/1, v/v) (checkers).
18. The product is stable for up to 6 months if stored at –20 °C (submitters).
19. (1*S*,3*R*,4*S*,5*R*,6*R*)-4,5-Bis(phenylmethoxy)-3-[(phenylmethoxy)methyl]-2-oxa-8-azabicyclo[4.2.0]octan-7-one: mp (uncorrected) = 72–75 °C; R_f = 0.33 (1:1 *n*-hexane : ethyl acetate, visualized by UV and anisaldehyde stain) (Figure 7); ^{1}H NMR (400 MHz, CDCl$_3$) δ: 3.34 (m, 1H), 3.57 (dd, *J* = 9.1, 6.8 Hz, 1H), 3.57–3.65 (m, 2H), 3.95 (m, 1H), 4.04 (dd, *J* = 6.8, 2.8 Hz, 1H), 4.42–4.53 (m, 4H), 4.62 (d, *J* = 11.9 Hz, 1H), 4.69 (d, *J* = 11.4 Hz, 1H), 5.44 (d, *J* = 4.1 Hz, 1H), 6.27 (brs, 1H), 7.10–7.13 (m, 2H), 7.19–7.34 (m, 13H). ^{13}C NMR (100 MHz, CDCl$_3$) δ: 54.8, 69.3, 69.6, 71.1, 73.4, 73.5, 75.6, 76.2, 99.8, 127.7 (2C), 127.8 (5C), 128.0 (2C), 128.3 (2C), 128.35 (2C), 128.43 (2C), 137.4, 137.7, 137.9, 167.0. IR (KBr film): 3289, 3030, 2868, 1768, 1495, 1453, 1362, 1263, 1208, 1109, 738 cm^{-1}. HRMS (ESI). *m/z* [M + Na]$^+$ calc'd for [C$_{28}$H$_{29}$NO$_5$Na]$^+$: 482.1938; found 482.1929. Purity of the lactam (28.04 mg) was assessed 97.9% by ^{1}H qNMR using duroquinone (10.82 mg) as the internal standard, which was purchased from Sigma Aldrich with 99.7% purity. Based on this purity, the corrected yield is 3.86 g (35%); The purity of the lactam was also assessed by elemental analysis (Calculated: C: 73.18, H: 6.36, N: 3.05; Observed: C: 72.95, H: 6.57, N: 3.10).

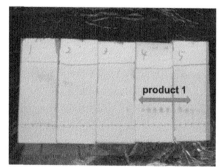
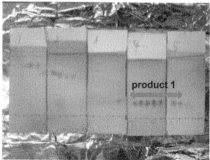

Figure 7. TLC analysis of column fractions (A) Spots on silica gel plate visualized under UV light (254 nm). (B) Anisaldehyde stain (photos provided by checkers)

20. The reaction was also performed on half scale and yielded 1.89 g (34%, 98.8% purity) of the lactam product.

Working with Hazardous Chemicals

The procedures in *Organic Syntheses* are intended for use only by persons with proper training in experimental organic chemistry. All hazardous materials should be handled using the standard procedures for work with chemicals described in references such as "Prudent Practices in the Laboratory" (The National Academies Press, Washington, D.C., 2011; the full text can be accessed free of charge at http://www.nap.edu/catalog.php?record_id=12654). All chemical waste should be disposed of in accordance with local regulations. For general guidelines for the management of chemical waste, see Chapter 8 of Prudent Practices.

In some articles in *Organic Syntheses*, chemical-specific hazards are highlighted in red "Caution Notes" within a procedure. It is important to recognize that the absence of a caution note does not imply that no significant hazards are associated with the chemicals involved in that procedure. Prior to performing a reaction, a thorough risk assessment should be carried out that includes a review of the potential hazards associated with each chemical and experimental operation on the scale that is planned for the procedure. Guidelines for carrying out a risk assessment and for analyzing the hazards associated with chemicals can be found in Chapter 4 of Prudent Practices.

The procedures described in *Organic Syntheses* are provided as published and are conducted at one's own risk. *Organic Syntheses, Inc.*, its Editors, and its Board of Directors do not warrant or guarantee the safety of individuals using these procedures and hereby disclaim any liability for any injuries or damages claimed to have resulted from or related in any way to the procedures herein.

Discussion

β-Lactams are of potential interest as precursors for anti-bacterial and anti-inflammatory agents,[2] building blocks for new biologically active compounds,[3] and as monomers for ring-opening polymerization.[4–10] β-Lactams of glycals are commonly used method for β-lactam synthesis, produces two new stereocenters. Therefore the product is often a mixture of stereoisomers.[15–17] Both the enolate-imine condensation reaction and the Kinugasa reaction rely on substrates that are not readily available or require harsh reaction conditions.[16] In contrast, the alkene-isocyanate cycloaddition method yields β-lactams with high regio- and stereo-selectivity and is compatible with glycal substrates. This reaction proceeds with electron-poor isocyanates (i.e., chlorosulfonyl isocyanate (CSI), p-tolunesulfonyl isocyanate, trifluoroacetyl isocyanate, and trichloroacetyl isocyanate (TCAI)) to yield stereoselective products. synthesized by the: Staudinger's ketene-imine [2+2] cycloaddition reaction;[11] enolate-imine condensation reaction; Kinugasa reaction;[12] and isocyanate cycloadditions.[13,14] Staudinger synthesis, the most

Chmielewski et al. reported synthetic methods to prepare stereoselective sugar vinyl ether β-lactams using TCAI[13] and CSI.[14] We perform both large scale TCAI and CSI cycloaddition reactions in our laboratory for the syntheses of β-lactam compounds from a variety of glycal structures (Table 1), including different derivatives of D-glucal (Table 1, entries 1-5), D-galactal (6), D-altral (7) and D-maltal (8). We then use these compounds or monomers for anionic ring-opening polymerization.

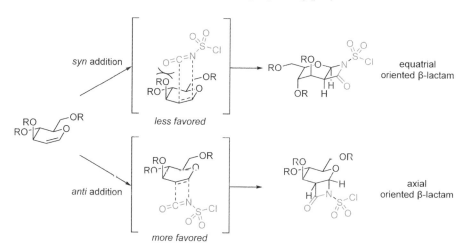

Figure 8. Rationale for stereoselectivity of cycloaddition reaction

The procedure described above offers a scalable, operationally simple method for large-scale synthesis of glycal β-lactams using CSI, which is the most widely used isocyanate cycloaddition reaction for β-lactam synthesis owing to its exceptionally high reactivity and, therefore, the extremely fast cycloaddition reaction.[18] Further, the CSI cycloaddition reaction is stereoselective due to the concerted addition of the sulfonyl isocyanate on the alkene (predicted by ab initio calculations)[19] as well as the unfavorable steric interactions between the substituent at the C3 position of the pyranose ring and the incoming CSI molecule (Figure 8).[14] The stereochemical control in this reaction leads to stereospecific syntheses of β-lactam monomers, ensuring a subsequent stereochemically controlled backbone post anionic ring-opening polymerization.[20,21]

Table 1. Substrate scope of CSI and TCAI cycloadditions

No.	Reactants	Products	Yield (Percentage)
1	tri-O-Bn-D-glucal		2.3 g from 10 g (21%)
2	6-O-PMB-3,4-di-O-Bn-D-glucal		3 g from 9.9 g (28%)
3	3,6-di-O-TBS-4-PMB-D-glucal		3.8 g from 5 g (69%)
4	6-O-TIPS-di-O-PMB-D-glucal		4 g from 9.5 g (39%)
5	6-N-Boc-N-Nosyl-amino-6-deoxy-di-O-Bn-D-glucal (TCAI)		2.6 g from 6.9 g (35%)
6	tri-O-Bn-D-galactal (TCAI)		2.2 g from 4.0 g (49%)
7	altrose		4.4 g from 9.5 g (42%)
8	maltose		5.3 g from 15.5 g (32%)

References

1. Department of Chemistry and Department of Biomedical Engineering, Boston University, Boston, MA, 02215, United States. Corresponding author: Mark W. Grinstaff, email: mgrin@bu.edu. ORCID: 0000-0002-5453-3668. This work was supported in part by BU and DoD Uniformed Services University (HU0001810012).
2. Kidwai, M.; Sapra, P.; Bhushan, K. R. *Curr. Med. Chem.* **1999**, *6*, 195–215.
3. Singh, G. S. *Mini-Reviews Med. Chem.* **2005**, *4*, 69–92.
4. Dane, E. L.; Grinstaff, M. W. *J. Am. Chem. Soc.* **2012**, *134*, 16255–16264.
5. Dane, E. L.; Chin, S. L.; Grinstaff, M. W. *ACS Macro Lett.* **2013**, *2*, 887–890.
6. Ghobril, C.; Heinrich, B.; Dane, E. L.; Grinstaff, M. W. *ACS Macro Lett.* **2014**, *3*, 359–363.
7. Xiao, R.; Dane, E. L.; Zeng, J.; McKnight, C. J.; Grinstaff, M. W. *J. Am. Chem. Soc.* **2017**, *139*, 14217–14223.
8. Balijepalli, A. S.; Sabatelle, R. C.; Chen, M.; Suki, B.; Grinstaff, M. W. *Angew. Chem. Int. Ed.* **2020**, *132*, 714–720.
9. Balijepalli, A. S.; Hamoud, A.; Grinstaff, M. W. *Polym. Chem.* **2020**, *11*, 1926–1936.
10. Balijepalli, A. S.; McNeely, J. H.; Hamoud, A.; Grinstaff, M. W. *J. Org. Chem.* **2020**, *85*, 12044–12057.
11. Hazelard, D.; Compain, P. *Org. Biomol. Chem.* **2017**, *15*, 3806–3827.
12. Popik, O.; Grzeszczyk, B.; Staszewska-Krajewska, O.; Furman, B.; Chmielewski, M. *Org. Biomol. Chem.* **2020**, *18*, 2852–2860.
13. Chmielewski, M.; Kałuża, Z. *Carbohydr. Res.* **1987**, *167*, 143–152.
14. Chmielewski, M.; Kałuża, Z. *J. Org. Chem.* **1986**, *51*, 2395–2397.
15. Jiao, L.; Liang, Y.; Xu, J. *J. Am. Chem. Soc.* **2006**, *128*, 6060–6069.
16. Hosseyni, S.; Jarrahpour, A. *Org. Biomol. Chem.* **2018**, *16*, 6840–6852.
17. Cossío, F. P.; Arrieta, A.; Sierra, M. A. *Acc. Chem. Res.* **2008**, *41*, 925–936.
18. Brandi, A.; Cicchi, S.; Cordero, F. M. *Chem. Rev.* **2008**, *108*, 3988–4035.
19. Cossío, F. P.; Lecea, B.; Lopez, X.; Roa, G.; Arrieta, A.; Ugalde, J. M. *J. Chem. Soc., Chem. Commun.* **1993**, *18*, 1450–1452.
20. Xiao, R.; Dane, E. L.; Zeng, J.; McKnight, C. J.; Grinstaff, M. W. *J. Am. Chem. Soc.* **2017**, *139*, 14217–14223.
21. Kałuża, Z.; Abramski, W.; Bełżecki, C.; Grodner, J.; Mostowicz, D.; Urbański, R.; Chmielewski, M. *Synlett* **1994**, *1994*, 539–541.

Appendix
Chemical Abstracts Nomenclature (Registry Number)

Sodium carbonate; (497-19-8)
Chlorosulfonyl isocyanate; (1189-71-5)
Tri-O-benzyl-D-glucal; (55628-54-1)
Red-Al® sodium bis(2-methoxyethoxy)aluminum hydride solution
(>60 wt% solution in toluene); (22722-98-1)
Sodium hydrogen carbonate; (144-55-8)
Sodium sulfate; (7757-82-6)
Hydrogen chloride; (7647-01-0)
Acetone; (67-64-1)
Methanol; (67-56-1)
Toluene; (108-88-3)

Maria Varghese received her combined B.S., M.S. degree (Chemistry major, Biology minor) from Indian Institute of Science Education and Research, Thiruvananthapuram (IISER, TVM), India, where she performed undergraduate research under Professor Kana M. Sureshan. She is currently a Ph.D. Candidate in Professor Mark W. Grinstaff's laboratory at Boston University. Her doctoral studies focus on the development of synthetic polysaccharide mimetics as synthetic anticoagulants.

Hannah E. Caputo received her Bachelor's degree from the University of New Hampshire in 2014. She completed her M.S. degree under the mentorship of Professors Mark W. Grinstaff and John E. Straub. Her research interests lie in the synthesis, characterization, and modeling of sulfated polysaccharide analogs.

Ruiqing Xiao received his B.S. degree in Material Chemistry (2009) and M.S. degree in Polymer Chemistry and Physics (2012) from Jilin University. His PhD studies focused on the synthesis of biologically active saccharide polymers at Boston University under the supervision of Prof. Mark W. Grinstaff, and he obtained his Ph.D. degree in Organic Chemistry in 2018. Currently, he is working as a post-doctoral researcher in Prof. Robert S. Langer's research group at the Massachusetts Institute of Technology. His research interests include the synthesis of novel immunostimulatory materials and polymeric materials for single-injection vaccine delivery and micronutrient delivery.

Anant Balijepalli attended the University of Michigan and earned a combined B.S.E / M.S.E in biomedical engineering performing research under Dr. Joel Swanson (UM Microbiology) and Dr. Mark Meyerhoff (UM Chemistry). He completed his Ph.D. in biomedical engineering under the mentorship of Dr. Mark W. Grinstaff at Boston University, focusing on the synthesis of cationic poly-amido-saccharides for drug delivery applications. He is currently a Scientist at Vivtex specializing in formulation and assay development to enable oral delivery of biologics.

Aladin Hamoud is a post-doctoral fellow in Prof. Mark W. Grinstaff research group at Boston University. He received his Bachelor's degree from the University of Bordeaux, France in 2013 followed by a Master's degree in organic chemistry in 2015 under the mentorship of Dr. Philippe Piexotto and Prof. Yannick Landais. He received his Ph.D. degree in 2018 under the mentorship of Dr. Valerie Desvergnes developing the aqueous Stetter reaction and the synthesis of nucleolipids for low molecular weight oleogels formation. His current research focuses on the development of mucoadhesive polysaccharide analogs and targeted cancer therapies.

Mark W. Grinstaff is the Distinguished Professor of Translational Research and a Professor of Biomedical Engineering, Chemistry, Materials Science and Engineering, and Medicine at Boston University. Mark received his A.B. from Occidental College, earned his Ph.D. under the mentorship of Kenneth S. Suslick at the University of Illinois, and completed his post-doctoral training with Harry B. Gray at the California Institute of Technology. His current research activities include the synthesis of new macromolecules and biomaterials, self-assembly chemistry, imaging contrast agents, drug delivery, and wound repair. He and his trainees have advanced ideas from the laboratory to the clinic with regulatory approved pharmaceutical and medical device products becoming the standard of care for patients. *www.grinstaff.org*

Kyohei Oga was born in Kanagawa, Japan. He graduated from Tokyo University of Pharmacy and Life Science in 2021 with B.S. in Life Science under the supervision of Professor Hisanaka Ito. In 2021, he began his graduate studies at the University of Tokyo under the supervision of Professor Masayuki Inoue. His research interests are in the area of the total synthesis of complex natural products.

Masanori Nagatomo completed his Ph.D. (2012) from the University of Tokyo under the supervision of Professor Masayuki Inoue. In 2012, he carried out visiting research with Professor Phil S. Baran at The Scripps Research Institute. In the same year, he was appointed as Assistant Professor in the Graduate School of Pharmaceutical Sciences at the University of Tokyo and was promoted to Lecturer in 2018. His research efforts focus on the development of novel synthetic methodology and applications to the multistep synthesis of complex molecules.

Preparation of 2-(Triethylsilyl)cyclohex-1-en-1-yl Trifluoromethanesulfonate as a Precursor to Cyclohexyne

Kazuki Inoue, Kengo Inoue, Atsunori Mori, and Kentaro Okano*[1]

Department of Chemical Science and Engineering, Kobe University, 1-1 Rokkodai, Nada, Kobe 657-8501, Japan

Checked by Zhixun Wang and Kevin Campos

Procedure (Note 1)

A. *2-(Triethylsilyl)cyclohex-1-en-1-yl trifluoromethanesulfonate* (**4**). An oven-dried 1-L three-necked flask equipped with two 125-mL pressure-equalizing dropping funnels with a rubber septum, a Teflon-coated magnetic stir bar (4.5 × 1.5 cm), and a rubber septum inserted with a nitrogen gas inlet and a thermocouple, is charged with *t*-BuOK (7.01 g, 62.5 mmol, 2.5 equiv) (Note 2) and anhydrous *n*-hexane (100 mL) (Note 3). A solution of LDA (1.0 M in THF/hexanes, 62.5 mL, 63 mmol, 2.5 equiv) (Note 4) is added by cannula to the one of the dropping funnels, and this LDA solution is added dropwise over 5 min to the round-bottomed flask. The resulting mixture is stirred at ambient temperature for 30 min (Figure 1).

Figure 1. A) LDA being transferred into the dropping funnel via cannula; B) reaction mixture after addition of LDA and stirring at ambient temperature for 30 min (photos provided by checkers)

(Cyclohex-1-en-1-yloxy)triethylsilane (**1**) (5.57 g of 95% purity, 25.0 mmol, 1 equiv) (Note 5) is added to the reaction mixture via a syringe through the septum over 1 min (Figure 2). After stirring at ambient temperature for 1.0–2.5 h (Note 6), distilled H_2O (675 µL, 37.5 mmol, 1.5 equiv) (Note 7) in anhydrous THF (100 mL) (Note 8) is added to the reaction mixture over 20 min (Note 9). After stirring at ambient temperature for 1 h, the reaction mixture is cooled to –78 °C with a dry ice-acetone bath. Comins' reagent (19.6 g, 50.0 mmol, 2.0 equiv) (Note 10) in anhydrous THF (75 mL) is added to the other pressure-equalizing dropping funnel and this solution is added dropwise to the round-bottomed flask over a 10-min period (Note 11) (Figure 3).

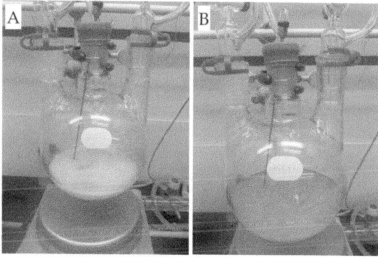

Figure 2. A) Reaction mixture after addition of (cyclohex-1-en-1-yloxy)triethylsilane (1) and stirring at ambient temperature for 1 h; B) reaction mixture after addition of water in THF and stirring at ambient temperature for 1 h (photos provided by checkers)

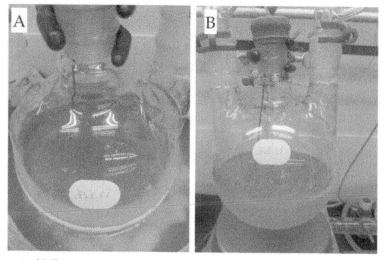

Figure 3. A) Reaction mixture after addition of Comins' reagent solution in THF at –78 °C; B) reaction mixture warming to ambient temperature (photos provided by checkers)

The solution is allowed to stir for another 10 min and then the cooling bath is removed. The reaction mixture warms to ambient temperature and is allowed to stir for 1 h (Note 12). The reaction is quenched with water (100 mL), which is added in one portion. The reaction mixture is transferred into a 1-L separatory funnel and extracted with *n*-hexane (3 × 100 mL). The combined organic extracts are washed with sat. aq. NaCl solution (1 × 150 mL), dried over Na_2SO_4 (25 g), and filtered. The filtrate is concentrated on a rotary evaporator under reduced pressure (15 °C, 10 mmHg), and the residue is dried *in vacuo* to afford 11.0–11.8 g of a crude product as a brown oil, which is purified by column chromatography on silica gel (elution: hexanes/Et_2O = 19:1) (Notes 13 and 14) to provide 2-(triethylsilyl)cyclohex-1-en-1-yl trifluoromethanesulfonate (**4**) as a colorless oil (7.00 g, 81%) (Note 15) (Figure 4).

Figure 4. 2-(Triethylsilyl)cyclohex-1-en-1-yl trifluoromethanesulfonate (4) (photo provided by checkers)

B. *9,10-Diphenyl-1,2,3,4,9,10-hexahydro-9,10-epoxyanthracene* (**7**). A 500-mL one-necked round-bottomed flask equipped with a Teflon-coated magnetic stir bar (4.5 × 1.5 cm) is charged with 1,3-diphenylisobenzofuran (**6**) (4.87 g, 18.0 mmol, 1.5 equiv) (Note 16), 2-(triethylsilyl)cyclohex-1-en-1-yl trifluoromethanesulfonate (**4**) (4.23 g of 98% purity, 12.0 mmol, 1 equiv), and THF (165 mL). A thermocouple is inserted into the reaction mixture. Tetrabutylammonium fluoride (1.0 M in THF, 18.0 mL, 18 mmol, 1.5 equiv) (Note 17) is added to the flask during 15 min via a syringe (Note 18), and after 1 h the thermocouple is removed (Figure 5).

Figure 5. Reaction mixture after addition of tetrabutylammonium fluoride (photo provided by submitters)

After stirring at ambient temperature for 2-4 h (Note 19), the reaction is quenched with water (120 mL). The reaction mixture is transferred into a 1-L separatory funnel and extracted with Et$_2$O (2 × 120 mL). The combined organic extracts are washed with sat. aq. NaCl solution (1 × 80 mL), dried over Na$_2$SO$_4$ (10 g), and filtered. The filtrate is concentrated on a rotary evaporator under reduced pressure (35 °C, 10 mmHg), and the residue is dried *in vacuo* to afford 10.3–12.2 g of a crude product as a yellow solid, which is purified by column chromatography on silica gel (Note 20) to provide 9,10-diphenyl-1,2,3,4,9,10-hexahydro-9,10-epoxyanthracene (**7**) as a pale yellowish green solid (4.05 g, 96%, uncorrected for purity).

The resulting solid is recrystallized from boiling ethanol (200 mL) in a 500-mL one-necked round-bottomed flask equipped with a reflux condenser and a Teflon-coated magnetic stir bar (4.5 × 1.5 cm). Crystallization is allowed to occur at ambient temperature over 2 h, then at 0 °C for 30 min. The crystals are collected by filtration and washed with cold (0 °C) ethanol (30 mL) to provide the title compound **7** as colorless crystals (3.55 g, 84%) (Note 21) (Figure 6).

Figure 6. A) Filtration of product 7; B) appearance of recrystallized product 7 (photos provided by checkers)

Notes

1. Prior to performing each reaction, a thorough hazard analysis and risk assessment should be carried out with regard to each chemical substance and experimental operation on the scale planned and in the context of the laboratory where the procedures will be carried out. Guidelines for carrying out risk assessments and for analyzing the hazards associated with chemicals can be found in references such as Chapter 4 of "Prudent Practices in the Laboratory" (The National Academies Press, Washington, D.C., 2011; the full text can be accessed free of charge at https://www.nap.edu/catalog/12654/prudent-practices-in-the-laboratory-handling-and-management-of-chemical. See also "Identifying and Evaluating Hazards in Research Laboratories" (American Chemical Society, 2015) which is available via the associated website "Hazard Assessment in Research Laboratories" at https://www.acs.org/content/acs/en/about/governance/committees/chemicalsafety/hazard-assessment.html. In the case of this procedure, the risk assessment should include (but not necessarily be limited to) an evaluation of the potential hazards associated with lithium

diisopropylamide, potassium *tert*-butoxide, *n*-hexane, tetrahydrofuran, Comins' reagent, sodium chloride, anhydrous sodium sulfate, diethyl ether, 1,3-diphenylisobenzofuran, tetrabutylammonium fluoride, ethyl acetate, silica gel, ethanol, deuterated chloroform, and 1,3,5-trimethoxybenzene, as well as the proper procedures for setting up experimental operations.
2. *t*-BuOK (>95.0%) was purchased from Tokyo Kasei Kogyo Co., Inc. by submitters and used as received without further purification. *t*-BuOK (>98%) was purchased from Sigma-Aldrich Co. by checkers and used as received without further purification.
3. Anhydrous *n*-hexane (>96.0%, water content: < 30 ppm) was purchased from Nacalai Tesque Co., Inc. by submitters and used as received without further purification. Anhydrous *n*-hexane (95%, water content: < 10 ppm) was purchased from Sigma-Aldrich Co. by checkers and used as received without further purification.
4. LDA (1.0 M in THF/hexanes) was purchased from Sigma-Aldrich Co. by both submitters and checkers, and the solution was used as received.
5. The starting (cyclohex-1-en-1-yloxy)triethylsilane (**1**) was prepared as reported in *Org. Synth.* **2021**, *98*, 407–429.
6. The reaction typically requires 1 h to consume all the silyl enol ether **1** and is monitored by TLC analysis on Merck silica gel 60 F_{254} plates developing with hexanes/ethyl acetate (9:1). The R_f values of (cyclohex-1-en-1-yloxy)triethylsilane (**1**) and 2-(triethylsilyl)cyclohexan-1-one are 0.86 and 0.45, respectively (stained with an ethanol solution of *p*-anisaldehyde). After dipping the TLC plate into the solution, the chromatogram is heated to reveal the stained compounds (Figure 7).

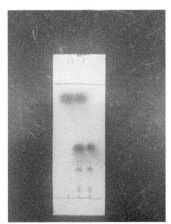

Figure 7. TLC of (cyclohex-1-en-1-yloxy)triethylsilane (1, left), reaction mixture after addition of 1 and stirring for 1 h (right), and their combination spot (middle). Eluent: hexanes/ethyl acetate (9:1, v/v) (photo provided by checkers)

7. Distilled water was purchased from Nacalai Tesque Co., Inc. by submitters and used as received without further purification. Water (UHPLC-MS grade) was purchased from Thermo Fisher Scientific Inc. by checkers and used as received without further purification.
8. THF (>99.5%, water content: < 10 ppm) was purchased from Kanto Chemical Co., Inc. by submitters and further dried by passing through a solvent purification system (Glass Contour) prior to use. Anhydrous THF (>99.9%, inhibitor-free, water content: < 20 ppm) was purchased from Sigma-Aldrich Co. by checkers and used as received without further purification.
9. An exotherm (from 22 to 29 °C) was observed during the 20 min addition time at the reported scale. A larger scale reaction than that described in this procedure may require a longer addition time to control the exotherm.
10. Comins' reagent (>99.0%) was purchased from Oakwood Products, Inc. by both submitters and checkers, and used as received without further purification.
11. An exotherm (from –72 to –65 °C) was observed during the 20 min addition time at the reported scale. A larger scale reaction than that described in this procedure may require a longer addition time to control the exotherm.

12. The reaction is monitored by TLC analysis on Merck silica gel 60 F_{254} plates developing with hexanes/ethyl acetate (9:1). The R_f values of 2-(triethylsilyl)cyclohexan-1-one and 2-(triethylsilyl)cyclohex-1-en-1-yl trifluoromethanesulfonate (4) are 0.45 and 0.82, respectively (stained with an ethanol solution of *p*-anisaldehyde). After dipping the TLC plate into the solution, the chromatogram is heated to reveal the stained compounds (Figure 8).

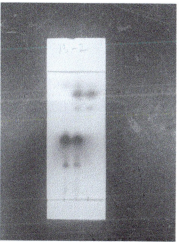

Figure 8. TLC of reaction mixture before addition of Comins' reagent (left), reaction mixture after addition of Comins' reagent and warming up to rt (right), and their combination spot (middle). Eluent: hexanes/ethyl acetate (9:1, v/v) (photo provided by checkers)

13. Silica gel (360 g, spherical, 100 µm, pH 6–8, purchased from TCI Chemicals by both submitters and checkers) and 0.9 L hexanes/Et_2O (19:1) were charged into a 1-L bottle, and shaken up to ensure mixing. The resulting silica gel slurry was loaded into the column (d = 6.5 cm, h = 20 cm). The crude product was suspended in 10 mL hexanes/Et_2O (19:1), and loaded onto the column, after which another 10 mL hexanes/Et_2O (19:1) was added to the flask to ensure a complete transfer. The column was eluted with hexanes/Et_2O (19:1). A total of 40 fractions (25 mL) were collected (total eluting time 15–20 min). Analysis by TLC (Note 12) revealed that the product was obtained in fractions 19–28, which were combined and concentrated on a rotary evaporator under reduced pressure (40 °C, 100 to 7 mmHg) (Figure 9).

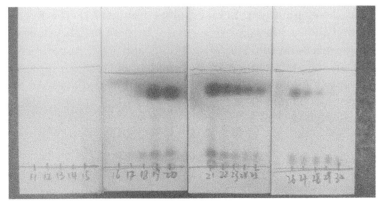

Figure 9. TLC of the fractions from flash column chromatography. Eluent: hexanes/Et$_2$O (19:1, v/v) (photo provided by checkers)

14. Both submitters and checkers observed that 2-(triethylsilyl)cyclohex-1-en-1-yl trifluoromethanesulfonate (**4**) undergoes decomposition on silica gel during column chromatography. Extensive optimization indicates that silica gel brands, silica gel/crude product ratio, and hexanes/Et$_2$O ratio in eluent (as shown in Note 13) are crucial to deliver satisfactory yield and purity.

15. 2-(Triethylsilyl)cyclohex-1-en-1-yl trifluoromethanesulfonate (**4**): colorless oil; R$_f$ = 0.79 (hexanes/Et$_2$O = 19:1); Merck silica gel 60 F$_{254}$ plates (stained with an ethanol solution of *p*-anisaldehyde). After dipping the TLC plate into the *p*-anisaldehyde solution, the chromatogram is stained by heating; IR (film, ATR): 2953, 1648, 1408, 1201, 1143, 986, 884, 850, 722 cm^{-1}; ^{1}H NMR (500 MHz, CDCl$_3$) δ: 0.73 (q, *J* = 7.9 Hz, 6H), 0.95 (t, *J* = 7.9 Hz, 9H), 1.51 – 1.63 (m, 2H), 1.69 – 1.81 (m, 2H), 2.19 (tt, *J* = 6.0, 2.5 Hz, 2H), 2.42 (tt, *J* = 6.4, 2.4 Hz, 2H); ^{13}C NMR (126 MHz, CDCl$_3$) δ: 2.9, 7.6, 21.9, 23.2, 28.5, 28.9, 118.4 (q, $^1J_{C-F}$ = 320 Hz), 125.6, 155.2; ^{19}F NMR (471 MHz, CDCl$_3$) δ: –74.77; The purity of the sample was determined to be 98% by ^{1}H qNMR using 37.42 mg of 1,3,5-trimethoxybenzene (Note 22) as an internal standard and 40.90 mg of product **4**. Based on the purity, 6.86 g (79%) of the product was prepared. A second reaction on the identical scale provided 7.19 g of the product with 94% purity.

16. 1,3-Diphenylisobenzofuran (**6**) (>97.0%) was purchased from Tokyo Kasei Kogyo Co., Inc. by submitters and used as received without

further purification. 1,3-Diphenylisobenzofuran (**6**) (97%) was purchased from Sigma-Aldrich Co. by checkers and used as received without further purification.

17. Tetrabutylammonium fluoride (1.0 M in THF) was purchased from Sigma-Aldrich Co. by both submitters and checkers, and the solution was used as received without further purification.
18. A small exotherm (+8 °C) was observed during 15 min addition time at current scale.
19. The reaction typically requires 2–4 h to consume all the 2-(triethylsilyl)cyclohex-1-en-1-yl trifluoromethanesulfonate (**4**) and is monitored by TLC analysis on Merck silica gel 60 F_{254} plates developing with *n*-hexane/CH_2Cl_2 (9:1). The R_f values of 2-(triethylsilyl)cyclohex-1-en-1-yl trifluoromethanesulfonate (**4**), 1,3-diphenylisobenzofuran (**6**), and 9,10-diphenyl-1,2,3,4,9,10-hexahydro-9,10-epoxyanthracene (**7**) are 0.68, 0.53, and 0.36, respectively (stained with an ethanol solution of *p*-anisaldehyde). After dipping the TLC plate to the solution, the chromatogram is stained by heating (Figure 10).

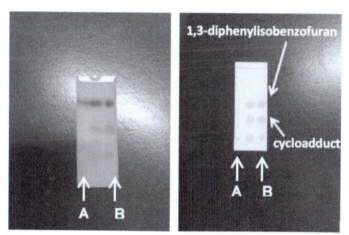

Figure 10. Left: TLC of 2-(triethylsilyl)cyclohex-1-en-1-yl trifluoromethanesulfonate (A) and the reaction mixture (B) (stained with an ethanol solution of *p*-anisaldehyde). Right: TLC of 2-(triethylsilyl)cyclohex-1-en-1-yl trifluoromethanesulfonate (A) and the reaction mixture (B) (visualized by UV lamp) (photos provided by submitters)

20. The crude product was dissolved in CH₂Cl₂ (40 mL) and 30-gram silica gel was added. The resulting slurry was evaporated to dryness on a rotary evaporator (35 °C, 100 to 10 mmHg). The crude product mixture was charged onto a RediSep RF column (diameter = 3.70 cm, height = 21.8 cm, 120-gram silica gel). The column was eluted with linear gradient from 0 to 25% CH₂Cl₂ in hexanes on a ISCO CombiFlash system, and 50-mL fractions were collected. Fractions 14–27 were combined and concentrated on a rotary evaporator under reduced pressure (40 °C, 100 to 10 mmHg).

21. 9,10-Diphenyl-1,2,3,4,9,10-hexahydro-9,10-epoxyanthracene (**7**): colorless crystals; R_f = 0.36 (*n*-hexane/CH₂Cl₂ = 9:1); Merck silica gel 60 F_{254} plates (stained with an ethanol solution of *p*-anisaldehyde). After dipping the TLC plate to the *p*-anisaldehyde solution, the chromatogram is stained by heating; mp 171–173 °C (ethanol); IR (film): 2920, 2853, 2826, 1601, 1446, 1307, 996, 742, 699 cm⁻¹; ¹H NMR (500 MHz, CDCl₃) δ: 1.40 – 1.51 (m, 2H), 1.55 – 1.69 (m, 2H), 1.99 – 2.16 (m, 2H), 2.22 – 2.38 (m, 2H), 6.99 (dd, *J* = 5.2, 3.0 Hz, 2H), 7.24 (dd, *J* = 5.2, 3.0 Hz, 2H), 7.40 – 7.46 (m, 2H), 7.49 – 7.57 (m, 4H), 7.73 – 7.80 (m, 4H); ¹³C NMR (126 MHz, CDCl₃) δ: 22.4, 23.4, 92.3, 119.0, 124.6, 126.3, 127.7, 128.4, 135.5, 150.2, 151.8; The purity of the sample was determined to be >99% by ¹H qNMR using 39.24 mg of 1,3,5-trimethoxybenzene (Note 22) as an internal standard and 47.23 mg of product **7**. A second reaction on the identical scale provided 3.99 g of crude product, which provided 3.48 g (83%, uncorrected for purity) of the product after recrystallization.

22. 1,3,5-Trimethoxybenzene (>99%) was purchased from Sigma-Aldrich Co. and used as received.

Working with Hazardous Chemicals

The procedures in *Organic Syntheses* are intended for use only by persons with proper training in experimental organic chemistry. All hazardous materials should be handled using the standard procedures for work with chemicals described in references such as "Prudent Practices in the Laboratory" (The National Academies Press, Washington, D.C., 2011; the full text can be accessed free of charge at http://www.nap.edu/catalog.php?record_id=12654). All chemical waste should be disposed of in accordance with local regulations.

For general guidelines for the management of chemical waste, see Chapter 8 of Prudent Practices.

In some articles in *Organic Syntheses*, chemical-specific hazards are highlighted in red "Caution Notes" within a procedure. It is important to recognize that the absence of a caution note does not imply that no significant hazards are associated with the chemicals involved in that procedure. Prior to performing a reaction, a thorough risk assessment should be carried out that includes a review of the potential hazards associated with each chemical and experimental operation on the scale that is planned for the procedure. Guidelines for carrying out a risk assessment and for analyzing the hazards associated with chemicals can be found in Chapter 4 of Prudent Practices.

The procedures described in *Organic Syntheses* are provided as published and are conducted at one's own risk. *Organic Syntheses, Inc.*, its Editors, and its Board of Directors do not warrant or guarantee the safety of individuals using these procedures and hereby disclaim any liability for any injuries or damages claimed to have resulted from or related in any way to the procedures herein.

Discussion

Strained molecules have attracted a great deal of attention because of their high reactivities. Cyclohexyne (**5**), in which a triple bond is present in a six-membered ring, is a short-lived intermediate and has been extensively investigated to construct the various molecular frameworks,[2] despite the limited number of reports compared to aryne[3] (Scheme 1).

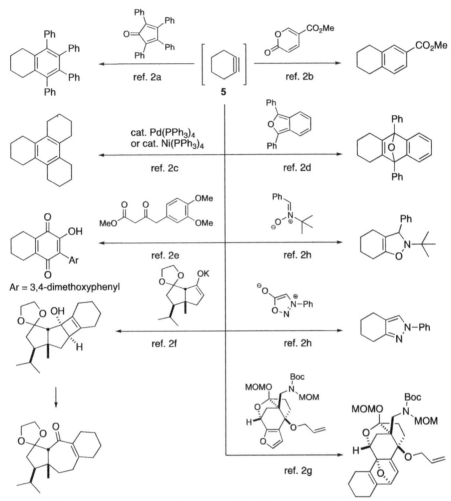

Scheme 1. Representative examples in constructing molecular frameworks using cyclohexyne (5)

Similar to the generation of 1,2-cyclohexadiene (**8**), cyclohexyne (**5**) can be generated from silylated cyclohexenyl triflate **9** with the aid of fluoride ion reported by Guitián[2b] (Scheme 2). Compared to other methods for generating cyclohexynes,[4] this fluoride ion-promoted conditions show the superior functional group compatibility. The only drawback of this method is the multi-step preparation of the silylated cyclohexenyl triflate **9**. The synthesis started with bromination of cyclohexenone (**10**), whose carbonyl group was protected as its acetal to provide **11**. The alkenyl bromide underwent halogen–lithium exchange followed by silylation. Aqueous workup gave 2-silylcyclohexenone **12**. The use of L-Selectride facilitated 1,4-reduction, and the resultant lithium enolate **13** was treated with PhNTf₂ to yield the silylated cyclohexenyl triflate **9**. In 2006, Peña reported generation of 1,2-cyclohexadiene (**8**) from silylated cyclohexenyl triflate **14** under the same reaction conditions. They first prepared 2-silylcyclohexanone **15** by protonation of lithium enolate **13**. Subsequent treatment with LDA led to the formation of trisubstituted lithium enolate **16** through kinetic deprotonation. The following triflation provided cyclohexenyl triflate **14** that is the precursor for 1,2-cyclohexadiene (**8**). Most of elaboration was dedicated to the introduction of the trimethylsilyl group, which is the key to shorten the synthesis of both silylated cyclohexenyl triflates.

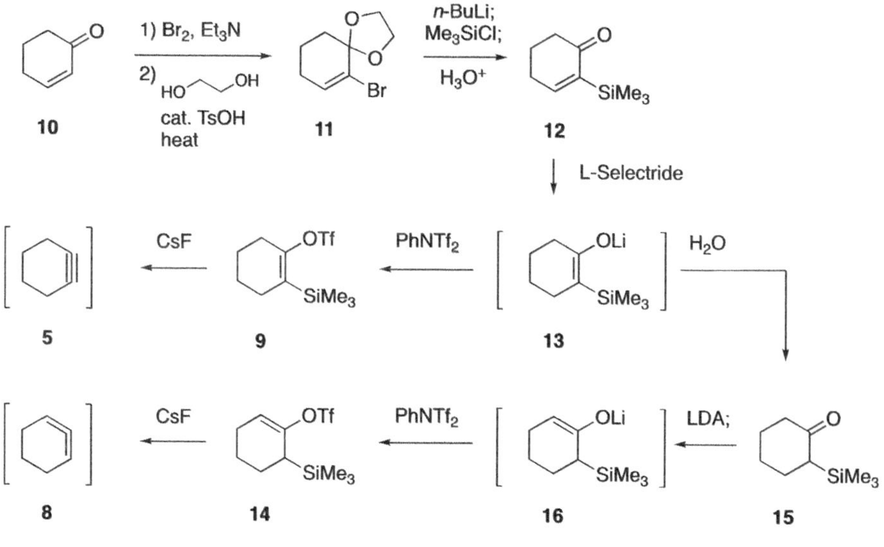

Scheme 2. Guitián and Peña's synthesis of the silylated cyclohexenyl triflates for cyclohexyne (**5**) and 1,2-cyclohexadiene (**8**)

As reported in our synthesis of 1,2-cyclohexadiene precursor **17**,[5] transfer of the triethylsilyl (TES) group in silyl enol ether **1**, which is easily prepared from cyclohexanone (**18**), is the powerful method to realize the facile synthesis of compound **17** (Scheme 3). Apart from the Guitián and Peña's method[2b,6a,6b] and recently reported Garg's method,[6c,6d] we first generate thermodynamically unstable trisubstituted silylated lithium enolate **19**, by rearrangement of allyllithium **20**. Subsequent triflation takes place smoothly with Comins' reagent[7] to provide 1,2-cyclohexadiene precursor **17** in 80–84% yield on a 7-gram scale from TES enol ether **1** in a single flask.[5c] We then devised an idea that isomerization of trisubstituted silylated lithium enolate **19** via 2-silylcyclohexanone **21** gives tetrasubstituted silylated lithium enolate **2**, which should be transformed to the corresponding cyclohexyne precursor **4**. We first performed the isomerization simply by prolonging reaction time and found that i-Pr$_2$NH, the conjugate acid of LDA, promoted the desired isomerization.[5a] We have also examined a variety of proton sources and found that water was the best additive to facilitate this isomerization of the double bond. The amount of water was carefully optimized; 1.5 equivalents of water were found to be best for the high-yielding process.

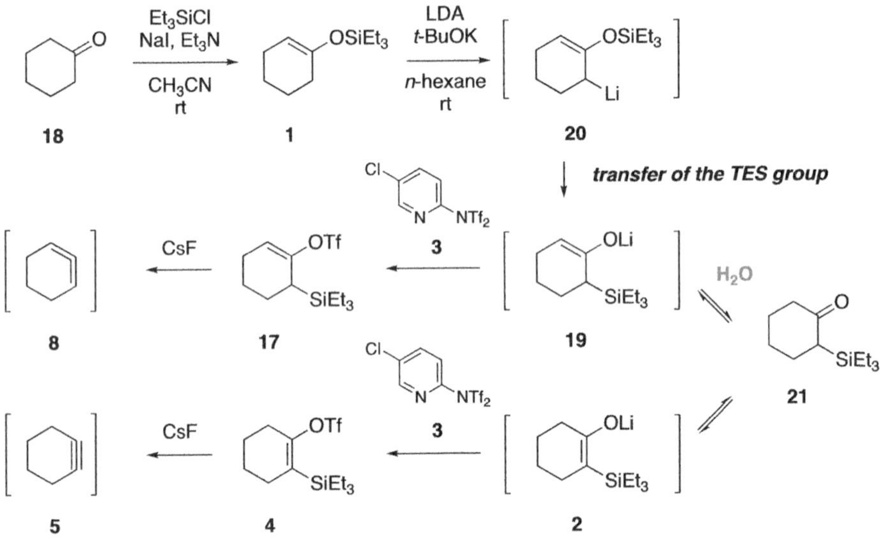

Scheme 3. Our synthetic strategy for the divergent synthesis of the silylated cyclohexenyl triflates for cyclohexyne (5) and 1,2-cyclohexadiene (8)

The amount of water was crucial for this isomerization (Table 1). In the absence of water, 1,2-cyclohexadiene precursor **17** was exclusively obtained (entry 1).[5c] As the amount of the water increased, the yields of cyclohexyne precursor **4** were improved; 1.5 equivalents proved most effective for the selective synthesis of silylated cyclohexenyl triflate **4** (entries 2–4). The excess amount of water resulted in significant reduction of the yields of compound **4** with concomitant formation of 2-silylcyclohexanone **21** by protonation (entries 5 and 6).

Table 1. Effects of water on the yields of the silylated cyclohexenyl triflates

entry	H₂O (equiv)	17 (%)[a]	4 (%)[a]
1	–	93	–[b]
2	0.5	71	13
3	1.0	34	54
4	1.5	–[b]	91
5	3.0	–[b]	76
6	5.0	–[b]	43

[a] Evaluated by ¹H NMR of the crude product.
[b] Not detected in the ¹H NMR spectrum of the crude product.

The established optimal conditions could be applied to various silylated cyclohexenyl triflates (Table 2). In addition to the TES enol ether, the TBS ether underwent the transfer of the silyl group and subsequent isomerization, which was then treated with Comins' reagent to give the desired product in moderate yield. The TIPS enol ether and TBDPS enol ether were converted to the corresponding products, albeit in low yields. The TES enol ether bearing the *tert*-butyl group at position 4 was transformed to the product in 59% yield as a sole regioisomer, whereas the TES enol ether bearing the *tert*-butyl group at position 3 was converted to the product in 44% yield.

Table 2. Scope of the reaction

substrate	product		isolated yield (%)
(cyclohexenyl-OSi)	(cyclohexenyl-OTf, Si)	Si = TES	85
		TBS	65
		TIPS	46
		TBDPS	16
4-*t*-Bu cyclohexenyl-OSiEt₃	4-*t*-Bu cyclohexenyl-OTf, SiEt₃		59
3-*t*-Bu cyclohexenyl-OSiEt₃	3-*t*-Bu cyclohexenyl-OTf, SiEt₃		44

In summary, the described protocol provides a reliable method for the synthesis of both silylated cyclohexenyl triflates that are precursors for cyclohexyne and 1,2-cyclohexadiene on multigram scales. The presented method is operationally simple and robust, and the starting cyclohexanone is widely available. This method can be applied to several silyl groups except for the trimethylsilyl group, which would promote the research on these reactive and strained synthetic intermediates for the synthesis of natural products, medicines, and functional organic materials.

References

1. Department of Chemical Science and Engineering, Kobe University, 1-1 Rokkodai, Nada-ku, Kobe 657-8501, Japan. E-mail: okano@harbor.kobe-u.ac.jp. ORCID: 0000-0003-2029-8505. These studies were supported by JSPS KAKENHI Grant Numbers JP16K05774 in Scientific Research (C), JP19H02717 in Scientific Research (B), Tonen General Sekiyu Research Development Encouragement & Scholarship Foundation, the Harmonic Ito Foundation, and Foundation for Interaction in Science & Technology.
2. (a) Barton, D. H. R.; Bashiardes, G.; Fourrey, J.-L. *Tetrahedron* **1988**, *44*, 147–162. (b) Atanes, N.; Escudero, S.; Pérez, D.; Guitián, E.; Castedo, L. *Tetrahedron Lett.* **1998**, *39*, 3039–3040. (c) Iglesias, B.; Peña, D.; Pérez, D.; Guitián, E.; Castedo, L. *Synlett* **2002**, 486–488. (d) Al-Omari, M.; Banert, K.; Hagedorn, M. *Angew. Chem. Int. Ed.* **2006**, *45*, 309–311. (e) Allan, K. M.; Hong, B. D.; Stoltz, B. M. *Org. Biomol. Chem.* **2009**, *7*, 4960–4964. (f) Gampe, C. M.; Carreira, E. M. *Angew. Chem. Int. Ed.* **2011**, *50*, 2962–2965. (g) Devlin, A. S.; Du Bois, J. *Chem. Sci.* **2013**, *4*, 1059–1063. (h) Medina, J. M.; McMahon, T. C.; Jiménez-Osés, G.; Houk, K. N.; Garg, N. K. *J. Am. Chem. Soc.* **2014**, *136*, 14706–14709. (i) Cho, S.; McLaren, E. J.; Wang, Q. *Angew. Chem. Int. Ed.* **2021**, *60*, 26332–26336. For a recent review, see: (j) Gampe, C. M.; Carreira, E. M. *Angew. Chem. Int. Ed.* **2012**, *51*, 3766–3778.
3. Recent reviews on aryne: (a) Sanz, R. *Org. Prep. Proced. Int.* **2008**, *40*, 215–291. (b) Kitamura, T. *Aust. J. Chem.* **2010**, *63*, 987–1001. (c) Tadross, P. M.; Stoltz, B. M. *Chem. Rev.* **2012**, *112*, 3550–3577. (d) Bhunia, A.; Yetra, S. R.; Biju, A. T. *Chem. Soc. Rev.* **2012**, *41*, 3140–3152. (e) Yoshida, H.; Takaki, K. *Heterocycles* **2012**, *85*, 1333–1349. (f) Pérez, D.; Peña, D.; Guitián, E. *Eur. J. Org. Chem.* **2013**, 5981–6013. (g) Wu, C.; Shi, F. *Asian J. Org. Chem.* **2013**, *2*, 116–125. (h) Shi, J.; Li, Y.; Li, Y. *Chem. Soc. Rev.* **2017**, *46*, 1707–1719. (i) Diamond, O. J.; Marder, T. B. *Org. Chem. Front.* **2017**, *4*, 891–910. (j) Takikawa, H.; Nishii, A.; Sakai, T.; Suzuki, K. *Chem. Soc. Rev.* **2018**, *47*, 8030–8056. (k) Nakamura, Y.; Yoshida, S.; Hosoya, T. *Heterocycles* **2019**, *98*, 1623–1677.
4. (a) Wittig, G.; Harborth, G. *Ber. Dtsch. Chem. Ges.* **1944**, *77*, 306–314. (b) Scardiglia, F.; Roberts, J. D. *Tetrahedron* **1957**, *1*, 343–344. (c) Wittig, G.; Pohlke, R. *Chem. Ber.* **1961**, *94*, 3276–3286. (d) Willey, F. G. *Angew. Chem. Int. Ed.* **1964**, *3*, 138. (e) Fujita, M.; Kim, W. H.; Sakanishi, Y.; Fujiwara, K.; Hirayama, S.; Okuyama, T.; Ohki, Y.; Tatsumi, K.; Yoshioka, Y. *J. Am. Chem. Soc.* **2004**, *126*, 7548–7558. (f) Yoshida, S.; Karaki, F.; Uchida, K.;

Hosoya, T. *Chem. Commun.* **2015**, *51*, 8745–8748. (g) Hioki, Y.; Okano, K.; Mori, A. *Chem. Commun.* **2017**, *53*, 2614–2617.
5. (a) Inoue, K.; Nakura, R.; Okano, K.; Mori, A. *Eur. J. Org. Chem.* **2018**, 3343–3347. (b) Nakura, R.; Inoue, K.; Okano, K.; Mori, A. *Synthesis* **2019**, *51*, 1561–1564. (c) Nakura, R.; Inoue, K.; Itoh, M.; Mori, A.; Okano, K. *Org. Synth.* **2021**, *98*, 407–429.
6. (a) Peña, D.; Iglesias, B.; Quintana, I.; Pérez, D.; Guitián, E.; Castedo, L. *Pure Appl. Chem.* **2006**, *78*, 451–455. (b) Quintana, I.; Peña, D.; Pérez, D.; Guitián, E. *Eur. J. Org. Chem.* **2009**, 5519–5524. (c) Barber, J. S.; Styduhar, E. D.; Pham, H. V.; McMahon, T. C.; Houk, K. N.; Garg, N. K. *J. Am. Chem. Soc.* **2016**, *138*, 2512–2515. (d) Chari, J. V.; Ippoliti, F. M.; Garg, N. K. *J. Org. Chem.* **2019**, *84*, 3652–3655.
7. (a) Comins, D. L.; Dehghani, A. *Tetrahedron Lett.* **1992**, *33*, 6299–6302. (b) Comins, D. L.; Dehghani, A.; Foti, C. J.; Joseph, S. P. *Org. Synth.* **1997**, *74*, 77–83.

Appendix
Chemical Abstracts Nomenclature (Registry Number)

t-BuOK: 2-Propanol, 2-methyl-, potassium salt (1:1); (865-47-4)
Hexane: hexane; (110-54-3)
LDA: 2-Propanamine, *N*-(1-methylethyl)-, lithium salt (1:1); (4111-54-0)
THF: Furan, tetrahydro-; (109-99-9)
Water; (7732-18-5)
Comins' reagent: Methanesulfonamide, *N*-(5-chloro-2-pyridinyl)-1,1,1-trifluoro-*N*-[(trifluoromethyl)sulfonyl]- (145100-51-2)
1,3-Diphenylisobenzofuran: Isobenzofuran, 1,3-diphenyl-; (5471-63-6)
Tetrabutylammonium fluoride: 1-Butanaminium, *N,N,N*-tributyl-, fluoride (1:1); (429-41-4)

Kazuki Inoue was born in Osaka in 1994. He received his B.S. in 2017 and M.S. in 2019 from Kobe University under the direction of Professors Atsunori Mori and Kentaro Okano. During his M.S. studies, he worked on generation of strained cycloalkynes and its application to the synthesis of functionalized poly(cycloalkyne)s. Currently, he works for Sumitomo Bakelite Co., Ltd.

Kengo Inoue was born in Fukuoka in 1997. He received his B.S. in 2020 from Kobe University under the direction of Professors Atsunori Mori and Kentaro Okano. During his M.S. studies, he is working on selective trapping of short-lived heteroaryllithiums in halogen dance and its application to the synthesis of multiply substituted heteroaromatic compounds.

Atsunori Mori was born in Aichi in 1959. He received B.S. (1982), M.S. (1984), and Ph.D. (1987) degrees from Nagoya University under the direction of Professor Hisashi Yamamoto. After postdoctoral study in U.C. Berkeley with Peter Vollhardt (1987–1988), he started his academic career as an Assistant Professor at the University of Tokyo (1988) and Japan Advanced Institute of Science and Technology (1993). He was appointed as an Associate Professor in 1995 at Chemical Resources Laboratory of Tokyo Institute of Technology. Since 2005, he has been Professor at the Graduate School of Engineering, Kobe University.

Kentaro Okano was born in Tokyo in 1979. He received his B.S. in 2003 from Kyoto University under the supervision of Professor Tamejiro Hiyama. He then moved to the laboratories of Professor Tohru Fukuyama at the University of Tokyo. In 2007, he joined the faculty at Tohoku University in Professor Hidetoshi Tokuyama's group. In 2014, he visited Professor Amir Hoveyda's laboratories at Boston College as a visiting researcher. In 2015, he moved to Kobe University, where he is currently Associate Professor of Department of Chemical Science and Engineering. His current research interest is natural product synthesis based on the development of new synthetic methodologies.

Zhixun Wang joined the Process Research Department of Merck & Co., Inc. in 2019. His research focuses on synthesize therapeutic candidates for large-scale production. He received his B. S. degree from Fudan University, and then obtained his Ph.D. under the supervision of Professor Liming Zhang at University of California, Santa Barbara with a research focus on homogeneous gold catalysis. Zhixun then moved to New Haven, CT, where he did postdoctoral work on synthesis of pleuromutilin analogs and structure elucidation of colibactin with Professor Seth Herzon in Yale University.

Late-stage C-H Functionalization with 2,3,7,8–Tetrafluorothianthrene: Preparation of a Tetrafluorothianthrenium-salt

Samira Speicher,[†] Matthew B. Plutschack,[†] and Tobias Ritter[†*1]

[†]Max-Planck-Institut für Kohlenforschung, Kaiser-Wilhelm-Platz 1, 45470 Mülheim an der Ruhr, Germany

Checked by Maurus Mathis, Jorge A. González, and Cristina Nevado

Procedure (Note 1)

A. *2,3,7,8-Tetrafluorothianthrene-S-oxide (2)*. In air, a three-necked 1000-mL round-bottomed flask (29/32) is charged with an egg-shaped Teflon coated magnetic stir bar (50 × 17 mm) and aluminum chloride (3.33 g, 25.0 mmol, 0.10 equiv) (Note 2). The central neck of the flask is fitted with a reflux condenser with the outlet passing through a round-bottomed flask (250 mL, 29/32) before entering a Drechsel bottle containing an aqueous sodium hydroxide solution (Note 3). Through a side neck, 1,2-difluorobenzene (0.250 L, 285 g, 2.50 mol, 10.0 equiv) (Note 4) is added to the 1000-mL round-bottomed flask. The flask is fitted with an adapter for the nitrogen inlet on the left neck (Note 5) and a 50 mL dropping funnel on the right neck. Disulfur dichloride (20.0 mL, 33.8 g, 0.250 mol, 1.00 equiv) (Note 6) is added to the addition funnel (Figure 1A), and then added dropwise at room temperature (Note 7) over a period of 10 min to the stirred reaction mixture (Note 8). Upon addition of the disulfur dichloride, the solution becomes black (Figure 1B) and evolves HCl gas. After complete addition of the disulfur dichloride, the reaction mixture is heated at a gentle reflux (105 °C, bath temp)) using a silicon oil bath (Note 9) for 1 h. The oil bath is then replaced by an ice water bath, and MeOH (200 mL) is added to the reaction mixture via the addition funnel (Note 10). The dark color quickly dissipates from the reaction mixture, resulting in a light yellow suspension with a colorless solid (2,3,7,8-tetrafluorothianthrene) (Figure 1C).

Figure 1. A) Reaction set-up, B) the reaction mixture after disulfur dichloride addition, C) reaction mixture after the methanol quench, and D) crude 2,3,7,8-tetrafluorothianthrene (photos A, B, and C were provided by the submitters; photo D was provided by the checkers)

The slurry is stirred for 30 min in the ice water bath, and the solid material is subsequently filtered by vacuum filtration (Note 11). The filter cake is washed with chilled methanol (2 × 50 mL) (Notes 10 and 12), and the resulting off-white solid is transferred into a 500-mL one-necked, round-bottomed flask (29/32) and dried under vacuum (Note 13) for 18 h at room temperature. The resulting 23.82 g off-white solid (Figure 1D) contained 74.3% of 2,3,7,8-tetrafluorothianthrene by weight (Note 14), which is used without any further purification in the subsequent reaction (Notes 15 and 16).

In air, the 500-mL one-necked, round-bottomed flask (29/32) containing 23.82 g of crude 2,3,7,8-tetrafluorothianthrene (wt% = 74.3%; Note 14) is charged with an egg-shaped Teflon-coated magnetic stir bar (27 × 10 mm), DCM (175 mL, c = 0.35 M), Fe(NO$_3$)$_3$·9H$_2$O (42.2 g, 105 mmol, 1.7 equiv),

NaBr (443 mg, 4.21 mmol, 0.07 equiv), and TFA (6.60 mL, 9.83 g, 86.2 mmol, 1.4 equiv) (Note 17). After addition, the flask is sealed with a rubber septum (Note 18). The resulting light brown suspension (Figure 2B) is stirred at room temperature for 20 h (Note 19). After this time, deionized water (150 mL) is added (Figure 2C).

Figure 2. The reaction mixture (A) before addition of TFA, (B) after addition of TFA, and (C) after addition of water (photos provided by submitters)

The suspension is filtered by vacuum filtration (Note 11), the filtrate is poured into a 500-mL separating funnel, and the layers are separated. The aqueous layer is extracted with DCM (2 × 150 mL). The filter cake is added to the combined organic layers, and the suspension is concentrated under reduced pressure (525 to 20 mmHg, 40 °C) (Note 20). MeCN (50 mL) (Note 21) is added to the residue, and the resulting suspension is filtered by vacuum filtration (Note 11). The solid is washed with MeCN (3 × 25 mL) (Note 12) (Figure 3A), which removes the majority of the red color. The solid is transferred into a 1000-mL one-necked, round-bottomed flask. Toluene (600 mL) (Note 22) is added, the flask is equipped with a reflux condenser, and the mixture is heated to reflux (110 °C) using a silicone oil bath. After 10 min at this temperature, heating and stirring is turned off and the mixture is left in the oil bath for 5 min. Subsequently, the mixture is decanted into a 1000-mL Erlenmeyer flask in order to remove the red-brown solid. Then the toluene solution is allowed to cool down to room temperature for 3 h. The obtained crystals (Figure 3B) are collected by vacuum filtration (Notes 11 and 23), washed with Et$_2$O (2 × 50 mL) (Notes

12 and 24) (Figure 3C), and transferred into a 500-mL one-necked, round-bottomed flask. The filtrate is concentrated via rotary evaporator (50 mmHg, 40 °C), and acetone (600 mL) is added to the resultant solid (Note 25). The mixture is heated at 60 °C using a silicone oil bath for 5 min. After this time the mixture is allowed to stand for 5 min. Subsequently, the mixture is decanted into a 1000-mL one-necked, round-bottomed flask (29/32). The solution is concentrated via rotary evaporator (400 mmHg, 40 °C). The product is transferred into a 250-mL one-necked, round-bottomed flask (29/32) and recrystallized from toluene (80 mL, 110 °C) using a silicone oil bath. After cooling to room temperature for 1 h the colorless crystals are collected by vacuum filtration (Note 11), washed with Et_2O (2 × 25 mL) (Note 12) (Figure 3D) and combined with the first batch of crystals in the 500-mL one-necked, round-bottomed flask (29/32). The combined batches are recrystallized again from toluene (240 mL, 110 °C) using a silicone oil bath. After cooling to room temperature for 2 h, the off-white crystals are collected by vacuum filtration (Note 11), washed with Et_2O (2 × 50 mL) (Note 12) and ground into a fine powder using a porcelain mortar. The resulting off-white powder is transferred into a 100-mL one-necked, round-bottomed flask and is dried under vacuum for 18 h at room temperature to afford 11.33 g of *2,3,7,8-tetrafluorothianthrene-S-oxide* (**2**) with a purity of 98.3% (15 % yield over two steps) (Figure 3E) (Notes 26 and 27).

Figure 3. A) Appearance of the crude product, B) the crystals during recrystallization, C) the product after filtration, D) second batch crystals, and E) final recrystallized product (photos A, B, and C were provided by the submitters; photos D and E were provided by the checkers)

B. *2-Phenethyl acetate-derived tetrafluorothianthrenium salt (4).* In air, a 250 mL one-necked, round-bottomed flask (29/32), equipped with a Teflon-coated magnetic stir bar (40 × 7 mm) is charged with 2-phenethyl acetate (6.98 mL, 7.19 g, 43.8 mmol, 1.00 equiv), acetonitrile (170 mL, c = 0.25 M) and TFAA (18.3 mL, 27.6 g, 131 mmol, 3.00 equiv) (Note 28). The reaction mixture is cooled to 0 to 5 °C with an ice water bath. *2,3,7,8-Tetrafluorothianthrene-S-oxide (2)* (13.32 g, 43.8 mmol, 1.00 equiv) (Note 26) is added, followed by the addition of tetrafluoroboric acid diethyl ether complex (8.9 mL, 10.6 g, 65.4 mmol, 1.49 equiv) (Note 29), resulting in a dark purple suspension (Figure 4B). The mixture is stirred at 0 to 5 °C for 1 h, after which the ice water bath is removed, and the reaction mixture is stirred for additional 23 h at room temperature (Note 30).

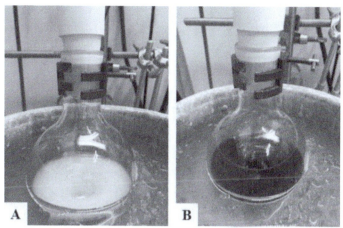

Figure 4. A) Reaction mixture before addition of tetrafluoroboric acid diethyl ether complex (A) and (B) after addition of tetrafluoroboric acid diethyl ether complex (photos provided by submitters)

After this time, the reaction mixture is concentrated on rotary evaporator (170 to 15 mmHg, 40 °C). *i*-PrOH (70 mL) (Note 31) is added in 10 mL portions during which the purple color disappears and a colorless solid precipitates (Figure 5A). The suspension is cooled to 0 to 5°C with an ice water bath, stirred at this temperature for 10 min, and then filtered (Note 11). The filter cake is washed with *i*-PrOH (15 mL) and with isohexane (100 mL) (Notes 12 and 32). The crude product is purified by flash chromatography on silica gel (Note 33), and the fractions containing the product are collected and concentrated via rotary evaporator (525 to 15 mmHg, 40 °C). The resulting colorless solid is ground into a fine powder using a porcelain mortar, transferred into a 100-mL one-necked, round-bottomed flask (29/32), and dried under vacuum at room temperature for 18 h to afford 17.9 g (76% yield) of 2-*phenethyl acetate-derived tetrafluorothianthrenium salt* (**4**) with a purity of 99.9% (Notes 34 and 35) (Figure 5B).

Figure 5. A) Appearance of the reaction mixture after the isopropanol quench, B) purified product (photos provided by submitters)

C. *[4-(4-Phenylphenyl)phenyl] ethyl acetate (5)*. A 500 mL two-necked, round-bottomed flask (29/32 and 14/23), equipped with a glass stopper on the small neck (14/23), a reflux condenser on the large neck (29/32) with a nitrogen inlet on top and a Teflon-coated magnetic stir bar (30 × 7 mm), is evacuated and flame-dried. After cooling down to room temperature again, the flask is filled with nitrogen. Under nitrogen (Note 36), the flask is charged via the small neck (14/23) with *2-phenethyl acetate derived tetrafluorothianthrenium salt (4)* (17.94 g, 33.3 mmol, 1.00 equiv), 4-biphenylboronic acid (7.26 g, 36.7 mmol, 1.10 equiv), Pd(dppf)Cl$_2$ (490 mg, 0.530 mmol, 2.00 mol%), K$_3$PO$_4$ (21.23 g, 100 mmol, 3.00 equiv), iPrOH (183 mL), and 1,4-dioxane (183 mL) (Notes 23 and 37) (Figure 6A). The flask is sealed with a glass stopper, and the reaction mixture is left stirring for 10 min under a nitrogen atmosphere at room temperature. After this time, the reaction mixture is heated to 50 °C using a silicone oil bath. During heating the suspension's color changes to bright orange (Figure 6B). The reaction mixture is stirred at 50 °C for 5 h (Note 38) under nitrogen atmosphere. After this time, the reaction mixture is cooled to room temperature. The solvent is removed via rotary evaporation (150 to 22 mmHg, 40 °C). Ethyl acetate (200 mL) (Note 39) is added, the mixture is stirred (300 rpm), and deionized water (200 mL) is added. The mixture is transferred into a 1 L separating funnel and additional ethyl acetate (100 mL) and deionized water (100 mL) are added. The layers are separated, and the aqueous layer is extracted with ethyl acetate (2 × 300 mL). The organic layers are combined, dried over 70 g of Na₂SO₄, filtered, and concentrated in vacuo (135 to 75 mmHg, 40 °C). The crude product is purified by flash chromatography on silica gel (Note 40). The fractions containing the product are collected and concentrated via rotary evaporator (525 to 15 mmHg, 40 °C). The off-white solid is transferred into a 250 mL one-necked, round-bottomed flask and recrystallized with 80 mL of ethyl acetate (77 °C) using a silicone oil bath as the heat source. The mixture is allowed to cool to room temperature for 1 hour and is subsequently filtered (Note 41) and washed with Et₂O (2 × 15 mL). The colorless crystals are grounded into a fine powder using a porcelain mortar, and the colorless powder is transferred into a 100-mL one-necked, round-bottomed flask. The powder is dried under vacuum at room temperature for 18 h to afford 8.16 g (77%) of [4-(4-phenylphenyl)phenyl] ethyl acetate (5) with a purity of 99.9% (Notes 42 and 43) (Figure 6C).

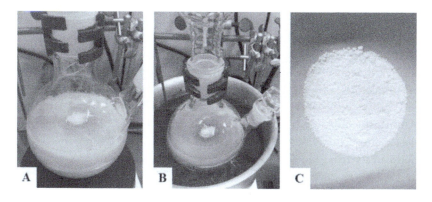

Figure 6. A) Reaction mixture at room temperature, B) after being heated to 50 °C, C) purified product (photos provided by submitters)

Notes

1. Prior to performing each reaction, a thorough hazard analysis and risk assessment should be carried out with regard to each chemical substance and experimental operation on the scale planned and in the context of the laboratory where the procedures will be carried out. Guidelines for carrying out risk assessments and for analyzing the hazards associated with chemicals can be found in references such as

Chapter 4 of "Prudent Practices in the Laboratory" (The National Academies Press, Washington, D.C., 2011; the full text can be accessed free of charge at http://www.nap.edu/catalog.php?record_id=12654). See also "Identifying and Evaluating Hazards in Research Laboratories" (American Chemical Society, 2015) which is available via the associated website "Hazard Assessment in Research Laboratories" at https://www.acs.org/content/acs/en/about/governance/committees/chemicalsafety/hazard-assessment.html. In the case of this procedure, the risk assessment should include (but not necessarily be limited to) an evaluation of the potential hazards associated with aluminium chloride, sodium hydroxide, 1,2-difluorobenzene, disulfur dichloride, methanol, 2,3,7,8-tetrafluorothianthrene, dichloromethane, Iron(III) nitrate nonahydrate, sodium bromide, trifluoroacetic acid, acetonitrile, toluene, diethyl ether, acetone, 2,3,7,8-tetrafluorothianthrene-S-oxide, 2-Phenylethyl acetate, trifluoroacetic anhydride, tetrafluoroboric acid diethyl ether complex, isopropyl alcohol, isohexane, 2-phenylethyl acetate derived tetrafluorothianthrenium salt, 4-biphenylboronic acid, [1,1'-bis(diphenylphosphino)ferrocene]dichloropalladium(II), tripotassium phosphate, 1,4-dioxane, ethyl acetate, [4-(4-phenylphenyl)-phenyl] ethyl acetate; as well as the proper procedures for quenching gaseous HCl which is evolving during a reaction.

2. Aluminum chloride (anhydrous, granular, 99%) was obtained from Alfa Aesar and used as received.
3. Sodium hydroxide (NaOH) (99%) was purchased from VWR chemicals and used as received. The Drechsel bottle was filled with aqueous NaOH solution (175 mL, 12.5 M). This HCl quenching setup using NaOH produces large amounts of heat and should be carried out with a glass container/flask rather than a plastic drying tube.
4. 1,2-Difluorobenzene (98%) was purchased from Fluorochem and used as received. The quantity of 1,2-difluorobenzene used in the reaction was consistent with its use as a solvent. When less 1,2-difluorobenzene was used, larger quantities of side-products resulted, which could not be removed by crystallization.
5. Nitrogen was used to purge HCl from the reaction flask and prevent backflow of the aqueous sodium hydroxide solution. Alternatively, argon can be used.
6. Disulfur dichloride (98%) was purchased from Sigma-Aldrich and used as received.

7. Room temperature throughout this manuscript refers to a temperature between 23 °C and 25 °C.
8. The stir plate was purchased from Heidolph Instruments GmbH & Co. KG. It has an input power of 230–240 V (50–60 Hz, 825 W), and the stirring range is from 100 rpm to 1400 rpm. Unless otherwise indicated, 500 rpm was used for stirring.
9. The silicone oil for the oil bath was purchased from abcr GmbH & Co. KG. The boiling point is over 205 °C. The oil in the oil bath should be level with the level of the reaction mixture in the reaction flask while heating. Unless otherwise indicated, the reported temperatures throughout this context refer to temperatures of oil in oil baths which were detected by the stirring plates' external temperature detectors. Approximately 15 min were necessary to increase temperature from 25 to 105 °C.
10. Methanol (MeOH, ≥ 99.8%) was purchased from Sigma-Aldrich and used as received. In order to precool the methanol, the required amount was placed in a 500 mL Erlenmeyer flask and cooled in an ice-water bath to 2–5 °C for 10 min. The addition was performed in 50 mL portions via the addition funnel.
11. Büchner funnel VitraPOR® with Por. 4 and a capacity of 125 mL was used.
12. To rinse the impurities out of the flask, manual stirring with a stainless spatula was applied to thoroughly mix the rinse solvent with the solid. When the filtration was complete, the filter cake was pressed with the top of a glass stopper to remove as much liquid as possible.
13. The vacuum pump was supplied by Vacuubrand GmbH & Co. KG. Vacuum refers to pressure lower than 0.075 mmHg.
14. Quantitative ^{19}F NMR (471 MHz, CDCl$_3$) was performed using 4,4'-difluorobenzophenone (≥ 99% purity, Sigma Aldrich) as an internal standard. The NMR sample was prepared by dissolving 4,4'-difluorobenzophenone (14.3 mg) and 2,3,7,8-tetrafluorothianthrene (10.3 mg) in CDCl$_3$ (0.5 mL) in a 5 mm NMR tube. The sample was subsequently analyzed by ^{19}F NMR spectroscopy with the following parameters: no. of scans: 32, D1: 20 s, rotation frequency: 15 Hz, spectral width: 49.7924 ppm, transmitter frequency offset: –121.3 ppm). Integration of the fluorine signals showed a 1:0.81 ratio, corresponding to 74.3 wt% of 2,3,7,8-tetrafluorothianthrene.

15. The off-white solid has the following characteristics: ^{1}H NMR (500 MHz, CDCl$_3$) δ: 7.32 (t, J = 8.5 Hz, 4H). ^{19}F NMR (471 MHz, 298 K, CDCl$_3$) δ: –136.86 (t, J = 8.6 Hz).

16. Purification of 2,3,7,8-tetrafluorothianthrene: A column (1 cm diameter) with a glass frit was charged with copper powder (2.9 g, 4.56 mmol) (powder, 99.999%, trace metal basis, Sigma Aldrich) The column was purged with argon, then, with the stopcock closed, 7 mL of concentrated HCl (37-38%, JT-Baker) was added to the copper powder, put under argon atmosphere, and gently agitated to liberate gas bubbles. After 5 min the stopcock was opened, and a positive pressure of argon was applied so that the concentrated HCl was eluted at a rate of about two drops·s^{-1} (total elution time about 2 min). The copper bed was washed twice with 10 mL of Milli-Q water by applying a positive pressure of argon (2 drops/s, about 2 min elution time for each wash). Nitrogen was allowed to flow through the column-bed for 5 min to remove as much water as possible. Then the copper was washed twice with 10 mL of anhydrous acetone using a positive pressure of argon (99.8%, ExtraDry, AcroSeal™, purchased from Fisher Scientific; about 2 min elution time for each wash). Again, argon was allowed to flow through the bed for 5 min to remove as much acetone as possible. Argon pressure was released, and the stopcock was closed. Then a solution of 320 mg of contaminated TFT in 10 mL of DCM was added to the column and allowed to sit with the stop cock closed for 10 min (all under argon atmosphere). At this point, the sample was allowed to elute into a 50 mL round-bottomed flask (approx. 10 min). The resulting column bed of black solid was washed with 10 mL of DCM using a positive pressure of argon and collected in the same 50 mL round-bottomed flask as the first DCM elution (about 2 min elution time). The combined DCM elutions were concentrated via rotary evaporator, and the colorless solid was dried under high vacuum at room temperature for 5 h yielding 225 mg of TFT with a purity of 98% (determined by quantitative ^{19}F-NMR). Purified 2,3,7,8-tetrafluorothianthrene (colorless solid) has the following characteristics: ^{1}H NMR (500 MHz, CDCl$_3$) δ: 7.32 (t, J = 8.5 Hz, 4H). ^{13}C NMR (126 MHz, CDCl$_3$) δ: 117.8 (dd, J = 13.8, 6.7 Hz), 131.5 (t, J = 5.0 Hz), 150.2 (dd, J = 254.3, 15.2 Hz). ^{19}F NMR (471 MHz, 298 K, CDCl$_3$) δ: –136.86 (t, J = 8.6 Hz). IR (neat): 3096, 3073, 3039, 2194, 2159, 2130, 2037, 2007, 1995, 1972, 1730, 1589, 1562, 1466, 1373, 1277, 1221, 1193, 1089, 969, 874, 852, 791, 679, 617, 587, 463, 440, 424 cm^{-1}.

EI-HRMS (*m/z*) calcd. for $C_{12}H_4S_2F_4$ [M]$^+$ 287.96850 found 287.96848. mp = 199.7 °C. The compound is bench stable.

17. Dichloromethane (DCM) (≥99%) was purchased from Sigma-Aldrich. Fe(NO$_3$)$_3$ · 9H$_2$O (98+%, metal basis) was purchased from Alfa Aesar. Sodium bromide (99.5%) was purchased from Riedel-de Haen. TFA (99%) was purchased from abcr GmbH & Co. They were all used as received.
18. The rubber septum was equipped with a short needle to prevent overpressure. The reaction was performed under air.
19. The progress of the reaction was followed by TLC analysis on silica gel (POLYGRAM® SIL G/UV with 0.20mm silica gel 60 with fluorescent indicator, purchased from Macherey-Nagel) with EtOAc-isohexane 1:10 (v/v) as eluent and visualized with KMnO$_4$-stain. The tetrafluorothianthrene starting material has R_f = 0.65 and the tetrafluorothianthrene-*S*-oxide product has R_f = 0.23.

Figure 7. TLC of starting material (SM) versus reaction spot of the end of reaction (P), a central co-spot (Co) and a spot for the product (R). (photo provided by submitters)

20. A BUCHI Vacuum Controller V-850 in combination with evaporator R-210 was used for rotary evaporation. Water was used in the heating bath. Unless otherwise indicated, the pressure and temperature were read from the controller.
21. Acetonitrile (MeCN) (≥ 99.9%) was purchased from Sigma-Aldrich and used as received.
22. Toluene (tech. grade) was purchased from OQEMA GmbH and used as received.

23. The filtrate was collected in a 1000-mL one-necked, round-bottomed flask.
24. Diethyl ether (Et$_2$O) (for analysis EMSURE® ACS, ISO, Reag. Ph Eur) was purchased from Sigma-Aldrich and dried over sodium-potassium alloy before use.
25. Acetone (tech. grade) was purchased from OQEMA GmbH and used as received.
26. The off-white solid has the following characteristics: ^{1}H NMR (400 MHz, CDCl$_3$) d: 7.49 (dd, J = 9.0, 6.5 Hz, 2H), 7.73 (dd, J = 8.8, 7.3 Hz, 2H). ^{13}C NMR (101 MHz, CDCl$_3$) d: 114.4 (dd, J = 21.3, 1.6 Hz), 118.9 (d, J = 20.4 Hz), 124.3 (dd, J = 7.2, 4.1 Hz), 138.4 (t, J = 3.7 Hz), 150.1 (dd, J = 24.4, 13.3 Hz), 152.7 (dd, J = 24.7, 13.3 Hz). ^{19}F NMR (471 MHz, CDCl$_3$) d: –132.82 (ddd, J = 20.2, 9.0, 7.3 Hz), –133.66 (ddd, J = 20.2, 8.8, 6.5 Hz). IR (neat): 3092, 3036, 1596, 1573, 1461, 1382, 1270, 1213, 1191, 1100, 1064, 955 cm^{-1}. EI-HRMS (m/z) calcd. for $C_{12}H_4O_1S_2F_4$ [M]$^+$ 303.9634, found 303.9634. mp = 254.8 °C. The compound is bench stable.

Quantitative ^{19}F NMR (471 MHz, CDCl$_3$) was performed using 4,4′-difluorobenzophenone (≥ 99% purity, Sigma Aldrich) as an internal standard. The NMR sample was prepared by dissolving 4,4′-difluorobenzophenone (7.4 mg) and 2,3,7,8-tetrafluorothianthrene-S-oxide (7.4 mg) in CDCl$_3$ (0.5 mL) in a 5 mm NMR tube. The sample was subsequently analyzed by ^{19}F NMR spectroscopy with the following parameters: no. of scans: 32, D1: 20 s, rotation frequency: 15 Hz, spectral width: 49.7923 ppm, transmitter frequency offset: –119.0 ppm). Integration of the fluorine signals showed a 1:0.705 ratio, corresponding to 97.6 wt% of 2,3,7,8-tetrafluorothianthrene-S-oxide.
27. The reaction (Step A) was also checked on half-scale and provided 5.94 g (15%) of the same product with purity of 97%.
28. 2-Phenylethyl acetate (98%) and TFAA (99+%) were purchased from Alfa Aesar and used as received.
29. Tetrafluoroboric acid diethyl ether complex (51.0 – 57.0 wt% HBF$_4$) was purchased from Sigma Aldrich and used as received.
30. The progress of the reaction was followed by TLC analysis on silica gel (POLYGRAM® SIL G/UV with 0.20mm silica gel 60 with fluorescent indicator, purchased from Macherey-Nagel) with DCM as eluent and visualized by 254 nm UV-light. The tetrafluorothianthrene-S-oxide starting material has R_f = 0.55 and 2-phenylethyl acetate has R_f = 0.65.

Figure 8. TLC of starting material tetrafluorothianthrene-S-oxide (SM TFTO, left) and starting material 2-phenylethyl actetae (SM, right) versus reaction spot of the end of reaction (P). A central co-spot (Co) for both is also shown (photo provided by submitters)

31. Isopropyl alcohol (iPrOH) (tech. grade) was purchased from OQEMA GmbH and used as received.
32. Isohexane (tech. grade) was purchased from OQEMA GmbH and distilled before use.
33. Flash column chromatography: A column (length: 26 cm, diameter: 8 cm, with a 1000 mL reservoir) was charged with 300 g SiO_2 (Geduran® Si 60 with pore size 40-63 μm from Merck KGaA) and 750 mL DCM. In order to dry load the crude 2-phenethyl acetate-derived tetrafluorothianthrenium salt, it was dissolved in 300 mL DCM at 40 °C using a water bath and 60 g of SiO_2 was added. The solvent was removed under reduced pressure (525 mmHg to 15 mmHg, 40 °C), and the dry residue was transferred on the column inside a fume hood. Sand (200 g) was added on top of the silica column bed. The product was eluted with 600 mL DCM. At that point, fraction collection was initiated, and elution was continued with 500 mL DCM/i-PrOH 100:1 (v/v), 8 L DCM/ i-PrOH 50:1 (v/v) and 2 L DCM/ i-PrOH 10:1. Fractions were collected in 50 mL test tubes. TLC analysis of the product was done with DCM/ i-PrOH 10:1 (v/v) as eluent and visualized with 254 nm UV light. Product could be found in fractions 41 to 219.
34. The colorless solid has the following characteristics: ^{1}H NMR (400 MHz, CD_3CN) δ: 1.91 (s, 3H), 2.96 (t, J = 6.5 Hz, 2H), 4.22 (t, J = 6.5 Hz, 2H), 7.11 – 7.17 (m, 2H), 7.38 – 7.42 (m, 2H), 7.95 (dd, J = 9.9, 7.0 Hz, 2H),

8.38 (dd, J = 9.1, 7.2 Hz, 2H). ^{13}C NMR (101 MHz, CD$_3$CN) δ: 21.0, 35.1, 64.5, 115.4 (dd, J = 7.1, 3.6 Hz), 121.2 (d, J = 21.9 Hz), 121.2, 125.5 (dd, J = 22.0, 2.6 Hz), 129.3, 132.2, 135.2 (dd, J = 8.0, 3.9 Hz), 146.3, 151.6 (dd, J = 255.4, 13.3 Hz), 154.8 (dd, J = 261.9, 13.2 Hz), 171.4. ^{19}F NMR (471 MHz, CD$_3$CN) δ: –125.42 (ddd, J = 20.2, 10.0, 7.1 Hz), –133.80 (ddd, J = 20.1, 9.1, 7.0 Hz), –151.70 (br, ^{10}B), –151.75 (br, ^{11}B). IR (neat): 3093, 3048, 1726, 1574, 1481, 1380,1284, 1267, 1236, 1199, 1055, 1026, 1008, 965, 904 cm^{-1}. ESI-HRMS: (m/z) calcd. for C$_{22}$H$_{15}$O$_2$S$_2$F$_4$ [M]$^+$ 451.0444, found 451.0444. mp = 207.0 °C. The compound is bench-stable.

Quantitative ^{19}F NMR (471 MHz, CD$_3$CN) was performed using 4,4'-difluorobenzophenone (≥ 99% purity, Sigma Aldrich) as an internal standard. The NMR sample was prepared by dissolving 4,4'-difluorobenzophenone (10.8 mg) and 2-phenethyl acetate derived tetrafluorothianthrenium salt (10.8 mg) in CD$_3$CN (0.5 mL) in a 5 mm NMR tube. The sample was subsequently analyzed by ^{19}F NMR spectroscopy with the following parameters: no. of scans: 32, D1: 20 s, rotation frequency: 15 Hz, spectral width: 49.7924 ppm, transmitter frequency offset: –121.3 ppm). Integration of the fluorine signals showed a 1:0.405 ratio, corresponding to 99.9 wt% of 2-phenethyl acetate derived tetrafluorothianthrenium salt.

35. The reaction (Step B) was also checked on half-scale and provided 6.75 g (64%) of the same product.
36. Alternatively, the reaction can be carried out under argon atmosphere.
37. 4-Biphenylboronic acid (95%) was purchased from Oxchem, Pd(dppf)Cl$_2$ (98%) was purchased from fluorochem, K$_3$PO$_4$ (97%) was purchased from Alfa Aesar, and 1,4-dioxane (≥ 99.5%) was purchased from Fisher Scientific. They were used as received.
38. The progress of the reaction was followed by TLC analysis on silica gel (POLYGRAM® SIL G/UV with 0.20mm silica gel 60 with fluorescent indicator, purchased from Macherey-Nagel) with DCM-iPrOH 10:1 (v/v) as eluent and visualized by 254 nm UV-light. The 2-phenethyl acetate derived tetrafluorothianthrenium salt starting material has R$_f$ = 0.47, and the Suzuki coupling product [4-(4-phenylphenyl)phenyl] ethyl acetate has R$_f$ = 0.80.

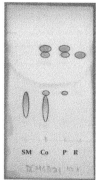

Figure 9. TLC of starting material (SM) versus reaction spot of the end of reaction (P). A central co-spot (Co) and a spot for the clean product (R) is also shown (photo provided by submitters)

39. Ethyl acetate (98-100%) was purchased from OQEMA GmbH and distilled before use.
40. Flash column chromatography: A column (length: 26 cm, diameter: 8 cm, with a 1000 mL reservoir) was charged with 300 g SiO_2 (Geduran® Si 60 with pore size 40-63 µm from Merck KGaA) and 800 mL Hex/DCM 8:1 (v/v). In order to dry load the crude [4-(4-phenylphenyl)phenyl] ethyl acetate was dissolved in 250 mL DCM at 40 °C using a water bath and 50 g of SiO_2 was added. The solvent was removed under reduced pressure (525 mmHg to 15 mmHg, 40 °C), and the dry residue was transferred on the column. Sand (200 g) was added on top of the column bed. The crude product is eluted with 1 L of a hexanes/DCM mixture (4:1 v/v) and 3.2 L (hexanes /DCM = 3:1 v/v). At that point, fraction collection is begun, and elution is continued with 4 L DCM/hexanes 1:1 (v/v) and 4 L DCM. Fractions were collected in 50 mL test tubes. TLC analysis of the product was done with DCM/i-PrOH 10:1 (v/v) as eluent and visualized with 254 nm UV light. Product could be found in fractions 44 to 120.
41. Büchner funnel VitraPOR® with Por. 4 and a capacity of 75 mL was used.
42. The off-white solid has the following characteristics: ^{1}H NMR (400 MHz, $CDCl_3$) δ: 2.08 (s, 3H), 3.00 (t, J = 7.1 Hz, 2H), 4.35 (t, J = 7.0 Hz, 2H), 7.31 – 7.35 (m, 2H), 7.35 – 7.41 (m, 1H), 7.47 (dd, J = 8.4, 6.9 Hz, 2H), 7.58 – 7.62 (m, 2H), 7.63 – 7.67 (m, 2H), 7.68 (s, 4H). ^{13}C NMR (101 MHz, $CDCl_3$) δ: 21.1, 34.9, 65.0, 127.2, 127.2, 127.5, 127.6, 127.6, 128.9, 129.5,

137.1, 139.1, 139.9, 140.2, 140.8, 171.2. IR (neat): 3032, 2955, 2894, 1732, 1484, 1400, 1383, 1366, 1233, 1030, 1001, 980, 829, 819, 691 cm^{-1}. ESI-HRMS (m/z) calcd. for $C_{22}H_{20}O_2Na_1$ [M + Na]$^+$ 339.1356, found 339.1357. mp = 168.1 °C. The compound is bench stable.

Quantitative ^{1}H NMR (500 MHz, CDCl$_3$) was performed using benzyl benzoate (certified reference material, TraceCERT®, Sigma Aldrich) as an internal standard. The NMR sample was prepared by dissolving benzyl benzoate (10.1 mg) and [4-(4-phenylphenyl)phenyl] ethyl acetate (9.80 mg) in CDCl$_3$ (0.5 mL) in a 5 mm NMR tube. Integration of the proton signals showed a 1:0.65 ratio, corresponding to 99.9 wt% of [4-(4-phenylphenyl)phenyl] ethyl acetate.

43. The reaction (Step C) was also checked on half-scale and provided 2.69 g (68%) of the same product.

Working with Hazardous Chemicals

The procedures in *Organic Syntheses* are intended for use only by persons with proper training in experimental organic chemistry. All hazardous materials should be handled using the standard procedures for work with chemicals described in references such as "Prudent Practices in the Laboratory" (The National Academies Press, Washington, D.C., 2011; the full text can be accessed free of charge at http://www.nap.edu/catalog.php?record_id=12654). All chemical waste should be disposed of in accordance with local regulations. For general guidelines for the management of chemical waste, see Chapter 8 of Prudent Practices.

In some articles in *Organic Syntheses*, chemical-specific hazards are highlighted in red "Caution Notes" within a procedure. It is important to recognize that the absence of a caution note does not imply that no significant hazards are associated with the chemicals involved in that procedure. Prior to performing a reaction, a thorough risk assessment should be carried out that includes a review of the potential hazards associated with each chemical and experimental operation on the scale that is planned for the procedure. Guidelines for carrying out a risk assessment and for analyzing the hazards associated with chemicals can be found in Chapter 4 of Prudent Practices.

The procedures described in *Organic Syntheses* are provided as published and are conducted at one's own risk. *Organic Syntheses, Inc.*, its Editors, and its Board of Directors do not warrant or guarantee the safety of individuals using these procedures and hereby disclaim any liability for any injuries or damages claimed to have resulted from or related in any way to the procedures herein.

Discussion

The C–H functionalization of arenes using 2,3,7,8–tetrafluorothianthrene-S-oxide accesses a versatile synthetic lynchpin (aryl tetrafluorothianthrenium salt) in a highly selective manner. While aryl halides and aryl boronic acids are proven synthetic lynchpins, no late-stage halogenation or borylation method currently has the same selectivity and scope which tetrafluorothianthrenation demonstrates.[2] Additionally, aryl tetrafluorothianthrenium salts are competent substrates for activation by palladium and photoredox catalysis. In contrast to mononuclear leaving groups such as bromide, complex leaving groups like tetrafluorothianthrene can raise the reduction potential of the aryl substrate and still possess the ability to engage in productive bond forming reactions.[3] We have shown that from aryl thianthrenium salts, C–F,[3] C–CF$_3$,[4] C–N,[5] and C–O[6] bonds can be formed using photoredox catalysis. A disadvantage of tetrafluoro-thianthrenium salts is their ability to undergo a nucleophilic aromatic substitution on the thianthrene scaffold by displacement of fluoride.[5] This problem was encountered in the coupling of amines, but this issue was overcome by using the non-fluorinated aryl thianthrenium analogs. The thianthrenation procedure entails the same protocol as reported here, however, with thianthrene-S-oxide. Thianthrenation has a similarly broad scope and functional group tolerance, but it cannot readily functionalize electron-poor aromatic rings. In addition to arene functionalization, we have shown that the regioselective thianthrenation of unactivated olefins produces versatile alkenyl electrophiles.[8] A minor drawback of the two thianthrene derivatives is their inability to cleanly react by direct substitution with nucleophiles. Direct substitution is important for applications such as ^{18}F-labeling with [^{18}F]fluoride. By employing dibenzothiophene-S-oxide derivatives, a light- and transition metal-free C–H to ^{18}F sequence is feasible.[7] The obvious disadvantage of the C–H functionalization reactions to form sulfonium salts is the low atom economy. Until equivalent C–H functionalizations are developed, stoichiometric waste is the price that must be paid.

References

1. Max-Planck-Institut für Kohlenforschung, Kaiser-Wilhelm-Platz 1, 45470, Muelheim an der Ruhr, Germany, E-mail: ritter@mpi-muelheim.mpg.de. ORCID: 0000-0002-6957-450X. Generous support by the Max-Planck-Society is gratefully acknowledged. The submitters acknowledge N. Haupt, D. Kampen, F. Kohler, S. Marcus and D. Margold for mass spectrometry analysis, D. Chamier Cieminski for collecting analytical data and J. Chen and Dr. F. Berger for helpful discussions.
2. Berger, F.; Plutschack, M. B.; Riegger, J.; Yu, W.; Speicher, S.; Ho, M.; Frank, N.; Ritter, T. *Nature* **2019**, *567*, 223–228.
3. Li, J.; Chen, J.; Sang, R.; Ham, W.-S.; Plutschack, M. B.; Berger, F.; Chabbra, S.; Schnegg, A.; Genicot, C.; Ritter, T. *Nat. Chem.* **2020**, *12*, 56–62.
4. Ye, F.; Berger, F.; Jia, H.; Ford, J.; Wortman, A.; Börgel, J.; Genicot, C.; Ritter, T. *Angew. Chem. Int. Ed.* **2019**, *58*, 14615–14619.
5. Engl, P. S.; Häring, A. P.; Berger, F.; Berger, G.; Pérez-Bitrián, A.; Ritter, T. *J. Am. Chem. Soc.* **2019**, *141*, 13346–13351.
6. Sang, R.; Korkis, S. E.; Su, W.; Ye, F.; Engl, P. S.; Berger, F.; Ritter, T. *Angew. Chem. Int. Ed.* **2019**, *58*, 16161–16166.
7. Xu, P.; Zhao, D; Berger, F.; Hamad, A.; Rickmeier, J.; Petzold, R.; Kondratiuk, M.; Bohdan, K.; Ritter, T. *Angew. Chem. Int. Ed.* **2020**, *59*, 1956–1960.
8. Chen, J.; Li, J.; Plutschack, M. B.; Berger, F.; Ritter, T. *Angew. Chem. Int. Ed.* **2020**, *59*, 5616–5620.

Appendix
Chemical Abstracts Nomenclature (Registry Number)

Aluminium chloride: aluminium chloride; (7446-70-0)
1,2-Difluorobenzene: 1,2-Difluorobenzene; (367-11-3)
Disulfur dichloride: disulfur dichloride (10025-67-9)
$Fe(NO_3)_3 \cdot 9H_2O$: iron(III) nitrate nonahydrate (7782-61-8)
NaBr: sodium bromide (7647-15-6)
TFA: trifluoroacetic acid (76-05-1)
2-Phenylethyl acetate: 2-Phenylethyl acetate (103-45-7)
TFAA: trifluoroacetic anhydride (407-25-0)

Tetrafluoroboric acid diethyl ether complex: tetrafluoroboric acid diethyl ether complex (67969-82-8)
4-Biphenylboronic acid: 4-Biphenylboronic acid (5122-94-1)
Pd(dppf)Cl$_2$: [1,1'-Bis-(diphenylphosphino)ferrocene]dichloropalladium(II) (72287-26-4)
K$_3$PO$_4$: potassium phosphate (7778-53-2)
1,4-Dioxane: 1,4-Dioxane (123-91-1)

Samira Speicher completed her apprenticeship as a technician in chemistry in 2017 (Max-Planck-Institut fuer Kohlenforschung, Muelheim an der Ruhr). During this time, she worked in the group of Prof. Benjamin List (homogenous catalysis) and in the group of Prof. Alois Fürstner (natural product synthesis). Since 2017, she is working in the group of Prof. Tobias Ritter on organic synthesis and radiochemistry.

Matthew B. Plutschack received his undergraduate degree from the University of Wisconsin – Madison, and he began conducting research under Prof. Howard E. Zimmerman. In 2013, he obtained his M.S. from Florida State University under the tutelage of Prof. D. Tyler McQuade. He conducted research at the Max-Planck-Institut fuer Kolloid- und Grenzflaechenforschung, and in 2017, he earned his doctorate from the Freie Universitaet Berlin under the mentorship of Prof. Peter H. Seeberger. Since 2017 he is a Post-doc in the lab of Prof. Tobias Ritter at the Max-Planck-Institut fuer Kohlenforschung.

Tobias Ritter received his undergraduate education in Braunschweig, Bordeaux, Lausanne, and Stanford. He has performed undergraduate research with Prof. Barry M. Trost, obtained his Ph.D. with Prof. Erick M. Carreira at ETH Zurich in 2004, and was a postdoc with Prof. Robert H. Grubbs at Caltech. In 2006, Tobias was appointed as Assistant Professor in the Department of Chemistry and Chemical Biology at Harvard, promoted to Associate Professor in 2010, and to Professor of Chemistry and Chemical Biology in 2012. Since 2015 he is director at the Max-Planck-Institut fuer Kohlenforschung. In 2011, Tobias founded SciFluor LifeScience, a pharmaceutical development company.

Maurus Mathis completed his apprenticeship in synthetic chemistry in 2018 while working in the group of Prof. Cristina Nevado at the University of Zürich. He is currently doing his Bachelor's Degree in chemistry at the UZH while still working as a laboratory technician in the Nevado group.

Jorge A. González was born in Xalapa, Mexico. He completed his Undergraduate Degree in Chemistry at the National and Autonomous University of Mexico in 2011. He obtained his PhD at the University of Edinburgh in 2016. He is currently a postdoctoral associate research associate in the group of Prof. Cristina Nevado.

Author Index Volume 98

Adler, M. J., **98**, 227
Altundas, B., **98**, 147
Alwedi, E., **98**, 147
Anandamurthy, A. S., **98**, 97
Aoki, Y., **98**, 84
Azam, F., **98**, 227

Balijepalli, A., **98**, 491
Ball, L. T., **98**, 289
Bao, H., **98**, 51
Baran, P. S., **98**, 97
Barber, T., **98**, 289
Bodnar, A. K., **98**, 263
Brodney, M. A., **98**, 131
Burman, J. S., **98**, 343
Butcher, W., **98**, 263

Caputo, H. E., **98**, 491
Cha, J. K., **98**, 430
Chaheine, C. M., **98**, 194
Chao, A., **98**, 147
Chiba, S., **98**, 363

Eastgate, M. D., **98**, 97

Fleming, F. F., **98**, 147
Fox, R. J., **98**, 97

Gair, J. J., **98**, 131
Galeote, O., **98**, 374
Garg, N. K., **98**, 68
Garner, P., **98**, 463
Giroux, S., **98**, 131
Gladen, P. T., **98**, 194
Grey, R. L., **98**, 131
Grinstaff, M. W., **98**, 491
Gryko, D. T., **98**, 242
Grzybowski, M., **98**, 242

Hamoud, A., **98**, 491
Hoblos, B., **98**, 391
Hu, J., **98**, 463
Huang, D., **98**, 263

Inoue, Ka., **98**, 407, 509
Inoue, Ke., **98**, 509
Ishihara, K., **98**, 1, 28
Ishizaki, S., **98**, 1, 28
Itoh, M., **98**, 407

Kaniskan, H. U., **98**, 463
Kawamura, S., **98**, 84
Kelleghan, A. V., **98**, 68
Kennington, S. C. D., **98**, 374
Kingsbury, J. S., **98**, 343
Knouse, K. W., **98**, 97
Krzeszewski, M., **98**, 242

Lew, J. K., **98**, 263
Lohbeck, J., **98**, 171, 315

Maity, P., **98**, 97
Mehta, M. M., **98**, 68
Mellado–Hidalgo, M., **98**, 374
Miller, A. K., **98**, 171, 315
Morgen, M., **98**, 171, 315
Mori, A., **98**, 407, 509

Nagorny, P., **98**, 446
Nakura, R., **98**, 407
Natrayan, M., **98**, 147
Newhouse, T. R., **98**, 263

Okano, K., **98**, 407, 509
Ong, D. Y., **98**, 363

Pang, J. H., **98**, 363
Parker, C. G., **98**, 463
Plutschack, M. B., **98**, 531

Rathnayake, U., **98**, 463
Rendina, V. L., **98**, 343
Ritter, T., **98**, 531
Romea, P., **98**, 374
Romo, D., **98**, 194

Sardini, S. R., **98**, 117
Schmidt, M. A., **98**, 97
Shekarappa, V., **98**, 97
Smolarski, B. A., **98**, 343
Sodeoka, M., **98**, 84
Song, C. J., **98**, 194
Speicher, S., **98**, 531
Stoltz, B. M., **98**, 117
Sun, S., **98**, 446

Tasior, M., **98**, 242
Turlik, A., **98**, 263

Urpí, F., **98**, 374
Uyanik, M., **98**, 1, 28

Vaidyanathan, R., **98**, 97
Vantourout, J. C., **98**, 97
Varghese, M., **98**, 491

Wan, W.-M., **98**, 51
Wengryniuk, S. E., **98**, 391

Xiao, R., **98**, 491

Zheng, B., **98**, 97
Zhu, J., **98**, 97
Zhu, N., **98**, 51
Ziminsky, Z. L., **98**, 147

Printed and bound by CPI Group (UK) Ltd, Croydon, CR0 4YY
05/03/2023

03197793-0001